Brief CALCULUS

An Applied Approach

Hybrid Edition

Brief CALCULUS
An Applied Approach

Ninth Edition
Hybrid Edition

Ron Larson
The Pennsylvania State University
The Behrend College

With the assistance of David C. Falvo
The Pennsylvania State University
The Behrend College

BROOKS/COLE
CENGAGE Learning™

Australia • Brazil • Japan • Korea • Mexico • Singapore • Spain • United Kingdom • United States

BROOKS/COLE
CENGAGE Learning™

Brief Calculus: An Applied Approach
Ninth Edition
Hybrid Edition

Ron Larson

Vice President, Editorial Director: P.J. Boardman

Publisher: Richard Stratton

Senior Development Editor: Laura Wheel

Senior Editorial Assistant: Haeree Chang

Associate Media Editor: Andrew Coppola

Senior Marketing Manager: Barb Bartoszek

Marketing Coordinator: Michael Ledesma

Marketing Communications Manager: Mary Anne Payumo

Content Project Manager: Jill Quinn

Senior Art Director: Jill Ort Haskell

Manufacturing Planner: Doug Bertke

Rights Acquisition Specialist: Shalice Shah-Caldwell

Text Designer: Larson Texts, Inc.

Cover Designer: Larson Texts, Inc.

Compositor: Larson Texts, Inc.

Cover Image: *Torus with Cross-Cap* by Helaman Ferguson.
Photograph by Ed Bernik photo/video.
Used with permission.

"The inscription on the base of my polished bronze *Torus with Cross-Cap* sculpture, $3x = x + h$ but $2x \neq h$, may seem a bit strange at first glance. The addition is of topological objects, h for a handle, x for a cross-cap. These are not numbers: the equation means that adding three cross-caps to a surface is equivalent to adding one cross-cap and a handle. In this topological "calculus" the usual cancellation law does not hold."

—Helaman Ferguson

For product information and technology assistance, contact us at
Cengage Learning Customer & Sales Support, 1-800-354-9706.
For permission to use material from this text or product, submit all requests online at **www.cengage.com/permissions.**
Further permissions questions can be emailed to
permissionrequest@cengage.com.

Library of Congress Control Number: 2011935165

Student Edition:
ISBN-13: 978-1-133-36514-3
ISBN-10: 1-133-36514-0

Brooks/Cole
20 Channel Center Street
Boston, MA 02210
USA

Cengage Learning is a leading provider of customized learning solutions with office locations around the globe, including Singapore, the United Kingdom, Australia, Mexico, Brazil, and Japan. Locate your local office at:
international.cengage.com/region

Cengage Learning products are represented in Canada by Nelson Education, Ltd.

For your course and learning solutions, visit **www.cengage.com.**
Purchase any of our products at your local college store or at our preferred online store *www.cengagebrain.com.*

Instructors: Please visit *login.cengage.com* and log in to access instructor-specific resources.

Printed in the United States of America
1 2 3 4 5 6 7 15 14 13 12 11

Contents

*Available at the text-specific website *www.cengagebrain.com*

Preface

Many traditional lecture-based courses are evolving into courses in which all homework and tests are delivered online. A hybrid text is designed to meet the needs of these courses through the integration of both print and online components. For this Hybrid Edition, the end-of-section exercises have been removed from the text and are available exclusively online in Enhanced WebAssign.

From the Author

Welcome to the Ninth Edition of *Brief Calculus: An Applied Approach*! I am always excited about a new edition, but with this edition, I am even more excited. I had a single goal in mind with this revision—to provide you with a book that is both real and relevant. This book has a bright business-oriented design that complements the multitude of business and life sciences applications found throughout.

The theme for the revision is **"IT'S ALL ABOUT YOU."** The pedagogy of the book is rock solid and is based on years of teaching, years of writing, and years of feedback from instructors and students. Please pay special attention to the study aids with a red **U**. These study aids will help you learn calculus, use technology, refresh your algebra skills, and prepare for tests. For an overview of these aids, check out CALCULUS & YOU on page 0.

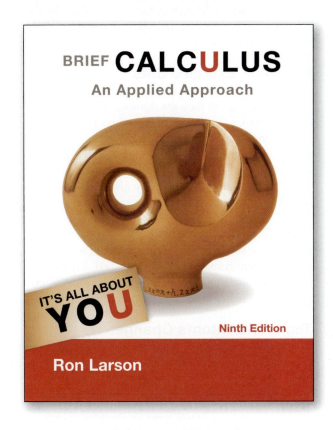

BRIEF **CALCULUS**
An Applied Approach

IT'S ALL ABOUT
YOU

Ninth Edition

Ron Larson

Example 5 on page 15 shows how the point of intersection of two graphs can be used to find the break-even point for a company manufacturing and selling a product.

New To This Edition

NEW Chapter Opener
Each *Chapter Opener* highlights a real-life problem from an example in the chapter, showing a graph related to the data and describing the math concept used to solve the problem.

NEW Section Opener
Each *Section Opener* highlights a real-life problem in the exercises, showing a graph for the situation with a description of how you will use the math of the section to solve the problem.

NEW SUMMARIZE
The *Summarize* feature at the end of each section helps you organize the lesson's key concepts into a concise summary, providing you with a valuable study tool.

NEW   HOW DO YOU SEE IT? Exercise

The *How Do You See It?* exercise in each section presents a real-life problem that you will solve by visual inspection using the concepts learned in the lesson. How Do You See It? exercises are in Cengage YouBook, which is available in Enhanced WebAssign.

REVISED Exercise Sets

The exercise sets have been carefully and extensively examined to ensure they are rigorous, relevant, and cover all topics suggested by our users. The exercises have been reorganized and titled so you can better see the connections between examples and exercises. Multi-step, real-life exercises reinforce problem-solving skills and mastery of concepts by giving you the opportunity to apply the concepts in real-life situations.

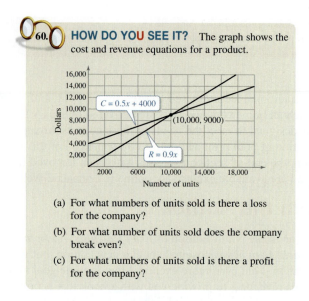

60. HOW DO YOU SEE IT? The graph shows the cost and revenue equations for a product.

$C = 0.5x + 4000$

(10,000, 9000)

$R = 0.9x$

(a) For what numbers of units sold is there a loss for the company?

(b) For what number of units sold does the company break even?

(c) For what numbers of units sold is there a profit for the company?

Calc Chat

For the past several years, an independent website—**CalcChat.com**—has been maintained to provide free solutions to all odd-numbered problems in the text. Thousands of students have visited the site for practice and help with their homework. For this edition, information from **CalcChat.com**, including which solutions students accessed most often, was used to help guide the revision of the exercises.

Table of Contents Changes

Chapter 0 (Precalculus Review) has been moved to Appendix A. Based on feedback from users, old Section 6.2 (Partial Fractions and Logistic Growth) has been removed.

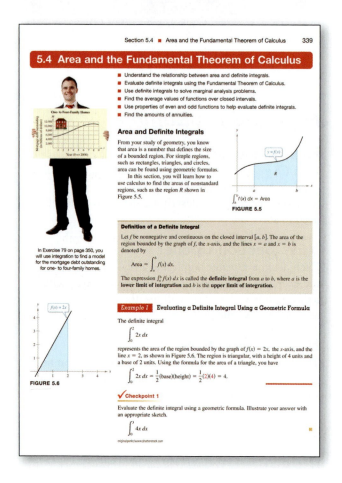

Trusted Features

Section Objectives

A bulleted list of learning objectives provides you the opportunity to preview what will be presented in the upcoming section.

Definitions and Theorems

All definitions and theorems are highlighted for emphasis and easy recognition.

Checkpoint

Paired with every example, the *Checkpoint* problems encourage immediate practice and check your understanding of the concepts presented in the example. Answers to all *Checkpoint* problems appear at the back of the text to reinforce understanding of the skill sets learned.

Business Capsule

Business Capsules appear at the end of selected sections in Cengage YouBook. These capsules and their accompanying research project highlight business situations related to the mathematical concepts covered in the chapter.

STUDY TIP

These hints and tips can be used to reinforce or expand upon concepts, help you learn how to study mathematics, caution you about common errors, address special cases, or show alternative or additional steps to a solution of an example.

TECH TUTOR

The *Tech Tutor* gives suggestions for effectively using tools such as calculators, graphing calculators, and spreadsheet programs to help deepen your understanding of concepts, ease lengthy calculations, and provide alternate solution methods for verifying answers obtained by hand.

ALGEBRA TUTOR

The *Algebra Tutor* appears throughout each chapter and offers algebraic support at point of use. This support is revisited in a two-page algebra review at the end of the chapter, where additional details of example solutions with explanations are provided.

Business Capsule

CitiKitty, Inc. was founded in 2005 by 26-year-old Rebecca Rescate after she moved into a small apartment in New York City with no place to hide her cat's litter box. Finding no easy-to-use cat toilet training kit, she created one, and CitiKitty was born with an initial investment of $20,000. Today the company flourishes with an expanded product line. Revenues in 2010 reached $350,000.

83. Research Project Use your school's library, the Internet, or some other reference source to find information about the start-up costs of beginning a business, such as the example above. Write a short paper about the company.

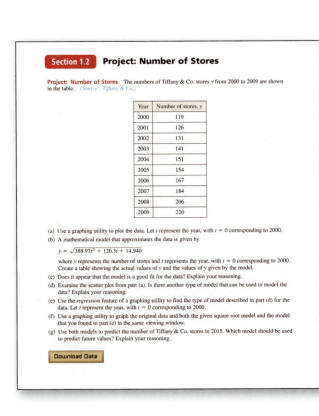

Section 1.2   **Project: Number of Stores**

Project: Number of Stores The numbers of Tiffany & Co. stores y from 2000 to 2009 are shown in the table. *(Source: Tiffany & Co.)*

Year	Number of stores, y
2000	119
2001	126
2002	131
2003	141
2004	151
2005	154
2006	167
2007	184
2008	206
2009	220

(a) Use a graphing utility to plot the data. Let t represent the year, with $t = 0$ corresponding to 2000.

(b) A mathematical model that approximates the data is given by

$$y = \sqrt{388.97t^2 + 120.3t + 14{,}940}$$

where y represents the number of stores and t represents the year, with $t = 0$ corresponding to 2000. Create a table showing the actual values of y and the values of y given by the model.

(c) Does it appear that the model is a good fit for the data? Explain your reasoning.

(d) Examine the scatter plot from part (a). Is there another type of model that can be used to model the data? Explain your reasoning.

(e) Use the *regression* feature of a graphing utility to find the type of model described in part (d) for the data. Let t represent the year, with $t = 0$ corresponding to 2000.

(f) Use a graphing utility to graph the original data and both the given square root model and the model that you found in part (e) in the same viewing window.

(g) Use both models to predict the number of Tiffany & Co. stores in 2015. Which model should be used to predict future values? Explain your reasoning.

Download Data

SKILLS WARM UP

The *Skills Warm Up* appears at the beginning of the exercise set for each section. These problems help you review previously learned skills that you will use in solving the section exercises.

Project

The projects for selected sections involve in-depth applied exercises in which you will work with large, real-life data sets, often creating or analyzing models. These projects are offered online in Enhanced WebAssign.

Instructor Resources

Print

Complete Solutions Manual
ISBN-13: 978-1-133-36432-0
The *Complete Solutions Manual* provides worked-out solutions for all exercises in the text, including Checkpoints, Quiz Yourself, Test Yourself, and Tech Tutors.

Media

PowerLecture
ISBN-13: 978-1-133-36473-3
This comprehensive CD-ROM provides dynamic media tools designed to help you teach. PowerLecture includes Solution Builder, Diploma Computerized Testing, Microsoft® Powerpoint® lecture slides, and all art from the text.

Solution Builder
www.cengage.com/solutionbuilder
This online instructor database offers complete worked-out solutions of all exercises in the text. Solution Builder allows you to create customized, secure solutions printouts (in PDF format) matched exactly to the problems you assign in class.

Diploma Computerized Testing
Diploma is an easy-to-use assessment software containing hundreds of algorithmic questions derived from the text exercises. With Diploma, you can quickly create, customize, and deliver tests in both print and online formats. Diploma is available on the PowerLecture CD.

www.webassign.net
WebAssign's homework delivery system lets you deliver, collect, grade, and record assignments via the web. Online exercises are assignable in free response, multiple choice, and multi-part formats. New enhancements include Cengage YouBook interactive eBook, Personal Study Plans, a Show My Work feature, an Answer Evaluator that accepts more mathematically equivalent answers, quizzes, videos, and more!

Cengage YouBook
YouBook is an interactive and customizable eBook! Containing all the content from the printed text, YouBook features a text edit tool that allows you to modify the textbook narrative as needed. With YouBook, you can quickly re-order entire sections and chapters or hide any content you don't teach to create an eBook that perfectly matches your syllabus. You can further customize the text by publishing web links. Additional media assets include: animated figures, video clips, highlighting, notes, and more! YouBook is available in Enhanced WebAssign.

Student Resources

Print

Student Solutions Manual
ISBN-13: 978-1-133-11279-2

The *Student Solutions Manual* provides complete worked-out solutions to all odd-numbered exercises in the text. In addition, the solutions of all Checkpoint, Quiz Yourself, Test Yourself, and Tech Tutor exercises are included.

Media

www.webassign.net

Enhanced WebAssign is an online homework system that lets instructors deliver, collect, grade, and record assignments via the web. Enhanced WebAssign includes Cengage YouBook interactive eBook, Personal Study Plans, a Show My Work feature, Answer Evaluator, quizzes, videos, and more!

CengageBrain.com

To access additional course materials and companion resources, please visit *www.cengagebrain.com*. At the CengageBrain.com home page, search for the ISBN of your title (from the back cover of your book) using the search box at the top of the page. This will take you to the product page where free companion resources can be found.

Acknowledgements

I would like to thank my colleagues who have helped me develop this program. Their encouragement, criticisms, and suggestions have been invaluable to me.

Reviewers

Nasri Abdel-Aziz, *State University of New York College of Environmental Sciences and Forestry*
Alejandro Acuna, *Central New Mexico Community College*
Dona Boccio, *Queensborough Community College*
George Bradley, *Duquesne University*
Andrea Marchese, *Pace University*
Benselamonyuy Ntatin, *Austin Peay State University*
Maijian Qian, *California State University, Fullerton*
Judy Smalling, *St. Petersburg College*
Eddy Stringer, *Tallahassee Community College*

I would also like to thank the following reviewers, who have given me many useful insights to this and previous editions.

Carol Achs, *Mesa Community College;* Lateef Adelani, *Harris-Stowe State University, Saint Louis;* Frederick Adkins, *Indiana University of Pennsylvania;* Polly Amstutz, *University of Nebraska at Kearney;* George Anastassiou, *University of Memphis;* Judy Barclay, *Cuesta College;* Jean Michelle Benedict, *Augusta State University;* David Bregenzer, *Utah State University;* Ben Brink, *Wharton County Junior College;* Mary Chabot, *Mt. San Antonio College;* Jimmy Chang, *St. Petersburg College;* Joseph Chance, *University of Texas—Pan American;* John Chuchel, *University of California;* Derron Coles, *Oregon State University;* Miriam E. Connellan, *Marquette University;* William Conway, *University of Arizona;* Karabi Datta, *Northern Illinois University;* Keng Deng, *University of Louisiana at Lafayette;* Roger A. Engle, *Clarion University of Pennsylvania;* David French, *Tidewater Community College;* Randy Gallaher, *Lewis & Clark Community College;* Perry Gillespie, *Fayetteville State University;* Jose Gimenez, *Temple University;* Betty Givan, *Eastern Kentucky University;* Walter J. Gleason, *Bridgewater State College;* Shane Goodwin, *Brigham Young University of Idaho;* Mark Greenhalgh, *Fullerton College;* Harvey Greenwald, *California Polytechnic State University;* Karen Hay, *Mesa Community College;* Raymond Heitmann, *University of Texas at Austin;* Larry Hoehn, *Austin Peay State University;* William C. Huffman, *Loyola University of Chicago;* Arlene Jesky, *Rose State College;* Raja Khoury, *Collin County Community College;* Ronnie Khuri, *University of Florida;* Bernadette Kocyba, *J. Sergeant Reynolds Community College;* Duane Kouba, *University of California—Davis;* James A. Kurre, *The Pennsylvania State University;* Melvin Lax, *California State University—Long Beach;* Norbert Lerner, *State University of New York at Cortland;* Yuhlong Lio, *University of South Dakota;* Peter J. Livorsi, *Oakton Community College;* Ivan Loy, *Front Range Community College;* Peggy Luczak, *Camden County College;* Lewis D. Ludwig, *Denison University;* Samuel A. Lynch, *Southwest Missouri State University;* Augustine Maison, *Eastern Kentucky University;* Kevin McDonald, *Mt. San Antonio College;* Earl H. McKinney, *Ball State University;* Randall McNiece, *San Jacinto College;* Philip R. Montgomery, *University of Kansas;* John Nardo, *Oglethorpe University;* Mike Nasab, *Long Beach City College;* Karla Neal, *Louisiana State University;* James Osterburg, *University of Cincinnati;* Darla Ottman, *Elizabethtown Community & Technical College;* William Parzynski, *Montclair State University;* Scott Perkins, *Lake Sumter Community College;* Laurie Poe, *Santa Clara University;*

Adelaida Quesada, *Miami Dade College—Kendall;* Brooke P. Quinlan, *Hillsborough Community College;* David Ray, *University of Tennessee at Martin;* Rita Richards, *Scottsdale Community College;* Stephen B. Rodi, *Austin Community College;* Carol Rychly, *Augusta State University;* Yvonne Sandoval-Brown, *Pima Community College;* Richard Semmler, *Northern Virginia Community College—Annandale;* Bernard Shapiro, *University of Massachusetts—Lowell;* Mike Shirazi, *Germanna Community College;* Rick Simon, *University of La Verne;* Jane Y. Smith, *University of Florida;* Marvin Stick, *University of Massachusetts—Lowell;* DeWitt L. Sumners, *Florida State University;* Devki Talwar, *Indiana University of Pennsylvania;* Linda Taylor, *Northern Virginia Community College;* Stephen Tillman, *Wilkes University;* Jay Wiestling, *Palomar College;* Jonathan Wilkin, *Northern Virginia Community College;* Carol G. Williams, *Pepperdine University;* John Williams, *St. Petersburg College;* Ted Williamson, *Montclair State University;* Melvin R. Woodard, *Indiana University of Pennsylvania;* Carlton Woods, *Auburn University at Montgomery;* Jan E. Wynn, *Brigham Young University;* Robert A. Yawin, *Springfield Technical Community College;* Charles W. Zimmerman, *Robert Morris College*

My thanks to Robert Hostetler, The Pennsylvania State University, The Behrend College, Bruce Edwards, University of Florida, and David Heyd, The Pennsylvania State University, The Behrend College, for their significant contributions to previous editions of this text.

I would also like to thank the staff at Larson Texts, Inc. who assisted with proofreading the manuscript, preparing and proofreading the art package, and checking and typesetting the supplements.

On a personal level, I am grateful to my spouse, Deanna Gilbert Larson, for her love, patience, and support. Also, a special thanks goes to R. Scott O'Neil.

If you have suggestions for improving this text, please feel free to write to me. Over the past two decades I have received many useful comments from both instructors and students, and I value these comments very highly.

Ron Larson, Ph.D.
Professor of Mathematics
Penn State University
www.RonLarson.com

CALCULUS & YOU

Every feature in this text is designed to help you learn calculus. Whenever you see a red **U**, pay special attention to the study aid. These study aids represent years of experience in teaching students *just like you*.   Ron Larson

STUDY TIP

The expressions for $f(g(x))$ and $g(f(x))$ are different in Example 5. In general, the composite of f with g is not the same as the composite of g with f.

The *Study Tips* occur at point of use throughout the text. They represent **common questions** that students ask me, **insights** into understanding concepts, and **alternative ways to look at concepts**. For instance, the *Study Tip* at the left provides insight into the importance of order when working with composite functions.

TECH TUTOR

If you have access to a symbolic differentiation utility, try using it to confirm the derivatives shown in this section.

The *Tech Tutors* give suggestions on how you can use various types of technology to help understand the material. This includes **graphing calculators**, **computer graphing programs**, and **spreadsheet programs** such as Excel. For instance, the *Tech Tutor* at the left points out that some calculators and some computer programs are capable of symbolic differentiation.

ALGEBRA TUTOR
xy

For help in evaluating the expressions in Example 2, see the review of order of operations on page 105.

Throughout years of teaching, I have found that the greatest stumbling block to success in calculus is a weakness in algebra. Each time you see an *Algebra Tutor*, please read it carefully. Then, flip ahead to the referenced page and give yourself a chance to enjoy a brief **algebra refresher**. It will be time well spent.

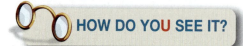 **HOW DO YOU SEE IT?**

The *How Do You See It?* question in each exercise set helps you **visually summarize concepts** without messy computations.

SUMMARIZE

The *Summarize* outline at the end of each section asks you to write each learning objective in **your own words**.

SKILLS WARM UP

The *Skills Warm Up* exercises that precede each exercise set will help you **review previously learned skills**.

SUMMARY AND STUDY STRATEGIES

The *Summary and Study Strategies*, coupled with the Review Exercises are designed to help you organize your thoughts as you **prepare for a chapter test**.

QUIZ YOURSELF

The *Quiz Yourself* occurs midway in each chapter. Take each of these quizzes as you would **take a quiz in class**.

TEST YOURSELF

The *Test Yourself* occurs at the end of each chapter. All questions are answered so you can **check your progress**.

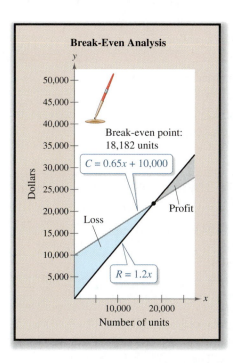

Break-Even Analysis

Break-even point:
18,182 units

$C = 0.65x + 10,000$

Profit

Loss

$R = 1.2x$

Dollars

Number of units

Example 5 on page 13 shows how the point of intersection of two graphs can be used to find the break-even point for a company manufacturing and selling a product.

1 Functions, Graphs, and Limits

1

1.1 The Cartesian Plane and the Distance Formula

In Exercise 29, you will use a line graph to estimate the Dow Jones Industrial Average.

■ Plot points in a coordinate plane and represent data graphically.
■ Find the distance between two points in a coordinate plane.
■ Find the midpoint of a line segment connecting two points.
■ Translate points in a coordinate plane.

The Cartesian Plane

Just as you can represent real numbers by points on a real number line, you can represent ordered pairs of real numbers by points in a plane called the **rectangular coordinate system,** or the **Cartesian plane,** after the French mathematician René Descartes (1596–1650).

The Cartesian plane is formed by using two real number lines intersecting at right angles, as shown in Figure 1.1. The horizontal real number line is usually called the **x-axis,** and the vertical real number line is usually called the **y-axis.** The point of intersection of these two axes is the **origin,** and the two axes divide the plane into four parts called **quadrants.**

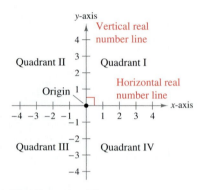

The Cartesian Plane
FIGURE 1.1

Each point in the plane corresponds to an **ordered pair** (x, y) of real numbers x and y, called **coordinates** of the point. The **x-coordinate** represents the directed distance from the y-axis to the point, and the **y-coordinate** represents the directed distance from the x-axis to the point, as shown in Figure 1.2.

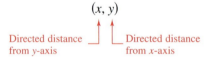

The notation (x, y) denotes both a point in the plane and an open interval on the real number line. The context will tell you which meaning is intended.

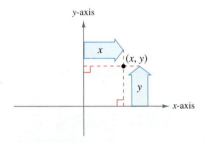

FIGURE 1.2

Example 1 Plotting Points in the Cartesian Plane

Plot the points

$$(-1, 2), \quad (3, 4), \quad (0, 0), \quad (3, 0), \quad \text{and} \quad (-2, -3).$$

SOLUTION   To plot the point

$$(-1, 2)$$

imagine a vertical line through -1 on the x-axis and a horizontal line through 2 on the y-axis. The intersection of these two lines is the point $(-1, 2)$. The other four points can be plotted in a similar way and are shown in Figure 1.3.

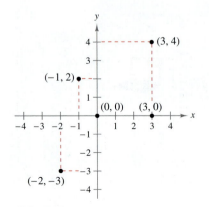

FIGURE 1.3

✓ Checkpoint 1

Plot the points

$$(-3, 2), \quad (4, -2), \quad (3, 1), \quad (0, -2), \quad \text{and} \quad (-1, -2).$$

Using a rectangular coordinate system allows you to visualize relationships between two variables. In Example 2, data is represented graphically by points plotted in a rectangular coordinate system. This type of graph is called a **scatter plot.**

 Example 2 **Sketching a Scatter Plot**

The numbers E (in millions of people) of private-sector employees in the United States from 2000 through 2009 are shown in the table, where t represents the year. Sketch a scatter plot of the data. *(Source: U.S. Bureau of Labor Statistics)*

t	2000	2001	2002	2003	2004	2005	2006	2007	2008	2009
E	111	111	109	108	110	112	114	115	114	108

STUDY TIP

In Example 2, $t = 1$ could have been used to represent the year 2000. In that case, the horizontal axis would not have been broken, and the tick marks would have been labeled 1 through 10 (instead of 2000 through 2009).

SOLUTION   To sketch a scatter plot of the data given in the table, you simply represent each pair of values by an ordered pair

(t, E)

and plot the resulting points, as shown in Figure 1.4. For instance, the first pair of values is represented by the ordered pair

$(2000, 111)$.

Note that the break in the t-axis indicates that the numbers between 0 and 2000 have been omitted.

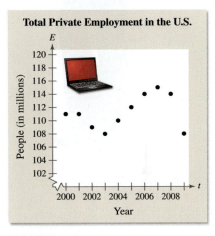

FIGURE 1.4

✓**Checkpoint 2**

The numbers E (in thousands of people) of federal employees in the United States from 2000 through 2009 are shown in the table, where t represents the year. Sketch a scatter plot of the data. *(Source: U.S. Bureau of Labor Statistics)*

t	2000	2001	2002	2003	2004	2005	2006	2007	2008	2009
E	2865	2764	2766	2761	2730	2732	2732	2734	2762	2828

The scatter plot in Example 2 is one way to represent the given data graphically. Another technique, a *bar graph*, is shown in Figure 1.5. Both graphical representations were created with a computer. If you have access to computer graphing software, try using it to represent graphically the data given in Example 2.

Another way to represent data is with a *line graph* (see Exercise 29).

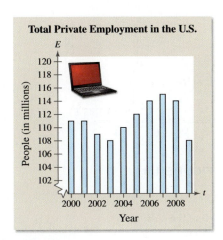

FIGURE 1.5

The Distance Formula

Recall from the Pythagorean Theorem that, for a right triangle with hypotenuse of length c and sides of lengths a and b, you have

$$a^2 + b^2 = c^2 \qquad \text{Pythagorean Theorem}$$

as shown in Figure 1.6. Note that the converse is also true. That is, if $a^2 + b^2 = c^2$, then the triangle is a right triangle.

Suppose you want to determine the distance d between two points

$$(x_1, y_1) \quad \text{and} \quad (x_2, y_2)$$

in the plane. With these two points, a right triangle can be formed, as shown in Figure 1.7. The length of the vertical side of the triangle is

$$|y_2 - y_1|$$

and the length of the horizontal side is

$$|x_2 - x_1|.$$

By the Pythagorean Theorem, you can write

$$d^2 = |x_2 - x_1|^2 + |y_2 - y_1|^2$$
$$d = \sqrt{|x_2 - x_1|^2 + |y_2 - y_1|^2}$$
$$d = \sqrt{(x_2 - x_1)^2 + (y_2 - y_1)^2}.$$

This result is the **Distance Formula.**

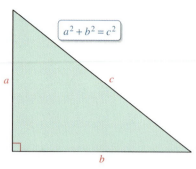

Pythagorean Theorem
FIGURE 1.6

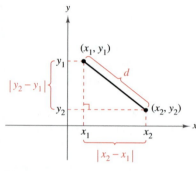

Distance Between Two Points
FIGURE 1.7

The Distance Formula

The distance d between the points (x_1, y_1) and (x_2, y_2) in the plane is

$$d = \sqrt{(x_2 - x_1)^2 + (y_2 - y_1)^2}.$$

Example 3   Finding a Distance

Find the distance between the points $(-2, 1)$ and $(3, 4)$.

SOLUTION Let $(x_1, y_1) = (-2, 1)$ and $(x_2, y_2) = (3, 4)$. Then apply the Distance Formula as shown.

$$
\begin{aligned}
d &= \sqrt{(x_2 - x_1)^2 + (y_2 - y_1)^2} && \text{Distance Formula} \\
&= \sqrt{[3 - (-2)]^2 + (4 - 1)^2} && \text{Substitute for } x_1, y_1, x_2, \text{ and } y_2. \\
&= \sqrt{(5)^2 + (3)^2} && \text{Simplify.} \\
&= \sqrt{34} \\
&\approx 5.83 && \text{Use a calculator.}
\end{aligned}
$$

So, the distance between the points is about 5.83 units. Note in Figure 1.8 that a distance of 5.83 looks about right.

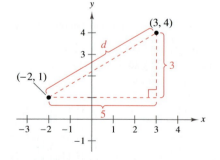

FIGURE 1.8

✓ Checkpoint 3

Find the distance between the points $(-2, 1)$ and $(2, 4)$.

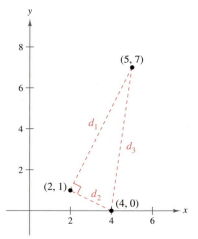

FIGURE 1.9

Example 4 **Verifying a Right Triangle**

Use the Distance Formula to show that the points

$$(2, 1), \quad (4, 0), \quad \text{and} \quad (5, 7)$$

are vertices of a right triangle.

SOLUTION The three points are plotted in Figure 1.9. Using the Distance Formula, you can find the lengths of the three sides as shown below.

$$d_1 = \sqrt{(5-2)^2 + (7-1)^2} = \sqrt{9+36} = \sqrt{45}$$
$$d_2 = \sqrt{(4-2)^2 + (0-1)^2} = \sqrt{4+1} = \sqrt{5}$$
$$d_3 = \sqrt{(5-4)^2 + (7-0)^2} = \sqrt{1+49} = \sqrt{50}$$

Because

$$d_1^2 + d_2^2 = 45 + 5 = 50 = d_3^2$$

you can apply the converse of the Pythagorean Theorem to conclude that the triangle must be a right triangle.

✓ **Checkpoint 4**

Use the Distance Formula to show that the points $(2, -1)$, $(5, 5)$, and $(6, -3)$ are vertices of a right triangle. ■

The figures provided with Examples 3 and 4 were not really essential to the solution. *Nevertheless*, it is strongly recommended that you develop the habit of including sketches with your solutions—even when they are not required.

Example 5 **Finding the Length of a Pass**

In a football game, a quarterback throws a pass from the 5-yard line, 20 yards from one sideline. The pass is caught by a wide receiver on the 45-yard line, 50 yards from the same sideline, as shown in Figure 1.10. How long was the pass?

SOLUTION You can find the length of the pass by finding the distance between the points $(20, 5)$ and $(50, 45)$.

$$d = \sqrt{(50-20)^2 + (45-5)^2} \qquad \text{\color{#B33}Distance Formula}$$
$$= \sqrt{900 + 1600}$$
$$= 50 \qquad\qquad\qquad\qquad \text{\color{#B33}Simplify.}$$

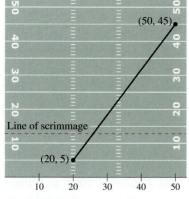

FIGURE 1.10

So, the pass was 50 yards long.

✓ **Checkpoint 5**

A quarterback throws a pass from the 10-yard line, 10 yards from one sideline. The pass is caught by a wide receiver on the 30-yard line, 25 yards from the same sideline. How long was the pass? ■

STUDY TIP

In Example 5, the scale along the goal line showing distance from the sideline does not normally appear on a football field. However, when you use coordinate geometry to solve real-life problems, you are free to place the coordinate system in any way that is convenient to the solution of the problem.

The Midpoint Formula

To find the **midpoint** of the line segment that joins two points in a coordinate plane, you can simply find the average values of the respective coordinates of the two endpoints.

The Midpoint Formula

The midpoint of the segment joining the points (x_1, y_1) and (x_2, y_2) is

$$\text{Midpoint} = \left(\frac{x_1 + x_2}{2}, \frac{y_1 + y_2}{2}\right).$$

Example 6 **Finding a Segment's Midpoint**

Find the midpoint of the line segment joining the points

$$(-5, -3) \quad \text{and} \quad (9, 3)$$

as shown in Figure 1.11.

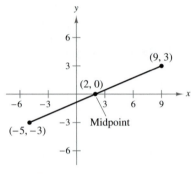

FIGURE 1.11

SOLUTION Let $(x_1, y_1) = (-5, -3)$ and $(x_2, y_2) = (9, 3)$.

$$\text{Midpoint} = \left(\frac{x_1 + x_2}{2}, \frac{y_1 + y_2}{2}\right) = \left(\frac{-5 + 9}{2}, \frac{-3 + 3}{2}\right) = (2, 0)$$

✓**Checkpoint 6**

Find the midpoint of the line segment joining the points $(-6, 2)$ and $(2, 8)$. ■

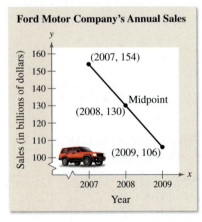

FIGURE 1.12

Example 7 **Estimating Annual Sales**

Ford Motor Company had annual sales of about $154 billion in 2007 and about $106 billion in 2009. Without knowing any additional information, what would you estimate the 2008 annual sales to have been? *(Source: Ford Motor Co.)*

SOLUTION One solution to the problem is to assume that sales followed a linear pattern. Then you can estimate the 2008 sales by finding the midpoint of the segment connecting the points (2007, 154) and (2009, 106).

$$\text{Midpoint} = \left(\frac{2007 + 2009}{2}, \frac{154 + 106}{2}\right) = (2008, 130)$$

So, you would estimate the 2008 sales to have been $130 billion, as shown in Figure 1.12. (The actual 2008 sales were $129 billion.)

✓**Checkpoint 7**

CVS Caremark Corporation had annual sales of about $76 billion in 2007 and about $99 billion in 2009. What would you estimate the 2008 annual sales to have been? *(Source: CVS Caremark Corp.)* ■

Translating Points in the Plane

Much of computer graphics consists of transformations of points in a coordinate plane. One type of transformation, a translation, is illustrated in Example 8. Other types of transformations include reflections, rotations, and stretches.

Example 8 Translating Points in the Plane

Figure 1.13(a) shows the vertices of a parallelogram. Find the vertices of the parallelogram after it has been translated four units to the right and two units down.

SOLUTION   To translate each vertex four units to the right, add 4 to each x-coordinate. To translate each vertex two units down, subtract 2 from each y-coordinate.

Original Point	*Translated Point*
$(1, 0)$	$(1 + 4, 0 - 2) = (5, -2)$
$(3, 2)$	$(3 + 4, 2 - 2) = (7, 0)$
$(3, 6)$	$(3 + 4, 6 - 2) = (7, 4)$
$(1, 4)$	$(1 + 4, 4 - 2) = (5, 2)$

The translated parallelogram is shown in Figure 1.13(b).

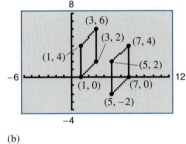

(a)   (b)

FIGURE 1.13

Many movies now use extensive computer graphics, much of which consists of transformations of points in two- and three-dimensional space. The photo above shows a character from *Avatar*. The movie's animators used computer graphics to design the scenery, characters, motion, and even the lighting throughout much of the film.

✓ **Checkpoint 8**

Find the vertices of the parallelogram in Example 8 after it has been translated two units to the left and four units down.

SUMMARIZE   (Section 1.1)

1. Describe the Cartesian plane *(page 2)*. For examples of plotting points in the Cartesian plane, see Examples 1 and 2.

2. State the Distance Formula *(page 4)*. For examples of using the Distance Formula to find the distance between two points, see Examples 3, 4, and 5.

3. State the Midpoint Formula *(page 6)*. For examples of using the Midpoint Formula to find the midpoint of a line segment, see Examples 6 and 7.

4. Describe how to translate points in the Cartesian plane *(page 7)*. For an example of translating points in the Cartesian plane, see Example 8.

SKILLS WARM UP 1.1 The following warm-up exercises involve skills that were covered in a previous course. You will use these skills in the exercise set for this section. For additional help, review Appendix Section A.3.

In Exercises 1–6, simplify the expression.

1. $\sqrt{(3-6)^2 + [1-(-5)]^2}$

2. $\sqrt{(-2-0)^2 + [-7-(-3)]^2}$

3. $\dfrac{5 + (-4)}{2}$

4. $\dfrac{-3 + (-1)}{2}$

5. $\sqrt{27} + \sqrt{12}$

6. $\sqrt{8} - \sqrt{18}$

In Exercises 7–10, solve for x or y.

7. $\sqrt{(3-x)^2 + (7-4)^2} = \sqrt{45}$

8. $\sqrt{(6-2)^2 + (-2-y)^2} = \sqrt{52}$

9. $\dfrac{x + (-5)}{2} = 7$

10. $\dfrac{-7 + y}{2} = -3$

1.2 Graphs of Equations

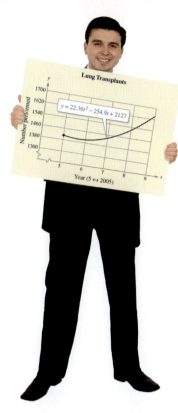

In Exercise 62, you will use a mathematical model to analyze the number of lung transplants in the United States.

■ Sketch graphs of equations by hand.
■ Find the *x*- and *y*-intercepts of graphs of equations.
■ Write the standard forms of equations of circles.
■ Find the points of intersection of two graphs.
■ Use mathematical models to model and solve real-life problems.

The Graph of an Equation

In Section 1.1, you used a coordinate system to represent graphically the relationship between two quantities. There, the graphical picture consisted of a collection of points in a coordinate plane (see Example 2 in Section 1.1).

Frequently, a relationship between two quantities is expressed as an equation. For instance, degrees on the Fahrenheit scale are related to degrees on the Celsius scale by the equation

$$F = \frac{9}{5}C + 32.$$

In this section, you will study some basic procedures for sketching the graphs of such equations. The **graph** of an equation is the set of all points that are solutions of the equation.

Example 1   Sketching the Graph of an Equation

Sketch the graph of $y = 7 - 3x$.

SOLUTION   The simplest way to sketch the graph of an equation is the *point-plotting method*. With this method, you construct a table of values that consists of several solution points of the equation, as shown in the table below. For instance, when $x = 0$,

$$y = 7 - 3(0) = 7$$

which implies that $(0, 7)$ is a solution point of the equation.

x	0	1	2	3	4
$y = 7 - 3x$	7	4	1	-2	-5

From the table, it follows that $(0, 7)$, $(1, 4)$, $(2, 1)$, $(3, -2)$, and $(4, -5)$ are solution points of the equation. After plotting these points, you can see that they appear to lie on a line, as shown in Figure 1.14. The graph of the equation is the line that passes through the five plotted points.

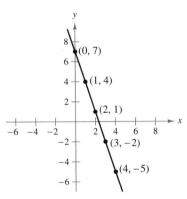

Solution Points for $y = 7 - 3x$
FIGURE 1.14

✓Checkpoint 1

Sketch the graph of $y = 2x - 1$.

STUDY TIP

Even though the sketch shown in Figure 1.14 is referred to as the graph of $y = 7 - 3x$, it actually represents only a *portion* of the graph. The entire graph is a line that would extend off the page.

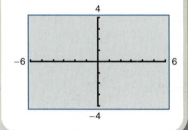
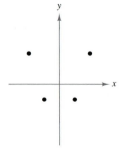
Example 2   **Sketching the Graph of an Equation**

Sketch the graph of $y = x^2 - 2$.

SOLUTION   Begin by constructing a table of values, as shown below.

x	-2	-1	0	1	2	3
$y = x^2 - 2$	2	-1	-2	-1	2	7

Next, plot the points given in the table, as shown in Figure 1.15(a). Finally, connect the points with a smooth curve, as shown in Figure 1.15(b).

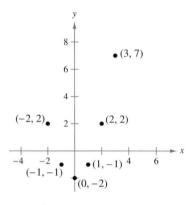

(a)

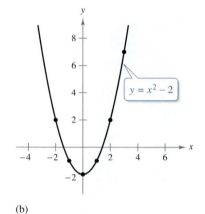

(b)

FIGURE 1.15

✓**Checkpoint 2**

Sketch the graph of $y = x^2 - 4$.

The graph shown in Example 2 is a **parabola.** The graph of any second-degree equation of the form

$$y = ax^2 + bx + c, \quad a \neq 0$$

has a similar shape. If $a > 0$, then the parabola opens upward, as shown in Figure 1.15(b), and if $a < 0$, then the parabola opens downward.

The point-plotting technique demonstrated in Examples 1 and 2 is easy to use, but it does have some shortcomings. With too few solution points, you can badly misrepresent the graph of a given equation. For instance, how would you connect the four points in Figure 1.16? Without further information, any one of the three graphs in Figure 1.17 would be reasonable.

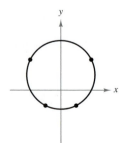

FIGURE 1.16

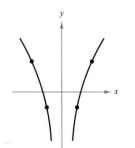

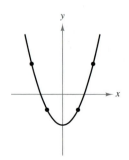

FIGURE 1.17

Intercepts of a Graph

Some solution points have zero as either the *x*-coordinate or the *y*-coordinate. These points are called **intercepts** because they are the points at which the graph intersects the *x*- or *y*-axis.

Some texts denote the *x*-intercept as the *x*-coordinate of the point

$$(a, 0)$$

rather than the point itself. Unless it is necessary to make a distinction, when the term *intercept* is used in this text, it will mean either the point or the coordinate.

A graph may have no intercepts or several intercepts, as shown in Figure 1.18.

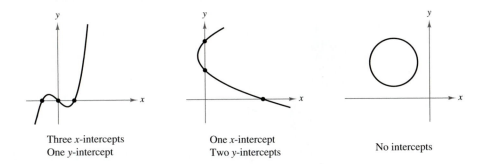

No *x*-intercept
One *y*-intercept

Three *x*-intercepts
One *y*-intercept

One *x*-intercept
Two *y*-intercepts

No intercepts

FIGURE 1.18

Finding Intercepts

1. To find **x-intercepts,** let *y* be zero and solve the equation for *x*.

2. To find **y-intercepts,** let *x* be zero and solve the equation for *y*.

Example 3   **Finding x- and y-Intercepts**

Find the *x*- and *y*-intercepts of the graph of $y = x^3 - 4x$.

SOLUTION   To find the *x*-intercepts, let *y* be zero and solve for *x*.

$$x^3 - 4x = 0 \qquad \text{Let } y \text{ be zero.}$$
$$x(x^2 - 4) = 0 \qquad \text{Factor out common monomial factor.}$$
$$x(x + 2)(x - 2) = 0 \qquad \text{Factor.}$$
$$x = 0, -2, \text{ or } 2 \qquad \text{Solve for } x.$$

Because this equation has three solutions, you can conclude that the graph has three *x*-intercepts:

$$(0, 0), \quad (-2, 0), \quad \text{and} \quad (2, 0). \qquad \text{\textit{x}-intercepts}$$

To find the *y*-intercepts, let *x* be zero and solve for *y*. Doing this produces

$$y = x^3 - 4x = 0^3 - 4(0) = 0.$$

This equation has only one solution, so the graph has one *y*-intercept:

$$(0, 0). \qquad \text{\textit{y}-intercept}$$

(See Figure 1.19.)

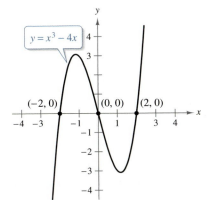

FIGURE 1.19

✓**Checkpoint 3**

Find the *x*- and *y*-intercepts of the graph of $y = x^2 - 2x - 3$.

Circles

Throughout this course, you will learn to recognize several types of graphs from their equations. For instance, you should recognize that the graph of a second-degree equation of the form

$$y = ax^2 + bx + c, \quad a \neq 0$$

is a parabola (see Example 2). Another easily recognized graph is that of a **circle.**

Consider the circle shown in Figure 1.20. A point (x, y) is on the circle if and only if its distance from the center (h, k) is r. By the Distance Formula,

$$\sqrt{(x - h)^2 + (y - k)^2} = r.$$

By squaring both sides of this equation, you obtain the **standard form of the equation of a circle.**

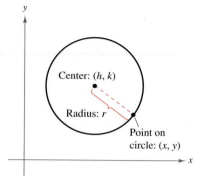

FIGURE 1.20

Standard Form of the Equation of a Circle

The **standard form of the equation of a circle** is

$$(x - h)^2 + (y - k)^2 = r^2. \qquad \text{Center at } (h, k)$$

The point (h, k) is the **center** of the circle, and the positive number r is the **radius** of the circle. The standard form of the equation of a circle whose center is the origin, $(h, k) = (0, 0)$, is

$$x^2 + y^2 = r^2. \qquad \text{Center at } (0, 0)$$

Example 4   Finding the Equation of a Circle

The point $(3, 4)$ lies on a circle whose center is at $(-1, 2)$. Find the standard form of the equation of this circle and sketch its graph.

SOLUTION The radius of the circle is the distance between $(-1, 2)$ and $(3, 4)$.

$$
\begin{aligned}
r &= \sqrt{[3 - (-1)]^2 + (4 - 2)^2} & \text{Distance Formula} \\
&= \sqrt{(4)^2 + (2)^2} & \text{Simplify.} \\
&= \sqrt{16 + 4} & \text{Simplify.} \\
&= \sqrt{20} & \text{Radius}
\end{aligned}
$$

Using $(h, k) = (-1, 2)$ and $r = \sqrt{20}$, the standard form of the equation of the circle is

$$
\begin{aligned}
(x - h)^2 + (y - k)^2 &= r^2 & \text{Equation of a circle} \\
[x - (-1)]^2 + (y - 2)^2 &= \left(\sqrt{20}\right)^2 & \text{Substitute for } h, k, \text{ and } r. \\
(x + 1)^2 + (y - 2)^2 &= 20. & \text{Write in standard form.}
\end{aligned}
$$

The graph of the equation of the circle is shown in Figure 1.21. _____

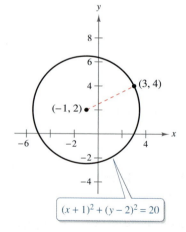

$$(x + 1)^2 + (y - 2)^2 = 20$$

FIGURE 1.21

✓ Checkpoint 4

The point $(1, 5)$ lies on a circle whose center is at $(-2, 1)$. Find the standard form of the equation of this circle and sketch its graph. ▪

Points of Intersection

An ordered pair that is a solution of two different equations is called a **point of intersection** of the graphs of the two equations. For instance, Figure 1.22 shows that the graphs of

$$y = x^2 - 3 \quad \text{and} \quad y = x - 1$$

have two points of intersection: $(2, 1)$ and $(-1, -2)$. To find the points analytically, set the two y-values equal to each other and solve the equation

$$x^2 - 3 = x - 1$$

for x.

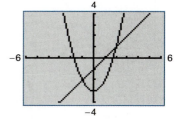

FIGURE 1.22

A common business application that involves points of intersection is **break-even analysis.** The marketing of a new product typically requires an initial investment. When sufficient units have been sold so that the total revenue has offset the total cost, the sale of the product has reached the **break-even point.** The **total cost** of producing x units of a product is denoted by C, and the **total revenue** from the sale of x units of the product is denoted by R. So, you can find the break-even point by setting the cost C equal to the revenue R and solving for x. In other words, the break-even point corresponds to the point of intersection of the cost and revenue graphs.

 Example 5   **Finding a Break-Even Point**

A company manufactures a product at a cost of \$0.65 per unit and sells the product for \$1.20 per unit. The company's initial investment to produce the product was \$10,000. Will the company break even when it sells 18,000 units? How many units must the company sell to break even?

SOLUTION   The total cost of producing x units of the product is given by

$$C = 0.65x + 10,000. \qquad \text{Cost equation}$$

The total revenue from the sale of x units is given by

$$R = 1.2x. \qquad \text{Revenue equation}$$

To find the break-even point, set the cost equal to the revenue and solve for x.

$$R = C \qquad \text{Set revenue equal to cost.}$$
$$1.2x = 0.65x + 10,000 \qquad \text{Substitute for } R \text{ and } C.$$
$$0.55x = 10,000 \qquad \text{Subtract } 0.65x \text{ from each side.}$$
$$x = \frac{10,000}{0.55} \qquad \text{Divide each side by 0.55.}$$
$$x \approx 18,182 \qquad \text{Use a calculator.}$$

So, the company will not break even when it sells 18,000 units. The company must sell 18,182 units before it breaks even. This result is shown graphically in Figure 1.23. Note in Figure 1.23 that sales less than 18,182 units correspond to a loss for the company ($R < C$), whereas sales greater than 18,182 units correspond to a profit for the company ($R > C$).

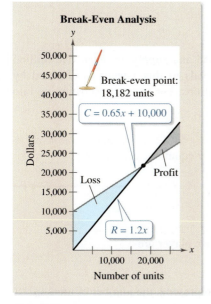

FIGURE 1.23

✓ **Checkpoint 5**

How many units must the company in Example 5 sell to break even when the selling price is \$1.45 per unit?

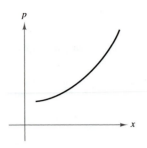

Supply Curve
FIGURE 1.24

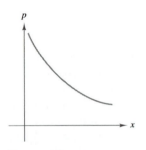

Demand Curve
FIGURE 1.25

Two types of equations that economists use to analyze a market are supply and demand equations. A **supply equation** shows the relationship between the unit price p of a product and the quantity supplied x. The graph of a supply equation is called a **supply curve.** (See Figure 1.24.) A typical supply curve rises because producers of a product want to sell more units when the unit price is higher.

A **demand equation** shows the relationship between the unit price p of a product and the quantity demanded x. The graph of a demand equation is called a **demand curve.** (See Figure 1.25.) A typical demand curve tends to show a decrease in the quantity demanded with each increase in price.

In an ideal situation, with no other factors present to influence the market, the production level should stabilize at the point of intersection of the graphs of the supply and demand equations. This point is called the **equilibrium point.** The x-coordinate of the equilibrium point is called the **equilibrium quantity** and the p-coordinate is called the **equilibrium price.** (See Figure 1.26.) You can find the equilibrium point by setting the demand equation equal to the supply equation and solving for x.

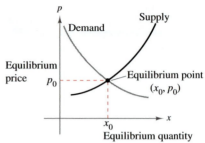

Equilibrium Point
FIGURE 1.26

Example 6   Finding the Equilibrium Point

The demand and supply equations for an e-book reader are given by

$$p = 195 - 5.8x \qquad \text{Demand equation}$$
$$p = 150 + 3.2x \qquad \text{Supply equation}$$

where p is the price in dollars and x represents the number of units in millions. Find the equilibrium point for this market.

SOLUTION   Begin by setting the demand equation equal to the supply equation.

$$195 - 5.8x = 150 + 3.2x \qquad \text{Set equations equal to each other.}$$
$$45 - 5.8x = 3.2x \qquad \text{Subtract 150 from each side.}$$
$$45 = 9x \qquad \text{Add } 5.8x \text{ to each side.}$$
$$5 = x \qquad \text{Divide each side by 9.}$$

So, the equilibrium point occurs when the demand and supply are each five million units. (See Figure 1.27.) The price that corresponds to this x-value is obtained by substituting $x = 5$ into either of the original equations. For instance, substituting into the demand equation produces

$$p = 195 - 5.8(5) = 195 - 29 = \$166.$$

Note that when you substitute $x = 5$ into the supply equation, you obtain

$$p = 150 + 3.2(5) = 150 + 16 = \$166.$$

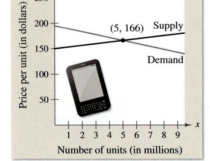

FIGURE 1.27

✓ Checkpoint 6

The demand and supply equations for a Blu-ray disc player are

$$p = 136 - 3.5x \quad \text{and} \quad p = 112 + 2.5x$$

respectively, where p is the price in dollars and x represents the number of units in millions. Find the equilibrium point for this market.

Mathematical Models

In this text, you will see many examples of the use of equations as **mathematical models** of real-life phenomena. In developing a mathematical model to represent actual data, you should strive for two (often conflicting) goals—accuracy and simplicity.

Example 7   Using Mathematical Models

The table shows the annual sales (in millions of dollars) for Dollar Tree and 99 Cents Only Stores from 2005 through 2009. In 2009, the publication *Value Line* listed the projected 2010 sales for the companies as $5770 million and $1430 million, respectively. How do you think these projections were obtained? *(Source: Dollar Tree, Inc. and 99 Cents Only Stores)*

Year	2005	2006	2007	2008	2009
t	5	6	7	8	9
Dollar Tree	3393.9	3969.4	4242.6	4644.9	5231.2
99 Cents Only Stores	1023.6	1104.7	1199.4	1302.9	1355.2

SOLUTION   The projections were obtained by using past sales to predict future sales. The past sales were modeled by equations that were found by a statistical procedure called least squares regression analysis. The models for the two companies are

$$S = 10.764t^2 + 284.31t + 1757.3, \quad 5 \le t \le 9 \qquad \text{Dollar Tree}$$

and

$$S = -3.486t^2 + 134.94t + 430.4, \quad 5 \le t \le 9. \qquad \text{99 Cents Only Stores}$$

Using $t = 10$ to represent 2010, you can predict the 2010 sales to be

$$S = 10.764(10)^2 + 284.31(10) + 1757.3 \approx 5676.8 \qquad \text{Dollar Tree}$$

and

$$S = -3.486(10)^2 + 134.94(10) + 430.4 \approx 1431.2. \qquad \text{99 Cents Only Stores}$$

These two projections are close to those projected by *Value Line*. The graphs of the two models are shown in Figure 1.28.

For help in evaluating the expressions in Example 7, see the review of order of operations on page 57.

ALGEBRA TUTOR xy

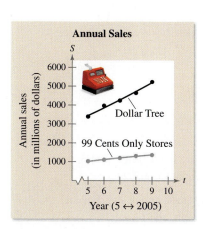

Annual Sales

FIGURE 1.28

✓ Checkpoint 7

The table shows the annual revenues (in millions of dollars) for BJ's Wholesale Club from 2005 through 2009. In 2009, the publication *Value Line* listed the projected 2010 revenues for BJ's Wholesale Club as $11,150 million. How does this projection compare with the projection obtained using the model below? *(Source: BJ's Wholesale Club, Inc.)*

$$S = -17.393t^2 + 845.59t + 4097.7, \quad 5 \le t \le 9$$

Year	2005	2006	2007	2008	2009
t	5	6	7	8	9
Revenues	7949.9	8480.3	9005.0	10,027.0	10,187.0

To test the accuracy of a model, you can compare the actual data with the values given by the model. Try doing this for each model in Example 7.

Much of your study of calculus will center around the behavior of the graphs of mathematical models. Figure 1.29 shows the graphs of six basic algebraic equations. Familiarity with these graphs will help you in the creation and use of mathematical models.

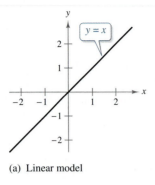

(a) Linear model

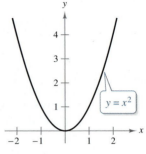

(b) Quadratic model

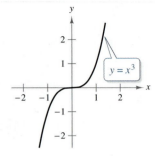

(c) Cubic model

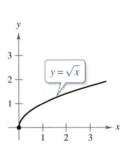

(d) Square root model

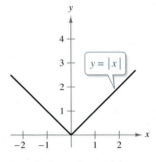

(e) Absolute value model

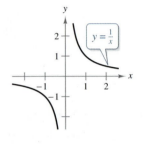

(f) Rational model

FIGURE 1.29

SUMMARIZE (Section 1.2)

1. Describe how to sketch the graph of an equation by hand *(page 9)*. For examples of sketching a graph by hand, see Examples 1 and 2.

2. Describe how to find the *x*- and *y*-intercepts of a graph *(page 11)*. For an example of finding the *x*- and *y*-intercepts of a graph, see Example 3.

3. State the standard form of the equation of a circle *(page 12)*. For an example of finding the standard form of the equation of a circle, see Example 4.

4. Describe how to find a point of intersection of the graphs of two equations *(page 13)*. For examples of finding points of intersection, see Examples 5 and 6.

5. Describe break-even analysis *(page 13)*. For an example of break-even analysis, see Example 5.

6. Describe supply equations and demand equations *(page 14)*. For examples of a supply equation and a demand equation, see Example 6.

7. Describe a mathematical model *(page 15)*. For an example of a mathematical model, see Example 7.

SKILLS WARM UP 1.2

The following warm-up exercises involve skills that were covered in a previous course. You will use these skills in the exercise set for this section. For additional help, review Appendix Sections A.3 and A.4.

In Exercises 1–6, solve for y.

1. $5y - 12 = x$

2. $-y = 15 - x$

3. $x^3y + 2y = 1$

4. $x^2 + x - y^2 - 6 = 0$

5. $(x - 2)^2 + (y + 1)^2 = 9$

6. $(x + 6)^2 + (y - 5)^2 = 81$

In Exercises 7–10, evaluate the expression for the given value of x.

Expression	*x-Value*
7. $y = 5x$	$x = -2$
8. $y = 3x - 4$	$x = 3$
9. $y = 2x^2 + 1$	$x = 2$
10. $y = x^2 + 2x - 7$	$x = -4$

In Exercises 11–14, factor the expression.

11. $x^2 - 3x + 2$

12. $x^2 + 5x + 6$

13. $y^2 - 3y + \frac{9}{4}$

14. $y^2 - 7y + \frac{49}{4}$

1.3 Lines in the Plane and Slope

■ Use the slope-intercept form of a linear equation to sketch graphs.
■ Find slopes of lines passing through two points.
■ Use the point-slope form to write equations of lines.
■ Find equations of parallel and perpendicular lines.
■ Use linear equations to model and solve real-life problems.

Using Slope

The simplest mathematical model for relating two variables is the **linear equation**

$$y = mx + b. \qquad \text{Linear equation}$$

This equation is called *linear* because its graph is a line. (In this text, the term *line* is used to mean *straight line*.) By letting $x = 0$, you can see that the line crosses the y-axis at

$$y = b$$

as shown in Figure 1.30. In other words, the y-intercept is $(0, b)$. The steepness or slope of the line is m.

$$y = mx + b$$

Slope ⟶   ⟵ y-intercept

The **slope** of a nonvertical line is the number of units the line rises (or falls) vertically for each unit of horizontal change from left to right, as shown in Figure 1.30.

In Exercise 87, you will use slope to analyze the average salaries of senior high school principals.

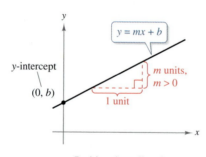

Positive slope, line rises.   Negative slope, line falls.

FIGURE 1.30

A linear equation that is written in the form $y = mx + b$ is said to be written in **slope-intercept form.**

The Slope-Intercept Form of the Equation of a Line

The graph of the equation

$$y = mx + b$$

is a line whose slope is m and whose y-intercept is $(0, b)$.

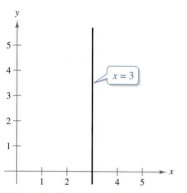

When the line is vertical, the slope is undefined.

FIGURE 1.31

A vertical line has an equation of the form

$$x = a. \qquad \text{Vertical line}$$

Because the equation of a vertical line cannot be written in the form $y = mx + b$, it follows that the slope of a vertical line is undefined, as indicated in Figure 1.31.

Once you have determined the slope and the y-intercept of a line, it is a relatively simple matter to sketch its graph.

Example 1   Graphing a Linear Equation

Sketch the graph of each linear equation.

a. $y = 2x + 1$

b. $y = 2$

c. $x + y = 2$

SOLUTION

a. This equation is written in slope-intercept form, $y = mx + b$. Because $b = 1$, the y-intercept is $(0, 1)$. Moreover, because the slope is $m = 2$, the line *rises* two units for each unit the line moves to the right, as shown in Figure 1.32(a).

b. By writing this equation in slope-intercept form

$$y = (0)x + 2$$

you can see that the y-intercept is $(0, 2)$ and the slope is zero. A zero slope implies that the line is horizontal—that is, it doesn't rise *or* fall, as shown in Figure 1.32(b).

c. By writing this equation in slope-intercept form

$x + y = 2$	Write original equation.
$y = -x + 2$	Subtract x from each side.
$y = (-1)x + 2$	Write in slope-intercept form.

you can see that the y-intercept is $(0, 2)$. Moreover, because the slope is $m = -1$, this line *falls* one unit for each unit the line moves to the right, as shown in Figure 1.32(c).

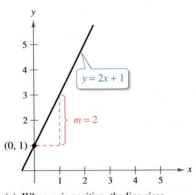

(a) When m is positive, the line rises from left to right.

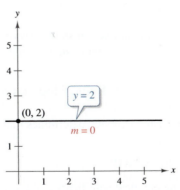

(b) When m is zero, the line is horizontal.

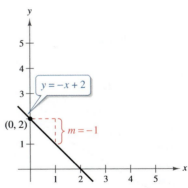

(c) When m is negative, the line falls from left to right.

FIGURE 1.32

✓ Checkpoint 1

Sketch the graph of each linear equation.

a. $y = 4x - 2$

b. $x = 1$

c. $2x + y = 6$

In real-life problems, the slope of a line can be interpreted as either a *ratio* or a *rate*. If the *x*-axis and *y*-axis have the same unit of measure, then the slope has no units and is a **ratio**. If the *x*-axis and *y*-axis have different units of measure, then the slope is a **rate** or **rate of change.**

 Example 2   **Using Slope as a Ratio**

The maximum recommended slope of a wheelchair ramp is $\frac{1}{12} \approx 0.083$. A business installs a wheelchair ramp that rises 22 inches over a horizontal length of 24 feet, as shown in Figure 1.33. Is the ramp steeper than recommended? *(Source: ADA Standards for Accessible Design)*

FIGURE 1.33

SOLUTION   The horizontal length of the ramp is 24 feet or $12(24) = 288$ inches. So, the slope of the ramp is

$$\text{Slope} = \frac{\text{vertical change}}{\text{horizontal change}}$$

$$= \frac{22 \text{ in.}}{288 \text{ in.}}$$

$$\approx 0.076.$$

So, the slope is not steeper than recommended.

✔**Checkpoint 2**

The business in Example 2 installs a second ramp that rises 27 inches over a horizontal length of 26 feet. Is the ramp steeper than recommended?

 Example 3   **Using Slope as a Rate of Change**

A manufacturing company determines that the total cost in dollars of producing *x* units of a product is

$$C = 25x + 3500.$$

Describe the practical significance of the *y*-intercept and slope of the line given by this equation.

SOLUTION   The *y*-intercept $(0, 3500)$ tells you that the cost of producing zero units is $3500. This is the **fixed cost** of production—it includes costs that must be paid regardless of the number of units produced. The slope, which is $m = 25$, tells you that the cost of producing each unit is $25, as shown in Figure 1.34. Economists call the cost per unit the **marginal cost.** If the production increases by one unit, then the "margin" or extra amount of cost is $25.

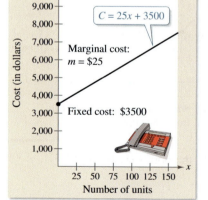

Production Cost

Cost (in dollars)

10,000
9,000
8,000
7,000
6,000
5,000
4,000
3,000
2,000
1,000

$C = 25x + 3500$

Marginal cost: $m = \$25$

Fixed cost: $3500

25 50 75 100 125 150

Number of units

FIGURE 1.34

✔**Checkpoint 3**

A small business determines that the value of a copier *t* years after its purchase is $V = -300t + 1500$. Describe the practical significance of the *y*-intercept and slope of the line given by this equation.

Finding the Slope of a Line

Given an equation of a nonvertical line, you can find its slope by writing the equation in slope-intercept form. When you are not given an equation, you can still find the slope of a line. For instance, suppose you want to find the slope of the line passing through the points (x_1, y_1) and (x_2, y_2), as shown in Figure 1.35. As you move from left to right along this line, a change of $(y_2 - y_1)$ units in the vertical direction corresponds to a change of $(x_2 - x_1)$ units in the horizontal direction. These two changes are denoted by the symbols

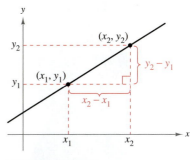

FIGURE 1.35

$$\Delta y = y_2 - y_1 = \text{the change in } y$$

and

$$\Delta x = x_2 - x_1 = \text{the change in } x.$$

(The symbol Δ is the Greek capital letter delta, and the symbols Δy and Δx are read as "delta y" and "delta x.") The ratio of Δy to Δx represents the slope of the line that passes through the points (x_1, y_1) and (x_2, y_2).

$$\text{Slope} = \frac{\Delta y}{\Delta x} = \frac{y_2 - y_1}{x_2 - x_1}$$

Be sure you see that Δx represents a single number, not the product of two numbers (Δ and x). The same is true for Δy.

The Slope of a Line Passing Through Two Points

The **slope** m of the nonvertical line passing through the points (x_1, y_1) and (x_2, y_2) is

$$m = \frac{\Delta y}{\Delta x} = \frac{y_2 - y_1}{x_2 - x_1}$$

where $x_1 \neq x_2$. Slope is not defined for vertical lines.

When this formula is used for slope, the *order of subtraction* is important. Given two points on a line, you may label either one of them as (x_1, y_1) and the other as (x_2, y_2). However, once you have done this, you must form the numerator and denominator using the same order of subtraction.

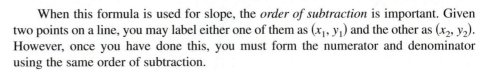

$$m = \frac{y_2 - y_1}{x_2 - x_1} \qquad m = \frac{y_1 - y_2}{x_1 - x_2} \qquad m = \frac{y_2 - y_1}{x_1 - x_2}$$

Correct   Correct   Incorrect

For instance, the slope of the line passing through the points $(3, 4)$ and $(5, 7)$ can be calculated as

$$m = \frac{7 - 4}{5 - 3} = \frac{3}{2}$$

or

$$m = \frac{4 - 7}{3 - 5} = \frac{-3}{-2} = \frac{3}{2}.$$

Example 4   **Finding the Slope of a Line**

Find the slope of the line passing through each pair of points.

a. $(-2, 0)$ and $(3, 1)$ **b.** $(-1, 2)$ and $(2, 2)$

c. $(0, 4)$ and $(1, -1)$ **d.** $(3, 4)$ and $(3, 1)$

SOLUTION

a. Letting $(x_1, y_1) = (-2, 0)$ and $(x_2, y_2) = (3, 1)$, you obtain a slope of

$$m = \frac{y_2 - y_1}{x_2 - x_1} = \frac{1 - 0}{3 - (-2)} = \frac{1}{5}$$

⟵ Difference in y-values
⟵ Difference in x-values

as shown in Figure 1.36(a).

b. The slope of the line passing through $(-1, 2)$ and $(2, 2)$ is

$$m = \frac{2 - 2}{2 - (-1)} = \frac{0}{3} = 0.$$

See Figure 1.36(b).

c. The slope of the line passing through $(0, 4)$ and $(1, -1)$ is

$$m = \frac{-1 - 4}{1 - 0} = \frac{-5}{1} = -5.$$

See Figure 1.36(c).

d. The slope of the vertical line passing through $(3, 4)$ and $(3, 1)$ is not defined because division by zero is undefined. [See Figure 1.36(d).]

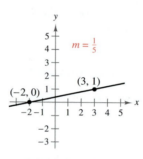

(a) Positive slope, line rises from left to right.

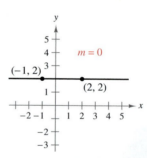

(b) Zero slope, line is horizontal.

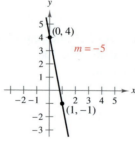

(c) Negative slope, line falls from left to right.

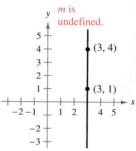

(d) Vertical line, undefined slope.

FIGURE 1.36

✓ **Checkpoint 4**

Find the slope of the line passing through each pair of points.

a. $(-3, 2)$ and $(5, 18)$

b. $(-2, 1)$ and $(-4, 2)$

c. $(2, -4)$ and $(-2, -4)$

Writing Linear Equations

When you know the slope of a line and the coordinates of one point on the line, you can find an equation for the line. For instance, in Figure 1.37, let (x_1, y_1) be a point on the line whose slope is m. If (x, y) is any other point on the line, then

$$\frac{y - y_1}{x - x_1} = m.$$

This equation, involving the variables x and y, can be rewritten in the form

$$y - y_1 = m(x - x_1)$$

which is the **point-slope form** of the equation of a line.

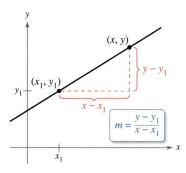

Any two points on a nonvertical line can be used to determine the slope of the line.

FIGURE 1.37

Point-Slope Form of the Equation of a Line

The equation of the line with slope m passing through the point (x_1, y_1) is

$$y - y_1 = m(x - x_1).$$

The point-slope form is most useful for *finding* the equation of a nonvertical line. You should remember this formula—it is used throughout the text.

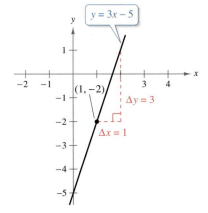

FIGURE 1.38

Example 5   **Using the Point-Slope Form**

Find the equation of the line that has a slope of 3 and passes through the point $(1, -2)$.

SOLUTION   Use the point-slope form with $m = 3$ and $(x_1, y_1) = (1, -2)$.

$$\begin{aligned}
y - y_1 &= m(x - x_1) & &\text{Point-slope form} \\
y - (-2) &= 3(x - 1) & &\text{Substitute for } m, x_1, \text{ and } y_1. \\
y + 2 &= 3x - 3 & &\text{Simplify.} \\
y &= 3x - 5 & &\text{Write in slope-intercept form.}
\end{aligned}$$

The slope-intercept form of the equation of the line is $y = 3x - 5$. The graph of this line is shown in Figure 1.38.

✓ **Checkpoint 5**

Find the equation of the line that has a slope of 2 and passes through the point $(-1, 2)$. ■

The point-slope form can be used to find an equation of the line passing through points (x_1, y_1) and (x_2, y_2). To do this, first find the slope of the line

$$m = \frac{y_2 - y_1}{x_2 - x_1}, \quad x_1 \neq x_2$$

and then use the point-slope form to obtain the equation

$$y - y_1 = \frac{y_2 - y_1}{x_2 - x_1}(x - x_1). \qquad \text{Two-point form}$$

This is sometimes called the **two-point form** of the equation of a line.

Example 6 Predicting Diluted Earnings Per Share

The diluted earnings per share (diluted EPS) for Tim Hortons, Inc. were $1.43 in 2007 and $1.55 in 2008. Using only this information, write a linear equation that gives the diluted EPS in terms of the year. Then predict the diluted EPS for 2009. *(Source: Tim Hortons, Inc.)*

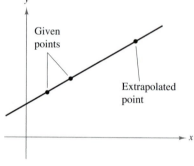

FIGURE 1.39

SOLUTION Let $t = 7$ represent 2007. Then the two given values are represented by the data points

$$(7, 1.43) \quad \text{and} \quad (8, 1.55).$$

The slope of the line through these points is

$$m = \frac{1.55 - 1.43}{8 - 7} = 0.12.$$

Using the point-slope form, you can find the equation that relates the diluted EPS y and the year t to be

$$y = 0.12t + 0.59.$$

Using $t = 9$ to represent 2009, you can predict the 2009 diluted EPS to be

$$y = 0.12(9) + 0.59 = 1.08 + 0.59 = 1.67.$$

According to this equation, the diluted EPS in 2009 was $1.67, as shown in Figure 1.39. (In this case, the prediction is fairly good—the actual diluted EPS in 2009 was $1.64.)

✓**Checkpoint 6**

The diluted EPS for Amazon.com was $0.45 in 2006 and $1.49 in 2008. Using only this information, write a linear equation that gives the diluted EPS in terms of the year. Then predict the diluted EPS for 2009. *(Source: Amazon.com)* ■

(a) Linear Extrapolation

The prediction method illustrated in Example 6 is called **linear extrapolation.** Note in Figure 1.40(a) that an extrapolated point does not lie between the given points. When the estimated point lies between two given points, as shown in Figure 1.40(b), the procedure is called **linear interpolation.**

Because the slope of a vertical line is not defined, its equation cannot be written in slope-intercept form. However, every line has an equation that can be written in the **general form**

$$Ax + By + C = 0 \qquad \text{General form}$$

where A and B are not both zero. For instance, the vertical line given by $x = a$ can be represented by the general form

$$x - a = 0. \qquad \text{General form of vertical line}$$

(b) Linear Interpolation

FIGURE 1.40

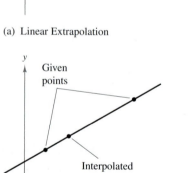

The five most common forms of equations of lines are summarized below.

Equations of Lines

1. General form: $Ax + By + C = 0$

2. Vertical line: $x = a$

3. Horizontal line: $y = b$

4. Slope-intercept form: $y = mx + b$

5. Point-slope form: $y - y_1 = m(x - x_1)$

Parallel and Perpendicular Lines

Slope can be used to decide whether two nonvertical lines in a plane are parallel, perpendicular, or neither.

Parallel and Perpendicular Lines

1. Two distinct nonvertical lines are **parallel** if and only if their slopes are equal. That is, $m_1 = m_2$.

2. Two nonvertical lines are **perpendicular** if and only if their slopes are negative reciprocals of each other. That is, $m_1 = -1/m_2$.

Example 7   Finding Parallel and Perpendicular Lines

Find equations of the lines that pass through the point $(2, -1)$ and are (a) parallel to the line $2x - 3y = 5$ and (b) perpendicular to the line $2x - 3y = 5$.

SOLUTION   By writing the given equation in slope-intercept form

$$2x - 3y = 5 \qquad \text{Write original equation.}$$

$$-3y = -2x + 5 \qquad \text{Subtract } 2x \text{ from each side.}$$

$$y = \frac{2}{3}x - \frac{5}{3} \qquad \text{Write in slope-intercept form.}$$

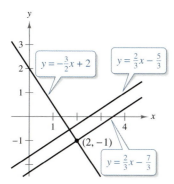

FIGURE 1.41

you can see that it has a slope of $m = \frac{2}{3}$, as shown in Figure 1.41.

a. Any line parallel to the given line must also have a slope of $\frac{2}{3}$. So, the line through $(2, -1)$ that is parallel to the given line has the following equation.

$$y - (-1) = \frac{2}{3}(x - 2) \qquad \text{Write in point-slope form.}$$

$$y + 1 = \frac{2}{3}x - \frac{4}{3} \qquad \text{Simplify.}$$

$$y = \frac{2}{3}x - \frac{4}{3} - 1 \qquad \text{Solve for } y.$$

$$y = \frac{2}{3}x - \frac{7}{3} \qquad \text{Write in slope-intercept form.}$$

b. Any line perpendicular to the given line must have a slope of $-\frac{3}{2}$ (because $-\frac{3}{2}$ is the negative reciprocal of $\frac{2}{3}$). So, the line through $(2, -1)$ that is perpendicular to the given line has the following equation.

$$y - (-1) = -\frac{3}{2}(x - 2) \qquad \text{Write in point-slope form.}$$

$$y + 1 = -\frac{3}{2}x + 3 \qquad \text{Simplify.}$$

$$y = -\frac{3}{2}x + 3 - 1 \qquad \text{Solve for } y.$$

$$y = -\frac{3}{2}x + 2 \qquad \text{Write in slope-intercept form.}$$

✓ Checkpoint 7

Find equations of the lines that pass through the point $(2, 1)$ and are (a) parallel to the line $2x - 4y = 5$ and (b) perpendicular to the line $2x - 4y = 5$. ■

TECH TUTOR

On a graphing utility, lines will not appear to have the correct slopes unless you use a viewing window that has a *square setting*. For instance, try graphing the lines in Example 7 using the standard setting $-10 \le x \le 10$ and $-10 \le y \le 10$. Then reset the viewing window with the square setting $-9 \le x \le 9$ and $-6 \le y \le 6$. On which setting do the lines $y = \frac{2}{3}x - \frac{5}{3}$ and $y = -\frac{3}{2}x + 2$ appear to be perpendicular?

Extended Application: Linear Depreciation

Most business expenses can be deducted the same year they occur. One exception to this is the cost of property that has a useful life of more than 1 year, such as buildings, cars, or equipment. Such costs must be **depreciated** over the useful life of the property. When the *same amount* is depreciated each year, the procedure is called **linear depreciation** or **straight-line depreciation.** The *book value* is the difference between the original value and the total amount of depreciation accumulated to date.

 Example 8 **Depreciating Equipment**

Your company has purchased a $12,000 machine that has a useful life of 8 years. The salvage value at the end of 8 years is $2000. Write a linear equation that describes the book value of the machine each year.

SOLUTION   Let V represent the value of the machine at the end of year t. You can represent the initial value of the machine by the ordered pair $(0, 12{,}000)$ and the salvage value of the machine by the ordered pair $(8, 2000)$. The slope of the line is

$$m = \frac{2000 - 12{,}000}{8 - 0} = \frac{-\$1250}{1 \text{ year}} \qquad m = \frac{V_2 - V_1}{t_2 - t_1}$$

which represents the annual depreciation in *dollars per year*. Using the point-slope form, you can write the equation of the line as shown.

$$V - 12{,}000 = -1250(t - 0) \qquad \text{Write in point-slope form.}$$
$$V = -1250t + 12{,}000 \qquad \text{Write in slope-intercept form.}$$

The graph of this equation is shown in Figure 1.42.

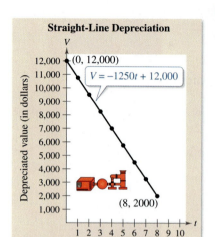

Straight-Line Depreciation

$(0, 12{,}000)$

$V = -1250t + 12{,}000$

$(8, 2000)$

Depreciated value (in dollars)

Number of years

FIGURE 1.42

✓**Checkpoint 8**

Write a linear equation for the machine in Example 8 when the salvage value at the end of 8 years is $1000. ■

SUMMARIZE   (Section 1.3)

1. State the slope-intercept form of the equation of a line *(page 18)*. For an example of an equation in slope-intercept form, see Example 1.

2. Explain how to decide whether the slope of a line is a ratio or a rate of change *(page 20)*. For an example of a slope that is a ratio, see Example 2. For an example of a slope that is a rate of change, see Example 3.

3. Explain how to find the slope of a line through two points *(page 21)*. For an example of finding the slope of a line through two points, see Example 4.

4. State the point-slope form of the equation of a line *(page 23)*. For examples of equations in point-slope form, see Examples 5 and 6.

5. Explain how to decide whether two lines are parallel, perpendicular, or neither *(page 25)*. For an example, see Example 7.

6. Describe a real-life example of how a linear equation can be used to analyze the depreciation of property *(page 26, Example 8)*.

SKILLS WARM UP 1.3 The following warm-up exercises involve skills that were covered in a previous course. You will use these skills in the exercise set for this section. For additional help, review Appendix Section A.3.

In Exercises 1 and 2, simplify the expression.

1. $\dfrac{5 - (-2)}{-3 - 4}$

2. $\dfrac{-7 - (-10)}{4 - 1}$

3. Evaluate $-\dfrac{1}{m}$ when $m = -3$.

4. Evaluate $-\dfrac{1}{m}$ when $m = \dfrac{6}{7}$.

In Exercises 5–10, solve for y in terms of x.

5. $-4x + y = 7$

6. $3x - y = 7$

7. $y - 2 = 3(x - 4)$

8. $y - (-5) = -1[x - (-2)]$

9. $y - (-3) = \dfrac{4 - (-3)}{2 - 1}(x - 2)$

10. $y - 1 = \dfrac{-3 - 1}{-7 - (-1)}[x - (-1)]$

ENHANCED **WebAssign**   Access end-of-section exercises online at **www.webassign.net**

QUIZ YOURSELF

Take this quiz as you would take a quiz in class. When you are done, check your work against the answers given in the back of the book.

In Exercises 1–3, (a) plot the points, (b) find the distance between the points, and (c) find the midpoint of the line segment joining the points.

1. $(3, -2), (-3, 1)$ **2.** $\left(\frac{1}{4}, -\frac{3}{2}\right), \left(\frac{1}{2}, 2\right)$ **3.** $(-12, 4), (6, -2)$

4. Use the Distance Formula to show that the points $(4, 0)$, $(2, 1)$, and $(-1, -5)$ are vertices of a right triangle.

5. The resident population of Georgia (in thousands) was 9534 in 2007 and 9829 in 2009. Use the Midpoint Formula to estimate the population in 2008. *(Source: U.S. Census Bureau)*

In Exercises 6–8, sketch the graph of the equation and label the intercepts.

6. $y = 5x + 2$ **7.** $y = x^2 + x - 6$ **8.** $y = |x - 3|$

In Exercises 9–11, find the standard form of the equation of the circle and sketch its graph.

9. Center: $(0, 0)$; radius: 9

10. Center: $(-1, 0)$; radius: 6

11. Center: $(2, -2)$; solution point: $(-1, 2)$

12. A business manufactures a product at a cost of $4.55 per unit and sells the product for $7.19 per unit. The company's initial investment to produce the product was $12,500. How many units must the company sell to break even?

In Exercises 13–15, find the equation of the line that passes through the given point and has the given slope. Then use the equation to sketch the line.

13. $(0, -3); m = \frac{1}{2}$ **14.** $(1, 1); m = 2$ **15.** $(6, 5); m = -\frac{1}{3}$

In Exercises 16–18, find the equation of the line that passes through the points. Then use the equation to sketch the line.

16. $(1, -1), (-4, 5)$ **17.** $(-2, 3), (-2, 2)$ **18.** $\left(\frac{5}{2}, 2\right), (0, 2)$

19. Find equations of the lines that pass through the point $(3, -5)$ and are

(a) parallel to the line $x + 4y = -2$.

(b) perpendicular to the line $x + 4y = -2$.

20. A company had sales of $1,330,000 in 2007 and $1,800,000 in 2011. The company's sales can be modeled by a linear equation. Predict the sales in 2015 and 2018.

21. A company reimburses its sales representatives $175 per day for lodging and meals, plus $0.55 per mile driven. Write a linear equation giving the daily cost C (in dollars) in terms of x, the number of miles driven.

22. Your annual salary was $34,600 in 2009 and $37,800 in 2011. Assume your salary can be modeled by a linear equation.

(a) Write a linear equation giving your salary S (in dollars) in terms of the year t. Let $t = 9$ represent 2009.

(b) Use the linear model to predict your salary in 2015.

1.4 Functions

- Decide whether the relationship between two variables is a function.
- Find the domains and ranges of functions.
- Use function notation and evaluate functions.
- Combine functions to create other functions.
- Find inverse functions algebraically.

Functions

In many common relationships between two variables, the value of one of the variables depends on the value of the other variable. Here are some examples.

1. The sales tax on an item depends on its selling price.

2. The distance an object moves in a given amount of time depends on its speed.

3. The area of a circle depends on its radius.

Consider the relationship between the total revenue R and the sale of x units of a product sold for $1.25 per unit. This relationship can be expressed by the equation

$$R = 1.25x.$$

In this equation, the value of R depends on the choice of x. Because of this, R is the **dependent variable** and x is the **independent variable.**

Most of the relationships that you will study in this course have the property that for a given value of the independent variable, there corresponds exactly one value of the dependent variable. Such a relationship is a **function.**

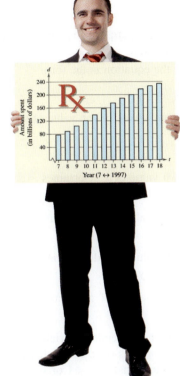

In Exercise 67, you will use a function to estimate the amounts spent on prescription drugs in the United States.

Definition of Function

A **function** is a relationship between two variables such that to each value of the independent variable there corresponds exactly one value of the dependent variable.

The **domain** of the function is the set of all values of the independent variable for which the function is defined. The **range** of the function is the set of all values taken on by the dependent variable.

As illustrated in Figure 1.43, a function can be thought of as a machine that inputs values of the independent variable and outputs values of the dependent variable.

Although functions can be described by various means such as tables, graphs, and diagrams, they are most often specified by formulas or equations. For instance, the equation

$$y = 4x^2 + 3$$

describes y as a function of x. For this function, x is the independent variable and y is the dependent variable.

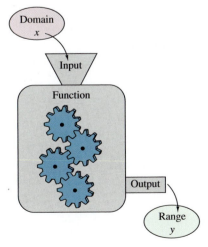

FIGURE 1.43

| Example 1 | Deciding Whether Equations Are Functions |

Decide whether each equation defines y as a function of x.

a. $x + y = 1$ **b.** $x^2 + y^2 = 1$

c. $x^2 + y = 1$ **d.** $x + y^2 = 1$

SOLUTION To decide whether an equation defines a function, it is helpful to isolate the dependent variable on the left side. For instance, to decide whether the equation $x + y = 1$ defines y as a function of x, write the equation in the form

$$y = 1 - x.$$

From this form, you can see that for any value of x, there is exactly one value of y. So, y is a function of x.

Original Equation	*Rewritten Equation*	*Test: Is y a function of x?*
a. $x + y = 1$	$y = 1 - x$	Yes, each value of x determines exactly one value of y.
b. $x^2 + y^2 = 1$	$y = \pm\sqrt{1 - x^2}$	No, some values of x determine two values of y.
c. $x^2 + y = 1$	$y = 1 - x^2$	Yes, each value of x determines exactly one value of y.
d. $x + y^2 = 1$	$y = \pm\sqrt{1 - x}$	No, some values of x determine two values of y.

Note that the equations that assign two values ($\pm$) to the dependent variable for a given value of the independent variable do not define functions of x. For instance, in part (b), when $x = 0$, the equation $y = \pm\sqrt{1 - x^2}$ indicates that $y = +1$ or $y = -1$. Figure 1.44 shows the graphs of the four equations.

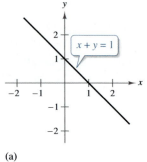 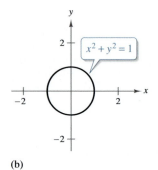 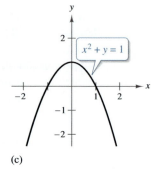 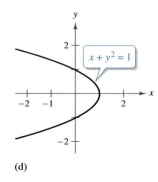

(a)   (b)   (c)   (d)

FIGURE 1.44

✓ Checkpoint 1

Decide whether each equation defines y as a function of x.

a. $x - y = 1$ **b.** $x^2 + y^2 = 4$ **c.** $y^2 + x = 2$ **d.** $x^2 - y = 0$

When the graph of a function is sketched, the standard convention is to let the horizontal axis represent the independent variable. When this convention is used, the test described in Example 1 has a nice graphical interpretation called the **Vertical Line Test.** This test states that if every vertical line intersects the graph of an equation at most once, then the equation defines y as a function of x. For instance, in Figure 1.44, the graphs in parts (a) and (c) pass the Vertical Line Test, but those in parts (b) and (d) do not.

TECH TUTOR

Many graphing utilities have an *equation editor* feature that requires an equation to be written in "$y =$" form in order to be entered. So, the procedure used in Example 1, isolating the dependent variable on the left side, is also useful for graphing equations with a graphing utility. Note that to graph an equation in which y is not a function of x, such as a circle, you usually have to enter two or more equations into the graphing utility.

The Domain and Range of a Function

The domain of a function may be described explicitly, or it may be *implied* by an equation used to define the function. For example, the function given by

$$y = \frac{1}{x^2 - 4}$$

has an implied domain that consists of all real numbers x except $x = \pm 2$. These two values are excluded from the domain because division by zero is undefined.

Another type of implied domain is that used to avoid even roots of negative numbers, as indicated in Example 2.

Example 2 — Finding the Domain and Range of a Function

Find the domain and range of each function.

a. $y = \sqrt{x - 1}$

b. $y = \begin{cases} 1 - x, & x < 1 \\ \sqrt{x - 1}, & x \geq 1 \end{cases}$

SOLUTION

a. Because $\sqrt{x - 1}$ is not defined for $x - 1 < 0$ (that is, for $x < 1$), it follows that the domain of the function is the interval

$$x \geq 1 \text{ or } [1, \infty).$$

To find the range, observe that $\sqrt{x - 1}$ is never negative. Moreover, as x takes on the various values in the domain, y takes on all nonnegative values. So, the range is the interval

$$y \geq 0 \text{ or } [0, \infty)$$

as shown in Figure 1.45(a).

b. Because this function is defined for $x < 1$ *and* for $x \geq 1$, the domain is the entire set of real numbers. This function is called a **piecewise-defined function** because it is defined by two or more equations over a specified domain. When $x \geq 1$, the function behaves as in part (a). For $x < 1$, the values of $1 - x$ are positive. So, the range of the function is

$$y \geq 0 \text{ or } [0, \infty)$$

as shown in Figure 1.45(b).

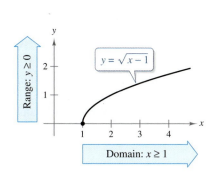

(a)

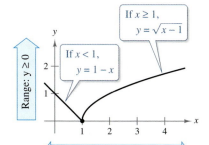

(b)

FIGURE 1.45

✓ Checkpoint 2

Find the domain and range of each function.

a. $y = \sqrt{x + 1}$ **b.** $y = \begin{cases} x^2, & x \leq 0 \\ \sqrt{x}, & x > 0 \end{cases}$

A function is **one-to-one** when to each value of the dependent variable in the range there corresponds exactly one value of the independent variable in the domain. For instance, the function in Example 2(a) is one-to-one, whereas the function in Example 2(b) is not one-to-one.

Geometrically, a function is one-to-one when every horizontal line intersects the graph of the function at most once. This geometrical interpretation is the **Horizontal Line Test** for one-to-one functions. So, a graph that represents a one-to-one function must satisfy *both* the Vertical Line Test and the Horizontal Line Test.

Function Notation

When using an equation to define a function, you generally isolate the dependent variable on the left. For instance, writing the equation $x + 2y = 1$ as

$$y = \frac{1 - x}{2}$$

indicates that y is the dependent variable. In **function notation,** this equation has the form

$$f(x) = \frac{1 - x}{2}. \qquad \text{\color{red} Function notation}$$

The independent variable is x, and the name of the function is "f." The symbol $f(x)$ is read as "f of x," and it denotes the value of the dependent variable. For instance, the value of f when $x = 3$ is

$$f(3) = \frac{1 - (3)}{2} = \frac{-2}{2} = -1.$$

The value $f(3)$ is called a **function value,** and it lies in the range of f. This means that the point $(3, f(3))$ lies on the graph of f. One of the advantages of function notation is that it allows you to be less wordy. For instance, instead of asking "What is the value of y when $x = 3$?" you can ask "What is $f(3)$?"

Example 3 Evaluating a Function

Find the values of the function

$$f(x) = 2x^2 - 4x + 1$$

when x is -1, 0, and 2. Is f one-to-one?

SOLUTION You can evaluate f at the given values of x as shown.

$$x = -1: \quad f(-1) = 2(-1)^2 - 4(-1) + 1 = 2 + 4 + 1 = 7$$
$$x = 0: \quad f(0) = 2(0)^2 - 4(0) + 1 = 0 - 0 + 1 = 1$$
$$x = 2: \quad f(2) = 2(2)^2 - 4(2) + 1 = 8 - 8 + 1 = 1$$

Because two different values of x yield the same value of $f(x)$, the function is *not* one-to-one, as shown in Figure 1.46.

STUDY TIP

You can use the Horizontal Line Test to determine whether the function in Example 3 is one-to-one. Because the line $y = 1$ intersects the graph of the function twice, the function is *not* one-to-one.

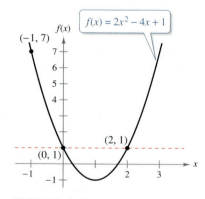

FIGURE 1.46

✓ Checkpoint 3

Find the values of $f(x) = x^2 - 5x + 1$ when x is 0, 1, and 4. Is f one-to-one?

Example 3 suggests that the role of the variable x in the equation

$$f(x) = 2x^2 - 4x + 1$$

is simply that of a placeholder. Informally, f could be defined by the equation

$$f() = 2()^2 - 4() + 1.$$

To evaluate $f(-2)$, simply place -2 in each set of parentheses.

$$f(-2) = 2(-2)^2 - 4(-2) + 1 = 8 + 8 + 1 = 17$$

Although f is often used as a convenient function name and x as the independent variable, you can use other symbols. For instance, the following equations all define the same function.

$$f(x) = x^2 - 4x + 7 \qquad \text{Function name is } f, \text{ independent variable is } x.$$
$$f(t) = t^2 - 4t + 7 \qquad \text{Function name is } f, \text{ independent variable is } t.$$
$$g(s) = s^2 - 4s + 7 \qquad \text{Function name is } g, \text{ independent variable is } s.$$

Example 4   Evaluating a Function

Given $f(x) = x^2 + 7$, evaluate each expression.

a. $f(x + \Delta x)$ **b.** $\dfrac{f(x + \Delta x) - f(x)}{\Delta x}$

SOLUTION

a. To evaluate f at $x + \Delta x$, substitute $x + \Delta x$ for x in the original function.

$$f(x + \Delta x) = (x + \Delta x)^2 + 7$$
$$= x^2 + 2x\,\Delta x + (\Delta x)^2 + 7$$

b. Using the result of part (a), you can write

$$\frac{f(x + \Delta x) - f(x)}{\Delta x} = \frac{[(x + \Delta x)^2 + 7] - (x^2 + 7)}{\Delta x}$$
$$= \frac{x^2 + 2x\,\Delta x + (\Delta x)^2 + 7 - x^2 - 7}{\Delta x}$$
$$= \frac{2x\,\Delta x + (\Delta x)^2}{\Delta x}$$
$$= \frac{\cancel{\Delta x}(2x + \Delta x)}{\cancel{\Delta x}}$$
$$= 2x + \Delta x, \quad \Delta x \neq 0.$$

✓ Checkpoint 4

Given $f(x) = x^2 + 3$, evaluate each expression.

a. $f(x + \Delta x)$ **b.** $\dfrac{f(x + \Delta x) - f(x)}{\Delta x}$

In Example 4(b), the expression

$$\frac{f(x + \Delta x) - f(x)}{\Delta x} \qquad \text{Difference quotient}$$

is called a *difference quotient* and has a special significance in calculus. You will learn more about this in Chapter 2.

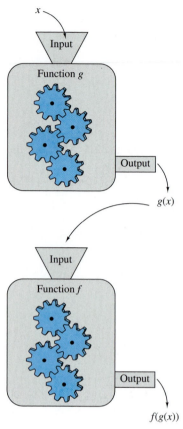

FIGURE 1.47

Combinations of Functions

Two functions can be combined in various ways to create new functions. For instance, given $f(x) = 2x - 3$ and $g(x) = x^2 + 1$, you can form the following functions.

$$f(x) + g(x) = (2x - 3) + (x^2 + 1) = x^2 + 2x - 2 \qquad \text{Sum}$$

$$f(x) - g(x) = (2x - 3) - (x^2 + 1) = -x^2 + 2x - 4 \qquad \text{Difference}$$

$$f(x)g(x) = (2x - 3)(x^2 + 1) = 2x^3 - 3x^2 + 2x - 3 \qquad \text{Product}$$

$$\frac{f(x)}{g(x)} = \frac{2x - 3}{x^2 + 1} \qquad \text{Quotient}$$

You can combine two functions in yet another way called a **composition**. The resulting function is called a **composite function**. For instance, given $f(x) = x^2$ and $g(x) = x + 1$, the composite of f with g is

$$f(g(x)) = f(x + 1) = (x + 1)^2.$$

This composition is denoted by $f \circ g$ and is read as "f composed with g."

Definition of Composite Function

Let f and g be functions. The function given by

$$(f \circ g)(x) = f(g(x))$$

is the **composite** of f with g. The **domain** of $f \circ g$ is the set of all x in the domain of g such that $g(x)$ is in the domain of f, as indicated in Figure 1.47.

The composite of f with g may not be equal to the composite of g with f, as shown in the next example.

Example 5   Forming Composite Functions

Given $f(x) = 2x - 3$ and $g(x) = x^2 + 1$, find each composite function.

a. $f(g(x))$ **b.** $g(f(x))$

SOLUTION

a. The composite of f with g is given by

$$\begin{aligned} f(g(x)) &= 2(g(x)) - 3 & \text{Evaluate } f \text{ at } g(x). \\ &= 2(x^2 + 1) - 3 & \text{Substitute } x^2 + 1 \text{ for } g(x). \\ &= 2x^2 + 2 - 3 & \text{Distributive Property} \\ &= 2x^2 - 1. & \text{Simplify.} \end{aligned}$$

b. The composite of g with f is given by

$$\begin{aligned} g(f(x)) &= (f(x))^2 + 1 & \text{Evaluate } g \text{ at } f(x). \\ &= (2x - 3)^2 + 1 & \text{Substitute } 2x - 3 \text{ for } f(x). \\ &= 4x^2 - 12x + 9 + 1 & \text{Expand.} \\ &= 4x^2 - 12x + 10. & \text{Simplify.} \end{aligned}$$

> **STUDY TIP**
>
> The expressions for $f(g(x))$ and $g(f(x))$ are different in Example 5. In general, the composite of f with g is not the same as the composite of g with f.

✔Checkpoint 5

Given $f(x) = 2x + 1$ and $g(x) = x^2 + 2$, find each composite function.

a. $f(g(x))$ **b.** $g(f(x))$ ■

Inverse Functions

Informally, the inverse function of f is another function g that "undoes" what f has done. For instance, subtraction can be used to undo addition, and division can be used to undo multiplication.

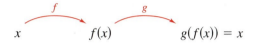

Definition of Inverse Function

Let f and g be two functions such that

$$f(g(x)) = x \text{ for each } x \text{ in the domain of } g$$

and

$$g(f(x)) = x \text{ for each } x \text{ in the domain of } f.$$

Under these conditions, the function g is the **inverse function** of f. The function g is denoted by f^{-1}, which is read as "f-inverse." So,

$$f(f^{-1}(x)) = x \quad \text{and} \quad f^{-1}(f(x)) = x.$$

The domain of f must be equal to the range of f^{-1}, and the range of f must be equal to the domain of f^{-1}.

Do not be confused by the use of the superscript -1 to denote the inverse function f^{-1}. In this text, whenever f^{-1} is written, it *always* refers to the inverse function of f and *not* to the reciprocal of $f(x)$.

Example 6 **Finding Inverse Functions Informally**

Find the inverse function of each function informally.

a. $f(x) = 2x$ **b.** $f(x) = x + 4$

SOLUTION

a. The function f *multiplies* each input by 2. To "undo" this function, you need to *divide* each input by 2. So, the inverse function of $f(x) = 2x$ is

$$f^{-1}(x) = \frac{x}{2}.$$

b. The function f *adds* 4 to each input. To "undo" this function, you need to *subtract* 4 from each input. So, the inverse function of $f(x) = x + 4$ is

$$f^{-1}(x) = x - 4.$$

Check that f and f^{-1} are inverse functions by showing that $f(f^{-1}(x)) = x$ and $f^{-1}(f(x)) = x.$

✓ **Checkpoint 6**

Find the inverse function of each function informally.

a. $f(x) = \frac{1}{5}x$ **b.** $f(x) = x - 6$

The graphs of f and f^{-1} are mirror images of each other (with respect to the line $y = x$), as shown in Figure 1.48. Try using a graphing utility with a *square setting* to confirm this for each of the functions given in Example 6.

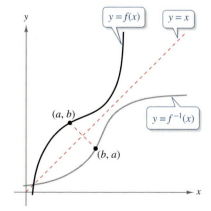

The graph of f^{-1} is a reflection of the graph of f in the line $y = x$.

FIGURE 1.48

The functions in Example 6 are simple enough so that their inverse functions can be found by inspection. The next example demonstrates a strategy for finding the inverse functions of more complicated functions.

Example 7   Finding an Inverse Function

Find the inverse function of $f(x) = \sqrt{2x - 3}$.

SOLUTION   Begin by replacing $f(x)$ with y. Then, interchange x and y and solve for y.

$$f(x) = \sqrt{2x - 3} \qquad \text{Write original function.}$$
$$y = \sqrt{2x - 3} \qquad \text{Replace } f(x) \text{ with } y.$$
$$x = \sqrt{2y - 3} \qquad \text{Interchange } x \text{ and } y.$$
$$x^2 = 2y - 3 \qquad \text{Square each side.}$$
$$x^2 + 3 = 2y \qquad \text{Add 3 to each side.}$$
$$\frac{x^2 + 3}{2} = y \qquad \text{Divide each side by 2.}$$

So, the inverse function has the form

$$f^{-1}(\boxed{}) = \frac{(\boxed{})^2 + 3}{2}.$$

Using x as the independent variable, you can write

$$f^{-1}(x) = \frac{x^2 + 3}{2}, \quad x \geq 0.$$

In Figure 1.49, note that the domain of f^{-1} coincides with the range of f.

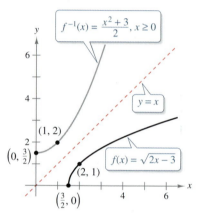

$f^{-1}(x) = \frac{x^2 + 3}{2}, x \geq 0$

$y = x$

$(1, 2)$

$\left(0, \frac{3}{2}\right)$

$(2, 1)$

$f(x) = \sqrt{2x - 3}$

$\left(\frac{3}{2}, 0\right)$

FIGURE 1.49

✓ Checkpoint 7

Find the inverse function of $f(x) = x^2 + 2$ for $x \geq 0$. ■

After you have found an inverse function, you should check your results. You can check your results *graphically* by observing that the graphs of f and f^{-1} are reflections of each other in the line

$$y = x.$$

You can check your results *algebraically* by evaluating $f(f^{-1}(x))$ and $f^{-1}(f(x))$—both should be equal to x.

Check that $f(f^{-1}(x)) = x$

$$f(f^{-1}(x)) = f\left(\frac{x^2 + 3}{2}\right)$$
$$= \sqrt{2\left(\frac{x^2 + 3}{2}\right) - 3}$$
$$= \sqrt{x^2}$$
$$= x, \quad x \geq 0$$

Check that $f^{-1}(f(x)) = x$

$$f^{-1}(f(x)) = f^{-1}\left(\sqrt{2x - 3}\right)$$
$$= \frac{\left(\sqrt{2x - 3}\right)^2 + 3}{2}$$
$$= \frac{2x}{2}$$
$$= x, \quad x \geq \frac{3}{2}$$

TECH TUTOR

A graphing utility can help you check that the graphs of f and f^{-1} are reflections of each other in the line $y = x$. To do this, graph $y = f(x)$, $y = f^{-1}(x)$, and $y = x$ in the same viewing window, using a square setting.

Not every function has an inverse function. In fact, for a function to have an inverse function, it must be one-to-one.

Example 8   A Function That Has No Inverse Function

Show that the function

$$f(x) = x^2 - 1$$

has no inverse function. (Assume that the domain of f is the set of all real numbers.)

SOLUTION   Begin by sketching the graph of f, as shown in Figure 1.50. Note that

$$f(2) = (2)^2 - 1 = 3$$

and

$$f(-2) = (-2)^2 - 1 = 3.$$

So, f does not pass the Horizontal Line Test, which implies that it is not one-to-one and therefore has no inverse function. The same conclusion can be obtained by trying to find the inverse function of f algebraically.

$f(x) = x^2 - 1$	Write original function.
$y = x^2 - 1$	Replace $f(x)$ with y.
$x = y^2 - 1$	Interchange x and y.
$x + 1 = y^2$	Add 1 to each side.
$\pm\sqrt{x + 1} = y$	Take square root of each side.

The last equation does not define y as a function of x, and so f has no inverse function.

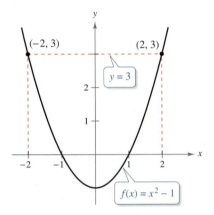

(−2, 3)   (2, 3)

$y = 3$

$f(x) = x^2 - 1$

f is not one-to-one and has no inverse function.

FIGURE 1.50

✓ Checkpoint 8

Show that the function

$$f(x) = x^2 + 4$$

has no inverse function.

SUMMARIZE   (Section 1.4)

1. State the definition of a function *(page 29)*. For an example of a function, see Example 1.

2. Explain the meanings of domain and range *(page 29)*. For an example of a domain and a range, see Example 2.

3. Explain the meaning of function notation *(page 32)*. For examples of function notation, see Examples 3 and 4.

4. State the definition of a composite function *(page 34)*. For an example of a composite function, see Example 5.

5. State the definition of an inverse function *(page 35)*. For examples of inverse functions, see Examples 6 and 7.

6. State when a function does not have an inverse function *(page 37)*. For an example of a function that does not have an inverse function, see Example 8.

SKILLS WARM UP 1.4   The following warm-up exercises involve skills that were covered in a previous course. You will use these skills in the exercise set for this section. For additional help, review Appendix Sections A.3 and A.5.

In Exercises 1–6, simplify the expression.

1. $5(-1)^2 - 6(-1) + 9$

2. $(-2)^3 + 7(-2)^2 - 10$

3. $(x - 2)^2 + 5x - 10$

4. $(3 - x) + (x + 3)^3$

5. $\dfrac{1}{1 - (1 - x)}$

6. $1 + \dfrac{x - 1}{x}$

In Exercises 7–12, solve for y in terms of x.

7. $2x + y - 6 = 11$

8. $5y - 6x^2 - 1 = 0$

9. $(y - 3)^2 = 5 + (x + 1)^2$

10. $y^2 - 4x^2 = 2$

11. $x = \dfrac{2y - 1}{4}$

12. $x = \sqrt[3]{2y - 1}$

WebAssign Access end-of-section exercises online at **www.webassign.net**

1.5 Limits

- Find limits of functions graphically and numerically.
- Understand the definition of the limit of a function and use the properties of limits to evaluate limits of functions.
- Use different analytic techniques to evaluate limits of functions.
- Evaluate one-sided limits.
- Recognize unbounded behavior of functions.

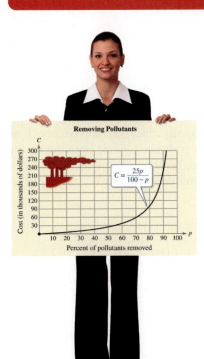

In Exercise 75, you will use a limit to analyze the cost of removing pollutants from a small lake.

The Limit of a Function

In everyday language, people refer to a speed limit, a wrestler's weight limit, the limit of one's endurance, or stretching a spring to its limit. These phrases all suggest that a limit is a bound, which on some occasions may not be reached but on other occasions may be reached or exceeded.

Consider a spring that will break only when a weight of 10 pounds or more is attached. To determine how far the spring will stretch without breaking, you could attach increasingly heavier weights and measure the spring length s for each weight w, as shown in Figure 1.51. If the spring length approaches a value of L, then it is said that "the limit of s as w approaches 10 is L." A mathematical limit is much like the limit of a spring. The notation for a limit is

$$\lim_{x \to c} f(x) = L$$

which is read as "the limit of $f(x)$ as x approaches c is L."

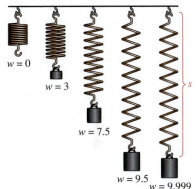

What is the limit of s as w approaches 10 pounds?

FIGURE 1.51

Example 1   Finding a Limit

Find the limit: $\lim_{x \to 1} (x^2 + 1)$.

SOLUTION Let $f(x) = x^2 + 1$. From the graph of f in Figure 1.52, it appears that $f(x)$ approaches 2 as x approaches 1 from either side, and you can write

$$\lim_{x \to 1} (x^2 + 1) = 2.$$

The table yields the same conclusion. Notice that as x gets closer and closer to 1, $f(x)$ gets closer and closer to 2.

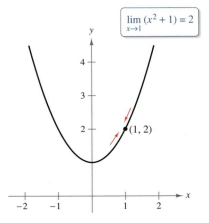

$\lim_{x \to 1} (x^2 + 1) = 2$

FIGURE 1.52

	x approaches 1.				x approaches 1.		
x	0.900	0.990	0.999	1.000	1.001	1.010	1.100
$f(x)$	1.810	1.980	1.998	2.000	2.002	2.020	2.210

$f(x)$ approaches 2. $f(x)$ approaches 2.

✓ Checkpoint 1

Find the limit: $\lim_{x \to 1} (2x + 4)$.

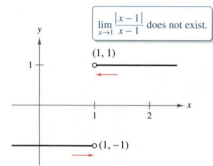

(a)

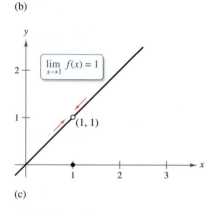

(b)

(c)

FIGURE 1.53

Example 2 Finding Limits Graphically and Numerically

Find the limit.

$$\lim_{x \to 1} f(x)$$

a. $f(x) = \dfrac{x^2 - 1}{x - 1}$ **b.** $f(x) = \dfrac{|x - 1|}{x - 1}$ **c.** $f(x) = \begin{cases} x, & x \neq 1 \\ 0, & x = 1 \end{cases}$

SOLUTION

a. From the graph of f in Figure 1.53(a), it appears that $f(x)$ approaches 2 as x approaches 1 from either side. A missing point is denoted by the open dot on the graph. This conclusion is reinforced by the table. Be sure you see that *it does not matter that $f(x)$ is undefined when $x = 1$.* The limit depends only on values of $f(x)$ near 1, not at 1.

	x approaches 1.				x approaches 1.		
x	0.900	0.990	0.999	1.000	1.001	1.010	1.100
$f(x)$	1.900	1.990	1.999	?	2.001	2.010	2.100
	$f(x)$ approaches 2.				$f(x)$ approaches 2.		

b. From the graph of f in Figure 1.53(b), you can see that $f(x) = -1$ for all values to the left of $x = 1$ and $f(x) = 1$ for all values to the right of $x = 1$. So, $f(x)$ is approaching a different value from the left of $x = 1$ than it is from the right of $x = 1$. In such situations, you can conclude that *the limit does not exist.* This conclusion is reinforced by the table.

	x approaches 1.				x approaches 1.		
x	0.900	0.990	0.999	1.000	1.001	1.010	1.100
$f(x)$	-1.000	-1.000	-1.000	?	1.000	1.000	1.000
	$f(x)$ approaches -1.				$f(x)$ approaches 1.		

c. From the graph of f in Figure 1.53(c), it appears that $f(x)$ approaches 1 as x approaches 1 from either side. This conclusion is reinforced by the table. It does not matter that $f(1) = 0$. The limit depends only on values of $f(x)$ near 1, not at 1.

	x approaches 1.				x approaches 1.		
x	0.900	0.990	0.999	1.000	1.001	1.010	1.100
$f(x)$	0.900	0.990	0.999	?	1.001	1.010	1.100
	$f(x)$ approaches 1.				$f(x)$ approaches 1.		

✓**Checkpoint 2**

Find the limit.

$$\lim_{x \to 2} f(x)$$

a. $f(x) = \dfrac{x^2 - 4}{x - 2}$ **b.** $f(x) = \dfrac{|x - 2|}{x - 2}$ **c.** $f(x) = \begin{cases} x^2, & x \neq 2 \\ 0, & x = 2 \end{cases}$

Definition of the Limit of a Function and Properties of Limits

There are three important ideas to learn from Examples 1 and 2.

1. Saying that the limit of $f(x)$ approaches L as x approaches c means that the value of $f(x)$ may be made *arbitrarily close* to the number L by choosing x closer and closer to c.

2. If $f(x)$ approaches the same number from *either side* of c, then the limit of $f(x)$ as x approaches c exists. However, if $f(x)$ approaches a different number from the right side of c than it does from the left side, then the limit of $f(x)$ as x approaches c *does not exist.* [See Example 2(b).]

3. The value of $f(x)$ when $x = c$ has no bearing on the existence or nonexistence of the limit of $f(x)$ as x approaches c. For instance, in Example 2(a), the limit of $f(x)$ exists as x approaches 1 even though the function f is not defined at $x = 1$.

> ### Definition of the Limit of a Function
>
> If $f(x)$ becomes arbitrarily close to a single number L as x approaches c from either side, then
>
> $$\lim_{x \to c} f(x) = L$$
>
> which is read as "the **limit** of $f(x)$ as x approaches c is L."

Many times the limit of $f(x)$ as x approaches c is simply $f(c)$, as shown in Example 1. Whenever the limit of $f(x)$ as x approaches c is

$$\lim_{x \to c} f(x) = f(c) \qquad \text{Substitute } c \text{ for } x.$$

the limit can be evaluated by **direct substitution.** (In the next section, you will learn that a function that has this property is *continuous at c.*) It is important that you learn to recognize the types of functions that have this property. Some basic ones are given in the following list.

> ### Some Basic Limits
>
> Let b and c be real numbers, and let n be a positive integer.
>
> **1.** $\lim_{x \to c} b = b$ **2.** $\lim_{x \to c} x = c$ **3.** $\lim_{x \to c} x^n = c^n$ **4.** $\lim_{x \to c} \sqrt[n]{x} = \sqrt[n]{c}$
>
> In Property 4, if n is even, then c must be positive.

Example 3   Evaluating Basic Limits

a. $\lim_{x \to 2} 3 = 3$ **b.** $\lim_{x \to -4} x = -4$

c. $\lim_{x \to 2} x^3 = 2^3 = 8$ **d.** $\lim_{x \to 9} \sqrt{x} = \sqrt{9} = 3$

✓ Checkpoint 3

Find the limit.

a. $\lim_{x \to 1} 5$ **b.** $\lim_{x \to 6} x$ **c.** $\lim_{x \to 5} x^2$ **d.** $\lim_{x \to -8} \sqrt[3]{x}$ ■

TECH TUTOR

You can use a graphing utility to estimate the limit of a function. One way is to estimate the limit numerically by creating a table of values using the *table* feature. Another way is to graph the function and use the *zoom* and *trace* features to estimate the limit graphically.

By combining the basic limits from the preceding page with the properties of limits shown below, you can find limits for a wide variety of algebraic functions.

Properties of Limits

Let b and c be real numbers, let n be a positive integer, and let f and g be functions with the following limits.

$$\lim_{x \to c} f(x) = L \quad \text{and} \quad \lim_{x \to c} g(x) = K$$

1. Scalar multiple: $\lim_{x \to c} [bf(x)] = bL$

2. Sum or difference: $\lim_{x \to c} [f(x) \pm g(x)] = L \pm K$

3. Product: $\lim_{x \to c} [f(x) \cdot g(x)] = LK$

4. Quotient: $\lim_{x \to c} \dfrac{f(x)}{g(x)} = \dfrac{L}{K}$, provided $K \neq 0$

5. Power: $\lim_{x \to c} [f(x)]^n = L^n$

6. Radical: $\lim_{x \to c} \sqrt[n]{f(x)} = \sqrt[n]{L}$

In Property 6, if n is even, then L must be positive.

TECH TUTOR

Symbolic computer algebra systems are capable of evaluating limits. Try using a computer algebra system to evaluate the limit given in Example 4.

Example 4   Finding the Limit of a Polynomial Function

Find the limit: $\lim\limits_{x \to 2} (x^2 + 2x - 3)$.

SOLUTION

$$
\begin{aligned}
\lim_{x \to 2} (x^2 + 2x - 3) &= \lim_{x \to 2} x^2 + \lim_{x \to 2} 2x - \lim_{x \to 2} 3 &&\text{Apply Property 2.}\\
&= \lim_{x \to 2} x^2 + 2 \lim_{x \to 2} x - \lim_{x \to 2} 3 &&\text{Apply Property 1.}\\
&= 2^2 + 2(2) - 3 &&\text{Use direct substitution.}\\
&= 5
\end{aligned}
$$

✓ Checkpoint 4

Find the limit: $\lim\limits_{x \to 1} (2x^2 - x + 4)$.   ■

In Example 4, note that the limit (as $x \to 2$) of the *polynomial function*

$$p(x) = x^2 + 2x - 3$$

is simply the value of p at $x = 2$.

$$\lim_{x \to 2} p(x) = p(2) = 2^2 + 2(2) - 3 = 4 + 4 - 3 = 5$$

This is an illustration of the following important result, which states that the limit of a polynomial function can be evaluated by direct substitution.

The Limit of a Polynomial Function

If p is a polynomial function and c is any real number, then

$$\lim_{x \to c} p(x) = p(c).$$

Techniques for Evaluating Limits

You have learned several techniques for evaluating limits. Another technique is shown in Example 5.

> **Example 5** **Finding the Limit of a Function**

Find the limit.

$$\lim_{x \to 1} \frac{x^3 - 1}{x - 1}$$

SOLUTION Note that the numerator and denominator are zero when $x = 1$. This implies that $x - 1$ is a factor of both, and you can divide out this common factor.

$$\frac{x^3 - 1}{x - 1} = \frac{(x - 1)(x^2 + x + 1)}{x - 1} \qquad \text{Factor numerator.}$$

$$= \frac{(x - 1)(x^2 + x + 1)}{x - 1} \qquad \text{Divide out common factor.}$$

$$= x^2 + x + 1, \quad x \neq 1 \qquad \text{Simplify.}$$

So, the rational function $(x^3 - 1)/(x - 1)$ and the polynomial function $x^2 + x + 1$ agree for all values of x other than $x = 1$, and you can apply the Replacement Theorem.

$$\lim_{x \to 1} \frac{x^3 - 1}{x - 1} = \lim_{x \to 1} (x^2 + x + 1) = 1^2 + 1 + 1 = 3$$

Figure 1.54 illustrates this result graphically. Note that the two graphs are identical except that the graph of g contains the point $(1, 3)$, whereas this point is missing on the graph of f. (In the graph of f in Figure 1.54, the missing point is denoted by an open dot.)

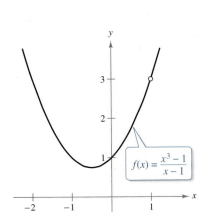
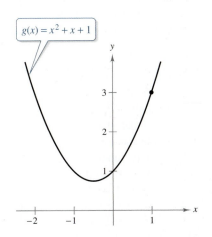

FIGURE 1.54

✓ **Checkpoint 5**

Find the limit.

$$\lim_{x \to 2} \frac{x^3 - 8}{x - 2}$$

The technique used to evaluate the limit in Example 5 is called the **dividing out** technique. The validity of this technique stems from the fact that when two functions agree at all but a single number c, the functions must have identical limit behavior at $x = c$. This technique is further demonstrated in the next example.

Example 6   Using the Dividing Out Technique

For $\lim\limits_{x \to -3} \dfrac{x^2 + x - 6}{x + 3}$, direct substitution fails because both the numerator and the denominator are zero when $x = -3$.

$$\lim\limits_{x \to -3} \dfrac{x^2 + x - 6}{x + 3} \qquad \longleftarrow \; \lim\limits_{x \to -3}(x^2 + x - 6) = 0$$
$$\longleftarrow \; \lim\limits_{x \to -3}(x + 3) = 0$$

Because the limits of both the numerator and the denominator are zero when $x = -3$, they must have a *common factor* of $x + 3$. So, for all $x \neq -3$, you can divide out this factor and find the limit as shown.

$$\lim\limits_{x \to -3} \dfrac{x^2 + x - 6}{x + 3} = \lim\limits_{x \to -3} \dfrac{(x - 2)(x + 3)}{x + 3} \qquad \text{Factor numerator.}$$

$$= \lim\limits_{x \to -3} \dfrac{(x - 2)(x + 3)}{x + 3} \qquad \text{Divide out common factor.}$$

$$= \lim\limits_{x \to -3}(x - 2) \qquad \text{Simplify.}$$

$$= -3 - 2 \qquad \text{Direct substitution}$$

$$= -5 \qquad \text{Simplify.}$$

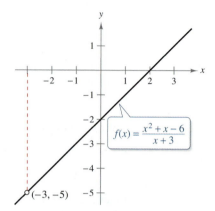

f is undefined when $x = -3$.

FIGURE 1.55

This result is shown graphically in Figure 1.55. Note that the graph of f coincides with the graph of $g(x) = x - 2$, except that the graph of f has a hole at $(-3, -5)$.

✓ Checkpoint 6

Find the limit: $\lim\limits_{x \to 3} \dfrac{x^2 + x - 12}{x - 3}$.

■

STUDY TIP

When you try to evaluate a limit and both the numerator and denominator are zero, remember that you must rewrite the fraction so that the new denominator does not have 0 as its limit. One way to do this is to divide out common factors, as shown in Example 6. Another technique is to rationalize the numerator, as shown in Example 7.

Example 7   Finding a Limit of a Function

Find the limit: $\lim\limits_{x \to 0} \dfrac{\sqrt{x + 1} - 1}{x}$.

SOLUTION   Direct substitution fails because both the numerator and the denominator are zero when $x = 0$. Rewrite the fraction by rationalizing the numerator.

$$\dfrac{\sqrt{x + 1} - 1}{x} = \left(\dfrac{\sqrt{x + 1} - 1}{x} \right) \left(\dfrac{\sqrt{x + 1} + 1}{\sqrt{x + 1} + 1} \right)$$

$$= \dfrac{(x + 1) - 1}{x\left(\sqrt{x + 1} + 1\right)}$$

$$= \dfrac{x}{x\left(\sqrt{x + 1} + 1\right)}$$

$$= \dfrac{1}{\sqrt{x + 1} + 1}, \quad x \neq 0$$

Now, using the Replacement Theorem, you can evaluate the limit as shown.

$$\lim\limits_{x \to 0} \dfrac{\sqrt{x + 1} - 1}{x} = \lim\limits_{x \to 0} \dfrac{1}{\sqrt{x + 1} + 1} = \dfrac{1}{\sqrt{0 + 1} + 1} = \dfrac{1}{1 + 1} = \dfrac{1}{2}$$

✓ Checkpoint 7

Find the limit: $\lim\limits_{x \to 0} \dfrac{\sqrt{x + 4} - 2}{x}$.

■

One-Sided Limits

In Example 2(b), you saw that one way in which a limit can fail to exist is when a function approaches a different value from the left of c than it approaches from the right of c. This type of behavior can be described more concisely with the concept of a **one-sided limit.**

$$\lim_{x \to c^-} f(x) = L \qquad \text{Limit from the left}$$

$$\lim_{x \to c^+} f(x) = L \qquad \text{Limit from the right}$$

The first of these two limits is read as "the limit of $f(x)$ as x approaches c from the left is L." The second is read as "the limit of $f(x)$ as x approaches c from the right is L."

Example 8 Finding One-Sided Limits

Find the limit as $x \to 0$ from the left and the limit as $x \to 0$ from the right for the function

$$f(x) = \frac{|2x|}{x}.$$

SOLUTION From the graph of f, shown in Figure 1.56, you can see that

$$f(x) = -2$$

for all $x < 0$. So, the limit from the left is

$$\lim_{x \to 0^-} \frac{|2x|}{x} = -2. \qquad \text{Limit from the left}$$

Because

$$f(x) = 2$$

for all $x > 0$, the limit from the right is

$$\lim_{x \to 0^+} \frac{|2x|}{x} = 2. \qquad \text{Limit from the right}$$

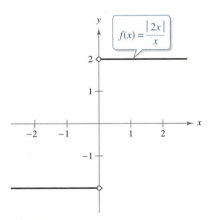

FIGURE 1.56

✓Checkpoint 8

Find each limit.

a. $\displaystyle \lim_{x \to 2^-} \frac{|x - 2|}{x - 2}$

b. $\displaystyle \lim_{x \to 2^+} \frac{|x - 2|}{x - 2}$

In Example 8, note that the function approaches different limits from the left and from the right. In such cases, the limit of $f(x)$ as $x \to c$ does not exist. For the limit of a function to exist as $x \to c$, *both* one-sided limits must exist and must be equal.

TECH TUTOR

On most graphing utilities, the absolute value function is denoted by *abs*. You can verify the result in Example 8 by graphing

$$y = \frac{\text{abs}(2x)}{x}$$

in the viewing window $-3 \le x \le 3$ and $-3 \le y \le 3$.

Existence of a Limit

If f is a function and c and L are real numbers, then

$$\lim_{x \to c} f(x) = L$$

if and only if both the left and right limits are equal to L.

Example 9   Finding One-Sided Limits

Find the limit of $f(x)$ as x approaches 1.

$$f(x) = \begin{cases} 4 - x, & x < 1 \\ 4x - x^2, & x > 1 \end{cases}$$

SOLUTION   Remember that you are concerned about the value of f near $x = 1$ rather than at $x = 1$. So, for $x < 1$, $f(x)$ is given by

$$4 - x$$

and you can use direct substitution to obtain

$$\lim_{x \to 1^-} f(x) = \lim_{x \to 1^-} (4 - x) = 4 - 1 = 3.$$

For $x > 1$, $f(x)$ is given by

$$4x - x^2$$

and you can use direct substitution to obtain

$$\lim_{x \to 1^+} f(x) = \lim_{x \to 1^+} (4x - x^2) = 4(1) - 1^2 = 4 - 1 = 3.$$

Because both one-sided limits exist and are equal to 3, it follows that

$$\lim_{x \to 1} f(x) = 3.$$

The graph in Figure 1.57 confirms this conclusion.

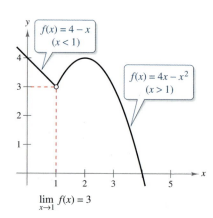

$f(x) = 4 - x$
$(x < 1)$

$f(x) = 4x - x^2$
$(x > 1)$

$\lim_{x \to 1} f(x) = 3$

FIGURE 1.57

✓ Checkpoint 9

Find the limit of $f(x)$ as x approaches 0.

$$f(x) = \begin{cases} x^2 + 1, & x < 0 \\ 2x + 1, & x > 0 \end{cases}$$

Example 10   Comparing One-Sided Limits

An overnight delivery service charges \$18 for the first pound and \$2 for each additional pound. Let x represent the weight of a parcel and let $f(x)$ represent the shipping cost.

$$f(x) = \begin{cases} 18, & 0 < x \le 1 \\ 20, & 1 < x \le 2 \\ 22, & 2 < x \le 3 \end{cases}$$

Show that the limit of $f(x)$ as $x \to 2$ does not exist.

SOLUTION   The graph of f is shown in Figure 1.58. The limit of $f(x)$ as x approaches 2 from the left is

$$\lim_{x \to 2^-} f(x) = 20$$

whereas the limit of $f(x)$ as x approaches 2 from the right is

$$\lim_{x \to 2^+} f(x) = 22.$$

Because these one-sided limits are not equal, the limit of $f(x)$ as $x \to 2$ does not exist.

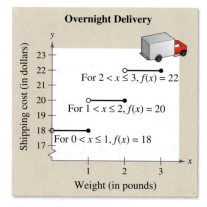

Overnight Delivery

Shipping cost (in dollars)

For $2 < x \le 3$, $f(x) = 22$

For $1 < x \le 2$, $f(x) = 20$

For $0 < x \le 1$, $f(x) = 18$

Weight (in pounds)

FIGURE 1.58

✓ Checkpoint 10

Show that the limit of $f(x)$ as $x \to 1$ does not exist in Example 10.

Unbounded Behavior

Example 10 shows a limit that fails to exist because the limits from the left and right differ. Another important way in which a limit can fail to exist is when $f(x)$ increases or decreases without bound as x approaches c.

Example 11 An Unbounded Function

Find the limit (if possible): $\lim\limits_{x \to 2} \dfrac{3}{x - 2}$.

SOLUTION From Figure 1.59, you can see that $f(x)$ decreases without bound as x approaches 2 from the left and $f(x)$ increases without bound as x approaches 2 from the right. Symbolically, you can write this as

$$\lim_{x \to 2^-} \frac{3}{x - 2} = -\infty$$

and

$$\lim_{x \to 2^+} \frac{3}{x - 2} = \infty.$$

Because f is unbounded as x approaches 2, the limit does not exist.

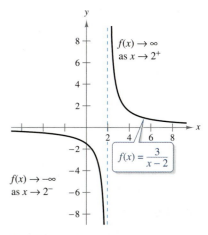

FIGURE 1.59

> **STUDY TIP**
>
> The equal sign in the statement $\lim\limits_{x \to c^+} f(x) = \infty$ does not mean that the limit exists. On the contrary, it tells you how the limit *fails to exist* by denoting the unbounded behavior of $f(x)$ as x approaches c.

✓ **Checkpoint 11**

Find the limit (if possible): $\lim\limits_{x \to -2} \dfrac{5}{x + 2}$.

SUMMARIZE (Section 1.5)

1. State the definition of the limit of a function *(page 41)*. For examples of limits, see Examples 1 and 2.

2. Make a list of the basic limits *(page 41)*. For examples of the basic limits, see Example 3.

3. Make a list of the properties of limits *(page 42)*. For an example of the use of these properties, see Example 4.

4. State the limit of a polynomial function *(page 42)*. For an example of the limit of a polynomial function, see Example 4.

5. Describe the dividing out technique *(page 43)*. For examples of the dividing out technique, see Examples 5 and 6.

6. Describe the rationalizing technique *(page 44)*. For an example of the rationalizing technique, see Example 7.

7. Describe a one-sided limit *(page 45)*. For examples of one-sided limits, see Examples 8, 9, and 10.

8. Describe the limit $\lim\limits_{x \to c} f(x)$ when $f(x)$ increases without bound as x approaches c *(page 47)*. For an example of an unbounded function, see Example 11.

SKILLS WARM UP 1.5   The following warm-up exercises involve skills that were covered in a previous course or in earlier sections. You will use these skills in the exercise set for this section. For additional help, review Appendix Section A.3, and Section 1.4.

In Exercises 1–4, simplify the expression by factoring.

1. $\dfrac{2x^3 + x^2}{6x}$

2. $\dfrac{x^5 + 9x^4}{x^2}$

3. $\dfrac{x^2 - 3x - 28}{x - 7}$

4. $\dfrac{x^2 + 11x + 30}{x + 5}$

In Exercises 5–8, evaluate the expression and simplify.

5. $f(x) = x^2 - 3x + 3$

    (a) $f(-1)$ (b) $f(c)$ (c) $f(x + h)$

6. $f(x) = \begin{cases} 2x - 2, & x < 1 \\ 3x + 1, & x \geq 1 \end{cases}$

    (a) $f(-1)$ (b) $f(3)$ (c) $f(t^2 + 1)$

7. $f(x) = x^2 - 2x + 2$ $\dfrac{f(1 + h) - f(1)}{h}$

8. $f(x) = 4x$ $\dfrac{f(2 + h) - f(2)}{h}$

In Exercises 9–12, find the domain and range of the function and sketch its graph.

9. $h(x) = -\dfrac{5}{x}$

10. $g(x) = \sqrt{25 - x^2}$

11. $f(x) = |x - 3|$

12. $f(x) = \dfrac{|x|}{x}$

In Exercises 13 and 14, determine whether y is a function of x.

13. $9x^2 + 4y^2 = 49$

14. $2x^2y + 8x = 7y$

ENHANCED **WebAssign**   Access end-of-section exercises online at **www.webassign.net**

1.6 Continuity

- ■ Determine the continuity of functions.
- ■ Determine the continuity of functions on a closed interval.
- ■ Use the greatest integer function to model and solve real-life problems.
- ■ Use compound interest models to solve real-life problems.

Continuity

In Exercise 67, you will examine the continuity of a function that represents an account balance.

In mathematics, the term "continuous" has much the same meaning as it has in everyday use. To say that a function is continuous at

$$x = c$$

means that there is no interruption in the graph of f at c. That is, the graph of f

1. is unbroken at c.

2. has no holes, jumps, or gaps.

As simple as this concept may seem, its precise definition eluded mathematicians for many years. In fact, it was not until the early 1800's that a precise definition was finally developed.

Before looking at this definition, consider the function whose graph is shown in Figure 1.60. This figure identifies three values of x at which the function f is not continuous.

1. At $x = c_1$, $f(c_1)$ is not defined.

2. At $x = c_2$, $\lim_{x \to c_2} f(x)$ does not exist.

3. At $x = c_3$, $f(c_3) \neq \lim_{x \to c_3} f(x)$.

At all other points in the interval (a, b), the graph of f is uninterrupted, which implies that the function f is continuous at all other points in the interval (a, b).

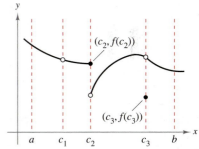

f is not continuous at $x = c_1, c_2, c_3$.

FIGURE 1.60

Definition of Continuity

Let c be a number in the interval (a, b), and let f be a function whose domain contains the interval (a, b). The function f is **continuous at the point c** when the following conditions are true.

1. $f(c)$ is defined.

2. $\lim_{x \to c} f(x)$ exists.

3. $\lim_{x \to c} f(x) = f(c)$.

If f is continuous at every point in the interval (a, b), then f is **continuous on the open interval (a, b).**

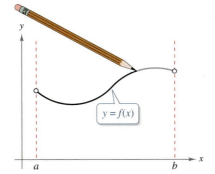

On the interval (a, b), the graph of f can be traced with a pencil.

FIGURE 1.61

Informally, you can say that a function is continuous on an interval when its graph on the interval can be traced using a pencil and paper without lifting the pencil from the paper, as shown in Figure 1.61.

TECH TUTOR

Most graphing utilities can draw graphs in two different modes: *connected mode* and *dot mode*. The *connected mode* works well as long as the function is continuous on the entire interval represented by the viewing window. If, however, the function is not continuous at one or more x-values in the viewing window, then the *connected mode* may try to "connect" parts of the graph that should not be connected. For instance, try graphing the function
$y_1 = (x + 3)/(x - 2)$
in the viewing window
$-8 \le x \le 8$ and
$-6 \le y \le 6$ in *connected mode* and then in *dot mode*.

Continuity of Polynomial and Rational Functions

1. A polynomial function is continuous at every real number.

2. A rational function is continuous at every number in its domain.

Example 1   Determining Continuity of a Polynomial Function

Discuss the continuity of each function.

a. $f(x) = x^2 - 2x + 3$ **b.** $f(x) = x^3 - x$ **c.** $f(x) = x^4 - 2x^2 + 1$

SOLUTION   Each of these functions is a *polynomial function*. So, each is continuous on the entire real number line, as indicated in Figure 1.62.

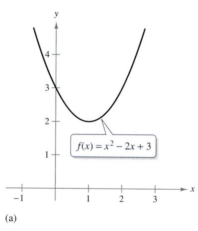

$f(x) = x^2 - 2x + 3$

(a)

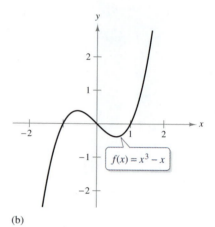

$f(x) = x^3 - x$

(b)

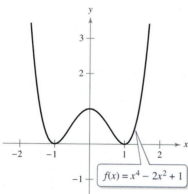

$f(x) = x^4 - 2x^2 + 1$

(c)

All three functions are continuous on $(-\infty, \infty)$.

FIGURE 1.62

✓Checkpoint 1

Discuss the continuity of each function.

a. $f(x) = x^2 + x + 1$ **b.** $f(x) = x^3 + x$ **c.** $f(x) = x^4$

Polynomial functions are one of the most important types of functions used in calculus. Be sure you see from Example 1 that the graph of a polynomial function is continuous on the entire real number line and therefore has no holes, jumps, or gaps. Rational functions, on the other hand, need not be continuous on the entire real number line, as shown in Example 2.

Example 2 **Determining Continuity of a Rational Function**

Discuss the continuity of each function.

a. $f(x) = \dfrac{1}{x}$ **b.** $f(x) = \dfrac{x^2 - 1}{x - 1}$ **c.** $f(x) = \dfrac{1}{x^2 + 1}$

SOLUTION Each of these functions is a rational function and is therefore continuous at every number in its domain.

a. The domain of $f(x) = 1/x$ consists of all real numbers except $x = 0$. So, this function is continuous on the intervals $(-\infty, 0)$ and $(0, \infty)$. [See Figure 1.63(a).]

b. The domain of $f(x) = (x^2 - 1)/(x - 1)$ consists of all real numbers except $x = 1$. So, this function is continuous on the intervals $(-\infty, 1)$ and $(1, \infty)$. [See Figure 1.63(b).]

c. The domain of $f(x) = 1/(x^2 + 1)$ consists of all real numbers. So, this function is continuous on the entire real number line. [See Figure 1.63(c).]

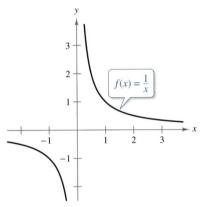

(a) Continuous on $(-\infty, 0)$ and $(0, \infty)$

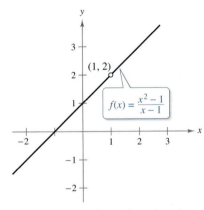

(b) Continuous on $(-\infty, 1)$ and $(1, \infty)$

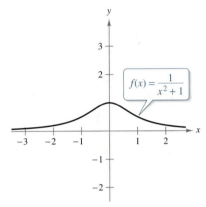

(c) Continuous on $(-\infty, \infty)$

FIGURE 1.63

TECH TUTOR

A graphing utility can give misleading information about the continuity of a function. For instance, try graphing the function from Example 2(b), $f(x) = (x^2 - 1)/(x - 1)$, in a standard viewing window. On most graphing utilities, the graph appears to be continuous at every real number. However, because $x = 1$ is not in the domain of f, you know that f is not continuous at $x = 1$. You can verify this on a graphing utility using the *trace* or *table* feature.

✓**Checkpoint 2**

Discuss the continuity of each function.

a. $f(x) = \dfrac{1}{x - 1}$

b. $f(x) = \dfrac{x^2 - 4}{x - 2}$

c. $f(x) = \dfrac{1}{x^2 + 2}$

Consider an open interval I that contains a real number c. If a function f is defined on I (except possibly at c), and f is not continuous at c, then f is said to have a **discontinuity** at c. Discontinuities fall into two categories: **removable** and **nonremovable.** A discontinuity at c is called removable when f can be made continuous by appropriately defining (or redefining) $f(c)$. For instance, the function in Example 2(b) has a removable discontinuity at $(1, 2)$. To remove the discontinuity, all you need to do is redefine the function so that $f(1) = 2$.

A discontinuity at $x = c$ is nonremovable when the function cannot be made continuous at $x = c$ by defining or redefining the function at $x = c$. For instance, the function in Example 2(a) has a nonremovable discontinuity at $x = 0$.

Continuity on a Closed Interval

The intervals discussed in Examples 1 and 2 are open. To discuss continuity on a closed interval, you can use the concept of one-sided limits, as defined in Section 1.5.

Definition of Continuity on a Closed Interval

Let f be defined on a closed interval $[a, b]$. If f is continuous on the open interval (a, b) and

$$\lim_{x \to a^+} f(x) = f(a) \quad \text{and} \quad \lim_{x \to b^-} f(x) = f(b)$$

then f is **continuous on the closed interval $[a, b]$**. Moreover, f is **continuous from the right** at a and **continuous from the left** at b.

Similar definitions can be made to cover continuity on intervals of the form $(a, b]$ and $[a, b)$, or on infinite intervals. For instance, the function

$$f(x) = \sqrt{x}$$

is continuous on the infinite interval $[0, \infty)$.

Example 3   **Examining Continuity at an Endpoint**

Discuss the continuity of

$$f(x) = \sqrt{3 - x}.$$

SOLUTION   Notice that the domain of f is the set $(-\infty, 3]$. Moreover, f is continuous from the left at $x = 3$ because

$$\begin{aligned}
\lim_{x \to 3^-} f(x) &= \lim_{x \to 3^-} \sqrt{3 - x} \\
&= \sqrt{3 - 3} \\
&= 0 \\
&= f(3).
\end{aligned}$$

For all $x < 3$, the function f satisfies the three conditions for continuity. So, you can conclude that f is continuous on the interval $(-\infty, 3]$, as shown in Figure 1.64.

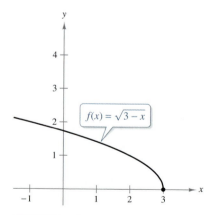

$$f(x) = \sqrt{3 - x}$$

FIGURE 1.64

✓ **Checkpoint 3**

Discuss the continuity of

$$f(x) = \sqrt{x - 2}.$$

Example 4 **Examining Continuity on a Closed Interval**

Discuss the continuity of

$$g(x) = \begin{cases} 5 - x, & -1 \le x \le 2 \\ x^2 - 1, & 2 < x \le 3 \end{cases}.$$

SOLUTION The polynomial functions

$$5 - x$$

and

$$x^2 - 1$$

are continuous on the intervals $[-1, 2]$ and $(2, 3]$, respectively. So, to conclude that g is continuous on the entire interval

$$[-1, 3]$$

you need only to check the behavior of g when $x = 2$. You can do this by taking the one-sided limits when $x = 2$.

$$\lim_{x \to 2^-} g(x) = \lim_{x \to 2^-} (5 - x) = 5 - 2 = 3 \qquad \text{Limit from the left}$$

and

$$\lim_{x \to 2^+} g(x) = \lim_{x \to 2^+} (x^2 - 1) = 2^2 - 1 = 3 \qquad \text{Limit from the right}$$

Because these two limits are equal,

$$\lim_{x \to 2} g(x) = g(2) = 3.$$

So, g is continuous at $x = 2$ and, consequently, it is continuous on the entire interval

$$[-1, 3].$$

The graph of g is shown in Figure 1.65.

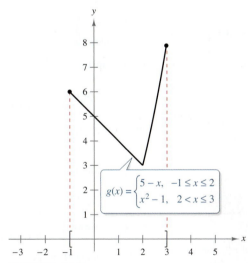

$$g(x) = \begin{cases} 5 - x, & -1 \le x \le 2 \\ x^2 - 1, & 2 < x \le 3 \end{cases}$$

FIGURE 1.65

✓ **Checkpoint 4**

Discuss the continuity of

$$f(x) = \begin{cases} x + 2, & -1 \le x < 3 \\ 14 - x^2, & 3 \le x \le 5 \end{cases}.$$

The Greatest Integer Function

Some functions that are used in business applications are **step functions.** For instance, the function in Example 10 in Section 1.5 is a step function. The **greatest integer function** is another example of a step function. This function is denoted by

$$[\![x]\!] = \text{greatest integer less than or equal to } x.$$

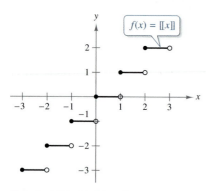

Greatest Integer Function
FIGURE 1.66

For example,

$$[\![-2.1]\!] = \text{greatest integer less than or equal to } -2.1 = -3$$
$$[\![-2]\!] = \text{greatest integer less than or equal to } -2 = -2$$
$$[\![1.5]\!] = \text{greatest integer less than or equal to } 1.5 = 1.$$

Note that the graph of the greatest integer function (Figure 1.66) jumps up one unit at each integer. This implies that the function is not continuous at each integer.

In real-life applications, the domain of the greatest integer function is often restricted to nonnegative values of x. In such cases, this function serves the purpose of **truncating** the decimal portion of x. For example, 1.345 is truncated to 1 and 3.57 is truncated to 3. That is,

$$[\![1.345]\!] = 1 \quad \text{and} \quad [\![3.57]\!] = 3.$$

Example 5 Modeling a Cost Function

A bookbinding company produces 10,000 books in an eight-hour shift. The fixed cost *per shift* amounts to $5000, and the unit cost per book is $3. Using the greatest integer function, you can write the cost of producing x books as

$$C = 5000\left(1 + \left[\!\left[\frac{x-1}{10,000}\right]\!\right]\right) + 3x.$$

Sketch the graph of this cost function.

SOLUTION Note that during the first eight-hour shift,

$$\left[\!\left[\frac{x-1}{10,000}\right]\!\right] = 0, \quad 1 \le x \le 10,000$$

which implies

$$C = 5000\left(1 + \left[\!\left[\frac{x-1}{10,000}\right]\!\right]\right) + 3x = 5000 + 3x.$$

During the second eight-hour shift,

$$\left[\!\left[\frac{x-1}{10,000}\right]\!\right] = 1, \quad 10,001 \le x \le 20,000$$

which implies

$$C = 5000\left(1 + \left[\!\left[\frac{x-1}{10,000}\right]\!\right]\right) + 3x$$
$$= 10,000 + 3x.$$

The graph of C is shown in Figure 1.67. Note the graph's discontinuities at $x = 10,000$, 20,000, and 30,000.

Cost of Producing Books

$$C = 5000\left(1 + \left[\!\left[\frac{x-1}{10,000}\right]\!\right]\right) + 3x$$

FIGURE 1.67

✓ Checkpoint 5

Use a graphing utility to graph the cost function in Example 5.

Extended Application: Compound Interest

Banks and other financial institutions differ on how interest is paid to an account. If the interest is added to the account so that future interest is paid on previously earned interest, then the interest is said to be **compounded.** For instance, a deposit of $10,000 is made in an account that pays 6% interest, compounded quarterly. Because the 6% is the annual interest rate, the quarterly rate is

$$\tfrac{1}{4}(0.06) = 0.015$$

or 1.5%. The balances during the first five quarters are shown below.

Quarter	Balance
1st	$10,000.00
2nd	$10,000.00 + (0.015)(10,000.00) = \$10,150.00$
3rd	$10,150.00 + (0.015)(10,150.00) = \$10,302.25$
4th	$10,302.25 + (0.015)(10,302.25) = \$10,456.78$
5th	$10,456.78 + (0.015)(10,456.78) = \$10,613.63$

Example 6   Graphing Compound Interest

Sketch the graph of the balance in the account described above.

SOLUTION   Let A represent the balance in the account and let t represent the time, in years. You can use the greatest integer function to represent the balance, as shown.

$$A = 10,000(1 + 0.015)^{[\![4t]\!]}$$

From the graph shown in Figure 1.68, notice that the function has a discontinuity at each quarter. That is, A has discontinuities at

$$t = \frac{1}{4}, \; t = \frac{1}{2}, \; t = \frac{3}{4}, \; t = 1,$$

and

$$t = \frac{5}{4}.$$

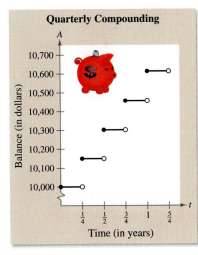

FIGURE 1.68

✓**Checkpoint 6**

Write an equation that gives the balance of the account in Example 6 when the annual interest rate is 3%. Then sketch the graph of the equation. ■

SUMMARIZE   (Section 1.6)

1. State the definition of continuity *(page 49)*. For an example of a function that is continuous at every real number, see Example 1.

2. State the definition of continuity on a closed interval *(page 52)*. For an example of a function that is continuous on a closed interval, see Example 4.

3. State the definition of the greatest integer function *(page 54)*. For real-life examples of the greatest integer function, see Examples 5 and 6.

SKILLS WARM UP 1.6   The following warm-up exercises involve skills that were covered in a previous course or in earlier sections. You will use these skills in the exercise set for this section. For additional help, review Appendix Sections A.4 and A.5, and Section 1.5.

In Exercises 1–4, simplify the expression.

1. $\dfrac{x^2 + 6x + 8}{x^2 - 6x - 16}$

2. $\dfrac{x^2 - 5x - 6}{x^2 - 9x + 18}$

3. $\dfrac{2x^2 - 2x - 12}{4x^2 - 24x + 36}$

4. $\dfrac{x^3 - 16x}{x^3 + 2x^2 - 8x}$

In Exercises 5–8, solve for x.

5. $x^2 + 7x = 0$

6. $x^2 + 4x - 5 = 0$

7. $3x^2 + 8x + 4 = 0$

8. $x^3 + 5x^2 - 24x = 0$

In Exercises 9 and 10, find the limit.

9. $\displaystyle\lim_{x \to 3} (2x^2 - 3x + 4)$

10. $\displaystyle\lim_{x \to -2} (3x^3 - 8x + 7)$

ALGEBRA TUTOR

Order of Operations

Much of the algebra in this chapter involves evaluation of algebraic expressions. When you evaluate an algebraic expression, you need to know the priorities assigned to different operations. These priorities are called the *order of operations*.

1. Perform operations inside *symbols of grouping or absolute value symbols,* starting with the innermost symbol.

2. Evaluate all *exponential* expressions.

3. Perform all *multiplications* and *divisions* from left to right.

4. Perform all *additions* and *subtractions* from left to right.

Example 1 Using Order of Operations

Evaluate each expression.

a. $20 - 2 \cdot 3^2$ **b.** $3 + 8 \div 2 \cdot 2$ **c.** $7 - [(5 \cdot 3) + 2^3]$

d. $[36 \div (3^2 \cdot 2)] + 6$ **e.** $36 - [3^2 \cdot (2 \div 6)]$ **f.** $10 - 2(8 + |5 - 7|)$

SOLUTION

a. $20 - 2 \cdot 3^2 = 20 - 2 \cdot 9$ Evaluate exponential expression.

$= 20 - 18$ Multiply.

$= 2$ Subtract.

b. $3 + 8 \div 2 \cdot 2 = 3 + 4 \cdot 2$ Divide.

$= 3 + 8$ Multiply.

$= 11$ Add.

c. $7 - [(5 \cdot 3) + 2^3] = 7 - [15 + 2^3]$ Multiply inside parentheses.

$= 7 - [15 + 8]$ Evaluate exponential expression.

$= 7 - 23$ Add inside brackets.

$= -16$ Subtract.

d. $[36 \div (3^2 \cdot 2)] + 6 = [36 \div (9 \cdot 2)] + 6$ Evaluate exponential expression inside parentheses.

$= [36 \div 18] + 6$ Multiply inside parentheses.

$= 2 + 6$ Divide inside brackets.

$= 8$ Add.

e. $36 - [3^2 \cdot (2 \div 6)] = 36 - \left[3^2 \cdot \frac{1}{3}\right]$ Divide inside parentheses.

$= 36 - \left[9 \cdot \frac{1}{3}\right]$ Evaluate exponential expression.

$= 36 - 3$ Multiply inside brackets.

$= 33$ Subtract.

f. $10 - 2(8 + |5 - 7|) = 10 - 2(8 + |-2|)$ Subtract inside absolute value symbols.

$= 10 - 2(8 + 2)$ Evaluate absolute value.

$= 10 - 2(10)$ Add inside parentheses.

$= 10 - 20$ Multiply.

$= -10$ Subtract.

TECH TUTOR

Most scientific and graphing calculators use the same order of operations listed above. Try entering the expressions in Example 1 into your calculator. Do you get the same results?

Solving Equations

A second algebraic skill used in this chapter is solving an equation in one variable.

1. To solve a *linear equation,* you can add or subtract the same quantity to or from each side of the equation. You can also multiply or divide each side of the equation by the same *nonzero* quantity.

2. To solve a *quadratic equation,* you can take the square root of each side, use factoring, or use the Quadratic Formula.

3. To solve a *radical equation,* isolate the radical on one side of the equation and square each side of the equation.

4. To solve an *absolute value equation,* use the definition of absolute value to rewrite the equation as two equations.

Example 2 Solving Equations

Solve each equation.

a. $3x - 3 = 5x - 7$ **b.** $2x^2 = 10$

c. $2x^2 + 5x - 6 = 6$ **d.** $\sqrt{2x - 7} = 5$

e. $|3x + 6| = 9$

SOLUTION

a. $3x - 3 = 5x - 7$ Write original (linear) equation.

$\qquad -3 = 2x - 7$ Subtract $3x$ from each side.

$\qquad\quad 4 = 2x$ Add 7 to each side.

$\qquad\quad 2 = x$ Divide each side by 2.

b. $2x^2 = 10$ Write original (quadratic) equation.

$\qquad x^2 = 5$ Divide each side by 2.

$\qquad x = \pm\sqrt{5}$ Take the square root of each side.

c. $\quad 2x^2 + 5x - 6 = 6$ Write original (quadratic) equation.

$\qquad 2x^2 + 5x - 12 = 0$ Write in general form.

$\qquad (2x - 3)(x + 4) = 0$ Factor.

$\qquad\qquad 2x - 3 = 0$ ⇨ $x = \tfrac{3}{2}$ Set first factor equal to zero.

$\qquad\qquad\quad x + 4 = 0$ ⇨ $x = -4$ Set second factor equal to zero.

d. $\sqrt{2x - 7} = 5$ Write original (radical) equation.

$\qquad 2x - 7 = 25$ Square each side.

$\qquad\quad 2x = 32$ Add 7 to each side.

$\qquad\quad x = 16$ Divide each side by 2.

e. $|3x + 6| = 9$ Write original (absolute value) equation.

$\qquad 3x + 6 = -9 \quad \text{or} \quad 3x + 6 = 9$ Rewrite equivalent equations.

$\qquad\quad 3x = -15 \quad \text{or} \qquad 3x = 3$ Subtract 6 from each side.

$\qquad\qquad x = -5 \quad \text{or} \qquad\quad x = 1$ Divide each side by 3.

STUDY TIP

Solving radical equations can sometimes lead to *extraneous solutions* (those that do not satisfy the original equation). For instance, squaring both sides of the following equation yields two possible solutions, one of which is extraneous.

$$\sqrt{x} = x - 2$$
$$x = x^2 - 4x + 4$$
$$0 = x^2 - 5x + 4$$
$$\quad = (x - 4)(x - 1)$$

$x - 4 = 0$ ⇨ $x = 4$
(solution)

$x - 1 = 0$ ⇨ $x = 1$
(extraneous)

SUMMARY AND STUDY STRATEGIES

After studying this chapter, you should have acquired the following skills.
The exercise numbers are keyed to the Review Exercises that begin on page 61.
Answers to odd-numbered Review Exercises are given in the back of the text.*

Section 1.1
                                                                        Review Exercises

■ Plot points in a coordinate plane.   *1, 2*

■ Find the distance between two points in a coordinate plane.   *3–8*

$$d = \sqrt{(x_2 - x_1)^2 + (y_2 - y_1)^2}$$

■ Find the midpoint of a line segment connecting two points.   *9–14*

$$\text{Midpoint} = \left(\frac{x_1 + x_2}{2}, \frac{y_1 + y_2}{2} \right)$$

■ Interpret real-life data that is presented graphically.   *15, 16*

■ Translate points in a coordinate plane.   *17, 18*

Section 1.2

■ Sketch graphs of equations by hand.   *19–28*

■ Find the x- and y-intercepts of graphs of equations.   *29–32*

■ Find the standard forms of equations of circles.   *33–36*

$$(x - h)^2 + (y - k)^2 = r^2$$

■ Find the points of intersection of two graphs.   *37–40*

■ Find the break-even point for a business.   *41, 42*

The break-even point occurs when the revenue R is equal to the cost C.

■ Find the equilibrium points of supply equations and demand equations.   *43*

The equilibrium point is the point of intersection of the graphs
of the supply and demand equations.

■ Use mathematical models to model and solve real-life problems.   *44*

Section 1.3

■ Use the slope-intercept form of a linear equation to sketch graphs of lines.   *45–52*

$$y = mx + b$$

■ Find slopes of lines passing through two points.   *53–56*

$$m = \frac{y_2 - y_1}{x_2 - x_1}$$

■ Use the point-slope form to write equations of lines and sketch graphs of lines.   *57–64*

$$y - y_1 = m(x - x_1)$$

---

* A wide range of valuable study aids are available to help you master the material in this chapter.
  The *Student Solutions Manual* includes step-by-step solutions to all odd-numbered exercises to
  help you review and prepare. The student website at *www.cengagebrain.com* offers algebra help
  and a *Graphing Technology Guide*, which contains step-by-step commands and instructions for a
  wide variety of graphing calculators.

Section 1.3 (continued)

Review Exercises

■ Find equations of parallel and perpendicular lines.

Parallel lines: $m_1 = m_2$

Perpendicular lines: $m_1 = -\dfrac{1}{m_2}$

65, 66

■ Use linear equations to solve real-life problems such as predicting future
sales or creating a linear depreciation schedule.

67, 68

Section 1.4

■ Use the Vertical Line Test to decide whether the relationship between
two variables is a function.

69–72

If every vertical line intersects the graph of an equation at most once,
then the equation defines y as a function of x.

■ Find the domains and ranges of functions.

73–78

■ Use function notation and evaluate functions.

79, 80

■ Combine functions to create other functions.

81, 82

■ Use the Horizontal Line Test to determine whether functions have inverse
functions. If they do, find the inverse functions.

83–88

A function is one-to-one when every horizontal line intersects the graph of
the function at most once. For a function to have an inverse function, it must
be one-to-one.

Section 1.5

■ Use a table to estimate limits.

89–92

■ Determine whether limits exist. If they do, find the limits.

93–110

Section 1.6

■ Determine whether functions are continuous at a point, on an open interval, and
on a closed interval.

111–120

■ Determine the constant such that f is continuous.

121, 122

■ Use analytic and graphical models of real-life data to solve real-life problems.

123–126

Study Strategies

■ **Use a Graphing Utility**   A graphing calculator or graphing software for a computer can help you in this course
in two important ways. As an *exploratory device,* a graphing utility allows you to learn concepts by allowing you
to compare graphs of equations. For instance, sketching the graphs of $y = x^2$, $y = x^2 + 1$, and $y = x^2 - 1$ helps
confirm that adding (or subtracting) a constant to (or from) a function shifts the graph of the function vertically. As
a *problem-solving tool,* a graphing utility frees you of some of the drudgery of sketching complicated graphs by hand.
The time that you save can be spent using mathematics to solve real-life problems.

■ **Use the Skills Warm-Up Exercises**   Each exercise set in this text begins with a set of skills warm-up exercises. You
should begin each homework session by quickly working all of these exercises. (All are answered in the back of the text.)
The "old" skills covered in these exercises are needed to master the "new" skills in the section exercise set. The skills
warm-up exercises remind you that mathematics is cumulative—to be successful in this course, you must retain "old" skills.

■ **Use the Additional Study Aids**   The additional study aids were prepared specifically to help you master the concepts
discussed in the text. They are the *Student Solutions Manual,* the student website, and the *Graphing Technology Guide.*

Review Exercises

Plotting Points in the Cartesian Plane   In Exercises 1 and 2, plot the points in the Cartesian plane.

1. $(2, 3), (0, 6), (-5, 1), (4, -3), (-3, -1)$
2. $(1, -4), (-1, -2), (6, -5), (-2, 0), (5, 5)$

Finding a Distance   In Exercises 3–8, find the distance between the two points.

3. $(0, 0), (5, 2)$
4. $(1, 2), (4, 3)$
5. $(-1, 3), (-4, 6)$
6. $(6, 8), (-3, 7)$
7. $\left(\frac{1}{4}, -8\right), \left(\frac{3}{4}, -6\right)$
8. $(-0.6, 3), (4, -1.8)$

Finding a Segment's Midpoint   In Exercises 9–14, find the midpoint of the line segment connecting the two points.

9. $(5, 6), (9, 2)$
10. $(0, 0), (-4, 8)$
11. $(-10, 4), (-6, 8)$
12. $(7, -9), (-3, 5)$
13. $\left(-1, \frac{1}{5}\right), \left(6, \frac{3}{5}\right)$
14. $(6, 1.2), (-3.2, 5)$

Revenues, Costs, and Profits   In Exercises 15 and 16, use the graph below, which gives the revenues, costs, and profits for Google from 2005 through 2009. *(Source: Google, Inc.)*

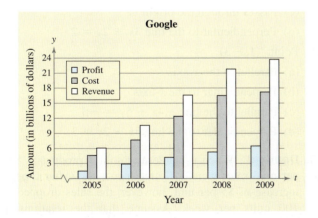

15. Write an equation that relates the revenue R, cost C, and profit P. Explain the relationship between the heights of the bars and the equation.
16. Estimate the revenue, cost, and profit for Google for each year.

Translating Points in the Plane   In Exercises 17 and 18, use the translation and the graph to find the vertices of the figure after it has been translated.

17. 3 units left and 4 units up

18. 4 units right and 1 unit down

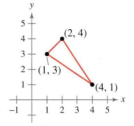

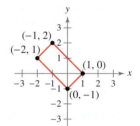

Sketching the Graph of an Equation   In Exercises 19–28, sketch the graph of the equation.

19. $y = 4x - 12$
20. $y = 4 - 3x$
21. $y = x^2 + 5$
22. $y = 1 - x^2$
23. $y = |4 - x|$
24. $y = |2x - 3|$
25. $y = x^3 + 4$
26. $y = 2x^3 - 1$
27. $y = \sqrt{4x + 1}$
28. $y = \sqrt{2x}$

Finding x- and y-Intercepts   In Exercises 29–32, find the x- and y-intercepts of the graph of the equation. Use a graphing utility to verify your results.

29. $4x + y + 3 = 0$
30. $3x - y + 6 = 0$
31. $y = x^2 + 2x - 8$
32. $y = (x - 1)^3 + 2(x - 1)^2$

Finding the Equation of a Circle   In Exercises 33–36, find the standard form of the equation of the circle and sketch its graph.

33. Center: $(0, 0)$; radius: 8
34. Center: $(-5, -2)$; radius: 6
35. Center: $(0, 0)$; solution point: $\left(2, \sqrt{5}\right)$
36. Center: $(3, -4)$; solution point: $(-1, -1)$

Finding Points of Intersection   In Exercises 37–40, find the point(s) of intersection (if any) of the graphs of the equations. Use a graphing utility to check your results.

37. $y = 2x + 13, \quad y = -5x - 1$
38. $y = x^2 - 5, \quad y = x + 1$
39. $y = x^3, \quad y = x$
40. $y = -x^2 + 4, \quad y = 2x - 1$

41. Break-Even Analysis A student organization wants to raise money by having a T-shirt sale. Each shirt costs $8. The silk screening costs $200 for the design, plus $2 per shirt. Each shirt will sell for $14.

(a) Find equations for the total cost C and the total revenue R for selling x shirts.

(b) Find the break-even point.

42. Break-Even Analysis You are starting a part-time business. You make an initial investment of $6000. The unit cost of the product is $6.50, and the selling price is $13.90.

(a) Find equations for the total cost C and the total revenue R for selling x units of the product.

(b) Find the break-even point.

43. Supply and Demand The demand and supply equations for a cordless screwdriver are given by

$p = 91.4 - 0.009x$ Demand equation

$p = 6.4 + 0.008x$ Supply equation

where p is the price in dollars and x represents the number of units. Find the equilibrium point for this market.

44. Wind Energy The table shows the annual amounts of U.S. consumption W (in trillion Btu) of wind energy for the years 2003 through 2008. *(Source: U.S. Energy Information Administration)*

Year	2003	2004	2005
Consumption	115	142	178

Year	2006	2007	2008
Consumption	264	341	546

A mathematical model for the data is given by

$W = 3.870t^3 - 45.04t^2 + 205.7t - 204$

where t represents the year, with $t = 3$ corresponding to 2003.

(a) Compare the actual consumptions with those given by the model. How well does the model fit the data? Explain your reasoning.

(b) Use the model to predict the consumption in 2014.

Finding the Slope and y-Intercept In Exercises 45–52, find the slope and y-intercept (if possible) of the equation of the line. Then sketch the graph of the equation.

45. $y = -x + 12$ **46.** $y = 3x - 1$

47. $3x + y = -2$ **48.** $2x - 4y = -8$

49. $y = -\frac{5}{3}$ **50.** $x = -3$

51. $-2x - 5y - 5 = 0$ **52.** $3.2x - 0.8y + 5.6 = 0$

Finding the Slope of a Line In Exercises 53–56, find the slope of the line passing through the pair of points.

53. $(0, 0), (7, 6)$ **54.** $(-1, 5), (-5, 7)$

55. $(10, 17), (-11, -3)$

56. $(-11, -3), (-1, -3)$

Using the Point-Slope Form In Exercises 57–60, find the equation of the line that passes through the given point and has the given slope. Then use the equation to sketch the line.

Point	Slope
57. $(3, -1)$	$m = -2$
58. $(-3, -3)$	$m = \frac{1}{2}$
59. $(1.5, -4)$	$m = 0$
60. $(8, 2)$	m is undefined.

Writing an Equation of a Line In Exercises 61–64, find the equation of the line that passes through the points. Then use the equation to sketch the line.

61. $(1, -7), (7, 5)$

62. $(2, 4), (8, 12)$

63. $(5, 7), (5, 14)$

64. $(4, -3), (-2, -3)$

Finding Parallel and Perpendicular Lines In Exercises 65 and 66, find the equation of the line passing through the given point and satisfying the given condition.

65. Point: $(-3, 6)$

(a) Slope is $\frac{7}{8}$.

(b) Parallel to the line $4x + 2y = 7$

(c) Passes through the origin

(d) Perpendicular to the line $3x - 2y = 2$

66. Point: $(1, -3)$

(a) Parallel to the x-axis

(b) Perpendicular to the x-axis

(c) Parallel to the line $-4x + 5y = -3$

(d) Perpendicular to the line $5x - 2y = 3$

67. Demand When a wholesaler sold a product at $32 per unit, sales were 750 units per week. After a price increase of $5 per unit, however, the sales dropped to 700 units per week. Assume that the relationship between the price p and the units sold per week x is linear.

(a) Write a linear equation expressing x in terms of p.

(b) Predict the number of units sold per week at a price of $34.50 per unit.

(c) Predict the number of units sold per week at a price of $42 per unit.

68. Linear Depreciation A printing company purchases an advanced color copier/printer for $117,000. After 9 years, the equipment will be obsolete and have no value.

(a) Write a linear equation giving the value v (in dollars) of the equipment in terms of the time t (in years), $0 \leq t \leq 9$.

(b) Use a graphing utility to graph the equation.

(c) Move the cursor along the graph and estimate (to two-decimal-place accuracy) the value of the equipment after 4 years.

(d) Move the cursor along the graph and estimate (to two-decimal-place accuracy) the time when the value of the equipment will be $84,000.

Vertical Line Test In Exercises 69–72, use the Vertical Line Test to determine whether y is a function of x.

69. $y = -x^2 + 2$

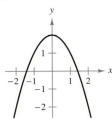

70. $x^2 + y^2 = 4$

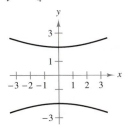

71. $y^2 - \frac{1}{4}x^2 = 4$

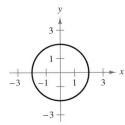

72. $y = |x + 4|$

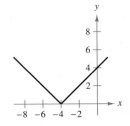

Finding the Domain and Range of a Function In Exercises 73–78, find the domain and range of the function. Use a graphing utility to verify your results.

73. $f(x) = x^3 + 2x^2 - x + 2$

74. $f(x) = 2$

75. $f(x) = \sqrt{x + 1}$

76. $f(x) = -|x| + 3$

77. $f(x) = \dfrac{x - 3}{x^2 + x - 12}$

78. $f(x) = \begin{cases} 6 - x, & x < 2 \\ 3x - 2, & x \geq 2 \end{cases}$

Evaluating a Function In Exercises 79 and 80, evaluate the function at the specified values of the independent variable. Simplify the result.

79. $f(x) = 3x + 4$

(a) $f(1)$ (b) $f(-5)$ (c) $f(x + 1)$

80. $f(x) = x^2 + 4x + 3$

(a) $f(0)$ (b) $f(3)$ (c) $f(x - 1)$

Combinations of Functions In Exercises 81 and 82, find (a) $f(x) + g(x)$, (b) $f(x) - g(x)$, (c) $f(x) \cdot g(x)$, (d) $f(x)/g(x)$, (e) $f(g(x))$, and (f) $g(f(x))$, if defined.

81. $f(x) = 1 + x^2$, $g(x) = 2x - 1$

82. $f(x) = 2x - 3$, $g(x) = \sqrt{x + 1}$

 Determine Whether a Function Is One-to-One In Exercises 83–88, use a graphing utility to graph the function. Then use the Horizontal Line Test to determine whether the function is one-to-one. If it is, find its inverse function.

83. $f(x) = 4x - 3$

84. $f(x) = \frac{3}{2}x$

85. $f(x) = -x^2 + \frac{1}{2}$

86. $f(x) = x^3 - 1$

87. $f(x) = |x + 1|$

88. $f(x) = 6$

Finding Limits Numerically In Exercises 89–92, complete the table and use the result to estimate the limit. Use a graphing utility to graph the function to confirm your result.

89. $\lim\limits_{x \to 1} (4x - 3)$

x	0.9	0.99	0.999	1	1.001	1.01	1.1
$f(x)$				?			

90. $\lim\limits_{x \to 3} \dfrac{x - 3}{x^2 - 2x - 3}$

x	2.9	2.99	2.999	3	3.001	3.01	3.1
$f(x)$				?			

91. $\lim\limits_{x \to 0} \dfrac{\sqrt{x + 6} - 6}{x}$

x	-0.1	-0.01	-0.001	0	0.001	0.01	0.1
$f(x)$				?			

92. $\lim\limits_{x \to 7} \dfrac{\dfrac{1}{x - 7} - \dfrac{1}{7}}{x}$

x	6.9	6.99	6.999	7	7.001	7.01	7.1
$f(x)$				?			

Finding Limits   In Exercises 93–110, find the limit (if it exists).

93. $\lim\limits_{x \to 3} 8$

94. $\lim\limits_{x \to 4} x^2$

95. $\lim\limits_{x \to 2} (5x - 3)$

96. $\lim\limits_{x \to 5} (2x + 4)$

97. $\lim\limits_{x \to -1} \dfrac{x + 3}{6x + 1}$

98. $\lim\limits_{t \to 3} \dfrac{t}{t + 5}$

99. $\lim\limits_{t \to 0} \dfrac{t^2 + 1}{t}$

100. $\lim\limits_{t \to 2} \dfrac{t + 1}{t - 2}$

101. $\lim\limits_{x \to -2} \dfrac{x + 2}{x^2 - 4}$

102. $\lim\limits_{x \to 3^-} \dfrac{x^2 - 9}{x - 3}$

103. $\lim\limits_{x \to 0^+} \left(x - \dfrac{1}{x} \right)$

104. $\lim\limits_{x \to 1/2} \dfrac{2x - 1}{6x - 3}$

105. $\lim\limits_{x \to 0} \dfrac{[1/(x - 2)] - 1}{x}$

106. $\lim\limits_{s \to 0} \dfrac{\left(1/\sqrt{1 + s}\right) - 1}{s}$

107. $\lim\limits_{x \to 0} f(x)$, where $f(x) = \begin{cases} x + 5, & x \neq 0 \\ 3, & x = 0 \end{cases}$

108. $\lim\limits_{x \to -2} f(x)$, where $f(x) = \begin{cases} \frac{1}{2}x + 5, & x < -2 \\ -x + 2, & x \geq -2 \end{cases}$

109. $\lim\limits_{\Delta x \to 0} \dfrac{(x + \Delta x)^3 - (x + \Delta x) - (x^3 - x)}{\Delta x}$

110. $\lim\limits_{\Delta x \to 0} \dfrac{1 - (x + \Delta x)^2 - (1 - x^2)}{\Delta x}$

Determining Continuity   In Exercises 111–120, describe the interval(s) on which the function is continuous. Explain why the function is continuous on the interval(s). If the function has a discontinuity, identify the conditions of continuity that are not satisfied.

111. $f(x) = x + 6$

112. $f(x) = x^2 + 3x + 2$

113. $f(x) = \dfrac{1}{(x + 4)^2}$

114. $f(x) = \dfrac{x + 2}{x}$

115. $f(x) = \dfrac{3}{x + 1}$

116. $f(x) = \dfrac{x + 1}{2x + 2}$

117. $f(x) = [\![x + 3]\!]$

118. $f(x) = [\![x]\!] - 2$

119. $f(x) = \begin{cases} x, & x \leq 0 \\ x + 1, & x > 0 \end{cases}$

120. $f(x) = \begin{cases} x, & x \leq 0 \\ x^2, & x > 0 \end{cases}$

Making a Function Continuous   In Exercises 121 and 122, find the constant a such that the function is continuous on the entire real number line.

121. $f(x) = \begin{cases} -x + 1, & x \leq 3 \\ ax - 8, & x > 3 \end{cases}$

122. $f(x) = \begin{cases} x + 1, & x < 1 \\ 2x + a, & x \geq 1 \end{cases}$

123. Consumer Awareness   The cost C (in dollars) of purchasing x bottles of vitamins at a whole foods store is shown below.

$$C(x) = \begin{cases} 5.99x, & 0 < x \leq 5 \\ 4.99x, & 5 < x \leq 10 \\ 3.99x, & 10 < x \leq 15 \\ 2.99x, & x > 15 \end{cases}$$

(a) Use a graphing utility to graph the function and then discuss its continuity. At what values is the function not continuous? Explain your reasoning.

(b) Find the cost of purchasing 10 bottles.

124. Salary Contract   A union contract guarantees a 10% salary increase yearly for 3 years. For a current salary of $28,000, the salaries S (in thousands of dollars) for the next 3 years are given by

$$S(t) = \begin{cases} 28.00, & 0 < t \leq 1 \\ 30.80, & 1 < t \leq 2 \\ 33.88, & 2 < t \leq 3 \end{cases}$$

where $t = 0$ represents the present year. Does the limit of S exist as t approaches 2? Explain your reasoning.

125. Consumer Awareness   A pay-as-you-go cellular phone charges $1 for the first minute and $0.10 for each additional minute or fraction thereof.

(a) Use the greatest integer function to create a model for the cost C of a phone call lasting t minutes.

(b) Use a graphing utility to graph the function and then discuss its continuity.

126. Recycling   A recycling center pays $0.50 for each pound of aluminum cans. Twenty-four aluminum cans weigh one pound. A mathematical model for the amount A paid by the recycling center is

$$A = \frac{1}{2} \left[\!\!\left[\frac{x}{24} \right]\!\!\right]$$

where x is the number of cans.

(a) Use a graphing utility to graph the function and then discuss its continuity.

(b) How much does the recycling center pay out for 1500 cans?

TEST YOURSELF

Take this test as you would take a test in class. When you are done, check your work against the answers given in the back of the book.

In Exercises 1–3, (a) find the distance between the points, (b) find the midpoint of the line segment joining the points, (c) find the slope of the line passing through the points, (d) find an equation of the line passing through the points, and (e) sketch the graph of the equation.

1. $(1, -1), (-4, 4)$ **2.** $\left(\frac{5}{2}, 2\right), (0, 2)$ **3.** $(2, 3), (-4, 1)$

4. The demand and supply equations for a product are $p = 65 - 2.1x$ and $p = 43 + 1.9x$, respectively, where p is the price (in dollars) and x represents the number of units, in thousands. Find the equilibrium point for this market.

In Exercises 5–7, find the slope and y-intercept (if possible) of the equation of the line. Then sketch the graph of the equation.

5. $y = \frac{1}{5}x - 2$ **6.** $x - \frac{7}{4} = 0$ **7.** $-x - 0.4y + 2.5 = 0$

8. Write an equation of the line that passes through $(-1, -7)$ and is perpendicular to $-4x + y = 8$.

9. Write an equation of the line that passes through $(2, 1)$ and is parallel to $5x - 2y = 8$.

In Exercises 10–12, (a) graph the function and label the intercepts, (b) determine the domain and range of the function, (c) find the value of the function when x is -3, -2, and 3, and (d) determine whether the function is one-to-one.

10. $f(x) = 2x + 5$ **11.** $f(x) = x^2 - x - 2$ **12.** $f(x) = \sqrt{x + 5}$

In Exercises 13 and 14, find the inverse function of f.

13. $f(x) = 4x + 6$ **14.** $f(x) = \sqrt[3]{8 - 3x}$

In Exercises 15–18, find the limit (if it exists).

15. $\lim\limits_{x \to 0} \dfrac{x - 2}{x + 2}$ **16.** $\lim\limits_{x \to 5} \dfrac{x + 5}{x - 5}$

17. $\lim\limits_{x \to -3} \dfrac{x^2 + 2x - 3}{x^2 + 4x + 3}$ **18.** $\lim\limits_{x \to 0} \dfrac{\sqrt{x + 9} - 3}{x}$

In Exercises 19–21, describe the interval(s) on which the function is continuous. Explain why the function is continuous on the interval(s). If the function has a discontinuity, identify the conditions of continuity that are not satisfied.

19. $f(x) = \dfrac{x^2 - 16}{x - 4}$ **20.** $f(x) = \sqrt{5 - x}$ **21.** $f(x) = \begin{cases} 1 - x, & x < 1 \\ x - x^2, & x \ge 1 \end{cases}$

22. The table lists the numbers of unemployed workers y (in thousands) in the United States for selected years. A mathematical model for the data is given by

$$y = 193.898t^3 - 3080.32t^2 + 15{,}478.5t - 16{,}925$$

where t represents the year, with $t = 4$ corresponding to 2004. *(Source: U.S. Bureau of Labor Statistics)*

(a) Compare the actual numbers of unemployed workers with those given by the model. How well does the model fit the data? Explain your reasoning.

(b) Use the model to predict the number of unemployed workers in 2014.

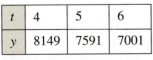

t	4	5	6
y	8149	7591	7001

t	7	8	9
y	7078	8924	14,265

Table for 22

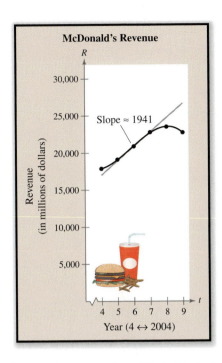

McDonald's Revenue

Slope ≈ 1941

Revenue (in millions of dollars)

Year (4 ↔ 2004)

Example 11 on page 85 shows how differentiation can be used to find the rate of change in a company's revenue.

2 Differentiation

67

2.1 The Derivative and the Slope of a Graph

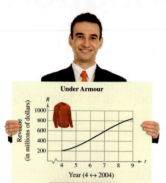

■ Identify tangent lines to a graph at a point.
■ Approximate the slopes of tangent lines to graphs at points.
■ Use the limit definition to find the slopes of graphs at points.
■ Use the limit definition to find the derivatives of functions.
■ Describe the relationship between differentiability and continuity.

In Exercise 13, you will estimate and interpret the slope of the graph of a revenue function.

Tangent Line to a Graph

Calculus is a branch of mathematics that studies rates of change of functions. In this course, you will learn that rates of change have many applications in real life. In Section 1.3, you learned how the slope of a line indicates the rate at which the line rises or falls. For a line, this rate (or slope) is the same at every point on the line. For graphs other than lines, the rate at which the graph rises or falls changes from point to point. For instance, in Figure 2.1, the parabola is rising more quickly at the point (x_1, y_1) than it is at the point (x_2, y_2). At the vertex (x_3, y_3), the graph levels off, and at the point (x_4, y_4), the graph is falling.

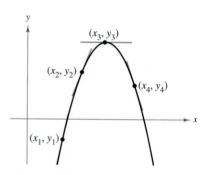

The slope of a nonlinear graph changes from one point to another.

FIGURE 2.1

To determine the rate at which a graph rises or falls at a *single point,* you can find the slope of the **tangent line** at the point. In simple terms, the tangent line to the graph of a function f at a point $P(x_1, y_1)$ is the line that best approximates the graph at that point, as shown in Figure 2.1. Figure 2.2 shows other examples of tangent lines.

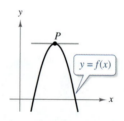 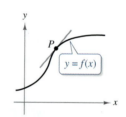 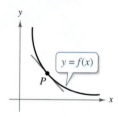

Tangent Line to a Graph at a Point

FIGURE 2.2

When Isaac Newton (1642–1727) was working on the "tangent line problem," he realized that it is difficult to define precisely what is meant by a tangent to a general curve. From geometry, you know that a line is tangent to a circle when the line intersects the circle at only one point, as shown in Figure 2.3. Tangent lines to a noncircular graph, however, can intersect the graph at more than one point. For instance, in the second graph in Figure 2.2, when the tangent line is extended, it intersects the graph at a point other than the point of tangency. In this section, you will see how the notion of a limit can be used to define a general tangent line.

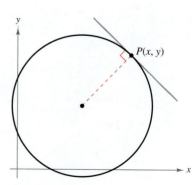

Tangent Line to a Circle

FIGURE 2.3

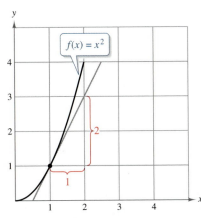

FIGURE 2.4

Slope of a Graph

Because a tangent line approximates the graph at a point, the problem of finding the slope of a graph at a point becomes one of finding the slope of the tangent line at the point.

Example 1   Approximating the Slope of a Graph

Use the graph in Figure 2.4 to approximate the slope of the graph of

$$f(x) = x^2$$

at the point $(1, 1)$.

SOLUTION   From the graph of

$$f(x) = x^2$$

you can see that the tangent line at $(1, 1)$ rises approximately two units for each unit change in x. So, the slope of the tangent line at $(1, 1)$ is given by

$$\text{Slope} = \frac{\Delta y}{\Delta x} = \frac{\text{change in } y}{\text{change in } x} \approx \frac{2}{1} = 2.$$

Because the tangent line at the point $(1, 1)$ has a slope of about 2, you can conclude that the graph has a slope of about 2 at the point $(1, 1)$.

✓ Checkpoint 1

STUDY TIP

When visually approximating the slope of a graph, note that the scales on the horizontal and vertical axes may differ. When this happens (as it frequently does in applications), the slope of the tangent line is distorted, and you must be careful to account for the difference in scales.

Use the graph to approximate the slope of the graph of

$$f(x) = x^3$$

at the point $(1, 1)$.

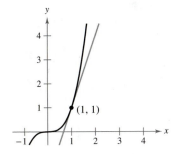

Example 2   Interpreting Slope

Figure 2.5 graphically depicts the average monthly temperature (in degrees Fahrenheit) in Duluth, Minnesota. Estimate the slope of this graph at the indicated point and give a physical interpretation of the result. *(Source: National Oceanic and Atmospheric Administration)*

SOLUTION   From the graph, you can see that the tangent line at the given point falls approximately 28 units for each two-unit change in x. So, you can estimate the slope at the given point to be

$$\text{Slope} = \frac{\Delta y}{\Delta x} = \frac{\text{change in } y}{\text{change in } x} \approx \frac{-28}{2} = -14 \text{ degrees per month.}$$

This means that you can expect the average daily temperatures in November to be about 14 degrees *lower* than the corresponding temperatures in October.

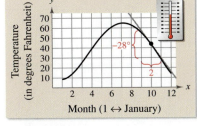

FIGURE 2.5

✓ Checkpoint 2

In Figure 2.5, for which months do the slopes of the tangent lines appear to be positive? Negative? Interpret these slopes in the context of the problem.

Slope and the Limit Process

In Examples 1 and 2, you approximated the slope of a graph at a point by making a careful graph and then "eyeballing" the tangent line at the point of tangency. A more precise method of approximating the slope of a tangent line makes use of a **secant line** through the point of tangency and a second point on the graph, as shown in Figure 2.6. If $(x, f(x))$ is the point of tangency and

$$(x + \Delta x, f(x + \Delta x))$$

is a second point on the graph of f, then the slope of the secant line through the two points is

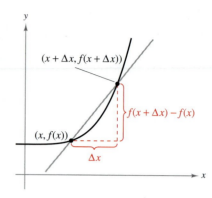

The Secant Line Through the Two Points $(x, f(x))$ and $(x + \Delta x, f(x + \Delta x))$

FIGURE 2.6

$$m = \frac{y_2 - y_1}{x_2 - x_1} \qquad \text{Formula for slope}$$

$$m_{\text{sec}} = \frac{f(x + \Delta x) - f(x)}{(x + \Delta x) - x} \qquad \frac{\text{Change in } y}{\text{Change in } x}$$

$$m_{\text{sec}} = \frac{f(x + \Delta x) - f(x)}{\Delta x}. \qquad \text{Slope of secant line}$$

The right side of this equation is called the **difference quotient.** The denominator Δx is the **change in x,** and the numerator is the **change in y.** The beauty of this procedure is that you obtain more and more accurate approximations of the slope of the tangent line by choosing points closer and closer to the point of tangency, as shown in Figure 2.7. Using the limit process, you can find the *exact* slope of the tangent line at $(x, f(x))$, which is also the slope of the graph of f at $(x, f(x))$.

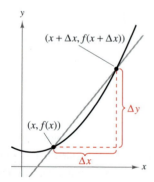
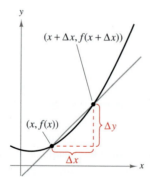
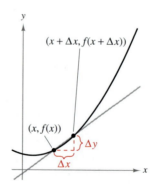
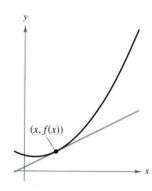

FIGURE 2.7   As Δx approaches 0, the secant lines approach the tangent line.

Definition of the Slope of a Graph

The **slope** m of the graph of f at the point

$$(x, f(x))$$

is equal to the slope of the tangent line to the graph of f at $(x, f(x))$, and is given by

$$m = \lim_{\Delta x \to 0} m_{\text{sec}} = \lim_{\Delta x \to 0} \frac{f(x + \Delta x) - f(x)}{\Delta x}$$

provided this limit exists.

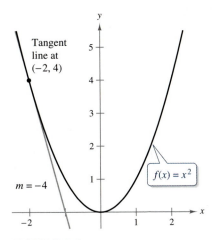

ALGEBRA TUTOR *xy*

For help in evaluating the expressions in Examples 3–6, see the review of simplifying fractional expressions on page 133.

FIGURE 2.8

Example 3 **Finding Slope by the Limit Process**

Find the slope of the graph of

$$f(x) = x^2$$

at the point $(-2, 4)$.

SOLUTION Begin by finding an expression that represents the slope of a secant line at the point $(-2, 4)$.

$$m_{sec} = \frac{f(-2 + \Delta x) - f(-2)}{\Delta x} \qquad \text{Set up difference quotient.}$$

$$= \frac{(-2 + \Delta x)^2 - (-2)^2}{\Delta x} \qquad \text{Use } f(x) = x^2.$$

$$= \frac{4 - 4\Delta x + (\Delta x)^2 - 4}{\Delta x} \qquad \text{Expand terms.}$$

$$= \frac{-4\Delta x + (\Delta x)^2}{\Delta x} \qquad \text{Simplify.}$$

$$= \frac{\Delta x(-4 + \Delta x)}{\Delta x} \qquad \text{Factor and divide out.}$$

$$= -4 + \Delta x, \quad \Delta x \neq 0 \qquad \text{Simplify.}$$

Next, take the limit of m_{sec} as $\Delta x \to 0$.

$$m = \lim_{\Delta x \to 0} m_{sec} = \lim_{\Delta x \to 0} (-4 + \Delta x) = -4 + 0 = -4$$

So, the graph of f has a slope of -4 at the point $(-2, 4)$, as shown in Figure 2.8.

✓Checkpoint 3

Find the slope of the graph of

$$f(x) = x^2$$

at the point $(2, 4)$. ■

Example 4 **Finding the Slope of a Graph**

Find the slope of the graph of $f(x) = -2x + 4$.

SOLUTION You know from your study of linear functions that the line given by $f(x) = -2x + 4$ has a slope of -2, as shown in Figure 2.9. This conclusion is consistent with the limit definition of slope.

$$m = \lim_{\Delta x \to 0} \frac{f(x + \Delta x) - f(x)}{\Delta x}$$

$$= \lim_{\Delta x \to 0} \frac{[-2(x + \Delta x) + 4] - (-2x + 4)}{\Delta x}$$

$$= \lim_{\Delta x \to 0} \frac{-2x - 2\Delta x + 4 + 2x - 4}{\Delta x}$$

$$= \lim_{\Delta x \to 0} \frac{-2\Delta x}{\Delta x}$$

$$= -2$$

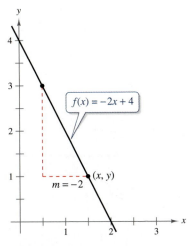

FIGURE 2.9

✓Checkpoint 4

Find the slope of the graph of $f(x) = 2x + 5$. ■

Example 5   Finding a Formula for the Slope of a Graph

Find a formula for the slope of the graph of $f(x) = x^2 + 1$. What are the slopes at the points $(-1, 2)$ and $(2, 5)$?

SOLUTION

$$
\begin{aligned}
m_{\text{sec}} &= \frac{f(x + \Delta x) - f(x)}{\Delta x} && \text{Set up difference quotient.} \\[2mm]
&= \frac{[(x + \Delta x)^2 + 1] - (x^2 + 1)}{\Delta x} && \text{Use } f(x) = x^2 + 1. \\[2mm]
&= \frac{x^2 + 2x\,\Delta x + (\Delta x)^2 + 1 - x^2 - 1}{\Delta x} && \text{Expand terms.} \\[2mm]
&= \frac{2x\,\Delta x + (\Delta x)^2}{\Delta x} && \text{Simplify.} \\[2mm]
&= \frac{\Delta x(2x + \Delta x)}{\Delta x} && \text{Factor and divide out.} \\[2mm]
&= 2x + \Delta x, \quad \Delta x \neq 0 && \text{Simplify.}
\end{aligned}
$$

Next, take the limit of m_{sec} as $\Delta x \to 0$.

$$
\begin{aligned}
m &= \lim_{\Delta x \to 0} m_{\text{sec}} \\
&= \lim_{\Delta x \to 0} (2x + \Delta x) \\
&= 2x + 0 \\
&= 2x
\end{aligned}
$$

Using the formula $m = 2x$, you can find the slopes at the specified points. At $(-1, 2)$ the slope is $m = 2(-1) = -2$, and at $(2, 5)$ the slope is $m = 2(2) = 4$. The graph of f is shown in Figure 2.10.

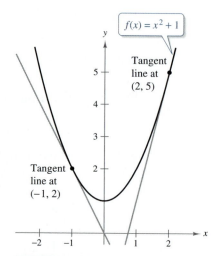

FIGURE 2.10

✓ Checkpoint 5

Find a formula for the slope of the graph of

$$f(x) = 4x^2 + 1.$$

What are the slopes at the points $(0, 1)$ and $(1, 5)$?

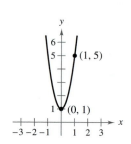

The Derivative of a Function

In Example 5, you started with the function

$$f(x) = x^2 + 1$$

and used the limit process to derive another function, $m = 2x$, that represents the slope of the graph of f at the point $(x, f(x))$. This derived function is called the **derivative** of f at x. It is denoted by $f'(x)$, which is read as "f prime of x."

Definition of the Derivative

The **derivative of f at x** is given by

$$f'(x) = \lim_{\Delta x \to 0} \frac{f(x + \Delta x) - f(x)}{\Delta x}$$

provided this limit exists. A function is **differentiable** at x when its derivative exists at x. The process of finding derivatives is called **differentiation.**

In addition to $f'(x)$, other notations can be used to denote the derivative of $y = f(x)$. The most common are

$$\frac{dy}{dx}, \quad y', \quad \frac{d}{dx}[f(x)], \quad \text{and} \quad D_x[y].$$

Example 6   Finding a Derivative

Find the derivative of

$$f(x) = 3x^2 - 2x.$$

SOLUTION

$$f'(x) = \lim_{\Delta x \to 0} \frac{f(x + \Delta x) - f(x)}{\Delta x}$$

$$= \lim_{\Delta x \to 0} \frac{[3(x + \Delta x)^2 - 2(x + \Delta x)] - (3x^2 - 2x)}{\Delta x}$$

$$= \lim_{\Delta x \to 0} \frac{3x^2 + 6x \Delta x + 3(\Delta x)^2 - 2x - 2 \Delta x - 3x^2 + 2x}{\Delta x}$$

$$= \lim_{\Delta x \to 0} \frac{6x \Delta x + 3(\Delta x)^2 - 2 \Delta x}{\Delta x}$$

$$= \lim_{\Delta x \to 0} \frac{\Delta x(6x + 3 \Delta x - 2)}{\Delta x}$$

$$= \lim_{\Delta x \to 0} (6x + 3 \Delta x - 2)$$

$$= 6x + 3(0) - 2$$

$$= 6x - 2$$

So, the derivative of $f(x) = 3x^2 - 2x$ is

$$f'(x) = 6x - 2.$$

✓Checkpoint 6

Find the derivative of

$$f(x) = x^2 - 5x.$$

In many applications, it is convenient to use a variable other than x as the independent variable. Example 7 shows a function that uses t as the independent variable.

Example 7   Finding a Derivative

Find the derivative of y with respect to t for the function

$$y = \frac{2}{t}.$$

SOLUTION   Consider $y = f(t)$, and use the limit process as shown.

$$\frac{dy}{dt} = \lim_{\Delta t \to 0} \frac{f(t + \Delta t) - f(t)}{\Delta t} \qquad \text{Set up difference quotient.}$$

$$= \lim_{\Delta t \to 0} \frac{\dfrac{2}{t + \Delta t} - \dfrac{2}{t}}{\Delta t} \qquad \text{Use } f(t) = 2/t.$$

$$= \lim_{\Delta t \to 0} \frac{\dfrac{2t - 2(t + \Delta t)}{t(t + \Delta t)}}{\Delta t} \qquad \text{Combine fractions in numerator.}$$

$$= \lim_{\Delta t \to 0} \frac{2t - 2t - 2\Delta t}{\Delta t(t)(t + \Delta t)} \qquad \text{Expand terms in numerator.}$$

$$= \lim_{\Delta t \to 0} \frac{-2 \cancel{\Delta t}}{\cancel{\Delta t}(t)(t + \Delta t)} \qquad \text{Factor and divide out.}$$

$$= \lim_{\Delta t \to 0} \frac{-2}{t(t + \Delta t)} \qquad \text{Simplify.}$$

$$= \frac{-2}{t(t + 0)} \qquad \text{Direct substitution}$$

$$= -\frac{2}{t^2} \qquad \text{Simplify.}$$

So, the derivative of y with respect to t is

$$\frac{dy}{dt} = -\frac{2}{t^2}.$$

✓ Checkpoint 7

Find the derivative of y with respect to t for the function $y = 4/t$.   ■

Remember that the derivative of a function gives you a formula for finding the slope of the tangent line at any point on the graph of the function. For instance, in Example 7 the slope of the tangent line to the graph of f at the point $(1, 2)$ is given by

$$f'(1) = -\frac{2}{1^2} = -2.$$

To find the slopes of the graph at other points, substitute the t-coordinate of the point into the derivative, as shown below.

Point	t-Coordinate	Slope
$(2, 1)$	$t = 2$	$m = f'(2) = -\dfrac{2}{2^2} = -\dfrac{1}{2}$
$(-2, -1)$	$t = -2$	$m = f'(-2) = -\dfrac{2}{(-2)^2} = -\dfrac{1}{2}$

TECH TUTOR

You can use a graphing utility to confirm the result given in Example 7. One way to do this is to choose a point on the graph of $y = 2/t$, such as $(1, 2)$, and find the equation of the tangent line at that point. Using the derivative found in the example, you know that the slope of the tangent line when $t = 1$ is $m = -2$. This means that the tangent line at the point $(1, 2)$ is

$$y - y_1 = m(t - t_1)$$
$$y - 2 = -2(t - 1)$$
$$y = -2t + 4.$$

By graphing $y = 2/t$ and $y = -2t + 4$ in the same viewing window, as shown below, you can confirm that the line is tangent to the graph at the point $(1, 2)$.

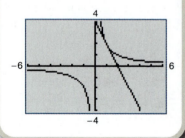

Differentiability and Continuity

Not every function is differentiable. Figure 2.11 shows some common situations in which a function will not be differentiable at a point—vertical tangent lines, discontinuities, and sharp turns in the graph. Each of the functions shown in Figure 2.11 is differentiable at every value of *x except x* = 0.

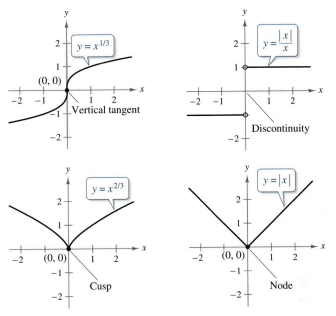

Functions That Are Not Differentiable at *x* = 0
FIGURE 2.11

In Figure 2.11, you can see that all but one of the functions are continuous at *x* = 0 but none are differentiable there. This shows that continuity is not a strong enough condition to guarantee differentiability. On the other hand, if a function is differentiable at a point, then it must be continuous at that point. This important result is stated in the following theorem.

Differentiability Implies Continuity

If a function *f* is differentiable at *x* = *c*, then *f* is continuous at *x* = *c*.

SUMMARIZE   (Section 2.1)

1. Describe a tangent line and how it can be used to approximate the slope of a graph at a point *(page 68)*. For an example of a tangent line, see Example 1.

2. State the definition of the slope of a graph using the limit process *(page 70)*. For examples of finding the slope of a graph using the limit process, see Examples 3, 4, and 5.

3. State the definition of the derivative of a function *(page 73)*. For examples of the derivative of a function, see Examples 6 and 7.

4. Describe the relationship between differentiability and continuity *(page 75)*. For an example showing that continuity does not guarantee differentiability, see Figure 2.11.

SKILLS WARM UP 2.1

The following warm-up exercises involve skills that were covered in earlier sections. You will use these skills in the exercise set for this section. For additional help, review Sections 1.3, 1.4, and 1.5.

In Exercises 1–4, find an equation of the line containing P and Q.

1. $P(2, 1)$, $Q(2, 4)$

2. $P(2, 2)$, $Q(-5, 2)$

3. $P(2, 0)$, $Q(3, -1)$

4. $P(3, 5)$, $Q(-1, -7)$

In Exercises 5–8, find the limit.

5. $\displaystyle \lim_{\Delta x \to 0} \frac{2x\Delta x + (\Delta x)^2}{\Delta x}$

6. $\displaystyle \lim_{\Delta x \to 0} \frac{3x^2\Delta x + 3x(\Delta x)^2 + (\Delta x)^3}{\Delta x}$

7. $\displaystyle \lim_{\Delta x \to 0} \frac{1}{x(x + \Delta x)}$

8. $\displaystyle \lim_{\Delta x \to 0} \frac{(x + \Delta x)^2 - x^2}{\Delta x}$

In Exercises 9–12, find the domain of the function.

9. $f(x) = 3x$

10. $f(x) = \dfrac{1}{x - 1}$

11. $f(x) = \dfrac{1}{5}x^3 - 2x^2 + \dfrac{1}{3}x - 1$

12. $f(x) = \dfrac{6x}{x^3 + x}$

2.2 Some Rules for Differentiation

■ Find the derivatives of functions using the Constant Rule.
■ Find the derivatives of functions using the Power Rule.
■ Find the derivatives of functions using the Constant Multiple Rule.
■ Find the derivatives of functions using the Sum and Difference Rules.
■ Use derivatives to answer questions about real-life situations.

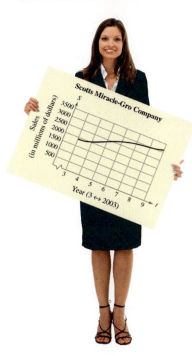

In Exercise 74, you will use differentiation to find the rate of change in a company's sales.

The Constant Rule

In Section 2.1, you found derivatives by the limit process. This process is tedious, even for simple functions, but fortunately there are rules that greatly simplify differentiation. These rules allow you to calculate derivatives without the *direct* use of limits.

The Constant Rule

The derivative of a constant function is zero. That is,

$$\frac{d}{dx}[c] = 0, \quad c \text{ is a constant.}$$

PROOF Let $f(x) = c$. Then, by the limit definition of the derivative, you can write

$$f'(x) = \lim_{\Delta x \to 0} \frac{f(x + \Delta x) - f(x)}{\Delta x} = \lim_{\Delta x \to 0} \frac{c - c}{\Delta x} = \lim_{\Delta x \to 0} 0 = 0.$$

So, $\frac{d}{dx}[c] = 0$.

Note in Figure 2.12 that the Constant Rule is equivalent to saying that the slope of a horizontal line is zero. An interpretation of the Constant Rule says that the tangent line to a constant function is the function itself. For instance, the equation of the tangent line to $f(x) = 4$ at $x = -1$ is

$$y = 4.$$

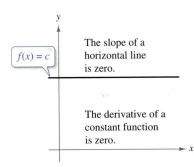

The slope of a horizontal line is zero.

The derivative of a constant function is zero.

FIGURE 2.12

Example 1   Finding Derivatives of Constant Functions

a. $\dfrac{d}{dx}[7] = 0$

b. If $f(x) = 0$, then $f'(x) = 0$.

c. If $y = 2$, then $\dfrac{dy}{dx} = 0$.

d. If $g(t) = -\dfrac{3}{2}$, then $g'(t) = 0$.

✓ Checkpoint 1

Find the derivative of each function.

a. $f(x) = -2$ **b.** $y = \pi$ **c.** $g(w) = \sqrt{5}$ **d.** $s(t) = 320.5$

The Power Rule

The binomial expansion process is used in proving a special case of the Power Rule.

$$(x + \Delta x)^2 = x^2 + 2x \, \Delta x + (\Delta x)^2$$

$$(x + \Delta x)^3 = x^3 + 3x^2 \, \Delta x + 3x(\Delta x)^2 + (\Delta x)^3$$

$$(x + \Delta x)^n = x^n + nx^{n-1} \, \Delta x + \underbrace{\frac{n(n-1)x^{n-2}}{2}(\Delta x)^2 + \cdots + (\Delta x)^n}_{(\Delta x)^2 \text{ is a factor of these terms.}}$$

The (Simple) Power Rule

$$\frac{d}{dx}[x^n] = nx^{n-1}, \quad n \text{ is any real number.}$$

PROOF This proof is limited to the case in which n is a positive integer. Let $f(x) = x^n$. Using the binomial expansion, you can write

$$f'(x) = \lim_{\Delta x \to 0} \frac{f(x + \Delta x) - f(x)}{\Delta x} \qquad \text{Definition of derivative}$$

$$= \lim_{\Delta x \to 0} \frac{(x + \Delta x)^n - x^n}{\Delta x}$$

$$= \lim_{\Delta x \to 0} \frac{x^n + nx^{n-1} \, \Delta x + \frac{n(n-1)x^{n-2}}{2}(\Delta x)^2 + \cdots + (\Delta x)^n - x^n}{\Delta x}$$

$$= \lim_{\Delta x \to 0} \left[nx^{n-1} + \frac{n(n-1)x^{n-2}}{2}(\Delta x) + \cdots + (\Delta x)^{n-1} \right]$$

$$= nx^{n-1} + 0 + \cdots + 0$$

$$= nx^{n-1}.$$

For the Power Rule, the case in which $n = 1$ is worth remembering as a separate differentiation rule. That is,

$$\frac{d}{dx}[x] = 1. \qquad \text{The derivative of } x \text{ is 1.}$$

This rule is consistent with the fact that the slope of the line given by $y = x$ is 1. (See Figure 2.13.)

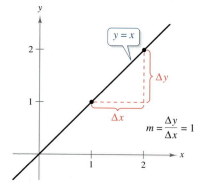

The slope of the line $y = x$ is 1.
FIGURE 2.13

Example 2   Applying the Power Rule

Original Function	Derivative
a. $f(x) = x^3$	$f'(x) = 3x^2$
b. $y = \dfrac{1}{x^2} = x^{-2}$	$\dfrac{dy}{dx} = (-2)x^{-3} = -\dfrac{2}{x^3}$
c. $g(t) = t$	$g'(t) = 1$

✓ Checkpoint 2

Find the derivative of each function.

a. $f(x) = x^4$ **b.** $y = \dfrac{1}{x^3}$ **c.** $g(w) = w^2$

In Example 2(b), note that *before* differentiating, you should rewrite $1/x^2$ as x^{-2}. Rewriting is the first step in *many* differentiation problems.

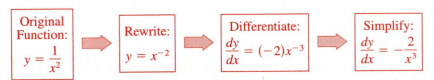

| Original Function: $y = \dfrac{1}{x^2}$ | Rewrite: $y = x^{-2}$ | Differentiate: $\dfrac{dy}{dx} = (-2)x^{-3}$ | Simplify: $\dfrac{dy}{dx} = -\dfrac{2}{x^3}$ |

Remember that the derivative of a function f is another function that gives the slope of the graph of f at any point at which f is differentiable. So, you can use the derivative to find slopes, as shown in Example 3.

Example 3 Finding the Slope of a Graph

Find the slopes of the graph of

$$f(x) = x^2$$

at $x = -2, -1, 0, 1,$ and 2.

SOLUTION Begin by using the Power Rule to find the derivative of f.

$$f'(x) = 2x \qquad \text{Derivative}$$

You can use the derivative to find the slopes of the graph of f, as shown.

x-Value	Slope of Graph of f
$x = -2$	$m = f'(-2) = 2(-2) = -4$
$x = -1$	$m = f'(-1) = 2(-1) = -2$
$x = 0$	$m = f'(0) = 2(0) = 0$
$x = 1$	$m = f'(1) = 2(1) = 2$
$x = 2$	$m = f'(2) = 2(2) = 4$

The graph of f is shown in Figure 2.14.

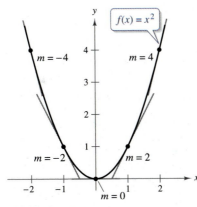

FIGURE 2.14

✓ Checkpoint 3

Find the slopes of the graph of

$$f(x) = x^3$$

at $x = -1, 0,$ and 1.

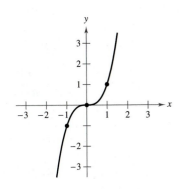

The Constant Multiple Rule

To prove the Constant Multiple Rule, the following property of limits is used.

$$\lim_{x \to a} cg(x) = c\left[\lim_{x \to a} g(x)\right]$$

The Constant Multiple Rule

If f is a differentiable function of x, and c is a real number, then

$$\frac{d}{dx}[cf(x)] = cf'(x), \quad c \text{ is a constant.}$$

PROOF   Apply the definition of the derivative to produce

$$\frac{d}{dx}[cf(x)] = \lim_{\Delta x \to 0} \frac{cf(x + \Delta x) - cf(x)}{\Delta x} \qquad \text{Definition of derivative}$$

$$= \lim_{\Delta x \to 0} c\left[\frac{f(x + \Delta x) - f(x)}{\Delta x}\right]$$

$$= c\left[\lim_{\Delta x \to 0} \frac{f(x + \Delta x) - f(x)}{\Delta x}\right]$$

$$= cf'(x).$$

Informally, the Constant Multiple Rule states that constants can be factored out of the differentiation process.

$$\frac{d}{dx}[cf(x)] = c\frac{d}{dx}[()f(x)] = cf'(x).$$

The usefulness of this rule is often overlooked, especially when the constant appears in the denominator, as shown below.

$$\frac{d}{dx}\left[\frac{f(x)}{c}\right] = \frac{d}{dx}\left[\frac{1}{c}f(x)\right]$$

$$= \frac{1}{c}\left(\frac{d}{dx}[()f(x)]\right)$$

$$= \frac{1}{c}f'(x).$$

To use the Constant Multiple Rule efficiently, look for constants that can be factored out *before* differentiating. For example,

$$\frac{d}{dx}[5x^2] = 5\frac{d}{dx}[x^2] \qquad \text{Factor out 5.}$$

$$= 5(2x) \qquad \text{Differentiate.}$$

$$= 10x \qquad \text{Simplify.}$$

and

$$\frac{d}{dx}\left[\frac{x^2}{5}\right] = \frac{1}{5}\left(\frac{d}{dx}[x^2]\right) \qquad \text{Factor out } \tfrac{1}{5}.$$

$$= \frac{1}{5}(2x) \qquad \text{Differentiate.}$$

$$= \frac{2}{5}x. \qquad \text{Simplify.}$$

TECH TUTOR

If you have access to a symbolic differentiation utility, try using it to confirm the derivatives shown in this section.

Example 4 **Using the Power and Constant Multiple Rules**

Find the derivative of (a) $y = 2x^{1/2}$ and (b) $f(t) = \dfrac{4t^2}{5}$.

SOLUTION

a. Using the Constant Multiple Rule and the Power Rule, you can write

$$\frac{dy}{dx} = \frac{d}{dx}[2x^{1/2}] = 2\underbrace{\frac{d}{dx}[x^{1/2}]}_{\text{Constant Multiple Rule}} = 2\underbrace{\left(\frac{1}{2}x^{-1/2}\right)}_{\text{Power Rule}} = x^{-1/2} = \frac{1}{\sqrt{x}}.$$

b. Begin by rewriting $f(t)$ as

$$f(t) = \frac{4t^2}{5} = \frac{4}{5}t^2.$$

Then, use the Constant Multiple Rule and the Power Rule to obtain

$$f'(t) = \frac{d}{dt}\left[\frac{4}{5}t^2\right] = \frac{4}{5}\left(\frac{d}{dt}[t^2]\right) = \frac{4}{5}(2t) = \frac{8}{5}t.$$

✓**Checkpoint 4**

Find the derivative of (a) $y = 4x^2$ and (b) $f(x) = 16x^{1/2}$. ■

You may find it helpful to combine the Constant Multiple Rule and the Power Rule into one combined rule.

$$\frac{d}{dx}[cx^n] = cnx^{n-1}, \quad n \text{ is a real number, } c \text{ is a constant.}$$

For instance, in Example 4(b), you can apply this combined rule to obtain

$$\frac{d}{dt}\left[\frac{4}{5}t^2\right] = \left(\frac{4}{5}\right)(2)(t) = \frac{8}{5}t.$$

The three functions in the next example are simple, yet errors are frequently made in differentiating functions involving constant multiples of the first power of x. Keep in mind that

$$\frac{d}{dx}[cx] = c, \quad c \text{ is a constant.}$$

Example 5 **Applying the Constant Multiple Rule**

Original Function	Derivative
a. $y = -\dfrac{3x}{2}$	$y' = -\dfrac{3}{2}$
b. $y = 3\pi x$	$y' = 3\pi$
c. $y = -\dfrac{x}{2}$	$y' = -\dfrac{1}{2}$

✓**Checkpoint 5**

Find the derivative of (a) $y = \dfrac{t}{4}$ and (b) $y = -\dfrac{2x}{5}$. ■

Parentheses can play an important role in the use of the Constant Multiple Rule and the Power Rule. In Example 6, be sure you understand the mathematical conventions involving the use of parentheses.

Example 6   Using Parentheses When Differentiating

Original Function	Rewrite	Differentiate	Simplify
a. $y = \dfrac{5}{2x^3}$	$y = \dfrac{5}{2}(x^{-3})$	$y' = \dfrac{5}{2}(-3x^{-4})$	$y' = -\dfrac{15}{2x^4}$
b. $y = \dfrac{5}{(2x)^3}$	$y = \dfrac{5}{8}(x^{-3})$	$y' = \dfrac{5}{8}(-3x^{-4})$	$y' = -\dfrac{15}{8x^4}$
c. $y = \dfrac{7}{3x^{-2}}$	$y = \dfrac{7}{3}(x^2)$	$y' = \dfrac{7}{3}(2x)$	$y' = \dfrac{14x}{3}$
d. $y = \dfrac{7}{(3x)^{-2}}$	$y = 63(x^2)$	$y' = 63(2x)$	$y' = 126x$

✔**Checkpoint 6**

Find the derivative of each function.

a. $y = \dfrac{9}{4x^2}$

b. $y = \dfrac{9}{(4x)^2}$

■

When differentiating functions involving radicals, you should rewrite the function with rational exponents. For instance, you should rewrite

$$y = \sqrt[3]{x} \quad \text{as} \quad y = x^{1/3}$$

and you should rewrite

$$y = \frac{1}{\sqrt[3]{x^4}} \quad \text{as} \quad y = x^{-4/3}.$$

Example 7   Differentiating Radical Functions

Original Function	Rewrite	Differentiate	Simplify
a. $y = \sqrt{x}$	$y = x^{1/2}$	$y' = \left(\dfrac{1}{2}\right)x^{-1/2}$	$y' = \dfrac{1}{2\sqrt{x}}$
b. $y = \dfrac{1}{2\sqrt[3]{x^2}}$	$y = \dfrac{1}{2}x^{-2/3}$	$y' = \dfrac{1}{2}\left(-\dfrac{2}{3}\right)x^{-5/3}$	$y' = -\dfrac{1}{3x^{5/3}}$
c. $y = \sqrt{2x}$	$y = \sqrt{2}(x^{1/2})$	$y' = \sqrt{2}\left(\dfrac{1}{2}\right)x^{-1/2}$	$y' = \dfrac{1}{\sqrt{2x}}$

✔**Checkpoint 7**

Find the derivative of each function.

a. $y = \sqrt{5x}$

b. $y = \sqrt[4]{x}$

■

The Sum and Difference Rules

To differentiate $y = 3x + 2x^3$, you would probably write

$$y' = 3 + 6x^2$$

without questioning your answer. The validity of differentiating a sum or difference of functions term by term is given by the Sum and Difference Rules.

The Sum and Difference Rules

The derivative of the sum or difference of two differentiable functions is the sum or difference of their derivatives.

$$\frac{d}{dx}[f(x) + g(x)] = f'(x) + g'(x) \qquad \text{Sum Rule}$$

$$\frac{d}{dx}[f(x) - g(x)] = f'(x) - g'(x) \qquad \text{Difference Rule}$$

PROOF   Let $h(x) = f(x) + g(x)$. Then, you can prove the Sum Rule as shown.

$$h'(x) = \lim_{\Delta x \to 0} \frac{h(x + \Delta x) - h(x)}{\Delta x} \qquad \text{Definition of derivative}$$

$$= \lim_{\Delta x \to 0} \frac{f(x + \Delta x) + g(x + \Delta x) - f(x) - g(x)}{\Delta x}$$

$$= \lim_{\Delta x \to 0} \frac{f(x + \Delta x) - f(x) + g(x + \Delta x) - g(x)}{\Delta x}$$

$$= \lim_{\Delta x \to 0} \left[\frac{f(x + \Delta x) - f(x)}{\Delta x} + \frac{g(x + \Delta x) - g(x)}{\Delta x} \right]$$

$$= \lim_{\Delta x \to 0} \frac{f(x + \Delta x) - f(x)}{\Delta x} + \lim_{\Delta x \to 0} \frac{g(x + \Delta x) - g(x)}{\Delta x}$$

$$= f'(x) + g'(x)$$

So,

$$\frac{d}{dx}[f(x) + g(x)] = f'(x) + g'(x).$$

The Difference Rule can be proved in a similar manner.

The Sum and Difference Rules can be extended to the sum or difference of any finite number of functions. For instance, if $y = f(x) + g(x) + h(x)$, then $y' = f'(x) + g'(x) + h'(x)$.

Example 8   Using the Sum and Difference Rules

Original Function	*Derivative*
a. $y = x^3 + 4x^2$	$y' = 3x^2 + 8x$
b. $f(x) = 3x^2 - 2x$	$f'(x) = 6x - 2$

✓ **Checkpoint 8**

Find the derivative of each function.

a. $f(x) = 2x^2 + 5x$ **b.** $y = x^4 - 2x$

STUDY TIP

Look back at Example 6 on page 82. Notice that the example asks for the derivative of the difference of two functions. Compare the result with the one obtained in Example 8(b) at the right.

R. Gino Santa Maria/Shutterstock.com

With the differentiation rules listed in this section, you can differentiate any polynomial function.

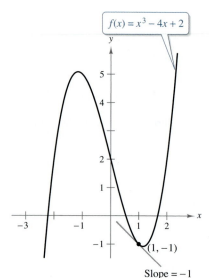

$f(x) = x^3 - 4x + 2$

FIGURE 2.15

Example 9 Finding the Slope of a Graph

Find the slope of the graph of $f(x) = x^3 - 4x + 2$ at the point $(1, -1)$.

SOLUTION The derivative of $f(x)$ is

$$f'(x) = 3x^2 - 4.$$

So, the slope of the graph of f at $(1, -1)$ is

$$\text{Slope} = f'(1) = 3(1)^2 - 4 = 3 - 4 = -1$$

as shown in Figure 2.15.

✔ **Checkpoint 9**

Find the slope of the graph of $f(x) = x^2 - 5x + 1$ at the point $(2, -5)$. ■

Example 9 illustrates the use of the derivative for determining the shape of a graph. A rough sketch of the graph of $f(x) = x^3 - 4x + 2$ might lead you to think that the point $(1, -1)$ is a minimum point of the graph. After finding the slope at this point to be -1, however, you can conclude that the minimum point (where the slope is 0) is farther to the right. (You will study techniques for finding minimum and maximum points in Section 3.2.)

Example 10 Finding an Equation of a Tangent Line

Find an equation of the tangent line to the graph of

$$g(x) = -\frac{1}{2}x^4 + 3x^3 - 2x$$

at the point $\left(-1, -\frac{3}{2}\right)$.

SOLUTION The derivative of $g(x)$ is $g'(x) = -2x^3 + 9x^2 - 2$, which implies that the slope of the graph at the point $\left(-1, -\frac{3}{2}\right)$ is

$$\begin{aligned}
\text{Slope} &= g'(-1) \\
&= -2(-1)^3 + 9(-1)^2 - 2 \\
&= 2 + 9 - 2 \\
&= 9
\end{aligned}$$

as shown in Figure 2.16. Using the point-slope form, you can write the equation of the tangent line at $\left(-1, -\frac{3}{2}\right)$ as shown.

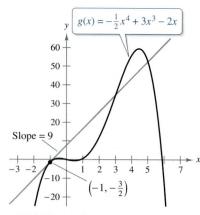

$g(x) = -\frac{1}{2}x^4 + 3x^3 - 2x$

FIGURE 2.16

$$y - \left(-\frac{3}{2}\right) = 9[x - (-1)] \qquad \text{Point-slope form}$$

$$y + \frac{3}{2} = 9x + 9 \qquad \text{Simplify.}$$

$$y = 9x + \frac{15}{2} \qquad \text{Equation of tangent line}$$

✔ **Checkpoint 10**

Find an equation of the tangent line to the graph of $f(x) = -x^2 + 3x - 2$ at the point $(2, 0)$. ■

Application

There are many applications of the derivative that you will study in this textbook. In Example 11, you will use a derivative to find the rate of change of a company's revenue with respect to time.

Example 11 Modeling Revenue

From 2004 through 2009, the revenue R (in millions of dollars) for McDonald's can be modeled by

$$R = -130.769t^3 + 2296.47t^2 - 11,493.5t + 35,493, \quad 4 \le t \le 9$$

where t represents the year, with $t = 4$ corresponding to 2004. At what rate was McDonald's revenue changing in 2006? *(Source: McDonald's Corporation)*

SOLUTION One way to answer this question is to find the derivative of the revenue model with respect to time.

$$\frac{dR}{dt} = -392.307t^2 + 4592.94t - 11,493.5, \quad 4 \le t \le 9$$

In 2006 (at $t = 6$), the rate of change of the revenue with respect to time is given by

$$\frac{dR}{dt} = -392.307(6)^2 + 4592.94(6) - 11,493.5 \approx 1941.$$

Because R is measured in millions of dollars and t is measured in years, it follows that the derivative dR/dt is measured in millions of dollars per year. So, at the end of 2006, McDonald's revenue was increasing at a rate of about $1941 million per year, as shown in Figure 2.17.

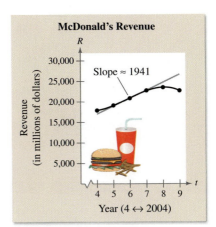

McDonald's Revenue

Slope ≈ 1941

FIGURE 2.17

✓ Checkpoint 11

From 2000 through 2010, the sales per share S (in dollars) for Microsoft Corporation can be modeled by

$$S = 0.0330t^2 + 0.208t + 2.13, \quad 0 \le t \le 10$$

where t represents the year, with $t = 0$ corresponding to 2000. At what rate was Microsoft's sales per share changing in 2002? *(Source: Microsoft Corporation)*

SUMMARIZE (Section 2.2)

1. State the Constant Rule *(page 77)*. For an example of the Constant Rule, see Example 1.

2. State the Power Rule *(page 78)*. For examples of the Power Rule, see Examples 2 and 3.

3. State the Constant Multiple Rule *(page 80)*. For examples of the Constant Multiple Rule, see Examples 4, 5, 6, and 7.

4. State the Sum Rule *(page 83)*. For an example of the Sum Rule, see Example 8.

5. State the Difference Rule *(page 83)*. For an example of the Difference Rule, see Example 8.

6. Describe a real-life example of how differentiation can be used to analyze the rate of change of a company's revenue *(page 85, Example 11)*.

SKILLS WARM UP 2.2

The following warm-up exercises involve skills that were covered in a previous course. You will use these skills in the exercise set for this section. For additional help, review Appendix Sections A.3 and A.4.

In Exercises 1 and 2, evaluate each expression when $x = 2$.

1. (a) $2x^2$ (b) $(2x)^2$ (c) $2x^{-2}$

2. (a) $\dfrac{1}{(3x)^2}$ (b) $\dfrac{1}{4x^3}$ (c) $\dfrac{(2x)^{-3}}{4x^{-2}}$

In Exercises 3–6, simplify the expression.

3. $4(3)x^3 + 2(2)x$

4. $\frac{1}{2}(3)x^2 - \frac{3}{2}x^{1/2}$

5. $\left(\frac{1}{4}\right)x^{-3/4}$

6. $\frac{1}{3}(3)x^2 - 2\left(\frac{1}{2}\right)x^{-1/2} + \frac{1}{3}x^{-2/3}$

In Exercises 7–10, solve the equation.

7. $3x^2 + 2x = 0$

8. $x^3 - x = 0$

9. $x^2 + 8x - 20 = 0$

10. $x^2 - 10x - 24 = 0$

ENHANCED WebAssign   Access end-of-section exercises online at **www.webassign.net**

2.3 Rates of Change: Velocity and Marginals

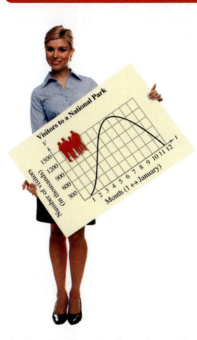

In Exercise 13, you will use the graph of a function to estimate the rate of change of the number of visitors to a national park.

■ Find the average rates of change of functions over intervals.
■ Find the instantaneous rates of change of functions at points.
■ Find the marginal revenues, marginal costs, and marginal profits for products.

Average Rate of Change

In Sections 2.1 and 2.2, you studied the two primary applications of derivatives.

1. **Slope** The derivative of f is a function that gives the slope of the graph of f at a point $(x, f(x))$.

2. **Rate of Change** The derivative of f is a function that gives the rate of change of $f(x)$ with respect to x at the point $(x, f(x))$.

In this section, you will see that there are many real-life applications of rates of change. A few are velocity, acceleration, population growth rates, unemployment rates, production rates, and water flow rates. Although rates of change often involve change with respect to time, you can investigate the rate of change of one variable with respect to any other related variable.

When determining the rate of change of one variable with respect to another, you must be careful to distinguish between *average* and *instantaneous* rates of change. The distinction between these two rates of change is comparable to the distinction between the slope of the secant line through two points on a graph and the slope of the tangent line at one point on the graph.

Definition of Average Rate of Change

If $y = f(x)$, then the **average rate of change** of y with respect to x on the interval $[a, b]$ is

$$\text{Average rate of change} = \frac{f(b) - f(a)}{b - a} = \frac{\Delta y}{\Delta x}.$$

Note that $f(a)$ is the value of the function at the *left* endpoint of the interval, $f(b)$ is the value of the function at the *right* endpoint of the interval, and $b - a$ is the width of the interval, as shown in Figure 2.18.

STUDY TIP

In real-life problems, it is important to list the units of measure for a rate of change. The units for $\Delta y / \Delta x$ are "y-units" per "x-units." For example, if y is measured in miles and x is measured in hours, then $\Delta y / \Delta x$ is measured in *miles per hour*.

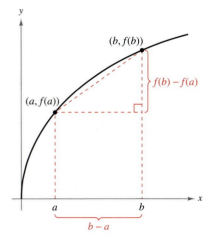

FIGURE 2.18

Example 1   Medicine

The concentration C (in milligrams per milliliter) of a drug in a patient's bloodstream is monitored over 10-minute intervals for 2 hours, where t is measured in minutes, as shown in the table.

t	0	10	20	30	40	50	60	70	80	90	100	110	120
C	0	2	17	37	55	73	89	103	111	113	113	103	68

Find the average rate of change of C over each interval.

a. $[0, 10]$

b. $[0, 20]$

c. $[100, 110]$

SOLUTION

a. For the interval $[0, 10]$, the average rate of change is

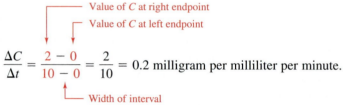

Value of C at right endpoint
Value of C at left endpoint

$$\frac{\Delta C}{\Delta t} = \frac{2 - 0}{10 - 0} = \frac{2}{10} = 0.2 \text{ milligram per milliliter per minute.}$$

Width of interval

b. For the interval $[0, 20]$, the average rate of change is

$$\frac{\Delta C}{\Delta t} = \frac{17 - 0}{20 - 0} = \frac{17}{20} = 0.85 \text{ milligram per milliliter per minute.}$$

c. For the interval $[100, 110]$, the average rate of change is

$$\frac{\Delta C}{\Delta t} = \frac{103 - 113}{110 - 100} = \frac{-10}{10} = -1 \text{ milligram per milliliter per minute.}$$

Notice in Figure 2.19 that the average rate of change is positive when the concentration increases and negative when the concentration decreases.

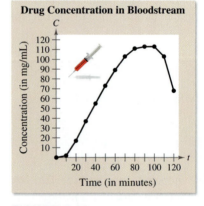

Drug Concentration in Bloodstream

FIGURE 2.19

✓ **Checkpoint 1**

Use the table in Example 1 to find the average rate of change of C over each interval.

a. $[0, 120]$

b. $[90, 100]$

c. $[90, 120]$

The rates of change in Example 1 are in milligrams per milliliter per minute because the concentration is measured in milligrams per milliliter and the time is measured in minutes.

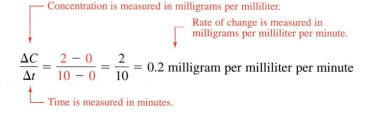

Concentration is measured in milligrams per milliliter.

Rate of change is measured in milligrams per milliliter per minute.

$$\frac{\Delta C}{\Delta t} = \frac{2 - 0}{10 - 0} = \frac{2}{10} = 0.2 \text{ milligram per milliliter per minute}$$

Time is measured in minutes.

A common application of an average rate of change is to find the **average velocity** of an object that is moving in a straight line. That is,

$$\text{Average velocity} = \frac{\text{change in distance}}{\text{change in time}}.$$

This formula is demonstrated in Example 2.

Example 2   Finding an Average Velocity

A free-falling object is dropped from a height of 100 feet. *Neglecting air resistance,* the height h (in feet) of the object at time t (in seconds) is given by

$$h = -16t^2 + 100. \qquad \text{(See Figure 2.20.)}$$

Find the average velocity of the object over each interval.

a. $[1, 2]$ **b.** $[1, 1.5]$ **c.** $[1, 1.1]$

SOLUTION You can use the position equation $h = -16t^2 + 100$ to determine the heights at

$$t = 1, 1.1, 1.5, \text{ and } 2$$

as shown in the table.

t (in seconds)	0	1	1.1	1.5	2
h (in feet)	100	84	80.64	64	36

a. For the interval $[1, 2]$, the object falls from a height of 84 feet to a height of 36 feet. So, the average velocity is

$$\frac{\Delta h}{\Delta t} = \frac{36 - 84}{2 - 1} = \frac{-48}{1} = -48 \text{ feet per second.}$$

b. For the interval $[1, 1.5]$, the average velocity is

$$\frac{\Delta h}{\Delta t} = \frac{64 - 84}{1.5 - 1} = \frac{-20}{0.5} = -40 \text{ feet per second.}$$

c. For the interval $[1, 1.1]$, the average velocity is

$$\frac{\Delta h}{\Delta t} = \frac{80.64 - 84}{1.1 - 1} = \frac{-3.36}{0.1} = -33.6 \text{ feet per second.}$$

✓**Checkpoint 2**

The height h (in feet) of a free-falling object at time t (in seconds) is given by

$$h = -16t^2 + 180.$$

Find the average velocity of the object over each interval.

a. $[0, 1]$ **b.** $[1, 2]$ **c.** $[2, 3]$ ■

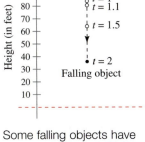

Some falling objects have considerable air resistance. Other falling objects have negligible air resistance. When modeling a falling-body problem, you must decide whether to account for air resistance or neglect it.

FIGURE 2.20

STUDY TIP

In Example 2, the average velocities are negative because the object is moving downward.

Instantaneous Rate of Change and Velocity

Suppose in Example 2 you wanted to find the rate of change of h at the instant $t = 1$ second. Such a rate is called an **instantaneous rate of change.** You can approximate the instantaneous rate of change at $t = 1$ by calculating the average rate of change over smaller and smaller intervals of the form $[1, 1 + \Delta t]$, as shown in the table. From the table, it seems reasonable to conclude that the instantaneous rate of change of the height at $t = 1$ is -32 feet per second.

Δt approaches 0.

Δt	1	0.5	0.1	0.01	0.001	0.0001	0
$\dfrac{\Delta h}{\Delta t}$	-48	-40	-33.6	-32.16	-32.016	-32.0016	-32

$\dfrac{\Delta h}{\Delta t}$ approaches -32.

Definition of Instantaneous Rate of Change

The **instantaneous rate of change** (or simply **rate of change**) of $y = f(x)$ at x is the limit of the average rate of change on the interval

$$[x, x + \Delta x]$$

as Δx approaches 0.

$$\lim_{\Delta x \to 0} \frac{\Delta y}{\Delta x} = \lim_{\Delta x \to 0} \frac{f(x + \Delta x) - f(x)}{\Delta x}$$

If y is a distance and x is time, then the rate of change is a **velocity.**

Example 3 **Finding an Instantaneous Rate of Change**

Find the velocity of the object in Example 2 at $t = 1$.

SOLUTION From Example 2, you know that the height of the falling object is given by

$$h = -16t^2 + 100. \qquad \text{Position function}$$

By taking the derivative of this position function, you obtain the velocity function.

$$h'(t) = -32t \qquad \text{Velocity function}$$

The velocity function gives the velocity at *any* time. So, at $t = 1$, the velocity is

$$h'(1) = -32(1)$$
$$= -32 \text{ feet per second.}$$

✓ **Checkpoint 3**

The height of the object in Checkpoint 2 is given by

$$h = -16t^2 + 180.$$

Find the velocities of the object at

a. $t = 1.75$.

b. $t = 2$.

The general **position function** for a free-falling object, neglecting air resistance, is

$$h = -16t^2 + v_0 t + h_0 \qquad \text{Position function}$$

where h is the height (in feet), t is the time (in seconds), v_0 is the initial velocity (in feet per second), and h_0 is the initial height (in feet). Remember that the model assumes that positive velocities indicate upward motion and negative velocities indicate downward motion. The derivative

$$h' = -32t^2 + v_0 \qquad \text{Velocity function}$$

is the **velocity function.** The absolute value of the velocity is the **speed** of the object.

 Example 4 **Finding the Velocity of a Diver**

At time $t = 0$, a diver jumps from a diving board that is 32 feet high, as shown in Figure 2.21. Because the diver's initial velocity is 16 feet per second, the position of the diver is given by

$$h = -16t^2 + 16t + 32. \qquad \text{Position function}$$

a. When does the diver hit the water?

b. What is the diver's velocity at impact?

SOLUTION

a. To find the time at which the diver hits the water, let $h = 0$ and solve for t.

$$-16t^2 + 16t + 32 = 0 \qquad \text{Set } h \text{ equal to 0.}$$
$$-16(t^2 - t - 2) = 0 \qquad \text{Factor out common factor.}$$
$$-16(t + 1)(t - 2) = 0 \qquad \text{Factor.}$$
$$t = -1 \text{ or } t = 2 \qquad \text{Solve for } t.$$

The solution $t = -1$ does not make sense in the problem because it would mean that the diver hits the water 1 second before jumping. So, you can conclude that the diver hits the water at $t = 2$ seconds.

b. The velocity at time t is given by the derivative

$$h' = -32t + 16. \qquad \text{Velocity function}$$

The velocity at time $t = 2$ is

$$h' = -32(2) + 16 = -48 \text{ feet per second.}$$

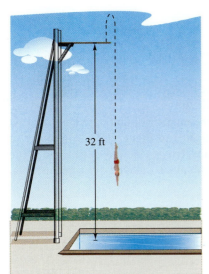

FIGURE 2.21

✓**Checkpoint 4**

At time $t = 0$, a diver jumps from a diving board that is 12 feet high with initial velocity 16 feet per second. The diver's position function is $h = -16t^2 + 16t + 12$.

a. When does the diver hit the water?

b. What is the diver's velocity at impact?

In Example 4, note that the diver's initial velocity is $v_0 = 16$ feet per second (upward) and the diver's initial height is $h_0 = 32$ feet.

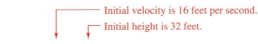

Initial velocity is 16 feet per second.
Initial height is 32 feet.

$$h = -16t^2 + 16t + 32$$

Rates of Change in Economics: Marginals

Another important use of rates of change is in the field of economics. Economists refer to *marginal profit, marginal revenue,* and *marginal cost* as the rates of change of the profit, revenue, and cost with respect to x, the number of units produced or sold. An equation that relates these three quantities is

$$P = R - C$$

where P, R, and C represent the following quantities.

$$P = \text{total profit}, \quad R = \text{total revenue}, \quad \text{and} \quad C = \text{total cost}$$

The derivatives of these quantities are called the **marginal profit, marginal revenue,** and **marginal cost,** respectively.

$$\frac{dP}{dx} = \text{marginal profit}$$

$$\frac{dR}{dx} = \text{marginal revenue}$$

$$\frac{dC}{dx} = \text{marginal cost}$$

In many business and economics problems, the number of units produced or sold is restricted to nonnegative integer values, as indicated in Figure 2.22. (Of course, it could happen that a sale involves half or quarter units, but it is hard to conceive of a sale involving $\sqrt{2}$ units.) The variable that denotes such units is called a **discrete variable.**

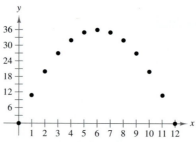

Function of a Discrete Variable
FIGURE 2.22

To analyze a function of a discrete variable x, you can temporarily assume that x is a **continuous variable** and is able to take on any real value in a given interval, as indicated in Figure 2.23. Then, you can use the methods of calculus to find the x-value that corresponds to the marginal revenue, maximum profit, minimum cost, or whatever is called for. Finally, you should round the solution to the nearest sensible x-value—cents, dollars, units, or days, depending on the context of the problem.

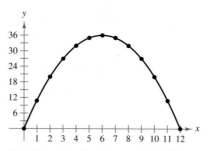

Function of a Continuous Variable
FIGURE 2.23

Example 5   **Finding the Marginal Profit**

The profit derived from selling x units of an alarm clock is given by

$$P = 0.0002x^3 + 10x.$$

a. Find the marginal profit for a production level of 50 units.

b. Compare the marginal profit with the actual gain in profit obtained by increasing the production level from 50 to 51 units.

SOLUTION

a. The profit is $P = 0.0002x^3 + 10x$. The marginal profit is given by the derivative

$$\frac{dP}{dx} = 0.0006x^2 + 10.$$

When $x = 50$, the marginal profit is

$$\frac{dP}{dx} = 0.0006(50)^2 + 10 \qquad \text{Substitute 50 for } x.$$
$$= 0.0006(2500) + 10$$
$$= 1.5 + 10$$
$$= \$11.50 \text{ per unit.} \qquad \text{Marginal profit for } x = 50$$

b. For $x = 50$, the actual profit is

$$P = 0.0002(50)^3 + 10(50) \qquad \text{Substitute 50 for } x.$$
$$= 0.0002(125,000) + 500$$
$$= 25 + 500$$
$$= \$525.00 \qquad \text{Actual profit for } x = 50$$

and for $x = 51$, the actual profit is

$$P = 0.0002(51)^3 + 10(51) \qquad \text{Substitute 51 for } x.$$
$$= 0.0002(132,651) + 510$$
$$\approx 26.53 + 510$$
$$= \$536.53. \qquad \text{Actual profit for } x = 51$$

So, the additional profit obtained by increasing production from 50 to 51 units is

$$536.53 - 525.00 = \$11.53. \qquad \text{Extra profit for one unit}$$

Note that the actual profit increase of \$11.53 (when x increases from 50 to 51 units) can be approximated by the marginal profit of \$11.50 per unit (when $x = 50$), as shown in Figure 2.24.

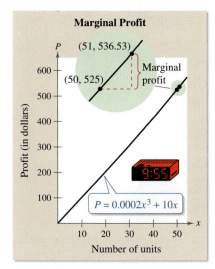

Marginal Profit

FIGURE 2.24

✓**Checkpoint 5**

Find the marginal profit in Example 5 for a production level of 100 units. Compare this with the actual gain in profit by increasing production from 100 to 101 units. ■

STUDY TIP

In Example 5, the marginal profit gives a good approximation of the actual change in profit because the graph of P is nearly straight over the interval $50 \leq x \leq 51$. You will study more about the use of marginals to approximate actual changes in Section 3.8.

The profit function in Example 5 is unusual in that the profit continues to increase as long as the number of units sold increases. In practice, it is more common to encounter situations in which sales can be increased only by lowering the price per item. Such reductions in price will ultimately cause the profit to decline.

The number of units x that consumers are willing to purchase at a given price per unit p is given by the **demand function**

$$p = f(x). \qquad \text{Demand function}$$

The total revenue R is then related to the price per unit and the quantity demanded (or sold) by the equation

$$R = xp. \qquad \text{Revenue function}$$

Example 6   Finding a Demand Function

The table shows the numbers x (in millions) of prerecorded high-definition DVDs sold in the United States and the average unit prices p (in dollars) from 2006 through 2009. Use this information to find the demand function and the total revenue function. *(Source: SNL Kagan)*

Year	2006	2007	2008	2009
x	1.2	9.8	22.7	54.0
p	23.75	23.38	22.21	20.43

SOLUTION   Begin by making a scatter plot of the data using the ordered pairs (x, p), as shown in Figure 2.25. From the graph, it appears that a linear model would be a good fit for the data. To find a linear model for the demand function, use any two points, such as $(1.2, 23.75)$ and $(54.0, 20.43)$. The slope of the line through these points is

$$m = \frac{20.43 - 23.75}{54.0 - 1.2}$$

$$= \frac{-3.32}{52.8}$$

$$\approx -0.063.$$

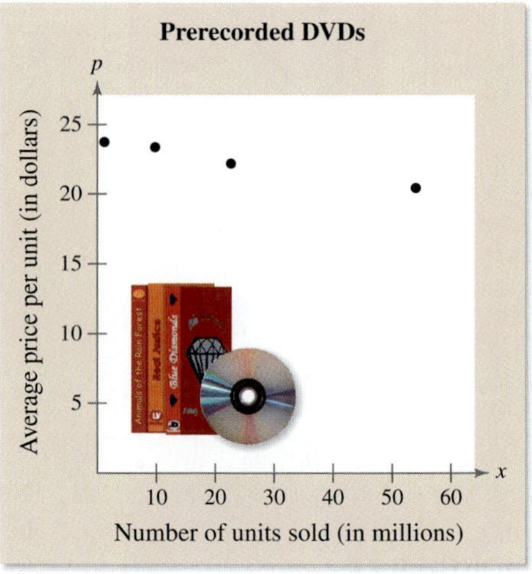

Using the point-slope form of a line, you can approximate the equation of the demand function to be $p = -0.063x + 23.83$. The total revenue function for prerecorded high-definition DVDs is

$$R = xp = x(-0.063x + 23.83) = -0.063x^2 + 23.83x.$$

Prerecorded DVDs

Average price per unit (in dollars) vs. Number of units sold (in millions)

FIGURE 2.25

✓ Checkpoint 6

Repeat Example 6 given that in 2010 an estimated 104.2 million prerecorded high-definition DVDs were sold at an average unit price of $17.88. To find a linear model for the demand function, use the points $(1.2, 23.75)$ and $(104.2, 17.88)$. *(Source: SNL Kagan)*

TECH TUTOR

Another way to find a linear model for the demand function in Example 6 is to use the *linear regression* feature of a graphing utility or a spreadsheet software program. Use a graphing utility or a spreadsheet software program to find the demand function and compare your results with those in Example 6. You will learn more about linear regression in Section 7.7. (Consult the user's manual of a graphing utility or a spreadsheet software program for specific instructions.)

Example 7   **Finding the Marginal Revenue**

A fast-food restaurant has determined that the monthly demand for its hamburgers is given by

$$p = \frac{60,000 - x}{20,000}.$$

Figure 2.26 shows that as the price decreases, the quantity demanded increases. The table shows the demands for hamburgers at various prices.

x	60,000	50,000	40,000	30,000	20,000	10,000	0
p	\$0.00	\$0.50	\$1.00	\$1.50	\$2.00	\$2.50	\$3.00

Find the increase in revenue per hamburger for monthly sales of 20,000 hamburgers. In other words, find the marginal revenue when $x = 20,000$.

SOLUTION   Because the demand is given by

$$p = \frac{60,000 - x}{20,000}$$

and the revenue is given by $R = xp$, you have

$$
\begin{aligned}
R &= xp & &\text{Formula for revenue} \\
&= x\left(\frac{60,000 - x}{20,000}\right) & &\text{Substitute for } p. \\
&= \frac{1}{20,000}(60,000x - x^2). & &\text{Revenue function}
\end{aligned}
$$

By differentiating, you can find the marginal revenue to be

$$\frac{dR}{dx} = \frac{1}{20,000}(60,000 - 2x).$$

So, at $x = 20,000$, the marginal revenue is

$$
\begin{aligned}
\frac{dR}{dx} &= \frac{1}{20,000}(60,000 - 2x) & &\text{Marginal revenue} \\
&= \frac{1}{20,000}[60,000 - 2(20,000)] & &\text{Substitute 20,000 for } x. \\
&= \frac{1}{20,000}(60,000 - 40,000) & &\text{Multiply.} \\
&= \frac{1}{20,000}(20,000) & &\text{Subtract.} \\
&= \$1 \text{ per unit.} & &\text{Marginal revenue when } x = 20,000
\end{aligned}
$$

So, for monthly sales of 20,000 hamburgers, you can conclude that the increase in revenue per hamburger is \$1.

✓**Checkpoint 7**

Find the revenue function and marginal revenue for a demand function of

$$p = 2000 - 4x.$$

Find the marginal revenue when $x = 250$.   ■

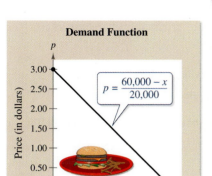

Demand Function

$$p = \frac{60,000 - x}{20,000}$$

Price (in dollars)

Number of hamburgers sold

As the price decreases, more hamburgers are sold.

FIGURE 2.26

STUDY TIP

Writing a demand function in the form $p = f(x)$ is a convention used in economics. From a consumer's point of view, it might seem more reasonable to think that the quantity demanded is a function of the price. Mathematically, however, the two points of view are equivalent because a typical demand function is one-to-one and so has an inverse function. For instance, in Example 7, you could write the demand function as $x = 60,000 - 20,000p$.

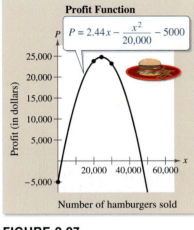

Profit Function

$$P = 2.44x - \frac{x^2}{20,000} - 5000$$

Profit (in dollars)

25,000
20,000
15,000
10,000
5,000

20,000 40,000 60,000   x

−5,000

Number of hamburgers sold

FIGURE 2.27

Example 8   Finding the Marginal Profit

For the fast-food restaurant in Example 7, the cost of producing x hamburgers is

$$C = 5000 + 0.56x, \quad 0 \le x \le 50,000.$$

Find the profit and the marginal profit for each production level.

a. $x = 20,000$ **b.** $x = 24,400$ **c.** $x = 30,000$

SOLUTION From Example 7, you know that the total revenue from selling x hamburgers is

$$R = \frac{1}{20,000}(60,000x - x^2).$$

Because the total profit is given by $P = R - C$, you have

$$P = \frac{1}{20,000}(60,000x - x^2) - (5000 + 0.56x)$$

$$= 3x - \frac{x^2}{20,000} - 5000 - 0.56x$$

$$= 2.44x - \frac{x^2}{20,000} - 5000. \qquad \textit{See Figure 2.27.}$$

So, the marginal profit is

$$\frac{dP}{dx} = 2.44 - \frac{x}{10,000}.$$

Using these formulas, you can compute the profit and marginal profit.

Production	Profit	Marginal Profit
a. $x = 20,000$	$P = \$23,800.00$	$2.44 - \dfrac{20,000}{10,000} = \$0.44$ per unit
b. $x = 24,400$	$P = \$24,768.00$	$2.44 - \dfrac{24,400}{10,000} = \$0.00$ per unit
c. $x = 30,000$	$P = \$23,200.00$	$2.44 - \dfrac{30,000}{10,000} = -\$0.56$ per unit

✓**Checkpoint 8**

From Example 8, compare the marginal profit when 10,000 units are produced with the actual increase in profit from 10,000 units to 10,001 units. ■

SUMMARIZE   (Section 2.3)

1. State the definition of average rate of change *(page 87)*. For examples of average rate of change, see Examples 1 and 2.

2. State the definition of instantaneous rate of change *(page 90)*. For examples of an instantaneous rate of change, see Examples 3 and 4.

3. Describe a real-life example of how rates of change can be used in the field of economics *(pages 93–96, Examples 5, 6, 7, and 8)*.

SKILLS WARM UP 2.3   The following warm-up exercises involve skills that were covered in earlier sections. You will use these skills in the exercise set for this section. For additional help, review Sections 2.1 and 2.2.

In Exercises 1–4, evaluate the expression.

1. $\dfrac{-63 - (-105)}{21 - 7}$

2. $\dfrac{-37 - 54}{16 - 3}$

3. $\dfrac{24 - 33}{9 - 6}$

4. $\dfrac{40 - 16}{18 - 8}$

In Exercises 5–12, find the derivative of the function.

5. $y = 4x^2 - 2x + 7$

6. $y = -3t^3 + 2t^2 - 8$

7. $s = -16t^2 + 24t + 30$

8. $y = -16x^2 + 54x + 70$

9. $A = \frac{1}{10}(-2r^3 + 3r^2 + 5r)$

10. $y = \frac{1}{9}(6x^3 - 18x^2 + 63x - 15)$

11. $y = 12x - \dfrac{x^2}{5000}$

12. $y = 138 + 74x - \dfrac{x^3}{10,000}$

2.4 The Product and Quotient Rules

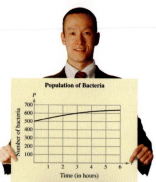

Population of Bacteria

In Exercise 63, you will use the Quotient Rule to find the rate of change of a population of bacteria.

- Find the derivatives of functions using the Product Rule.
- Find the derivatives of functions using the Quotient Rule.
- Use derivatives to answer questions about real-life situations.

The Product Rule

In Section 2.2, you saw that the derivative of a sum or difference of two functions is simply the sum or difference of their derivatives. The rules for the derivative of a product or quotient of two functions are not as simple.

The Product Rule

The derivative of the product of two differentiable functions is equal to the first function times the derivative of the second plus the second function times the derivative of the first.

$$\frac{d}{dx}[f(x)g(x)] = f(x)g\,'(x) + g(x)f'(x)$$

PROOF Some mathematical proofs, such as the proof of the Sum Rule, are straightforward. Others involve clever steps that may not appear to follow clearly from a prior step. The proof below involves such a step—adding and subtracting the same quantity. (This step is shown in color.) Let $F(x) = f(x)g(x)$.

$$F'(x) = \lim_{\Delta x \to 0} \frac{F(x + \Delta x) - F(x)}{\Delta x}$$

$$= \lim_{\Delta x \to 0} \frac{f(x + \Delta x)g(x + \Delta x) - f(x)g(x)}{\Delta x}$$

$$= \lim_{\Delta x \to 0} \frac{f(x + \Delta x)g(x + \Delta x) - f(x + \Delta x)g(x) + f(x + \Delta x)g(x) - f(x)g(x)}{\Delta x}$$

$$= \lim_{\Delta x \to 0} \left[f(x + \Delta x)\frac{g(x + \Delta x) - g(x)}{\Delta x} + g(x)\frac{f(x + \Delta x) - f(x)}{\Delta x} \right]$$

$$= \lim_{\Delta x \to 0} f(x + \Delta x)\frac{g(x + \Delta x) - g(x)}{\Delta x} + \lim_{\Delta x \to 0} g(x)\frac{f(x + \Delta x) - f(x)}{\Delta x}$$

$$= \lim_{\Delta x \to 0} f(x + \Delta x) \cdot \lim_{\Delta x \to 0} \frac{g(x + \Delta x) - g(x)}{\Delta x} + \lim_{\Delta x \to 0} g(x) \cdot \lim_{\Delta x \to 0} \frac{f(x + \Delta x) - f(x)}{\Delta x}$$

$$= f(x)g\,'(x) + g(x)f'(x)$$

STUDY TIP

Rather than trying to remember the formula for the Product Rule, it can be more helpful to remember its verbal statement:

the first function times the derivative of the second plus the second function times the derivative of the first.

Example 1   Using the Product Rule

Find the derivative of $y = (3x - 2x^2)(5 + 4x)$.

SOLUTION   Using the Product Rule, you can write

First | Derivative of second | Second | Derivative of first

$$\frac{dy}{dx} = (3x - 2x^2)\frac{d}{dx}[5 + 4x] + (5 + 4x)\frac{d}{dx}[3x - 2x^2]$$

$$= (3x - 2x^2)(4) + (5 + 4x)(3 - 4x)$$

$$= (12x - 8x^2) + (15 - 8x - 16x^2)$$

$$= 15 + 4x - 24x^2.$$

✓ Checkpoint 1

Find the derivative of $y = (4x + 3x^2)(6 - 3x)$.   ■

In general, the derivative of the product of two functions is not equal to the product of the derivatives of the two functions. To see this, compare the product of the derivatives of

$$f(x) = 3x - 2x^2 \quad \text{and} \quad g(x) = 5 + 4x$$

with the derivative found in Example 1.

In the next example, notice that the first step in differentiating is *rewriting the original function*.

Example 2   Using the Product Rule

Find the derivative of $f(x) = \left(\frac{1}{x} + 1\right)(x - 1)$.

SOLUTION   Rewrite the function. Then use the Product Rule to find the derivative.

$$f(x) = \left(\frac{1}{x} + 1\right)(x - 1)$$ Write original function.

$$= (x^{-1} + 1)(x - 1)$$ Rewrite function.

$$f'(x) = (x^{-1} + 1)\frac{d}{dx}[x - 1] + (x - 1)\frac{d}{dx}[x^{-1} + 1]$$ Product Rule

$$= (x^{-1} + 1)(1) + (x - 1)(-x^{-2})$$

$$= \frac{1}{x} + 1 - \frac{x - 1}{x^2}$$

$$= \frac{x + x^2 - x + 1}{x^2}$$ Write with common denominator.

$$= \frac{x^2 + 1}{x^2}$$ Simplify.

✓ Checkpoint 2

Find the derivative of

$$f(x) = \left(\frac{1}{x} + 1\right)(2x + 1).$$ ■

You now have two differentiation rules that deal with products—the Constant Multiple Rule and the Product Rule. The difference between these two rules is that the Constant Multiple Rule is used when one of the factors is a constant

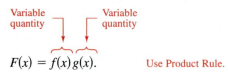

$$F(x) = c\,f(x)$$ Use Constant Multiple Rule.

whereas the Product Rule is used when both of the factors are variable quantities

$$F(x) = f(x)\,g(x).$$ Use Product Rule.

The next example compares these two rules.

Example 3   Comparing Differentiation Rules

Find the derivative of each function.

a. $y = 2x(x^2 + 3x)$

b. $y = 2(x^2 + 3x)$

SOLUTION

a. Because both factors are variable quantities, use the Product Rule.

$$y = 2x(x^2 + 3x)$$

$$\frac{dy}{dx} = (2x)\frac{d}{dx}[x^2 + 3x] + (x^2 + 3x)\frac{d}{dx}[2x]$$ Product Rule

$$= (2x)(2x + 3) + (x^2 + 3x)(2)$$

$$= 4x^2 + 6x + 2x^2 + 6x$$

$$= 6x^2 + 12x$$

b. Because one of the factors is a constant, use the Constant Multiple Rule.

$$y = 2(x^2 + 3x)$$

$$\frac{dy}{dx} = 2\frac{d}{dx}[x^2 + 3x]$$ Constant Multiple Rule

$$= 2(2x + 3)$$

$$= 4x + 6$$

> **STUDY TIP**
>
> You could calculate the derivative in Example 3(a) without the Product Rule. For instance,
>
> $$y = 2x(x^2 + 3x) = 2x^3 + 6x^2$$
>
> and
>
> $$\frac{dy}{dx} = 6x^2 + 12x.$$

✓ Checkpoint 3

Find the derivative of each function.

a. $y = 3x(2x^2 + 5x)$

b. $y = 3(2x^2 + 5x)$ ■

The Product Rule can be extended to products that have more than two factors. For example, if f, g, and h are differentiable functions of x, then

$$\frac{d}{dx}[f(x)g(x)h(x)] = f'(x)g(x)h(x) + f(x)g'(x)h(x) + f(x)g(x)h'(x).$$

The Quotient Rule

In Section 2.2, you saw that by using the Constant Rule, the Power Rule, the Constant Multiple Rule, and the Sum and Difference Rules, you were able to differentiate any polynomial function. By combining these rules with the Quotient Rule, you can now differentiate any *rational* function.

STUDY TIP

As suggested for the Product Rule, it can be more helpful to remember the verbal statement of the Quotient Rule rather than trying to remember the formula for the rule.

The Quotient Rule

The derivative of the quotient of two differentiable functions is equal to the denominator times the derivative of the numerator minus the numerator times the derivative of the denominator, all divided by the square of the denominator.

$$\frac{d}{dx}\left[\frac{f(x)}{g(x)}\right] = \frac{g(x)f'(x) - f(x)g'(x)}{[g(x)]^2}, \quad g(x) \neq 0$$

PROOF Begin by letting

$$F(x) = \frac{f(x)}{g(x)}.$$

As in the proof of the Product Rule, a key step in this proof is adding and subtracting the same quantity.

$$F'(x) = \lim_{\Delta x \to 0} \frac{F(x + \Delta x) - F(x)}{\Delta x}$$

$$= \lim_{\Delta x \to 0} \frac{\dfrac{f(x + \Delta x)}{g(x + \Delta x)} - \dfrac{f(x)}{g(x)}}{\Delta x}$$

$$= \lim_{\Delta x \to 0} \left[\left(\frac{g(x)f(x + \Delta x)}{g(x)g(x + \Delta x)} - \frac{f(x)g(x + \Delta x)}{g(x)g(x + \Delta x)}\right) \div \Delta x\right]$$

$$= \lim_{\Delta x \to 0} \left[\frac{g(x)f(x + \Delta x) - f(x)g(x + \Delta x)}{g(x)g(x + \Delta x)} \cdot \frac{1}{\Delta x}\right]$$

$$= \lim_{\Delta x \to 0} \frac{g(x)f(x + \Delta x) - f(x)g(x + \Delta x)}{\Delta x g(x)g(x + \Delta x)}$$

$$= \lim_{\Delta x \to 0} \frac{g(x)f(x + \Delta x) - f(x)g(x) + f(x)g(x) - f(x)g(x + \Delta x)}{\Delta x g(x)g(x + \Delta x)}$$

$$= \frac{\displaystyle\lim_{\Delta x \to 0} \frac{g(x)[f(x + \Delta x) - f(x)]}{\Delta x} - \lim_{\Delta x \to 0} \frac{f(x)[g(x + \Delta x) - g(x)]}{\Delta x}}{\displaystyle\lim_{\Delta x \to 0} [g(x)g(x + \Delta x)]}$$

$$= \frac{g(x)\left[\displaystyle\lim_{\Delta x \to 0} \frac{f(x + \Delta x) - f(x)}{\Delta x}\right] - f(x)\left[\displaystyle\lim_{\Delta x \to 0} \frac{g(x + \Delta x) - g(x)}{\Delta x}\right]}{\displaystyle\lim_{\Delta x \to 0} [g(x)g(x + \Delta x)]}$$

$$= \frac{g(x)f'(x) - f(x)g'(x)}{[g(x)]^2}$$

From the Quotient Rule, you can see that the derivative of a quotient is not, in general, the quotient of the derivatives. That is,

$$\frac{d}{dx}\left[\frac{f(x)}{g(x)}\right] \neq \frac{f'(x)}{g'(x)}.$$

Example 4   **Using the Quotient Rule**

Find the derivative of $y = \dfrac{x - 1}{2x + 3}$.

SOLUTION   Apply the Quotient Rule, as shown.

$$\frac{dy}{dx} = \frac{(2x + 3)\dfrac{d}{dx}[x - 1] - (x - 1)\dfrac{d}{dx}[2x + 3]}{(2x + 3)^2}$$

$$= \frac{(2x + 3)(1) - (x - 1)(2)}{(2x + 3)^2}$$

$$= \frac{2x + 3 - 2x + 2}{(2x + 3)^2}$$

$$= \frac{5}{(2x + 3)^2}$$

✓ **Checkpoint 4**

Find the derivative of $y = \dfrac{x + 4}{5x - 2}$.

Example 5   **Finding an Equation of a Tangent Line**

Find an equation of the tangent line to the graph of

$$y = \frac{2x^2 - 4x + 3}{2 - 3x}$$

at $x = 1$.

SOLUTION   Apply the Quotient Rule, as shown.

$$\frac{dy}{dx} = \frac{(2 - 3x)\dfrac{d}{dx}[2x^2 - 4x + 3] - (2x^2 - 4x + 3)\dfrac{d}{dx}[2 - 3x]}{(2 - 3x)^2}$$

$$= \frac{(2 - 3x)(4x - 4) - (2x^2 - 4x + 3)(-3)}{(2 - 3x)^2}$$

$$= \frac{-12x^2 + 20x - 8 - (-6x^2 + 12x - 9)}{(2 - 3x)^2}$$

$$= \frac{-12x^2 + 20x - 8 + 6x^2 - 12x + 9}{(2 - 3x)^2}$$

$$= \frac{-6x^2 + 8x + 1}{(2 - 3x)^2}$$

When $x = 1$, the value of the function is $y = -1$ and the slope is $m = 3$. Using the point-slope form of a line, you can find the equation of the tangent line to be $y = 3x - 4$. The graph of the function and the tangent line is shown in Figure 2.28.

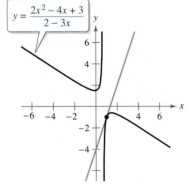

$y = \dfrac{2x^2 - 4x + 3}{2 - 3x}$

FIGURE 2.28

✓ **Checkpoint 5**

Find an equation of the tangent line to the graph of

$$y = \frac{x^2 - 4}{2x + 5}$$

at $x = 0$. Sketch the line tangent to the graph at $x = 0$.

STUDY TIP

Note in Example 6 that much of the work in obtaining the final form of the derivative occurs *after* applying the Quotient Rule. In general, direct application of differentiation rules often yields results that are not in simplified form. Note that two characteristics of simplified form are the absence of negative exponents and the combining of like terms.

Example 6   **Rewriting Before Differentiating**

Find the derivative of

$$y = \frac{3 - (1/x)}{x + 5}.$$

SOLUTION   Begin by rewriting the function. Then apply the Quotient Rule and simplify the result.

$$y = \frac{3 - (1/x)}{x + 5} \qquad \text{Write original function.}$$

$$= \frac{x[3 - (1/x)]}{x(x + 5)} \qquad \text{Multiply numerator and denominator by } x.$$

$$= \frac{3x - 1}{x^2 + 5x} \qquad \text{Rewrite.}$$

$$\frac{dy}{dx} = \frac{(x^2 + 5x)(3) - (3x - 1)(2x + 5)}{(x^2 + 5x)^2} \qquad \text{Apply Quotient Rule.}$$

$$= \frac{(3x^2 + 15x) - (6x^2 + 13x - 5)}{(x^2 + 5x)^2}$$

$$= \frac{-3x^2 + 2x + 5}{(x^2 + 5x)^2} \qquad \text{Simplify.}$$

✓Checkpoint 6

Find the derivative of $y = \dfrac{3 - (2/x)}{x + 4}$. ■

Not every quotient needs to be differentiated by the Quotient Rule. For instance, each quotient in the next example can be considered as the product of a constant and a function of x. In such cases, the Constant Multiple Rule is more efficient than the Quotient Rule.

STUDY TIP

To see the benefit of using the Constant Multiple Rule for some quotients, try using the Quotient Rule to differentiate the functions in Example 7. You should obtain the same results, but with more work.

Example 7   **Using the Constant Multiple Rule**

Original Function	*Rewrite*	*Differentiate*	*Simplify*
a. $y = \dfrac{x^2 + 3x}{6}$	$y = \dfrac{1}{6}(x^2 + 3x)$	$y' = \dfrac{1}{6}(2x + 3)$	$y' = \dfrac{1}{3}x + \dfrac{1}{2}$
b. $y = \dfrac{5x^4}{8}$	$y = \dfrac{5}{8}x^4$	$y' = \dfrac{5}{8}(4x^3)$	$y' = \dfrac{5}{2}x^3$
c. $y = \dfrac{-3(3x - 2x^2)}{7x}$	$y = -\dfrac{3}{7}(3 - 2x)$	$y' = -\dfrac{3}{7}(-2)$	$y' = \dfrac{6}{7}$
d. $y = \dfrac{9}{5x^2}$	$y = \dfrac{9}{5}(x^{-2})$	$y' = \dfrac{9}{5}(-2x^{-3})$	$y' = -\dfrac{18}{5x^3}$

✓Checkpoint 7

Find the derivative of each function.

a. $y = \dfrac{x^2 + 4x}{5}$ **b.** $y = \dfrac{3x^4}{4}$ ■

Application

 Example 8   Rate of Change of Systolic Blood Pressure

As blood moves from the heart through the major arteries out to the capillaries and back through the veins, the systolic blood pressure continuously drops. Consider a person whose systolic blood pressure P (in millimeters of mercury) is given by

$$P = \frac{25t^2 + 125}{t^2 + 1}, \quad 0 \le t \le 10$$

where t is measured in seconds. At what rate is the blood pressure changing 5 seconds after blood leaves the heart?

SOLUTION   Begin by applying the Quotient Rule.

$$P = \frac{25t^2 + 125}{t^2 + 1}$$ Write original function.

$$\frac{dP}{dt} = \frac{(t^2 + 1)(50t) - (25t^2 + 125)(2t)}{(t^2 + 1)^2}$$ Quotient Rule

$$= \frac{50t^3 + 50t - 50t^3 - 250t}{(t^2 + 1)^2}$$

$$= -\frac{200t}{(t^2 + 1)^2}$$ Simplify.

When $t = 5$, the rate of change is

$$\frac{dP}{dt} = -\frac{200(5)}{26^2} \approx -1.48 \text{ millimeters per second.}$$

So, the pressure is *dropping* at a rate of 1.48 millimeters per second at $t = 5$ seconds.

✓**Checkpoint 8**

In Example 8, find the rate at which systolic blood pressure is changing at each time shown in the table below. Describe the changes in blood pressure as the blood moves away from the heart.

t	0	1	2	3	4	5	6	7
$\dfrac{dP}{dt}$								

SUMMARIZE   (Section 2.4)

1. State the Product Rule *(page 98)*. For examples of the Product Rule, see Examples 1, 2, and 3.

2. State the Quotient Rule *(page 101)*. For examples of the Quotient Rule, see Examples 4, 5, and 6.

3. Describe a real-life example of how the Quotient Rule can be used to analyze the rate of change of systolic blood pressure *(page 104, Example 8)*.

SKILLS WARM UP 2.4
The following warm-up exercises involve skills that were covered in a previous course or earlier sections. You will use these skills in the exercise set for this section. For additional help, review Appendix Sections A.4 and A.5, and Section 2.2.

In Exercises 1–10, simplify the expression.

1. $(x^2 + 1)(2) + (2x + 7)(2x)$

2. $(2x - x^3)(8x) + (4x^2)(2 - 3x^2)$

3. $x(4)(x^2 + 2)^3(2x) + (x^2 + 4)(1)$

4. $x^2(2)(2x + 1)(2) + (2x + 1)^4(2x)$

5. $\dfrac{(2x + 7)(5) - (5x + 6)(2)}{(2x + 7)^2}$

6. $\dfrac{(x^2 - 4)(2x + 1) - (x^2 + x)(2x)}{(x^2 - 4)^2}$

7. $\dfrac{(x^2 + 1)(2) - (2x + 1)(2x)}{(x^2 + 1)^2}$

8. $\dfrac{(1 - x^4)(4) - (4x - 1)(-4x^3)}{(1 - x^4)^2}$

9. $(x^{-1} + x)(2) + (2x - 3)(-x^{-2} + 1)$

10. $\dfrac{(1 - x^{-1})(1) - (x - 4)(x^{-2})}{(1 - x^{-1})^2}$

In Exercises 11–14, find $f'(2)$.

11. $f(x) = 3x^2 - x + 4$

12. $f(x) = -x^3 + x^2 + 8x$

13. $f(x) = \dfrac{1}{x}$

14. $f(x) = x^2 - \dfrac{1}{x^2}$

QUIZ YOURSELF

Take this quiz as you would take a quiz in class. When you are done, check your work against the answers given in the back of the book.

In Exercises 1–3, use the limit definition to find the derivative of the function. Then find the slope of the tangent line to the graph of f at the given point.

1. $f(x) = 5x + 3; (-2, -7)$

2. $f(x) = \sqrt{x + 3}; (1, 2)$

3. $f(x) = x^2 - 2x; (3, 3)$

In Exercises 4–13, find the derivative of the function.

4. $f(x) = 12$

5. $f(x) = 19x + 9$

6. $f(x) = 5 - 3x^2$

7. $f(x) = 12x^{1/4}$

8. $f(x) = 4x^{-2}$

9. $f(x) = 2\sqrt{x}$

10. $f(x) = \dfrac{2x + 3}{3x + 2}$

11. $f(x) = (x^2 + 1)(-2x + 4)$

12. $f(x) = (x^2 + 3x + 4)(5x - 2)$

13. $f(x) = \dfrac{4x}{x^2 + 3}$

 In Exercises 14–17, use a graphing utility to graph the function and find its average rate of change on the given interval. Compare this rate with the instantaneous rates of change at the endpoints of the interval.

14. $f(x) = x^2 - 3x + 1; [0, 3]$

15. $f(x) = 2x^3 + x^2 - x + 4; [-1, 1]$

16. $f(x) = \dfrac{1}{2x}; [2, 5]$

17. $f(x) = \sqrt[3]{x}; [8, 27]$

18. The profit P (in dollars) from selling x units of a product is given by

$$P = -0.0125x^2 + 16x - 600.$$

(a) Find the additional profit when sales increase from 175 to 176 units.

(b) Find the marginal profit when $x = 175$.

(c) Compare the results of parts (a) and (b).

 In Exercises 19 and 20, find an equation of the tangent line to the graph of f at the given point. Then use a graphing utility to graph the function and the equation of the tangent line in the same viewing window.

19. $f(x) = 5x^2 + 6x - 1; (-1, -2)$

20. $f(x) = (x^2 + 1)(4x - 3); (1, 2)$

21. From 2003 through 2009, the sales per share S (in dollars) for Columbia Sportswear can be modeled by

$$S = -0.13556t^3 + 1.8682t^2 - 4.351t + 23.52, \quad 3 \le t \le 9$$

where t represents the year, with $t = 3$ corresponding to 2003. *(Source: Columbia Sportswear Company)*

(a) Find the rate of change of the sales per share with respect to the year.

(b) At what rate were the sales per share changing in 2004? in 2007? in 2008?

2.5 The Chain Rule

■ Find derivatives using the Chain Rule.
■ Find derivatives using the General Power Rule.
■ Write derivatives in simplified form.
■ Use derivatives to answer questions about real-life situations.
■ Review the basic differentiation rules for algebraic functions.

The Chain Rule

In this section, you will study one of the most powerful rules of differential calculus—the **Chain Rule.** This differentiation rule deals with composite functions and adds versatility to the rules presented in Sections 2.2 and 2.4. For example, compare the functions below. Those on the left can be differentiated without the Chain Rule, whereas those on the right are best done with the Chain Rule.

Without the Chain Rule	*With the Chain Rule*
$y = x^2 + 1$	$y = \sqrt{x^2 + 1}$
$y = x + 1$	$y = (x + 1)^{-1/2}$
$y = 3x + 2$	$y = (3x + 2)^5$
$y = \dfrac{x + 5}{x^2 + 2}$	$y = \left(\dfrac{x + 5}{x^2 + 2}\right)^2$
$y = \dfrac{x + 1}{x}$	$y = \sqrt{\dfrac{x + 1}{x}}$

In Exercise 71, you will use the General Power Rule to find the rate of change of the balance in an account.

The Chain Rule

If $y = f(u)$ is a differentiable function of u, and $u = g(x)$ is a differentiable function of x, then $y = f(g(x))$ is a differentiable function of x, and

$$\frac{dy}{dx} = \frac{dy}{du} \cdot \frac{du}{dx}$$

or, equivalently,

$$\frac{d}{dx}[f(g(x))] = f'(g(x))g'(x).$$

Basically, the Chain Rule states that if y changes dy/du times as fast as u, and u changes du/dx times as fast as x, then y changes

$$\frac{dy}{du} \cdot \frac{du}{dx}$$

times as fast as x, as illustrated in Figure 2.29. One advantage of the

$$\frac{dy}{dx}$$

notation for derivatives is that it helps you remember differentiation rules, such as the Chain Rule. For instance, in the formula

$$\frac{dy}{dx} = \frac{dy}{du} \cdot \frac{du}{dx}$$

you can imagine that the du's divide out.

x

Rate of change of u with respect to x is $\dfrac{du}{dx}$.

Input

Function g Output

$u = g(x)$

u

Rate of change of y with respect to u is $\dfrac{dy}{du}$.

Input

Function f Output

$y = f(u) = f(g(x))$

Rate of change of y with respect to x is $\dfrac{dy}{dx} = \dfrac{dy}{du}\dfrac{du}{dx}$.

FIGURE 2.29

When applying the Chain Rule, it is helpful to think of the composite function $y = f(g(x))$ or $y = f(u)$ as having two parts—an *inside* and an *outside*—as illustrated below.

Inside

$$y = f(g(x)) = f(u)$$

Outside

The Chain Rule tells you that the derivative of $y = f(u)$ is the derivative of the outer function (at the inner function u) *times* the derivative of the inner function. That is,

$$y' = f'(u) \cdot u'.$$

Example 1   Decomposing Composite Functions

Write each function as the composition of two functions.

a. $y = \dfrac{1}{x + 1}$ **b.** $y = \sqrt{3x^2 - x + 1}$

SOLUTION   There is more than one correct way to decompose each function. One way for each is shown below.

$y = f(g(x))$	$u = g(x)$ *(inside)*	$y = f(u)$ *(outside)*
a. $y = \dfrac{1}{x + 1}$	$u = x + 1$	$y = \dfrac{1}{u}$
b. $y = \sqrt{3x^2 - x + 1}$	$u = 3x^2 - x + 1$	$y = \sqrt{u}$

✓ **Checkpoint 1**

Write each function as the composition of two functions, where $y = f(g(x))$.

a. $y = \dfrac{1}{\sqrt{x + 1}}$ **b.** $y = (x^2 + 2x + 5)^3$ ■

Example 2   Using the Chain Rule

Find the derivative of $y = (x^2 + 1)^3$.

SOLUTION   To apply the Chain Rule, you need to identify the inside function u.

$$y = \overbrace{(x^2 + 1)}^{u}{}^3 = u^3$$

The inside function is $u = x^2 + 1$. By the Chain Rule, you can write the derivative as shown.

$$\frac{dy}{dx} = \overbrace{3(x^2 + 1)^2}^{\frac{dy}{du}}\overbrace{(2x)}^{\frac{du}{dx}} = 6x(x^2 + 1)^2$$

STUDY TIP

Try checking the result of Example 2 by expanding the function to obtain

$$y = x^6 + 3x^4 + 3x^2 + 1$$

and finding the derivative. Do you obtain the same answer?

✓ **Checkpoint 2**

Find the derivative of $y = (x^3 + 1)^2$.   ■

The General Power Rule

The function in Example 2 illustrates one of the most common types of composite functions—a power function of the form

$$y = [u(x)]^n.$$

The rule for differentiating such functions is called the **General Power Rule,** and it is a special case of the Chain Rule.

The General Power Rule

If $y = [u(x)]^n$, where u is a differentiable function of x and n is a real number, then

$$\frac{dy}{dx} = n[u(x)]^{n-1}\frac{du}{dx}$$

or, equivalently,

$$\frac{d}{dx}[u^n] = nu^{n-1}u'.$$

PROOF   Apply the Chain Rule and the Simple Power Rule as shown.

$$\frac{dy}{dx} = \frac{dy}{du} \cdot \frac{du}{dx}$$

$$= \frac{d}{du}[u^n]\frac{du}{dx}$$

$$= nu^{n-1}\frac{du}{dx}$$

TECH TUTOR

If you have access to a symbolic differentiation utility, try using it to confirm the result of Example 3.

Example 3   **Using the General Power Rule**

Find the derivative of

$$y = (3x - 2x^2)^3.$$

SOLUTION   To apply the General Power Rule, you need to identify the inside function u.

$$y = \overbrace{(3x - 2x^2)}^{u}{}^3 = u^3$$

The inside function is

$$u = 3x - 2x^2.$$

So, by the General Power Rule,

$$\frac{dy}{dx} = \overbrace{3}^{n}\overbrace{(3x - 2x^2)^2}^{u^{n-1}}\overbrace{\frac{d}{dx}[3x - 2x^2]}^{u'}$$

$$= 3(3x - 2x^2)^2(3 - 4x).$$

✓ Checkpoint 3

Find the derivative of

$$y = (x^2 + 3x)^4.$$

Example 4   Finding an Equation of a Tangent Line

Find an equation of the tangent line to the graph of

$$y = \sqrt[3]{(x^2 + 4)^2}$$

at $x = 2$.

SOLUTION   Begin by rewriting the function in rational exponent form.

$$y = (x^2 + 4)^{2/3}$$ Rewrite original function.

Then, using the inside function, $u = x^2 + 4$, apply the General Power Rule.

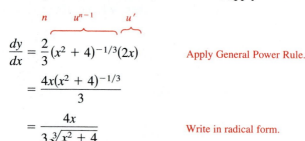

$$\frac{dy}{dx} = \frac{2}{3}(x^2 + 4)^{-1/3}(2x)$$ Apply General Power Rule.

$$= \frac{4x(x^2 + 4)^{-1/3}}{3}$$

$$= \frac{4x}{3\sqrt[3]{x^2 + 4}}$$ Write in radical form.

When $x = 2$, $y = 4$ and the slope of the line tangent to the graph at $(2, 4)$ is $\frac{4}{3}$. Using the point-slope form, you can find the equation of the tangent line to be $y = \frac{4}{3}x + \frac{4}{3}$. The graph of the function and the tangent line is shown in Figure 2.30.

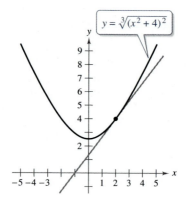

FIGURE 2.30

✓**Checkpoint 4**

Find an equation of the tangent line to the graph of $y = \sqrt[3]{(x + 4)^2}$ at $x = 4$. ▪

STUDY TIP

The derivative of a quotient can sometimes be found more easily with the General Power Rule than with the Quotient Rule. This is especially true when the numerator is a constant, as shown in Example 5.

Example 5   Differentiating a Quotient with a Constant Numerator

Find the derivative of

$$y = \frac{5}{(4x - 3)^2}.$$

SOLUTION   Begin by rewriting the function in rational exponent form.

$$y = 5(4x - 3)^{-2}$$ Rewrite original function.

Then, using the inside function, $u = 4x - 3$, apply the General Power Rule.

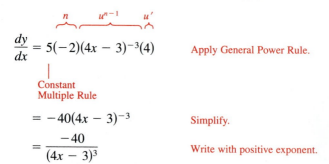

$$\frac{dy}{dx} = 5(-2)(4x - 3)^{-3}(4)$$ Apply General Power Rule.

Constant
Multiple Rule

$$= -40(4x - 3)^{-3}$$ Simplify.

$$= \frac{-40}{(4x - 3)^3}$$ Write with positive exponent.

✓**Checkpoint 5**

Find the derivative of $y = \dfrac{4}{2x + 1}.$ ▪

Simplification Techniques

Throughout this chapter, writing derivatives in simplified form has been emphasized. The reason for this is that most applications of derivatives require a simplified form. The next two examples illustrate some useful simplification techniques.

Example 6 Simplifying by Factoring Out Least Powers

Find the derivative of $y = x^2\sqrt{1 - x^2}$.

SOLUTION

$$
\begin{aligned}
y &= x^2\sqrt{1 - x^2} & &\text{Write original function.}\\
&= x^2(1 - x^2)^{1/2} & &\text{Rewrite function.}\\
y' &= x^2\frac{d}{dx}\left[(1 - x^2)^{1/2}\right] + (1 - x^2)^{1/2}\frac{d}{dx}\left[x^2\right] & &\text{Product Rule}\\
&= x^2\left[\frac{1}{2}(1 - x^2)^{-1/2}(-2x)\right] + (1 - x^2)^{1/2}(2x) & &\text{General Power Rule}\\
&= -x^3(1 - x^2)^{-1/2} + 2x(1 - x^2)^{1/2} & &\text{Simplify.}\\
&= x(1 - x^2)^{-1/2}\left[-x^2(1) + 2(1 - x^2)\right] & &\text{Factor.}\\
&= x(1 - x^2)^{-1/2}(2 - 3x^2) & &\text{Simplify.}\\
&= \frac{x(2 - 3x^2)}{\sqrt{1 - x^2}} & &\text{Write in radical form.}
\end{aligned}
$$

✓ **Checkpoint 6**

Find and simplify the derivative of $y = x^2\sqrt{x^2 + 1}$. ■

Example 7 Differentiating a Quotient Raised to a Power

Find the derivative of

$$f(x) = \left(\frac{3x - 1}{x^2 + 3}\right)^2.$$

SOLUTION

$$
\begin{aligned}
f'(x) &= 2\left(\frac{3x - 1}{x^2 + 3}\right)\frac{d}{dx}\left[\frac{3x - 1}{x^2 + 3}\right] & &\text{General Power Rule}\\
&= \left[\frac{2(3x - 1)}{x^2 + 3}\right]\left[\frac{(x^2 + 3)(3) - (3x - 1)(2x)}{(x^2 + 3)^2}\right] & &\text{Quotient Rule}\\
&= \frac{2(3x - 1)(3x^2 + 9 - 6x^2 + 2x)}{(x^2 + 3)^3} & &\text{Multiply.}\\
&= \frac{2(3x - 1)(-3x^2 + 2x + 9)}{(x^2 + 3)^3} & &\text{Simplify.}
\end{aligned}
$$

where n corresponds to the outer exponent, u^{n-1} and u' label the inner terms.

✓ **Checkpoint 7**

Find the derivative of

$$f(x) = \left(\frac{x + 1}{x - 5}\right)^2.$$ ■

ALGEBRA TUTOR xy

In Example 6, note that you subtract exponents when factoring. That is, when $(1 - x^2)^{-1/2}$ is factored out of $(1 - x^2)^{1/2}$, the *remaining* factor has an exponent of $\frac{1}{2} - \left(-\frac{1}{2}\right) = 1$. So, $(1 - x^2)^{1/2}$ is equal to the product of $(1 - x^2)^{-1/2}$ and $(1 - x^2)^1$. For help in simplifying expressions like the one in Example 6, see the *Chapter 2 Algebra Tutor* on pages 133 and 134.

STUDY TIP

In Example 7, try to find $f'(x)$ by applying the Quotient Rule to

$$f(x) = \frac{(3x - 1)^2}{(x^2 + 3)^2}.$$

Which method do you prefer?

Application

Example 8   **Finding Rates of Change**

From 2000 through 2009, the revenue per share R (in dollars) for U.S. Cellular can be modeled by

$$R = (-0.009t^2 + 0.39t + 4.3)^2, \quad 0 \le t \le 9$$

where t is the year, with $t = 0$ corresponding to 2000. Use the model to approximate the rates of change in the revenue per share in 2001, 2002, and 2005. Would U.S. Cellular stockholders have been satisfied with the performance of this stock from 2000 through 2009?   *(Source: U.S. Cellular)*

SOLUTION   The rate of change in R is given by the derivative dR/dt. You can use the General Power Rule to find the derivative.

$$\frac{dR}{dt} = 2(-0.009t^2 + 0.39t + 4.3)(-0.018t + 0.39)$$

$$= (-0.036t + 0.78)(-0.009t^2 + 0.39t + 4.3)$$

In 2001, the revenue per share was changing at a rate of

$$[-0.036(1) + 0.78][-0.009(1)^2 + 0.39(1) + 4.3] \approx \$3.48 \text{ per year.}$$

In 2002, the revenue per share was changing at a rate of

$$[-0.036(2) + 0.78][-0.009(2)^2 + 0.39(2) + 4.3] \approx \$3.57 \text{ per year.}$$

In 2005, the revenue per share was changing at a rate of

$$[-0.036(5) + 0.78][-0.009(5)^2 + 0.39(5) + 4.3] \approx \$3.62 \text{ per year.}$$

The graph of the revenue per share function R is shown in Figure 2.31. So, most stockholders would have been satisfied with the performance of this stock.

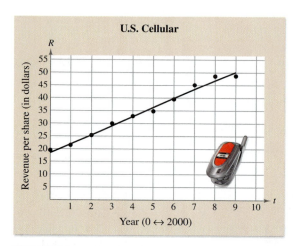

FIGURE 2.31

✓Checkpoint 8

From 2000 through 2009, the sales per share S (in dollars) for Dollar Tree can be modeled by

$$S = (0.010t^2 + 0.27t + 3.1)^2, \quad 0 \le t \le 9$$

where t is the year, with $t = 0$ corresponding to 2000. Use the model to approximate the rate of change of sales per share in 2005.   *(Source: Dollar Tree, Inc.)*

Review of Basic Differentiation Rules

You now have all the rules you need to differentiate *any* algebraic function. For your convenience, they are summarized below.

Summary of Basic Differentiation Rules

Let u and v be differentiable functions of x.

1. Constant Rule
$$\frac{d}{dx}[c] = 0, \quad c \text{ is a constant.}$$

2. Constant Multiple Rule
$$\frac{d}{dx}[cu] = c\frac{du}{dx}, \quad c \text{ is a constant.}$$

3. Sum and Difference Rules
$$\frac{d}{dx}[u \pm v] = \frac{du}{dx} \pm \frac{dv}{dx}$$

4. Product Rule
$$\frac{d}{dx}[uv] = u\frac{dv}{dx} + v\frac{du}{dx}$$

5. Quotient Rule
$$\frac{d}{dx}\left[\frac{u}{v}\right] = \frac{v\frac{du}{dx} - u\frac{dv}{dx}}{v^2}$$

6. Power Rules
$$\frac{d}{dx}[x^n] = nx^{n-1}$$

$$\frac{d}{dx}[u^n] = nu^{n-1}\frac{du}{dx}$$

7. Chain Rule
$$\frac{dy}{dx} = \frac{dy}{du} \cdot \frac{du}{dx}$$

SUMMARIZE (Section 2.5)

1. State the Chain Rule *(page 107)*. For an example of the Chain Rule, see Example 2.

2. State the General Power Rule *(page 109)*. For examples of the General Power Rule, see Examples 3, 4, and 5.

3. Describe a real-life example of how the General Power Rule can be used to analyze the rate of change of a company's revenue per share *(page 112, Example 8)*.

4. Use the Summary of Basic Differentiation Rules to identify the differentiation rules illustrated by (a)–(f) below *(page 113)*.

(a) $\dfrac{d}{dx}[2x] = 2\dfrac{d}{dx}[x]$ (b) $\dfrac{d}{dx}[x^4] = 4x^3$

(c) $\dfrac{d}{dx}[8] = 0$ (d) $\dfrac{d}{dx}[x^2 + x] = \dfrac{d}{dx}[x^2] + \dfrac{d}{dx}[x]$

(e) $\dfrac{d}{dx}[x - x^3] = \dfrac{d}{dx}[x] - \dfrac{d}{dx}[x^3]$

(f) $\dfrac{d}{dx}[x(x+1)] = (x)\dfrac{d}{dx}[x+1] + (x+1)\dfrac{d}{dx}[x]$

SKILLS WARM UP 2.5   The following warm-up exercises involve skills that were covered in a previous course. You will use these skills in the exercise set for this section. For additional help, review Appendix Sections A.3 and A.4.

In Exercises 1–6, rewrite the expression with rational exponents.

1. $\sqrt[5]{(1 - 5x)^2}$

2. $\sqrt[4]{(2x - 1)^3}$

3. $\dfrac{1}{\sqrt{4x^2 + 1}}$

4. $\dfrac{1}{\sqrt[3]{x - 6}}$

5. $\dfrac{\sqrt{x}}{\sqrt[3]{1 - 2x}}$

6. $\dfrac{\sqrt{(3 - 7x)^3}}{2x}$

In Exercises 7–10, factor the expression.

7. $3x^3 - 6x^2 + 5x - 10$

8. $5x\sqrt{x} - x - 5\sqrt{x} + 1$

9. $4(x^2 + 1)^2 - x(x^2 + 1)^3$

10. $-x^5 + 3x^3 + x^2 - 3$

ENHANCED WebAssign   Access end-of-section exercises online at **www.webassign.net**

2.6 Higher-Order Derivatives

■ Find higher-order derivatives.

■ Find and use a position function to determine the velocity and acceleration of a moving object.

Second, Third, and Higher-Order Derivatives

The "standard" derivative f' is often called the **first derivative** of f. The derivative of f' is the **second derivative** of f and is denoted by f''.

$$\frac{d}{dx}[f'(x)] = f''(x)$$ Second derivative

The derivative of f'' is the **third derivative** of f and is denoted by f'''.

$$\frac{d}{dx}[f''(x)] = f'''(x)$$ Third derivative

By continuing this process, you obtain **higher-order derivatives** of f. Higher-order derivatives are denoted as follows.

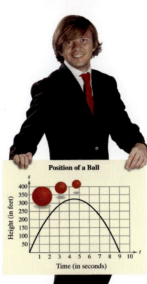

Position of a Ball

In Exercise 35, you will use derivatives to find the velocity function and the acceleration function of a ball.

Notation for Higher-Order Derivatives

1. 1st derivative:	y',	$f'(x)$,	$\dfrac{dy}{dx}$,	$\dfrac{d}{dx}[f(x)]$,	$D_x[y]$
2. 2nd derivative:	y'',	$f''(x)$,	$\dfrac{d^2y}{dx^2}$,	$\dfrac{d^2}{dx^2}[f(x)]$,	$D_x^2[y]$
3. 3rd derivative:	y''',	$f'''(x)$,	$\dfrac{d^3y}{dx^3}$,	$\dfrac{d^3}{dx^3}[f(x)]$,	$D_x^3[y]$
4. 4th derivative:	$y^{(4)}$,	$f^{(4)}(x)$,	$\dfrac{d^4y}{dx^4}$,	$\dfrac{d^4}{dx^4}[f(x)]$,	$D_x^4[y]$
5. nth derivative:	$y^{(n)}$,	$f^{(n)}(x)$,	$\dfrac{d^ny}{dx^n}$,	$\dfrac{d^n}{dx^n}[f(x)]$,	$D_x^n[y]$

Example 1   **Finding Higher-Order Derivatives**

Find the first five derivatives of

$$f(x) = 2x^4 - 3x^2.$$

SOLUTION

$f(x) = 2x^4 - 3x^2$	Write original function.
$f'(x) = 8x^3 - 6x$	First derivative
$f''(x) = 24x^2 - 6$	Second derivative
$f'''(x) = 48x$	Third derivative
$f^{(4)}(x) = 48$	Fourth derivative
$f^{(5)}(x) = 0$	Fifth derivative

✓**Checkpoint 1**

Find the first four derivatives of $f(x) = 6x^3 - 2x^2 + 1$.

Example 2   Finding Higher-Order Derivatives

Find the value of $g'''(2)$ for the function

$$g(t) = -t^4 + 2t^3 + t + 4.$$

SOLUTION   Begin by differentiating three times.

$$g'(t) = -4t^3 + 6t^2 + 1 \qquad \text{First derivative}$$
$$g''(t) = -12t^2 + 12t \qquad \text{Second derivative}$$
$$g'''(t) = -24t + 12 \qquad \text{Third derivative}$$

Then, evaluate the third derivative of g at $t = 2$.

$$g'''(2) = -24(2) + 12$$
$$= -36 \qquad \text{Value of third derivative}$$

✓ **Checkpoint 2**

Find the value of $g'''(1)$ for $g(x) = x^4 - x^3 + 2x$.   ■

Examples 1 and 2 show how to find higher-order derivatives of *polynomial* functions. Note that with each successive differentiation, the degree of the polynomial drops by one. Eventually, higher-order derivatives of polynomial functions degenerate to a constant function. Specifically, the *n*th-order derivative of an *n*th-degree polynomial function

$$f(x) = a_n x^n + a_{n-1} x^{n-1} + \cdots + a_1 x + a_0$$

is the constant function

$$f^{(n)}(x) = n! a_n$$

where $n! = 1 \cdot 2 \cdot 3 \cdots n$. Each derivative of order higher than n is the zero function.

TECH TUTOR

Higher-order derivatives of nonpolynomial functions can be difficult to find by hand. If you have access to a symbolic differentiation utility, try using it to find higher-order derivatives.

Example 3   Finding Higher-Order Derivatives

Find the first four derivatives of $y = x^{-1}$.

SOLUTION

$$y = x^{-1} = \frac{1}{x} \qquad \text{Write original function.}$$

$$y' = (-1)x^{-2} = -\frac{1}{x^2} \qquad \text{First derivative}$$

$$y'' = (-1)(-2)x^{-3} = \frac{2}{x^3} \qquad \text{Second derivative}$$

$$y''' = (-1)(-2)(-3)x^{-4} = -\frac{6}{x^4} \qquad \text{Third derivative}$$

$$y^{(4)} = (-1)(-2)(-3)(-4)x^{-5} = \frac{24}{x^5} \qquad \text{Fourth derivative}$$

✓ **Checkpoint 3**

Find the fourth derivative of

$$y = \frac{1}{x^2}.$$ ■

David Davis/Shutterstock.com

Acceleration

In Section 2.3, you saw that the velocity of a free-falling object (neglecting air resistance) is given by the derivative of its position function. In other words, the rate of change of the position with respect to time is defined to be the velocity. In a similar way, the rate of change of the velocity with respect to time is defined to be the **acceleration** of the object.

$$s = f(t)$$ Position function

$$\frac{ds}{dt} = f'(t)$$ Velocity function

$$\frac{d^2s}{dt^2} = f''(t)$$ Acceleration function

To find the position, velocity, or acceleration at a particular time t, substitute the given value of t into the appropriate function, as illustrated in Example 4.

 Example 4   **Finding Acceleration**

A ball is thrown upward from the top of a 160-foot cliff, as shown in Figure 2.32. The initial velocity of the ball is 48 feet per second, which implies that the position function is

$$s = -16t^2 + 48t + 160$$

where the time t is measured in seconds. Find the height, velocity, and acceleration of the ball at $t = 3$.

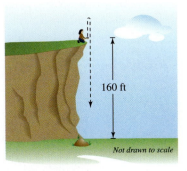

SOLUTION   Begin by differentiating to find the velocity and acceleration functions.

FIGURE 2.32

$$s = -16t^2 + 48t + 160$$ Position function

$$\frac{ds}{dt} = -32t + 48$$ Velocity function

$$\frac{d^2s}{dt^2} = -32$$ Acceleration function

To find the height, velocity, and acceleration at $t = 3$, substitute $t = 3$ into each of the functions above.

Height $= -16(3)^2 + 48(3) + 160 = 160$ feet

Velocity $= -32(3) + 48 = -48$ feet per second

Acceleration $= -32$ feet per second squared

✓**Checkpoint 4**

A ball is thrown upward from the top of an 80-foot cliff with an initial velocity of 64 feet per second, which implies that the position function is

$$s = -16t^2 + 64t + 80$$

where the time t is measured in seconds. Find the height, velocity, and acceleration of the ball at $t = 2$. ■

The acceleration due to gravity on the surface of the moon is only about one-sixth that exerted on the surface of Earth.

In Example 4, notice that the acceleration of the ball is -32 feet per second squared at any time t. This constant acceleration is due to the gravitational force of Earth and is called the **acceleration due to gravity.** Note that the negative value indicates that the ball is being pulled *down*—toward Earth.

Although the acceleration exerted on a falling object is relatively constant near Earth's surface, it varies greatly throughout our solar system. Large planets exert a much greater gravitational pull than do small planets or moons. The next example describes the motion of a free-falling object on the moon.

Example 5   Finding Acceleration on the Moon

An astronaut standing on the surface of the moon throws a rock upward. The height s (in feet) of the rock is given by

$$s = -\frac{27}{10}t^2 + 27t + 6$$

where t is measured in seconds. How does the acceleration due to gravity on the moon compare with that on Earth?

SOLUTION

$$s = -\frac{27}{10}t^2 + 27t + 6 \qquad \text{Position function}$$

$$\frac{ds}{dt} = -\frac{27}{5}t + 27 \qquad \text{Velocity function}$$

$$\frac{d^2s}{dt^2} = -\frac{27}{5} \qquad \text{Acceleration function}$$

So, the acceleration at any time is

$$-\frac{27}{5} = -5.4 \text{ feet per second squared}$$

—about one-sixth of the acceleration due to gravity on Earth. ──────

✓ Checkpoint 5

The position function on Earth, where s is measured in meters, t is measured in seconds, v_0 is the initial velocity in meters per second, and h_0 is the initial height in meters, is

$$s = -4.9t^2 + v_0t + h_0.$$

An object is thrown upward with an initial velocity of 2.2 meters per second from an initial height of 3.6 meters. What is the acceleration due to gravity on Earth in meters per second squared?   ■

The position function described in Example 5 neglects air resistance, which is appropriate because the moon has no atmosphere—and *no air resistance.* This means that the position function for any free-falling object on the moon is given by

$$s = -\frac{27}{10}t^2 + v_0t + h_0$$

where s is the height (in feet), t is the time (in seconds), v_0 is the initial velocity (in feet per second), and h_0 is the initial height (in feet). For instance, the rock in Example 5 was thrown upward with an initial velocity of 27 feet per second and had an initial height of 6 feet. This position function is valid for all objects, whether heavy ones such as hammers or light ones such as feathers.

Example 6   **Finding Velocity and Acceleration**

The velocity v (in feet per second) of a certain automobile starting from rest is

$$v = \frac{80t}{t + 5}$$ Velocity function

where t is the time (in seconds). The positions of the automobile at 10-second intervals are shown in Figure 2.33. Find the velocity and acceleration of the automobile at 10-second intervals from $t = 0$ to $t = 60$.

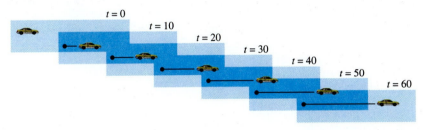

FIGURE 2.33

SOLUTION   To find the acceleration function, differentiate the velocity function.

$$\frac{dv}{dt} = \frac{(t + 5)(80) - (80t)(1)}{(t + 5)^2}$$ Apply Quotient Rule.

$$= \frac{400}{(t + 5)^2}$$ Acceleration function

t (seconds)	0	10	20	30	40	50	60
v (ft/sec)	0	53.3	64.0	68.6	71.1	72.7	73.8
$\frac{dv}{dt}$ (ft/sec²)	16	1.78	0.64	0.33	0.20	0.13	0.09

In the table, note that the acceleration approaches zero as the velocity levels off. This observation should agree with your experience—when riding in an accelerating automobile, you do not feel the velocity, but you do feel the acceleration. In other words, you feel changes in velocity.

✓**Checkpoint 6**

Use a graphing utility to graph the velocity function and acceleration function in Example 6 in the same viewing window. Compare the graphs with the table in Example 6. As the velocity levels off, what does the acceleration approach?   ■

SUMMARIZE   (Section 2.6)

1. State the meaning of each derivative listed below *(page 115)*. For examples of higher-order derivatives, see Examples 1, 2, and 3.

    (a) y'' (b) $f^{(4)}(x)$ (c) $\dfrac{d^3y}{dx^3}$

2. Describe a real-life example of how higher-order derivatives can be used to analyze the velocity and acceleration of an object *(page 117, Examples 4, 5, and 6)*.

SKILLS WARM UP 2.6

The following warm-up exercises involve skills that were covered in earlier sections. You will use these skills in the exercise set for this section. For additional help, review Sections 1.4 and 2.4.

In Exercises 1–4, solve the equation.

1. $-16t^2 + 24t = 0$
2. $-16t^2 + 80t + 224 = 0$
3. $-16t^2 + 128t + 320 = 0$
4. $-16t^2 + 9t + 1440 = 0$

In Exercises 5–8, find dy/dx.

5. $y = x^2(2x + 7)$
6. $y = (x^2 + 3x)(2x^2 - 5)$
7. $y = \dfrac{x^2}{2x + 7}$
8. $y = \dfrac{x^2 + 3x}{2x^2 - 5}$

In Exercises 9 and 10, find the domain and range of f.

9. $f(x) = x^2 - 4$
10. $f(x) = \sqrt{x - 7}$

ENHANCED
WebAssign   Access end-of-section exercises online at **www.webassign.net**

2.7 Implicit Differentiation

■ Find derivatives explicitly.
■ Find derivatives implicitly.
■ Use derivatives to answer questions about real-life situations.

Explicit and Implicit Functions

So far in this text, most functions involving two variables have been expressed in the **explicit form**

$$y = f(x). \qquad \text{Explicit form}$$

That is, one of the two variables has been explicitly given in terms of the other. For example, in the equation

$$y = 3x - 5$$

the variable y is explicitly written as a function of x. Some functions, however, are not given explicitly and are only implied by a given equation, as shown in Example 1.

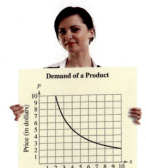

Demand of a Product

In Exercise 43, you will use implicit differentiation to find the rate of change for a demand function.

Example 1 Finding a Derivative Explicitly

Find dy/dx for the equation

$$xy = 1.$$

SOLUTION In this equation, y is **implicitly** defined as a function of x. One way to find dy/dx is first to solve the equation for y, then differentiate as usual.

$xy = 1$	Write original equation.
$y = \dfrac{1}{x}$	Solve for y.
$= x^{-1}$	Rewrite.
$\dfrac{dy}{dx} = -x^{-2}$	Differentiate with respect to x.
$= -\dfrac{1}{x^2}$	Simplify.

✓ Checkpoint 1

Find dy/dx for the equation

$$x^2 y = 1.$$

The procedure shown in Example 1 works well whenever you can easily write the given function explicitly. You cannot, however, use this procedure when you are unable to solve for y as a function of x. For instance, how would you find dy/dx for the equations

$$x^2 - 2y^3 + 4y = 2$$

and

$$x^2 + 2xy - y^3 = 5$$

where it is very difficult to express y as a function of x explicitly? To differentiate such equations, you can use a procedure called **implicit differentiation.**

Implicit Differentiation

To understand how to find dy/dx implicitly, you must realize that the differentiation is taking place *with respect to x*. This means that when you differentiate terms involving x alone, you can differentiate as usual. *But* when you differentiate terms involving y, you must apply the Chain Rule because you are assuming that y is defined implicitly as a differentiable function of x. Study the next example carefully. Note in particular how the Chain Rule is used to introduce the dy/dx factors in Examples 2(b) and 2(d).

Example 2   Applying the Chain Rule

Differentiate each expression with respect to x.

a. $3x^2$ **b.** $2y^3$

c. $x + 3y$ **d.** xy^2

SOLUTION

a. The only variable in this expression is x. So, to differentiate with respect to x, you can use the Simple Power Rule and the Constant Multiple Rule to obtain

$$\frac{d}{dx}[3x^2] = 6x.$$

b. This case is different. The variable in the expression is y, and yet you are asked to differentiate with respect to x. To do this, assume that y is a differentiable function of x and use the Chain Rule.

$$\frac{d}{dx}[2y^3] = \overbrace{2}^{c} \quad \underset{|}{(3)}^{\,n} \quad \overbrace{y^2}^{u^{n-1}} \quad \overbrace{\frac{dy}{dx}}^{u'} \qquad \text{Chain Rule}$$

$$= 6y^2\frac{dy}{dx}$$

c. This expression involves both x and y. By the Sum Rule and the Constant Multiple Rule, you can write

$$\frac{d}{dx}[x + 3y] = 1 + 3\frac{dy}{dx}.$$

d. By the Product Rule and the Chain Rule, you can write

$$\frac{d}{dx}[xy^2] = x\frac{d}{dx}[y^2] + y^2\frac{d}{dx}[x] \qquad \text{Product Rule}$$

$$= x\left(2y\frac{dy}{dx}\right) + y^2(1) \qquad \text{Chain Rule}$$

$$= 2xy\frac{dy}{dx} + y^2.$$

✓Checkpoint 2

Differentiate each expression with respect to x.

a. $4x^3$

b. $3y^2$

c. $x + 5y$

d. xy^3

Guidelines for Implicit Differentiation

Consider an equation involving x and y in which y is a differentiable function of x. You can use the steps below to find dy/dx.

1. Differentiate both sides of the equation *with respect to x*.

2. Write the result so that all terms involving dy/dx are on the left side of the equation and all other terms are on the right side of the equation.

3. Factor dy/dx out of the terms on the left side of the equation.

4. Solve for dy/dx by dividing both sides of the equation by the left-hand factor that does not contain dy/dx.

In Example 3, note that implicit differentiation can produce an expression for dy/dx that contains both x and y.

Example 3   Using Implicit Differentiation

Find dy/dx for the equation $y^3 + y^2 - 5y - x^2 = -4$.

SOLUTION

1. Differentiate both sides of the equation with respect to x.

$$\frac{d}{dx}[y^3 + y^2 - 5y - x^2] = \frac{d}{dx}[-4]$$

$$\frac{d}{dx}[y^3] + \frac{d}{dx}[y^2] - \frac{d}{dx}[5y] - \frac{d}{dx}[x^2] = \frac{d}{dx}[-4]$$

$$3y^2\frac{dy}{dx} + 2y\frac{dy}{dx} - 5\frac{dy}{dx} - 2x = 0$$

2. Collect the dy/dx terms on the left side of the equation and move all other terms to the right side of the equation.

$$3y^2\frac{dy}{dx} + 2y\frac{dy}{dx} - 5\frac{dy}{dx} = 2x$$

3. Factor dy/dx out of the left side of the equation.

$$\frac{dy}{dx}(3y^2 + 2y - 5) = 2x$$

4. Solve for dy/dx by dividing by $(3y^2 + 2y - 5)$.

$$\frac{dy}{dx} = \frac{2x}{3y^2 + 2y - 5}$$

✓ Checkpoint 3

Find dy/dx for the equation $y^2 + x^2 - 2y - 4x = 4$.

To see how you can use an implicit derivative, consider the graph shown in Figure 2.34. The derivative found in Example 3 gives a formula for the slope of the tangent line at a point on this graph. For instance, the slope at the point $(1, -3)$ is

$$\frac{dy}{dx} = \frac{2(1)}{3(-3)^2 + 2(-3) - 5} = \frac{1}{8}.$$

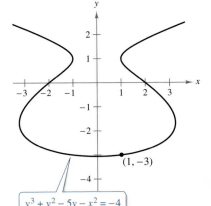

$y^3 + y^2 - 5y - x^2 = -4$

FIGURE 2.34

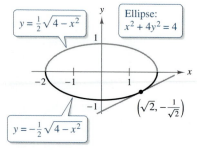

$y = \frac{1}{2}\sqrt{4 - x^2}$

Ellipse: $x^2 + 4y^2 = 4$

$y = -\frac{1}{2}\sqrt{4 - x^2}$

$\left(\sqrt{2}, -\frac{1}{\sqrt{2}}\right)$

Slope of tangent line is $\frac{1}{2}$.

FIGURE 2.35

STUDY TIP

To see the benefit of implicit differentiation, try reworking Example 4 using the explicit function

$$y = -\frac{1}{2}\sqrt{4 - x^2}.$$

The graph of this function is the lower half of the ellipse.

Example 4   Finding the Slope of a Graph Implicitly

Find the slope of the tangent line to the ellipse given by $x^2 + 4y^2 = 4$ at the point $\left(\sqrt{2}, -1/\sqrt{2}\right)$, as shown in Figure 2.35.

SOLUTION

$x^2 + 4y^2 = 4$	Write original equation.
$\dfrac{d}{dx}[x^2 + 4y^2] = \dfrac{d}{dx}[4]$	Differentiate with respect to x.
$2x + 8y\dfrac{dy}{dx} = 0$	Implicit differentiation
$8y\dfrac{dy}{dx} = -2x$	Subtract $2x$ from each side.
$\dfrac{dy}{dx} = \dfrac{-2x}{8y}$	Divide each side by $8y$.
$\dfrac{dy}{dx} = -\dfrac{x}{4y}$	Simplify.

To find the slope at the given point, substitute $x = \sqrt{2}$ and $y = -1/\sqrt{2}$ into the derivative, as shown below.

$$\frac{dy}{dx} = -\frac{\sqrt{2}}{4\left(-1/\sqrt{2}\right)}$$

$$= \frac{1}{2}$$

✓**Checkpoint 4**

Find the slope of the tangent line to the circle $x^2 + y^2 = 25$ at the point $(3, -4)$. ■

Example 5   Finding the Slope of a Graph Implicitly

Find the slope of the graph of $2x^2 - y^2 = 1$ at the point $(1, 1)$.

SOLUTION   Begin by finding dy/dx implicitly.

$2x^2 - y^2 = 1$	Write original equation.
$4x - 2y\dfrac{dy}{dx} = 0$	Differentiate with respect to x.
$-2y\dfrac{dy}{dx} = -4x$	Subtract $4x$ from each side.
$\dfrac{dy}{dx} = \dfrac{2x}{y}$	Divide each side by $-2y$.

At the point $(1, 1)$, the slope of the graph is

$$\frac{dy}{dx} = \frac{2(1)}{1}$$

$$= 2$$

$2x^2 - y^2 = 1$

Hyperbola

FIGURE 2.36

as shown in Figure 2.36. The graph is called a **hyperbola.**

✓**Checkpoint 5**

Find the slope of the graph of $x^2 - 9y^2 = 16$ at the point $(5, 1)$. ■

Application

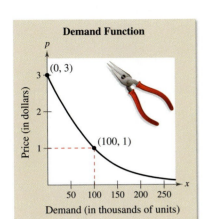

Demand Function

FIGURE 2.37

Example 6   Using a Demand Function

The demand function for a product is modeled by

$$p = \frac{3}{0.000001x^3 + 0.01x + 1}$$

where p is measured in dollars and x is measured in thousands of units, as shown in Figure 2.37. Find the rate of change of the demand x with respect to the price p when $x = 100$.

SOLUTION   To simplify the differentiation, begin by rewriting the function. Then, differentiate *with respect to p*.

$$p = \frac{3}{0.000001x^3 + 0.01x + 1}$$

$$0.000001x^3 + 0.01x + 1 = \frac{3}{p}$$

$$0.000003x^2 \frac{dx}{dp} + 0.01\frac{dx}{dp} = -\frac{3}{p^2}$$

$$(0.000003x^2 + 0.01)\frac{dx}{dp} = -\frac{3}{p^2}$$

$$\frac{dx}{dp} = -\frac{3}{p^2(0.000003x^2 + 0.01)}$$

When $x = 100$, the price is

$$p = \frac{3}{0.000001(100)^3 + 0.01(100) + 1} = \$1.$$

So, when $x = 100$ and $p = 1$, the rate of change of the demand with respect to the price is

$$\frac{dx}{dp} = -\frac{3}{(1)^2[0.000003(100)^2 + 0.01]} = -75.$$

This means that when $x = 100$, the demand is dropping at the rate of 75 thousand units for each dollar increase in price.

✓**Checkpoint 6**

The demand function for a product is given by

$$p = \frac{2}{0.001x^2 + x + 1}.$$

Find dx/dp implicitly.

SUMMARIZE   (Section 2.7)

1. State the guidelines for implicit differentiation *(page 123)*. For examples of implicit differentiation, see Examples 2, 3, 4, and 5.

2. Describe a real-life example of how implicit differentiation can be used to analyze the rate of change of a product's demand *(page 125, Example 6)*.

SKILLS WARM UP 2.7 The following warm-up exercises involve skills that were covered in a previous course. You will use these skills in the exercise set for this section. For additional help, review Appendix Section A.3.

In Exercises 1–6, solve the equation for y.

1. $x - \dfrac{y}{x} = 2$

2. $\dfrac{4}{x - 3} = \dfrac{1}{y}$

3. $xy - x + 6y = 6$

4. $12 + 3y = 4x^2 + x^2y$

5. $x^2 + y^2 = 5$

6. $x = \pm\sqrt{6 - y^2}$

In Exercises 7–9, evaluate the expression at the given point.

7. $\dfrac{3x^2 - 4}{3y^2}, \quad (2, 1)$

8. $\dfrac{x^2 - 2}{1 - y}, \quad (0, -3)$

9. $\dfrac{5x}{3y^2 - 12y + 5}, \quad (-1, 2)$

ENHANCED
WebAssign Access end-of-section exercises online at **www.webassign.net**

2.8 Related Rates

■ Examine related variables.
■ Solve related-rate problems.

Related Variables

In this section, you will study problems involving variables that are changing with respect to time. If two or more such variables are related to each other, then their rates of change with respect to time are also related.

For instance, suppose that x and y are related by the equation

$$y = 2x.$$

If both variables are changing with respect to time, then their rates of change will also be related.

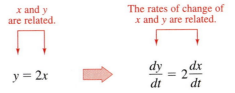

In this simple example, you can see that because y always has twice the value of x, it follows that the rate of change of y with respect to time is always twice the rate of change of x with respect to time.

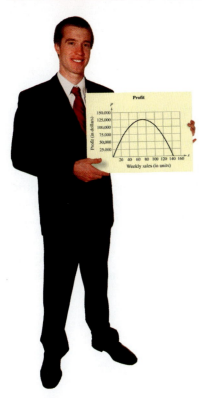

In Exercise 25, you will use related rates to find the rate of change of the sales for a product.

Example 1   Examining Two Rates That Are Related

The variables x and y are differentiable functions of t and are related by the equation

$$y = x^2 + 3.$$

When $x = 1$, $dx/dt = 2$. Find dy/dt when $x = 1$.

SOLUTION   Use the Chain Rule to differentiate both sides of the equation with respect to t.

$$y = x^2 + 3 \qquad \text{Write original equation.}$$

$$\frac{d}{dt}[y] = \frac{d}{dt}[x^2 + 3] \qquad \text{Differentiate with respect to } t.$$

$$\frac{dy}{dt} = 2x\frac{dx}{dt} \qquad \text{Apply Chain Rule.}$$

When $x = 1$ and $dx/dt = 2$, you have

$$\frac{dy}{dt} = 2(1)(2)$$

$$= 4.$$

✓Checkpoint 1

The variables x and y are differentiable functions of t and are related by the equation

$$y = x^3 + 2.$$

When $x = 1$, $dx/dt = 3$. Find dy/dt when $x = 1$.

Solving Related-Rate Problems

In Example 1, you were *given* the mathematical model.

Given equation: $y = x^2 + 3$

Given rate: $\dfrac{dx}{dt} = 2$ at $x = 1$

Find: $\dfrac{dy}{dt}$ at $x = 1$

In the next example, you must *create* a mathematical model from a verbal description.

Example 2 Changing Area

A pebble is dropped into a calm pool of water, causing ripples in the form of concentric circles, as shown in the photo. The radius r of the outer ripple is increasing at a constant rate of 1 foot per second. When the radius is 4 feet, at what rate is the total area A of the disturbed water changing?

SOLUTION   The variables r and A are related by the equation for the area of a circle, $A = \pi r^2$. To solve this problem, use the fact that the rate of change of the radius is given by dr/dt.

Equation: $A = \pi r^2$

Given rate: $\dfrac{dr}{dt} = 1$ at $r = 4$

Find: $\dfrac{dA}{dt}$ at $r = 4$

Using this model, you can proceed as in Example 1.

$A = \pi r^2$	Write original equation.
$\dfrac{d}{dt}[A] = \dfrac{d}{dt}[\pi r^2]$	Differentiate with respect to t.
$\dfrac{dA}{dt} = 2\pi r \dfrac{dr}{dt}$	Apply Chain Rule.

When $r = 4$ and $dr/dt = 1$, you have

$$\dfrac{dA}{dt} = 2\pi(4)(1) = 8\pi \qquad \text{Substitute 4 for } r \text{ and 1 for } dr/dt.$$

When the radius is 4 feet, the area is changing at a rate of 8π square feet per second.

✓ Checkpoint 2

As in Example 2, a pebble is dropped into the pool, but this time the radius r of the outer ripple is increasing at a rate of 2 feet per second. At what rate is the total area changing when the radius is 3 feet? ■

In Example 2, note that the radius changes at a *constant* rate ($dr/dt = 1$ for all t), but the area changes at a *nonconstant* rate.

When $r = 1$ ft	When $r = 2$ ft	When $r = 3$ ft	When $r = 4$ ft
$\dfrac{dA}{dt} = 2\pi$ ft²/sec	$\dfrac{dA}{dt} = 4\pi$ ft²/sec	$\dfrac{dA}{dt} = 6\pi$ ft²/sec	$\dfrac{dA}{dt} = 8\pi$ ft²/sec

Total area increases as the outer radius increases.

The solution shown in Example 2 illustrates the steps for solving a related-rate problem.

Guidelines for Solving a Related-Rate Problem

1. Identify all *given* quantities and all quantities *to be determined*. If possible, make a sketch and label the quantities.

2. Write an equation that relates all variables whose rates of change are either given or to be determined.

3. Use the Chain Rule to differentiate both sides of the equation *with respect to time*.

4. *After* completing Step 3, substitute into the resulting equation all known values for the variables and their rates of change. Then solve for the required rate of change.

STUDY TIP

Be sure you notice the order of Steps 3 and 4 in the guidelines. Do not substitute the known values for the variables until after you have differentiated.

In Step 2 of the guidelines, note that you must write an equation that relates the given variables. To help you with this step, reference tables that summarize many common formulas are included in the appendices. For instance, the volume of a sphere of radius r is given by the formula

$$V = \frac{4}{3}\pi r^3$$

as listed in Appendix D.

The table below lists examples of the mathematical models for some common rates of change that can be used in the first step of the solution of a related-rate problem.

Verbal statement	Mathematical model
The velocity of a car after traveling for 1 hour is 50 miles per hour.	x = distance traveled $\frac{dx}{dt} = 50$ when $t = 1$
Water is being pumped into a swimming pool at a rate of 10 cubic feet per minute.	V = volume of water in pool $\frac{dV}{dt} = 10$ ft³/min
A population of bacteria is increasing at a rate of 2000 per hour.	x = number in population $\frac{dx}{dt} = 2000$ bacteria per hour
Revenue is increasing at a rate of $4000 per month.	R = revenue $\frac{dR}{dt} = 4000$ dollars per month
Profit is decreasing at a rate of $2500 per day.	P = profit $\frac{dP}{dt} = -2500$ dollars per day

Example 3   Analyzing a Profit Function

A company's profit P (in dollars) from selling x units of a product can be modeled by

$$P = 500x - \frac{1}{4}x^2. \qquad \text{Model for profit}$$

The sales are increasing at a rate of 10 units per day. Find the rate of change in the profit (in dollars per day) when 500 units have been sold.

SOLUTION   Because the sales are increasing at a rate of 10 units per day, you know that at time t the rate of change is $dx/dt = 10$. So, the problem can be stated as shown.

Given rate: $\dfrac{dx}{dt} = 10$

Find: $\dfrac{dP}{dt}$ when $x = 500$

To find the rate of change of the profit, use the model for profit that relates the profit P and the units of the product sold x.

Equation: $P = 500x - \dfrac{1}{4}x^2$

By differentiating both sides of the equation with respect to t, you obtain

$$\frac{d}{dt}[P] = \frac{d}{dt}\left[500x - \frac{1}{4}x^2\right] \qquad \text{Differentiate with respect to } t.$$

$$\frac{dP}{dt} = \left(500 - \frac{1}{2}x\right)\frac{dx}{dt}. \qquad \text{Apply Chain Rule.}$$

When $x = 500$ units and $dx/dt = 10$, the rate of change in the profit is

$$\frac{dP}{dt} = \left[500 - \frac{1}{2}(500)\right](10) = (500 - 250)(10) = 250(10) = \$2500 \text{ per day.}$$

The graph of the profit function (in terms of x) is shown in Figure 2.38.

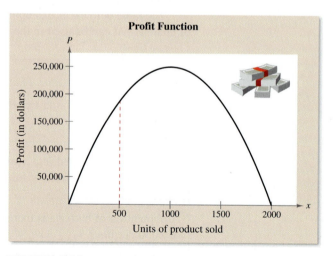

FIGURE 2.38

✓ **Checkpoint 3**

Find the rate of change in profit (in dollars per day) when 50 units have been sold, sales have increased at a rate of 10 units per day, and $P = 200x - \frac{1}{2}x^2$. ∎

Example 4   Increasing Production

A company is increasing the production of a product at the rate of 200 units per week. The weekly demand function is modeled by

$$p = 100 - 0.001x$$

where p is the price per unit and x is the number of units produced in a week. Find the rate of change of the revenue with respect to time when the weekly production is 2000 units. Will the rate of change of the revenue be greater than $20,000 per week?

SOLUTION   Because production is increasing at a rate of 200 units per week, you know that at time t the rate of change is $dx/dt = 200$. So, the problem can be stated as shown.

$$\text{Given rate: } \frac{dx}{dt} = 200$$

$$\text{Find: } \frac{dR}{dt} \text{ when } x = 2000$$

To find the rate of change of the revenue, you must find an equation that relates the revenue R and the number of units produced x.

$$\text{Equation: } R = xp = x(100 - 0.001x) = 100x - 0.001x^2$$

By differentiating both sides of the equation with respect to t, you obtain

$$R = 100x - 0.001x^2 \qquad \text{Write original equation.}$$

$$\frac{d}{dt}[R] = \frac{d}{dt}[100x - 0.001x^2] \qquad \text{Differentiate with respect to } t.$$

$$\frac{dR}{dt} = (100 - 0.002x)\frac{dx}{dt}. \qquad \text{Apply Chain Rule.}$$

Using $x = 2000$ and $dx/dt = 200$, you have

$$\frac{dR}{dt} = [100 - 0.002(2000)](200)$$

$$= \$19,200 \text{ per week.}$$

No, the rate of change of the revenue will not be greater than $20,000 per week.

✓ **Checkpoint 4**

Find the rate of change of the revenue with respect to time for the company in Example 4 when the weekly demand function is

$$p = 150 - 0.002x.$$

SUMMARIZE   (Section 2.8)

1. Give a description of related variables (*page 127*). For an example of two related variables, see Example 1.

2. State the guidelines for solving a related-rate problem (*page 129*). For examples of solving related-rate problems, see Examples 2, 3, and 4.

3. Describe a real-life example of how related rates can be used to analyze the rate of change of a company's revenue (*page 131, Example 4*).

SKILLS WARM UP 2.8   The following warm-up exercises involve skills that were covered in earlier sections. You will use these skills in the exercise set for this section. For additional help, review Section 2.7.

In Exercises 1–6, write a formula for the given quantity.

1. Area of a circle

2. Volume of a sphere

3. Surface area of a cube

4. Volume of a cube

5. Volume of a cone

6. Area of a triangle

In Exercises 7–10, find dy/dx by implicit differentiation.

7. $x^2 + y^2 = 9$

8. $3xy - x^2 = 6$

9. $x^2 + 2y + xy = 12$

10. $x + xy^2 - y^2 = xy$

ENHANCED WebAssign   Access end-of-section exercises online at **www.webassign.net**

ALGEBRA TUTOR

Simplifying Algebraic Expressions

To be successful in using derivatives, you must be good at simplifying algebraic expressions. Here are some helpful simplification techniques.

1. Combine *like terms*. This may involve expanding an expression by multiplying factors.

2. Divide out *common factors* in the numerator and denominator of an expression.

3. Factor an expression.

4. Rationalize a denominator.

5. Add, subtract, multiply, or divide fractions.

Example 1   Simplifying Fractional Expressions

a. $\dfrac{[3(x + \Delta x) + 5] - (3x + 5)}{\Delta x} = \dfrac{3x + 3\Delta x + 5 - 3x - 5}{\Delta x}$ Multiply factors and remove parentheses.

$= \dfrac{3\Delta x}{\Delta x}$ Combine like terms.

$= 3, \quad \Delta x \neq 0$ Divide out common factors.

b. $\dfrac{(x + \Delta x)^2 - x^2}{\Delta x} = \dfrac{x^2 + 2x(\Delta x) + (\Delta x)^2 - x^2}{\Delta x}$ Expand terms.

$= \dfrac{2x(\Delta x) + (\Delta x)^2}{\Delta x}$ Combine like terms.

$= \dfrac{\Delta x(2x + \Delta x)}{\Delta x}$ Factor.

$= 2x + \Delta x, \quad \Delta x \neq 0$ Divide out common factors.

c. $\dfrac{(x^2 - 1)(-2 - 2x) - (3 - 2x - x^2)(2)}{(x^2 - 1)^2}$

$= \dfrac{(-2x^2 - 2x^3 + 2 + 2x) - (6 - 4x - 2x^2)}{(x^2 - 1)^2}$ Expand terms.

$= \dfrac{-2x^2 - 2x^3 + 2 + 2x - 6 + 4x + 2x^2}{(x^2 - 1)^2}$ Remove parentheses.

$= \dfrac{-2x^3 + 6x - 4}{(x^2 - 1)^2}$ Combine like terms.

d. $2\left(\dfrac{2x + 1}{3x}\right)\left[\dfrac{3x(2) - (2x + 1)(3)}{(3x)^2}\right]$

$= 2\left(\dfrac{2x + 1}{3x}\right)\left[\dfrac{6x - (6x + 3)}{(3x)^2}\right]$ Multiply factors.

$= \dfrac{2(2x + 1)(6x - 6x - 3)}{(3x)^3}$ Multiply fractions and remove parentheses.

$= \dfrac{2(2x + 1)(-3)}{3(9)x^3}$ Combine like terms and factor.

$= \dfrac{-2(2x + 1)}{9x^3}$ Divide out common factors.

Example 2   **Simplifying Expressions with Powers**

Simplify each expression.

a. $(2x + 1)^2(6x + 1) + (3x^2 + x)(2)(2x + 1)(2)$

b. $(-1)(3x^2 - 2x)^{-2}(6x - 2)$

c. $(x)\left(\frac{1}{2}\right)(2x + 3)^{-1/2} + (2x + 3)^{1/2}(1)$

d. $\dfrac{x^2\left(\frac{1}{2}\right)(x^2 + 1)^{-1/2}(2x) - (x^2 + 1)^{1/2}(2x)}{x^4}$

SOLUTION

a. $(2x + 1)^2(6x + 1) + (3x^2 + x)(2)(2x + 1)(2)$

$$= (2x + 1)[(2x + 1)(6x + 1) + (3x^2 + x)(2)(2)] \qquad \text{Factor.}$$
$$= (2x + 1)[12x^2 + 8x + 1 + (12x^2 + 4x)] \qquad \text{Multiply factors.}$$
$$= (2x + 1)(12x^2 + 8x + 1 + 12x^2 + 4x) \qquad \text{Remove parentheses.}$$
$$= (2x + 1)(24x^2 + 12x + 1) \qquad \text{Combine like terms.}$$

b. $(-1)(3x^2 - 2x)^{-2}(6x - 2)$

$$= \frac{(-1)(6x - 2)}{(3x^2 - 2x)^2} \qquad \text{Rewrite as a fraction.}$$
$$= \frac{(-1)(2)(3x - 1)}{(3x^2 - 2x)^2} \qquad \text{Factor.}$$
$$= \frac{-2(3x - 1)}{(3x^2 - 2x)^2} \qquad \text{Multiply factors.}$$

c. $(x)\left(\frac{1}{2}\right)(2x + 3)^{-1/2} + (2x + 3)^{1/2}(1)$

$$= (2x + 3)^{-1/2}\left(\tfrac{1}{2}\right)[x + (2x + 3)(2)] \qquad \text{Factor.}$$
$$= \frac{x + 4x + 6}{(2x + 3)^{1/2}(2)} \qquad \text{Rewrite as a fraction.}$$
$$= \frac{5x + 6}{2(2x + 3)^{1/2}} \qquad \text{Combine like terms.}$$

d. $\dfrac{x^2\left(\frac{1}{2}\right)(x^2 + 1)^{-1/2}(2x) - (x^2 + 1)^{1/2}(2x)}{x^4}$

$$= \frac{(x^3)(x^2 + 1)^{-1/2} - (x^2 + 1)^{1/2}(2x)}{x^4} \qquad \text{Multiply factors.}$$
$$= \frac{(x^2 + 1)^{-1/2}(x)[x^2 - (x^2 + 1)(2)]}{x^4} \qquad \text{Factor.}$$
$$= \frac{x[x^2 - (2x^2 + 2)]}{(x^2 + 1)^{1/2}x^4} \qquad \text{Write with positive exponents.}$$
$$= \frac{x^2 - 2x^2 - 2}{(x^2 + 1)^{1/2}x^3} \qquad \text{Divide out common factors and remove parentheses.}$$
$$= \frac{-x^2 - 2}{(x^2 + 1)^{1/2}x^3} \qquad \text{Combine like terms.}$$

STUDY TIP

All but one of the expressions in this Algebra Tutor are derivatives. Can you see what the original function is? Explain your reasoning.

SUMMARY AND STUDY STRATEGIES

After studying this chapter, you should have acquired the following skills.
The exercise numbers are keyed to the Review Exercises that begin on page 137.
Answers to odd-numbered Review Exercises are given in the back of the text.*

Section 2.1   Review Exercises

- Approximate the slope of the tangent line to a graph at a point.   *1–4*
- Interpret the slope of a graph in a real-life setting.   *5–8*
- Use the limit definition to find the slope of a graph at a point and the derivative   *9–24*
  of a function.

$$f'(x) = \lim_{\Delta x \to 0} \frac{f(x + \Delta x) - f(x)}{\Delta x}$$

- Use the graph of a function to recognize points at which the function is   *25–28*
  not differentiable.

Section 2.2

- Use the Constant Rule for differentiation.   *29, 30*

$$\frac{d}{dx}[c] = 0$$

- Use the Power Rule for differentiation.   *31, 32*

$$\frac{d}{dx}[x^n] = nx^{n-1}$$

- Use the Constant Multiple Rule for differentiation.   *33–36*

$$\frac{d}{dx}[cf(x)] = cf'(x)$$

- Use the Sum and Difference Rules for differentiation.   *37–40*

$$\frac{d}{dx}[f(x) \pm g(x)] = f'(x) \pm g'(x)$$

- Use derivatives to find the slope of a graph.   *41–44*
- Use derivatives to write equations of tangent lines.   *45–48*
- Use derivatives to answer questions about real-life situations.   *49, 50*

Section 2.3

- Find the average rate of change of a function over an interval and the instantaneous   *51–54*
  rate of change at a point.

  Average rate of change: $\dfrac{f(b) - f(a)}{b - a}$; Instantaneous rate of change: $\lim\limits_{\Delta x \to 0} \dfrac{f(x + \Delta x) - f(x)}{\Delta x}$

- Use derivatives to find the velocities of objects.   *55, 56*
- Find the marginal revenues, marginal costs, and marginal profits for products.   *57–66*
- Use derivatives to answer questions about real-life situations.   *67, 68*

---

* A wide range of valuable study aids are available to help you master the material in this chapter.
  The *Student Solutions Manual* includes step-by-step solutions to all odd-numbered exercises to help
  you review and prepare. The student website at *www.cengagebrain.com* offers algebra help and a
  Graphing Technology Guide, which contains step-by-step commands and instructions for a wide
  variety of graphing calculators.

Section 2.4

■ Use the Product Rule for differentiation.

$$\frac{d}{dx}[f(x)g(x)] = f(x)g'(x) + g(x)f'(x)$$

■ Use the Quotient Rule for differentiation.

$$\frac{d}{dx}\left[\frac{f(x)}{g(x)}\right] = \frac{g(x)f'(x) - f(x)g'(x)}{[g(x)]^2}$$

Section 2.5

■ Use the General Power Rule for differentiation.

$$\frac{d}{dx}[u^n] = nu^{n-1}u'$$

■ Use differentiation rules efficiently to find the derivative of any algebraic function, then simplify the result.

■ Use derivatives to answer questions about real-life situations.   (Sections 2.1–2.5)

Section 2.6

■ Find higher-order derivatives.

■ Find and use a position function to determine the velocity and acceleration of a moving object.

Section 2.7

■ Find derivatives implicitly.

■ Use implicit differentiation to write equations of tangent lines.

Section 2.8

■ Solve related-rate problems.

Study Strategies

■ **Simplify Your Derivatives** You may ask if you have to simplify your derivatives. The answer is "Yes, if you expect to use them." In the next chapter, you will see that almost all applications of derivatives require that the derivatives be written in simplified form. It is not difficult to see the advantage of a derivative in simplified form. Consider, for instance, the derivative of

$$f(x) = \frac{x}{\sqrt{x^2 + 1}}.$$

The "raw form" produced by the Quotient and Chain Rules

$$f'(x) = \frac{(x^2 + 1)^{1/2}(1) - (x)\left(\frac{1}{2}\right)(x^2 + 1)^{-1/2}(2x)}{\left(\sqrt{x^2 + 1}\right)^2}$$

is obviously much more difficult to use than the simplified form

$$f'(x) = \frac{1}{(x^2 + 1)^{3/2}}.$$

■ **List Units of Measure in Applied Problems** When using derivatives in real-life applications, be sure to list the units of measure for each variable. For instance, if R is measured in dollars and t is measured in years, then the derivative dR/dt is measured in dollars per year.

Review Exercises

Approximating the Slope of a Graph In Exercises 1–4, estimate the slope of the graph at the point (x, y). (Each square on the grid is 1 unit by 1 unit.)

1.
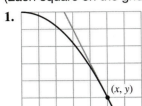

2.
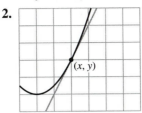

3.
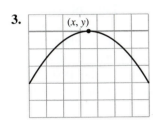

4.
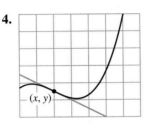

5. Sales The graph represents the sales S (in millions of dollars) for Tractor Supply Company for the years 2003 through 2009, where t represents the year, with $t = 3$ corresponding to 2003. Estimate and interpret the slopes of the graph for the years 2004 and 2007. *(Source: Tractor Supply Company)*

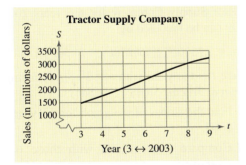

6. Farms The graph represents the amount of farm land L (in millions of acres) in the United States for the years 2004 through 2009, where t represents the year, with $t = 4$ corresponding to 2004. Estimate and interpret the slopes of the graph for the years 2005 and 2008. *(Source: U.S. Department of Agriculture)*

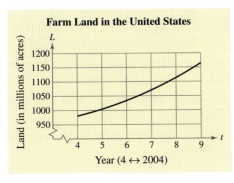

7. Consumer Trends The graph shows the number of visitors V (in thousands) to a national park during a one-year period, where $t = 1$ corresponds to January. Estimate and interpret the slopes of the graph at $t = 1, 8$, and 12.

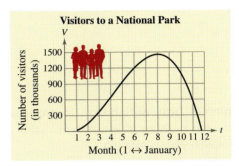

8. White-Water Rafting Two white-water rafters leave a campsite simultaneously and start downstream on a 9-mile trip. Their distances from the campsite are given by $s = f(t)$ and $s = g(t)$, where s is measured in miles and t is measured in hours.

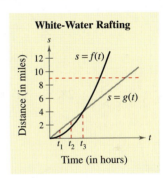

(a) Which rafter is traveling at a greater rate at t_1?

(b) What can you conclude about their rates at t_2?

(c) What can you conclude about their rates at t_3?

(d) Which rafter finishes the trip first? Explain your reasoning.

Finding the Slope of a Graph In Exercises 9–16, use the limit definition to find the slope of the tangent line to the graph of f at the given point.

9. $f(x) = -3x - 5; (-2, 1)$

10. $f(x) = 7x + 3; (-1, -4)$

11. $f(x) = x^2 - 4x; (1, -3)$

12. $f(x) = x^2 + 10; (2, 14)$

13. $f(x) = \sqrt{x + 9}; (-5, 2)$

14. $f(x) = \sqrt{x - 1}; (10, 3)$

15. $f(x) = \dfrac{1}{x - 5}; (6, 1)$ **16.** $f(x) = \dfrac{1}{x + 4}; (-3, 1)$

Finding a Derivative In Exercises 17–24, use the limit definition to find the derivative of the function.

17. $f(x) = 9x + 1$

18. $f(x) = 1 - 4x$

19. $f(x) = -\frac{1}{2}x^2 + 2x$

20. $f(x) = 4 - x^2$

21. $f(x) = \sqrt{x - 5}$

22. $f(x) = \sqrt{x} + 3$

23. $f(x) = \frac{5}{x}$

24. $f(x) = \frac{1}{x + 4}$

Determining Differentiability In Exercises 25–28, determine the x-values at which the function is differentiable. Explain your reasoning.

25. $y = \dfrac{x + 1}{x - 1}$

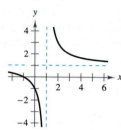

26. $y = -|x| + 3$

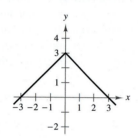

27. $y = \begin{cases} -x - 2, & x \leq 0 \\ x^3 + 2, & x > 0 \end{cases}$

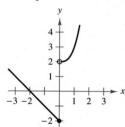

28. $y = (x + 1)^{2/3}$

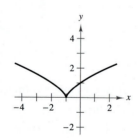

Finding Derivatives In Exercises 29–40, find the derivative of the function.

29. $y = -6$

30. $f(x) = 5$

31. $f(x) = x^3$

32. $h(x) = \dfrac{1}{x^4}$

33. $f(x) = 4x^2$

34. $g(t) = 6t^5$

35. $f(x) = \dfrac{2x^4}{5}$

36. $y = 3x^{2/3}$

37. $g(x) = 2x^4 + 3x^2$

38. $f(x) = 6x^2 - 4x$

39. $y = x^2 + 6x - 7$

40. $y = 2x^4 - 3x^3 + x$

Finding the Slope of a Graph In Exercises 41–44, find the slope of the graph of the function at the given point.

41. $f(x) = 2x^{-1/2}; \ (4, 1)$

42. $y = \dfrac{3}{2x} + 3; \ \left(\dfrac{1}{2}, 6\right)$

43. $g(x) = x^3 - 4x^2 - 6x + 8; \ (-1, 9)$

44. $y = 2x^4 - 5x^3 + 6x^2 - x; \ (1, 2)$

 Finding an Equation of a Tangent Line In Exercises 45–48, (a) find an equation of the tangent line to the graph of the function at the given point, (b) use a graphing utility to graph the function and its tangent line at the point, and (c) use the *derivative* feature of a graphing utility to confirm your results.

45. $f(x) = 2x^2 - 3x + 1; \ (2, 3)$

46. $y = 11x^4 - 5x^2 + 1; \ (-1, 7)$

47. $f(x) = \sqrt{x} - \dfrac{1}{\sqrt{x}}; \ (1, 0)$

48. $f(x) = -x^2 - 4x - 4; \ (-4, -4)$

49. Sales The annual sales S (in millions of dollars) for Tractor Supply Company for the years 2003 through 2009 can be modeled by

$$S = -0.7500t^4 + 13.278t^3 - 74.50t^2 + 440.2t + 523$$

where t is the year, with $t = 3$ corresponding to 2003. *(Source: Tractor Supply Company)*

(a) Find the slopes of the graph for the years 2004 and 2007.

(b) Compare your results with those obtained in Exercise 5.

(c) Interpret the slope of the graph in the context of the problem.

50. Farms The amount of farm land L (in millions of acres) in the United States for the years 2004 through 2009 can be modeled by

$$L = 0.0991t^3 + 1.512t^2 + 4.01t + 933.9$$

where t is the year, with $t = 4$ corresponding to 2004. *(Source: U.S. Department of Agriculture)*

(a) Find the slopes of the graph for the years 2005 and 2008.

(b) Compare your results with those obtained in Exercise 6.

(c) Interpret the slope of the graph in the context of the problem.

Finding Rates of Change In Exercises 51–54, use a graphing utility to graph the function and find its average rate of change over the interval. Compare this rate with the instantaneous rates of change at the endpoints of the interval.

51. $f(t) = 4t + 3; [-3, 1]$

52. $f(x) = x^{2/3}; [1, 8]$

53. $f(x) = x^2 + 3x - 4; [0, 1]$

54. $f(x) = x^3 + x; [-2, 2]$

55. Velocity The height s (in feet) at time t (in seconds) of a ball thrown upward from the top of a 300-foot building with an initial velocity of 24 feet per second is given by

$$s = -16t^2 + 24t + 300.$$

(a) Find the average velocity on the interval $[1, 2]$.

(b) Find the instantaneous velocities when $t = 1$ and $t = 3$.

(c) How long will it take the ball to hit the ground?

(d) Find the velocity of the ball when it hits the ground.

56. Velocity A rock is dropped from a tower on the Brooklyn Bridge, 276 feet above the East River. Let t represent the time in seconds.

(a) Find the position and velocity functions for the rock.

(b) Find the average velocity over the interval $[0, 2]$.

(c) Find the instantaneous velocities at $t = 2$ and $t = 3$.

(d) How long will it take the rock to hit the water?

(e) Find the velocity of the rock when it hits the water.

Marginal Cost In Exercises 57–60, find the marginal cost for producing x units. (The cost is measured in dollars.)

57. $C = 2500 + 320x$

58. $C = 3x^3 + 24,000$

59. $C = 370 + 2.55\sqrt{x}$

60. $C = 475 + 5.25x^{2/3}$

Marginal Revenue In Exercises 61–64, find the marginal revenue for producing x units. (The revenue is measured in dollars.)

61. $R = 150x - 0.6x^2$

62. $R = 150x - \frac{3}{4}x^2$

63. $R = -4x^3 + 2x^2 + 100x$

64. $R = 4x + 10\sqrt{x}$

Marginal Profit In Exercises 65 and 66, find the marginal profit for producing x units. (The profit is measured in dollars.)

65. $P = -0.0002x^3 + 6x^2 - x - 2000$

66. $P = -\frac{1}{15}x^3 + 4000x^2 - 120x - 144,000$

67. Marginal Profit The profit P (in dollars) from selling x units of a product is given by

$$P = -0.05x^2 + 20x - 1000.$$

(a) Find the additional profit when the sales increase from 100 to 101 units.

(b) Find the marginal profit when $x = 100$ units.

(c) Compare the results of parts (a) and (b).

68. Population Growth The population P (in millions) of Brazil from 1980 through 2010 can be modeled by

$$P = -0.007t^2 + 2.78t + 123.6$$

where t represents the year, with $t = 0$ corresponding to 1980. *(Source: U.S. Census Bureau)*

(a) Evaluate P for $t = 0$, 5, 10, 15, 20, 25, and 30. Explain these values.

(b) Determine the population growth rate, dP/dt.

(c) Evaluate dP/dt for the same values as in part (a). Explain your results.

Finding Derivatives In Exercises 69–90, find the derivative of the function. State which differentiation rule(s) you used to find the derivative.

69. $f(x) = x^3(5 - 3x^2)$

70. $y = (3x^2 + 7)(x^2 - 2x)$

71. $y = (4x - 3)(x^3 - 2x^2)$

72. $s = \left(4 - \frac{1}{t^2}\right)(t^2 - 3t)$

73. $g(x) = \dfrac{x}{x + 3}$

74. $f(x) = \dfrac{2 - 5x}{3x + 1}$

75. $f(x) = \dfrac{6x - 5}{x^2 + 1}$

76. $f(x) = \dfrac{x^2 + x - 1}{x^2 - 1}$

77. $f(x) = (5x^2 + 2)^3$

78. $f(x) = \sqrt[3]{x^2 - 1}$

79. $h(x) = \dfrac{2}{\sqrt{x + 1}}$

80. $g(x) = \sqrt{x^6 - 12x^3 + 9}$

81. $g(x) = x\sqrt{x^2 + 1}$ **82.** $g(t) = \dfrac{t}{(1 - t)^3}$

83. $f(x) = x(1 - 4x^2)^2$ **84.** $f(x) = \left(x^2 + \dfrac{1}{x}\right)^5$

85. $h(x) = [x^2(2x + 3)]^3$ **86.** $f(x) = [(x - 2)(x + 4)]^2$

87. $f(x) = x^2(x - 1)^5$ **88.** $f(s) = s^3(s^2 - 1)^{5/2}$

89. $h(t) = \dfrac{\sqrt{3t + 1}}{(1 - 3t)^2}$ **90.** $g(x) = \dfrac{(3x + 1)^2}{(x^2 + 1)^2}$

91. Physical Science The temperature T (in degrees Fahrenheit) of food placed in a freezer can be modeled by

$$T = \frac{1300}{t^2 + 2t + 25}$$

where t is the time (in hours).

(a) Find the rates of change of T at $t = 1, 3, 5,$ and 10.

(b) Graph the model on a graphing utility and describe the rate at which the temperature is changing.

92. Forestry According to the *Doyle Log Rule*, the volume V (in board-feet) of a log of length L (in feet) and diameter D (in inches) at the small end is

$$V = \left(\frac{D - 4}{4}\right)^2 L.$$

Find the rates at which the volume is changing with respect to D for a 12-foot-long log whose smallest diameter is (a) 8 inches, (b) 16 inches, (c) 24 inches, and (d) 36 inches.

Finding Higher-Order Derivatives In Exercises 93–100, find the higher-order derivative.

Given	Derivative
93. $f(x) = 3x^2 + 7x + 1$	$f''(x)$
94. $f'(x) = 5x^4 - 6x^2 + 2x$	$f'''(x)$
95. $f'''(x) = -\dfrac{6}{x^4}$	$f^{(5)}(x)$
96. $f(x) = \sqrt{x}$	$f^{(4)}(x)$
97. $f'(x) = 7x^{5/2}$	$f''(x)$
98. $f(x) = x^2 + \dfrac{3}{x}$	$f''(x)$
99. $f''(x) = 6\sqrt[3]{x}$	$f'''(x)$
100. $f'''(x) = 20x^4 - \dfrac{2}{x^3}$	$f^{(5)}(x)$

101. Athletics A person dives from a 30-foot platform with an initial velocity of 5 feet per second (upward).

(a) Find the position function of the diver.

(b) How long will it take the diver to hit the water?

(c) What is the diver's velocity at impact?

(d) What is the diver's acceleration at impact?

102. Velocity and Acceleration The position function of a particle is given by

$$s = \frac{1}{t^2 + 2t + 1}$$

where s is the height (in feet) and t is the time (in seconds). Find the velocity and acceleration functions.

Finding Derivatives In Exercises 103–106, use implicit differentiation to find dy/dx.

103. $x^2 + 3xy + y^3 = 10$

104. $x^2 + 9xy + y^2 = 0$

105. $y^2 - x^2 + 8x - 9y - 1 = 0$

106. $y^2 + x^2 - 6y - 2x - 5 = 0$

Finding an Equation of a Tangent Line In Exercises 107–110, use implicit differentiation to find an equation of the tangent line to the graph of the function at the given point.

	Equation	*Point*
107.	$y^2 = x - y$	$(2, 1)$
108.	$2\sqrt[3]{x} + 3\sqrt{y} = 10$	$(8, 4)$
109.	$y^2 - 2x = xy$	$(1, 2)$
110.	$y^3 - 2x^2y + 3xy^2 = -1$	$(0, -1)$

111. Area The radius r of a circle is increasing at a rate of 2 inches per minute. Find the rate of change of the area at (a) $r = 3$ inches and (b) $r = 10$ inches.

112. Moving Point A point is moving along the graph of $y = \sqrt{x}$ such that dx/dt is 3 centimeters per second. Find dy/dt for (a) $x = \frac{1}{4}$, (b) $x = 1$, and (c) $x = 4$.

113. Water Level A swimming pool is 40 feet long, 20 feet wide, 4 feet deep at the shallow end, and 9 feet deep at the deep end (see figure). Water is being pumped into the pool at the rate of 10 cubic feet per minute. How fast is the water level rising when there is 4 feet of water in the deep end?

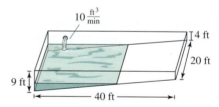

114. Profit The demand and cost functions for a product can be modeled by

$$p = 211 - 0.002x$$

and

$$C = 30x + 1{,}500{,}000$$

where x is the number of units produced.

(a) Write the profit function for this product.

(b) Find the marginal profit when 80,000 units are produced.

(c) Graph the profit function on a graphing utility and use the graph to determine the price you would charge for the product. Explain your reasoning.

TEST YOURSELF

Take this test as you would take a test in class. When you are done, check your work against the answers given in the back of the book.

In Exercises 1 and 2, use the limit definition to find the derivative of the function. Then find the slope of the tangent line to the graph of *f* at the given point.

1. $f(x) = x^2 + 1; (2, 5)$ **2.** $f(x) = \sqrt{x} - 2; (4, 0)$

In Exercises 3–11, find the derivative of the function.

3. $f(t) = t^3 + 2t$ **4.** $f(x) = 4x^2 - 8x + 1$ **5.** $f(x) = x^{3/2}$

6. $f(x) = (x + 3)(x^2 + 2x)$ **7.** $f(x) = -3x^{-3}$ **8.** $f(x) = \sqrt{x}(5 + x)$

9. $f(x) = (3x^2 + 4)^2$ **10.** $f(x) = \sqrt{1 - 2x}$ **11.** $f(x) = \dfrac{(5x - 1)^3}{x}$

12. Find an equation of the tangent line to the graph of

$$f(x) = x - \frac{1}{x}$$

at the point $(1, 0)$. Then use a graphing utility to graph the function and the tangent line in the same viewing window.

13. The annual sales *S* (in billions of dollars) of CVS Caremark for the years 2004 through 2009 can be modeled by

$$S = -1.3241t^3 + 26.562t^2 - 155.81t + 314.3$$

where *t* represents the year, with $t = 4$ corresponding to 2004. *(Source: CVS Caremark Corporation)*

(a) Approximate the average rate of change for the years from 2005 through 2008.

(b) Find the instantaneous rates of change of the model for the years 2005 and 2008.

(c) Interpret the results of parts (a) and (b) in the context of the problem.

14. The monthly demand and cost functions for a product are given by

$$p = 1700 - 0.016x \quad \text{and} \quad C = 715{,}000 + 240x.$$

(a) Write the profit function for this product.

(b) Find the rate of change of the profit when the monthly sales are $x = 700$ units.

In Exercises 15–17, find the third derivative of the function.

15. $f(x) = 2x^2 + 3x + 1$ **16.** $f(x) = \sqrt{3 - x}$ **17.** $f(x) = \dfrac{2x + 1}{2x - 1}$

18. A ball is thrown straight upward from a height of 75 feet above the ground with an initial velocity of 30 feet per second. Write the position, velocity, and acceleration functions of the ball. Find the height, velocity, and acceleration when $t = 2$.

In Exercises 19–21, use implicit differentiation to find dy/dx.

19. $x + xy = 6$ **20.** $y^2 + 2x - 2y + 1 = 0$ **21.** $x^2 - 2y^2 = 4$

22. The radius *r* of a right circular cylinder is increasing at a rate of 0.25 centimeter per minute. The height *h* of the cylinder is related to the radius by $h = 20r$. Find the rate of change of the volume when (a) $r = 0.5$ centimeter and (b) $r = 1$ centimeter.

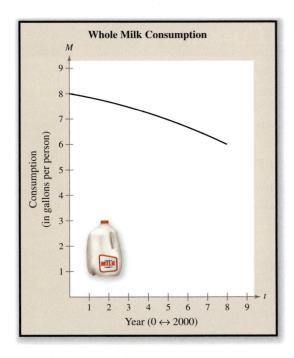

Whole Milk Consumption

Consumption (in gallons per person) vs. Year (0 ↔ 2000)

Example 2 on page 145 shows how the derivative can be used to show that milk consumption decreased in the United States from 2000 through 2008.

3 Applications of the Derivative

3.1 Increasing and Decreasing Functions

■ Test for increasing and decreasing functions.
■ Find the critical numbers of functions and find the open intervals on which functions are increasing or decreasing.
■ Use increasing and decreasing functions to model and solve real-life problems.

Increasing and Decreasing Functions

A function is **increasing** when its graph moves up as x moves to the right and **decreasing** when its graph moves down as x moves to the right. The following definition states this more formally.

In Exercise 47, you will use derivatives and critical numbers to find the intervals on which the profit from selling popcorn is increasing and decreasing.

Definition of Increasing and Decreasing Functions

A function f is **increasing** on an interval when, for any two numbers x_1 and x_2 in the interval,

$$x_2 > x_1 \quad \text{implies} \quad f(x_2) > f(x_1).$$

A function f is **decreasing** on an interval when, for any two numbers x_1 and x_2 in the interval,

$$x_2 > x_1 \quad \text{implies} \quad f(x_2) < f(x_1).$$

The function in Figure 3.1 is decreasing on the interval $(-\infty, a)$, constant on the interval (a, b), and increasing on the interval (b, ∞). Actually, from the definition of increasing and decreasing functions, the function shown in Figure 3.1 is decreasing on the interval $(-\infty, a]$ and increasing on the interval $[b, \infty)$. This text restricts the discussion to finding *open* intervals on which a function is increasing or decreasing.

The derivative of a function can be used to determine whether the function is increasing or decreasing on an interval.

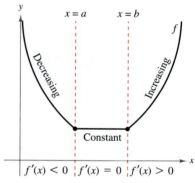

FIGURE 3.1

Test for Increasing and Decreasing Functions

Let f be differentiable on the interval (a, b).

1. If $f'(x) > 0$ for all x in (a, b), then f is increasing on (a, b).

2. If $f'(x) < 0$ for all x in (a, b), then f is decreasing on (a, b).

3. If $f'(x) = 0$ for all x in (a, b), then f is constant on (a, b).

STUDY TIP

The conclusions in the first two cases of testing for increasing and decreasing functions are valid even when $f'(x) = 0$ at a finite number of x-values in (a, b).

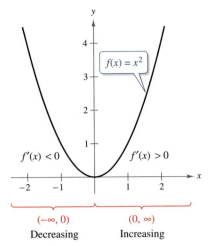

FIGURE 3.2

Example 1 Testing for Increasing and Decreasing Functions

Show that the function $f(x) = x^2$ is decreasing on the open interval $(-\infty, 0)$ and increasing on the open interval $(0, \infty)$.

SOLUTION The derivative of f is

$$f'(x) = 2x.$$

On the open interval $(-\infty, 0)$, the fact that x is negative implies that $f'(x) = 2x$ is also negative. So, by the test for a decreasing function, you can conclude that f is *decreasing* on this interval. Similarly, on the open interval $(0, \infty)$, the fact that x is positive implies that $f'(x) = 2x$ is also positive. So, it follows that f is *increasing* on this interval, as shown in Figure 3.2.

✓ **Checkpoint 1**

Show that the function $f(x) = x^4$ is decreasing on the open interval $(-\infty, 0)$ and increasing on the open interval $(0, \infty)$. ■

Example 2 Modeling Consumption

From 2000 through 2008, the consumption M of whole milk in the United States (in gallons per person) can be modeled by

$$M = -0.015t^2 + 0.13t + 8.0, \quad 0 \le t \le 8$$

where $t = 0$ corresponds to 2000 (see Figure 3.3). Show that the consumption of whole milk was decreasing from 2000 to 2008. *(Source: U.S. Department of Agriculture)*

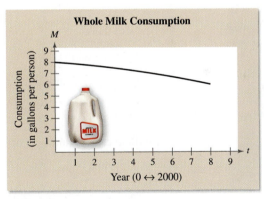

FIGURE 3.3

SOLUTION The derivative of this model is $dM/dt = -0.030t + 0.13$. For the open interval $(0, 8)$, the derivative is negative. So, the function is decreasing, which implies that the consumption of whole milk was decreasing during the given time period.

✓ **Checkpoint 2**

From 2003 through 2008, the consumption F of fresh fruit in the United States (in pounds per person) can be modeled by

$$F = -0.7674t^2 + 2.872t + 277.87, \quad 3 \le t \le 8$$

where $t = 3$ corresponds to 2003. Show that the consumption of fresh fruit was decreasing from 2003 to 2008. *(Source: U.S. Department of Agriculture)* ■

Vladimir Wrangel/Shutterstock.com

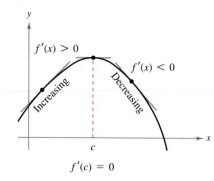

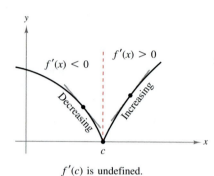

FIGURE 3.4

Critical Numbers and Their Use

In Example 1, you were given two intervals: one on which the function was decreasing and one on which it was increasing. Suppose you had been asked to determine these intervals. To do this, you could have used the fact that for a continuous function, $f'(x)$ can change signs only at x-values for which $f'(x) = 0$ or at x-values for which $f'(x)$ is undefined, as shown in Figure 3.4. These two types of numbers are called the **critical numbers** of f.

Definition of a Critical Number

If f is defined at c, then c is a critical number of f when $f'(c) = 0$ or when $f'(c)$ is undefined.

Example 3   **Finding Critical Numbers**

Find the critical numbers of

$$f(x) = 2x^3 - 9x^2.$$

SOLUTION   Begin by differentiating the function.

$$f(x) = 2x^3 - 9x^2 \qquad \text{Write original function.}$$
$$f'(x) = 6x^2 - 18x \qquad \text{Differentiate.}$$

To find the critical numbers of f, you must find all x-values for which $f'(x) = 0$ and all x-values for which $f'(x)$ is undefined.

$$6x^2 - 18x = 0 \qquad \text{Set } f'(x) \text{ equal to 0.}$$
$$6x(x - 3) = 0 \qquad \text{Factor.}$$
$$x = 0, x = 3 \qquad \text{Critical numbers}$$

Because there are no x-values for which f' is undefined, you can conclude that

$$x = 0 \quad \text{and} \quad x = 3$$

are the only critical numbers of f.

✓**Checkpoint 3**

Find the critical numbers of

$$f(x) = x^2 - x.$$

To determine the intervals on which a continuous function is increasing or decreasing, you can use the guidelines below.

Guidelines for Applying the Increasing/Decreasing Test

1. Find the derivative of f.

2. Locate the critical numbers of f and use these numbers to determine test intervals. That is, find all x for which $f'(x) = 0$ or $f'(x)$ is undefined.

3. Determine the sign of $f'(x)$ at one test value in each of the intervals.

4. Use the test for increasing and decreasing functions to decide whether f is increasing or decreasing on each interval.

Example 4 Intervals on Which *f* Is Increasing or Decreasing

Find the open intervals on which the function is increasing or decreasing.

$$f(x) = x^3 - \frac{3}{2}x^2$$

SOLUTION Begin by finding the derivative of *f*. Then set the derivative equal to zero and solve for the critical numbers.

$f'(x) = 3x^2 - 3x$	Differentiate original function.
$3x^2 - 3x = 0$	Set derivative equal to 0.
$3x(x - 1) = 0$	Factor.
$x = 0, x = 1$	Critical numbers

Because there are no *x*-values for which f' is undefined, it follows that $x = 0$ and $x = 1$ are the only critical numbers. So, the intervals that need to be tested are

$$(-\infty, 0), (0, 1), \quad \text{and} \quad (1, \infty). \qquad \text{Test intervals}$$

The table summarizes the testing of these three intervals.

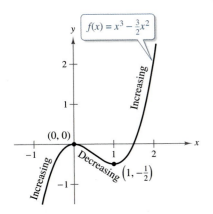

$f(x) = x^3 - \frac{3}{2}x^2$

(0, 0)

Increasing

Decreasing $\left(1, -\frac{1}{2}\right)$

Increasing

FIGURE 3.5

Interval	$-\infty < x < 0$	$0 < x < 1$	$1 < x < \infty$
Test value	$x = -1$	$x = \frac{1}{2}$	$x = 2$
Sign of $f'(x)$	$f'(-1) = 6 > 0$	$f'\left(\frac{1}{2}\right) = -\frac{3}{4} < 0$	$f'(2) = 6 > 0$
Conclusion	Increasing	Decreasing	Increasing

The graph of *f* is shown in Figure 3.5. Note that the test values in the intervals were chosen for convenience—other *x*-values could have been used.

✓**Checkpoint 4**

Find the open intervals on which the function $f(x) = x^3 - 12x$ is increasing or decreasing.

TECH TUTOR

You can use the *trace* feature of a graphing utility to confirm the result of Example 4. Begin by graphing the function, as shown below. Then use the *trace* feature and move the cursor from left to right. In intervals on which the function is increasing, note that the *y*-values increase as the *x*-values increase, whereas in intervals on which the function is decreasing, the *y*-values decrease as the *x*-values increase.

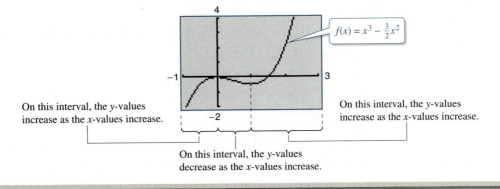

$f(x) = x^3 - \frac{3}{2}x^2$

On this interval, the *y*-values increase as the *x*-values increase.

On this interval, the *y*-values increase as the *x*-values increase.

On this interval, the *y*-values decrease as the *x*-values increase.

Not only is the function in Example 4 continuous on the entire real number line, it is also differentiable there. For such functions, the only critical numbers are those for which $f'(x) = 0$. The next example considers a continuous function that has *both* types of critical numbers—those for which $f'(x) = 0$ and those for which $f'(x)$ is undefined.

ALGEBRA TUTOR

For help on the algebra in Example 5, see Example 2(d) in the *Chapter 3 Algebra Tutor*, on page 207.

Example 5   Intervals on Which *f* Is Increasing or Decreasing

Find the open intervals on which the function

$$f(x) = (x^2 - 4)^{2/3}$$

is increasing or decreasing.

SOLUTION   Begin by finding the derivative of the function.

$$f'(x) = \frac{2}{3}(x^2 - 4)^{-1/3}(2x) \qquad \text{Differentiate.}$$

$$= \frac{4x}{3(x^2 - 4)^{1/3}} \qquad \text{Simplify.}$$

From this, you can see that the derivative is zero when $x = 0$ and the derivative is undefined when $x = \pm 2$. So, the critical numbers are

$$x = -2, \quad x = 0, \quad \text{and} \quad x = 2. \qquad \text{Critical numbers}$$

This implies that the test intervals are

$$(-\infty, -2), \quad (-2, 0), \quad (0, 2), \quad \text{and} \quad (2, \infty). \qquad \text{Test intervals}$$

The table summarizes the testing of these four intervals, and the graph of the function is shown in Figure 3.6.

Interval	$-\infty < x < -2$	$-2 < x < 0$	$0 < x < 2$	$2 < x < \infty$
Test value	$x = -3$	$x = -1$	$x = 1$	$x = 3$
Sign of $f'(x)$	$f'(-3) < 0$	$f'(-1) > 0$	$f'(1) < 0$	$f'(3) > 0$
Conclusion	Decreasing	Increasing	Decreasing	Increasing

STUDY TIP

To test the intervals in the table in Example 5, it is not necessary to *evaluate* $f'(x)$ at each test value—you only need to determine its sign. For instance, you can determine the sign of $f'(-3)$ as shown.

$$f'(-3) = \frac{4(-3)}{3(9 - 4)^{1/3}}$$

$$= \frac{\text{negative}}{\text{positive}}$$

$$= \text{negative}$$

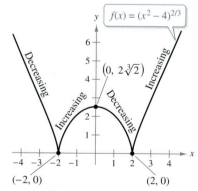

FIGURE 3.6

✓**Checkpoint 5**

Find the open intervals on which the function

$$f(x) = x^{2/3}$$

is increasing or decreasing.

The functions in Examples 1 through 5 are continuous on the entire real number line. If there are isolated x-values at which a function is not continuous, then these x-values should be used along with the critical numbers to determine the test intervals.

Example 6   Testing a Function That Is Not Continuous

The function

$$f(x) = \frac{x^4 + 1}{x^2}$$

is not continuous at $x = 0$. Because the derivative of f

$$f'(x) = \frac{2(x^4 - 1)}{x^3}$$

is zero at $x = \pm 1$, you should use the following numbers to determine the test intervals.

$x = -1, x = 1$ Critical numbers

$x = 0$ Discontinuity

After testing $f'(x)$, you can determine that f is decreasing on the intervals $(-\infty, -1)$ and $(0, 1)$, and increasing on the intervals $(-1, 0)$ and $(1, \infty)$, as shown in Figure 3.7.

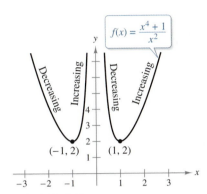

FIGURE 3.7

✔ Checkpoint 6

Find the open invervals on which the function $f(x) = \dfrac{x^2 + 1}{x}$ is increasing or decreasing. ■

The converse of the test for increasing and decreasing functions is *not* true. For instance, it is possible for a function to be increasing on an interval even though its derivative is not positive at every point in the interval.

Example 7   Testing an Increasing Function

Show that $f(x) = x^3 - 3x^2 + 3x$ is increasing on the entire real number line.

SOLUTION   From the derivative of f

$$f'(x) = 3x^2 - 6x + 3 = 3(x - 1)^2$$

you can see that the only critical number is $x = 1$. So, the test intervals are $(-\infty, 1)$ and $(1, \infty)$. The table summarizes the testing of these two intervals. From Figure 3.8, you can see that f is increasing on the entire real number line, even though $f'(1) = 0$. To convince yourself of this, look back at the definition of an increasing function.

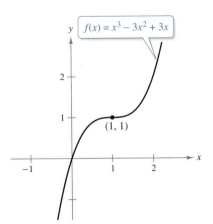

FIGURE 3.8

Interval	$-\infty < x < 1$	$1 < x < \infty$
Test value	$x = 0$	$x = 2$
Sign of $f'(x)$	$f'(0) = 3(-1)^2 > 0$	$f'(2) = 3(1)^2 > 0$
Conclusion	Increasing	Increasing

✔ Checkpoint 7

Show that $f(x) = -x^3 + 2$ is decreasing on the entire real number line. ■

Application

 Example 8   **Profit Analysis**

A national toy distributor determines the cost and revenue models for one of its games.

$$C = 2.4x - 0.0002x^2, \quad 0 \le x \le 6000$$
$$R = 7.2x - 0.001x^2, \quad 0 \le x \le 6000$$

Determine the interval on which the profit function is increasing.

SOLUTION   The profit for producing x games is

$$P = R - C$$
$$= (7.2x - 0.001x^2) - (2.4x - 0.0002x^2)$$
$$= 4.8x - 0.0008x^2.$$

To find the interval on which the profit is increasing, set the marginal profit P' equal to zero and solve for x.

$P' = 4.8 - 0.0016x$	Differentiate profit function.
$4.8 - 0.0016x = 0$	Set P' equal to 0.
$-0.0016x = -4.8$	Subtract 4.8 from each side.
$x = \dfrac{-4.8}{-0.0016}$	Divide each side by -0.0016.
$x = 3000$ games	Simplify.

On the interval $(0, 3000)$, P' is positive and the profit is *increasing*. On the interval $(3000, 6000)$, P' is negative and the profit is *decreasing*. The graphs of the cost, revenue, and profit functions are shown in Figure 3.9.

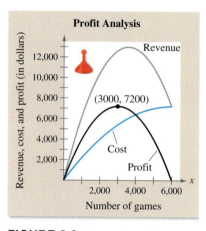

Profit Analysis

FIGURE 3.9

✓ **Checkpoint 8**

A national distributor of pet toys determines the cost and revenue functions for one of its toys.

$$C = 1.2x - 0.0001x^2, \quad 0 \le x \le 6000$$
$$R = 3.6x - 0.0005x^2, \quad 0 \le x \le 6000$$

Determine the interval on which the profit function is increasing.   ■

SUMMARIZE   (Section 3.1)

1. State the test for increasing and decreasing functions *(page 144)*. For an example of testing for increasing and decreasing functions, see Example 1.

2. State the definition of a critical number *(page 146)*. For an example of finding a critical number, see Example 3.

3. State the guidelines for determining the intervals on which a continuous function is increasing or decreasing *(page 146)*. For examples of finding the intervals on which a function is increasing or decreasing, see Examples 4, 5, and 7.

4. Describe a real-life example of how testing for increasing and decreasing functions can be used to analyze the profit of a company *(page 150, Example 8)*.

SKILLS WARM UP 3.1 The following warm-up exercises involve skills that were covered in a previous course or earlier sections. You will use these skills in the exercise set for this section. For additional help, review Appendix Section A.3, and Section 1.4.

In Exercises 1–4, solve the equation.

1. $x^2 = 8x$

2. $15x = \dfrac{5}{8}x^2$

3. $\dfrac{x^2 - 25}{x^3} = 0$

4. $\dfrac{2x}{\sqrt{1 - x^2}} = 0$

In Exercises 5–8, find the domain of the function.

5. $y = \dfrac{x + 3}{x - 3}$

6. $y = \dfrac{2}{\sqrt{1 - x}}$

7. $y = \dfrac{2x + 1}{x^2 - 3x - 10}$

8. $y = \dfrac{3x}{\sqrt{9 - 3x^2}}$

In Exercises 9–12, evaluate the expression when $x = -2$, 0, and 2.

9. $-2(x + 1)(x - 1)$

10. $4(2x + 1)(2x - 1)$

11. $\dfrac{2x + 1}{(x - 1)^2}$

12. $\dfrac{-2(x + 1)}{(x - 4)^2}$

ENHANCED WebAssign Access end-of-section exercises online at **www.webassign.net**

3.2 Extrema and the First-Derivative Test

■ Recognize the occurrence of relative extrema of functions.
■ Use the First-Derivative Test to find the relative extrema of functions.
■ Find absolute extrema of continuous functions on a closed interval.
■ Find minimum and maximum values of real-life models and interpret the results in context.

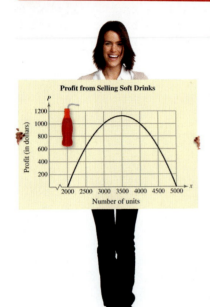

Profit from Selling Soft Drinks

In Exercise 49, you will use the First-Derivative Test to find the price of a soft drink that yields a maximum profit.

Relative Extrema

You have used the derivative to determine the intervals on which a function is increasing or decreasing. In this section, you will examine the points at which a function changes from increasing to decreasing, or vice versa. At such a point, the function has a **relative extremum.** (The plural of extremum is *extrema*.) The **relative extrema** of a function include the **relative minima** and **relative maxima** of the function. For instance, the function shown in Figure 3.10 has a relative maximum at the left point and a relative minimum at the right point.

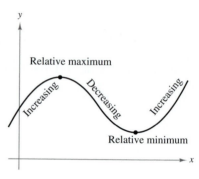

FIGURE 3.10

Definition of Relative Extrema

Let f be a function defined at c.

1. $f(c)$ is a **relative maximum** of f when there exists an interval (a, b) containing c such that $f(x) \leq f(c)$ for all x in (a, b).
2. $f(c)$ is a **relative minimum** of f when there exists an interval (a, b) containing c such that $f(x) \geq f(c)$ for all x in (a, b).

If $f(c)$ is a relative extremum of f, then the relative extremum is said to occur at $x = c$.

For a continuous function, the relative extrema must occur at critical numbers of the function, as shown in Figure 3.11.

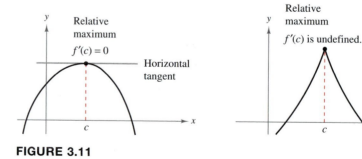

FIGURE 3.11

Occurrences of Relative Extrema

If f has a relative minimum or relative maximum at $x = c$, then c is a critical number of f. That is, either $f'(c) = 0$ or $f'(c)$ is undefined.

The First-Derivative Test

The discussion on the preceding page implies that in your search for relative extrema of a continuous function, you need to test only the critical numbers of the function. Once you have determined that c is a critical number of a function f, the **First-Derivative Test** for relative extrema enables you to classify $f(c)$ as a relative minimum, a relative maximum, or neither.

First-Derivative Test for Relative Extrema

Let f be continuous on the interval (a, b) in which c is the only critical number. If f is differentiable on the interval (except possibly at c), then $f(c)$ can be classified as a relative minimum, a relative maximum, or neither, as shown.

1. On the interval (a, b), if $f'(x)$ is negative to the left of $x = c$ and positive to the right of $x = c$, then $f(c)$ is a relative minimum.

2. On the interval (a, b), if $f'(x)$ is positive to the left of $x = c$ and negative to the right of $x = c$, then $f(c)$ is a relative maximum.

3. On the interval (a, b), if $f'(x)$ is positive on both sides of $x = c$ or negative on both sides of $x = c$, then $f(c)$ is neither a relative minimum nor a relative maximum.

A graphical interpretation of the First-Derivative Test is shown in Figure 3.12.

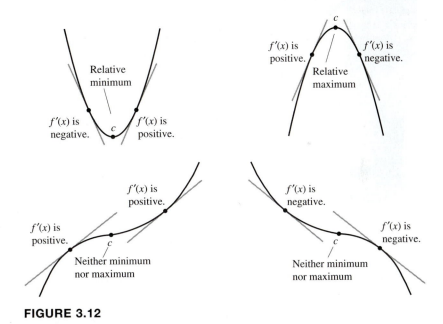

FIGURE 3.12

Guidelines for Finding Relative Extrema

1. Find the derivative of f.

2. Locate the critical numbers of f and use these numbers to determine the test intervals.

3. Test the sign of $f'(x)$ at an arbitrary number in each of the test intervals.

4. For each critical number c, use the First-Derivative Test to decide whether $f(c)$ is a relative minimum, a relative maximum, or neither.

Example 1   Finding Relative Extrema

Find all relative extrema of the function

$$f(x) = 2x^3 - 3x^2 - 36x + 14.$$

SOLUTION   Begin by finding the derivative of f.

$$f'(x) = 6x^2 - 6x - 36 \qquad \text{Differentiate.}$$

Next, find the critical numbers of f.

$$6x^2 - 6x - 36 = 0 \qquad \text{Set derivative equal to 0.}$$
$$6(x^2 - x - 6) = 0 \qquad \text{Factor out common factor.}$$
$$6(x - 3)(x + 2) = 0 \qquad \text{Factor.}$$
$$x = -2, x = 3 \qquad \text{Critical numbers}$$

Because $f'(x)$ is defined for all x, the only critical numbers of f are

$$x = -2 \quad \text{and} \quad x = 3. \qquad \text{Critical numbers}$$

Using these numbers, you can form the three test intervals

$$(-\infty, -2), (-2, 3), \quad \text{and} \quad (3, \infty). \qquad \text{Test intervals}$$

The testing of the three intervals is shown in the table.

Interval	$-\infty < x < -2$	$-2 < x < 3$	$3 < x < \infty$
Test value	$x = -3$	$x = 0$	$x = 4$
Sign of $f'(x)$	$f'(-3) = 36 > 0$	$f'(0) = -36 < 0$	$f'(4) = 36 > 0$
Conclusion	Increasing	Decreasing	Increasing

Using the First-Derivative Test, you can conclude that the critical number -2 yields a relative maximum [$f'(x)$ changes sign from positive to negative], and the critical number 3 yields a relative minimum [$f'(x)$ changes sign from negative to positive]. The graph of f is shown in Figure 3.13. The relative maximum is

$$f(-2) = 58$$

and the relative minimum is

$$f(3) = -67.$$

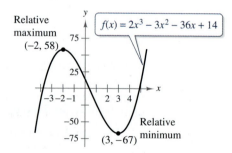

FIGURE 3.13

✓ Checkpoint 1

Find all relative extrema of

$$f(x) = 2x^3 - 6x + 1.$$

In Example 1, both critical numbers yielded relative extrema. In the next example, only one of the two critical numbers yields a relative extremum.

Example 2   Finding Relative Extrema

Find all relative extrema of the function $f(x) = x^4 - x^3$.

SOLUTION   From the derivative of the function

$$f'(x) = 4x^3 - 3x^2 = x^2(4x - 3)$$

you can see that the function has only two critical numbers: $x = 0$ and $x = \frac{3}{4}$. These numbers produce the test intervals $(-\infty, 0)$, $\left(0, \frac{3}{4}\right)$, and $\left(\frac{3}{4}, \infty\right)$, which are tested in the table.

ALGEBRA TUTOR *xy*

For help on the algebra in Example 2, see Example 2(c) in the *Chapter 3 Algebra Tutor*, on page 207.

Interval	$-\infty < x < 0$	$0 < x < \frac{3}{4}$	$\frac{3}{4} < x < \infty$
Test value	$x = -1$	$x = \frac{1}{2}$	$x = 1$
Sign of $f'(x)$	$f'(-1) = -7 < 0$	$f'\left(\frac{1}{2}\right) = -\frac{1}{4} < 0$	$f'(1) = 1 > 0$
Conclusion	Decreasing	Decreasing	Increasing

By the First-Derivative Test, it follows that f has a relative minimum at $x = \frac{3}{4}$, as shown in Figure 3.14. The relative minimum is

$$f\left(\frac{3}{4}\right) = -\frac{27}{256}.$$

Note that the critical number $x = 0$ does not yield a relative extremum because $f'(x)$ is negative on both sides of $x = 0$.

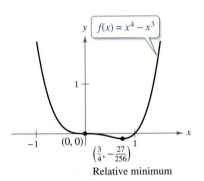

FIGURE 3.14

✓ Checkpoint 2

Find all relative extrema of $f(x) = x^4 - 4x^3$.

Example 3   Finding Relative Extrema

Find all relative extrema of the function

$$f(x) = 2x - 3x^{2/3}.$$

SOLUTION   From the derivative of the function

$$f'(x) = 2 - \frac{2}{x^{1/3}} = \frac{2(x^{1/3} - 1)}{x^{1/3}}$$

you can see that $f'(1) = 0$ and f' is undefined at $x = 0$. So, the function has two critical numbers: $x = 1$ and $x = 0$. These numbers produce the test intervals $(-\infty, 0)$, $(0, 1)$, and $(1, \infty)$. By testing these intervals, you can conclude that f has a relative maximum at $(0, 0)$ and a relative minimum at $(1, -1)$, as shown in Figure 3.15.

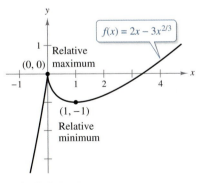

FIGURE 3.15

✓ Checkpoint 3

Find all relative extrema of $f(x) = 3x^{2/3} - 2x$.

Absolute Extrema

The terms *relative minimum* and *relative maximum* describe the *local* behavior of a function. To describe the *global* behavior of the function on an entire interval, you can use the terms **absolute maximum** and **absolute minimum.**

Definition of Absolute Extrema

Let f be defined on an interval I containing c.

1. $f(c)$ is an **absolute minimum of** f on I when $f(c) \leq f(x)$ for every x in I.

2. $f(c)$ is an **absolute maximum of** f on I when $f(c) \geq f(x)$ for every x in I.

The absolute minimum and absolute maximum values of a function on an interval are sometimes called simply the **minimum** and **maximum** of f on I.

Be sure that you understand the distinction between relative extrema and absolute extrema. For instance, in Figure 3.16, the function has a relative minimum that also happens to be an absolute minimum on the interval $[a, b]$. The relative maximum of f, however, is not the absolute maximum on the interval $[a, b]$. The next theorem points out that if a continuous function has a closed interval as its domain, then it *must* have both an absolute minimum and an absolute maximum on the interval. From Figure 3.16, note that these extrema can occur at the endpoints of the interval.

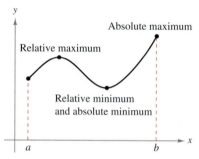

FIGURE 3.16

Extreme Value Theorem

If f is continuous on a closed interval $[a, b]$, then f has both a minimum value and a maximum value on $[a, b]$.

Although a continuous function has just one minimum and one maximum value on a closed interval, either of these values can occur for more than one x-value. For instance, on the interval $[-3, 3]$, the function

$$f(x) = 9 - x^2$$

has a minimum value of zero at $x = -3$ *and* at $x = 3$, as shown in Figure 3.17.

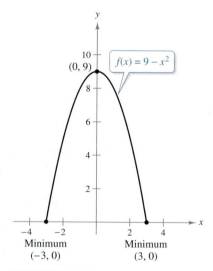

FIGURE 3.17

When looking for extrema of a function on a *closed* interval, remember that you must consider the values of the function at the endpoints as well as at the critical numbers of the function. You can use the guidelines below to find extrema on a closed interval.

Guidelines for Finding Extrema on a Closed Interval

To find the extrema of a continuous function f on a closed interval $[a, b]$, use the following steps.

1. Find the critical numbers of f in the open interval (a, b).
2. Evaluate f at each of its critical numbers in (a, b).
3. Evaluate f at each endpoint, a and b.
4. The least of these values is the minimum, and the greatest is the maximum.

Example 4   Finding Extrema on a Closed Interval

Find the minimum and maximum values of

$$f(x) = x^2 - 6x + 2$$

on the interval $[0, 5]$.

SOLUTION   Begin by differentiating the function.

$f(x) = x^2 - 6x + 2$	Write original function.
$f'(x) = 2x - 6$	Differentiate.

Next, find the critical numbers of f.

$2x - 6 = 0$	Set derivative equal to 0.
$2x = 6$	Add 6 to each side.
$x = 3$	Solve for x.

Because f' is defined for all x, you can conclude that the only critical number of f is $x = 3$. Because this number lies in the interval $[0, 5]$, you should evaluate f at this number *and* at the endpoints of the interval, as shown in the table.

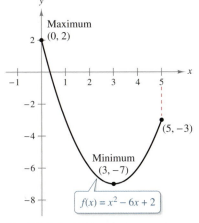

FIGURE 3.18

x-value	Endpoint: $x = 0$	Critical number: $x = 3$	Endpoint: $x = 5$
$f(x)$	$f(0) = 2$	$f(3) = -7$	$f(5) = -3$
Conclusion	Maximum is 2.	Minimum is -7.	Neither maximum nor minimum

From the table, you can see that the minimum of f on the interval $[0, 5]$ is $f(3) = -7$. Moreover, the maximum of f on the interval $[0, 5]$ is $f(0) = 2$. This is confirmed by the graph of f, as shown in Figure 3.18.

✓**Checkpoint 4**

Find the minimum and maximum values of

$$f(x) = x^2 - 8x + 10$$

on the interval $[0, 7]$. Sketch the graph of f and label the minimum and maximum values.

Application

Finding the minimum and maximum values of a function is one of the most common applications of calculus.

 Example 5   Finding the Maximum Profit

Recall the fast-food restaurant in Examples 7 and 8 in Section 2.3. The restaurant's profit function for hamburgers is given by

$$P = 2.44x - \frac{x^2}{20{,}000} - 5000, \quad 0 \le x \le 50{,}000.$$

Find the sales level that yields a maximum profit.

SOLUTION   To begin, find an equation for marginal profit.

$$\frac{dP}{dx} = 2.44 - \frac{x}{10{,}000} \qquad \text{Find marginal profit.}$$

Next, set the marginal profit equal to zero and solve for x.

$$2.44 - \frac{x}{10{,}000} = 0 \qquad \text{Set marginal profit equal to 0.}$$

$$-\frac{x}{10{,}000} = -2.44 \qquad \text{Subtract 2.44 from each side.}$$

$$x = 24{,}400 \text{ hamburgers} \qquad \text{Critical number}$$

In Figure 3.19, you can see that the critical number $x = 24{,}400$ corresponds to the sales level that yields a maximum profit. To find the maximum profit, substitute $x = 24{,}400$ into the profit function.

$$P = 2.44x - \frac{x^2}{20{,}000} - 5000$$

$$= 2.44(24{,}400) - \frac{(24{,}400)^2}{20{,}000} - 5000$$

$$= \$24{,}768$$

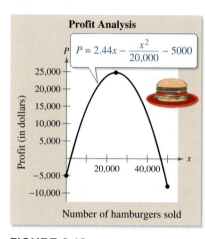

Profit Analysis

$P = 2.44x - \dfrac{x^2}{20{,}000} - 5000$

Profit (in dollars)

Number of hamburgers sold

FIGURE 3.19

✔ **Checkpoint 5**

Verify the results of Example 5 by completing the table.

x (units)	24,000	24,200	24,300	24,400	24,500	24,600	24,800
P (profit)							

SUMMARIZE   (Section 3.2)

1. State the First-Derivative Test *(page 153)*. For examples in which the First-Derivative Test is used, see Examples 1, 2, and 3.

2. State the guidelines for finding extrema on a closed interval *(page 157)*. For an example of finding the extrema of a function on a closed interval, see Example 4.

3. Describe a real-life example of how the First-Derivative Test can be used to find the sales level that yields a maximum profit for a company *(page 158, Example 5)*.

SKILLS WARM UP 3.2   The following warm-up exercises involve skills that were covered in earlier sections. You will use these skills in the exercise set for this section. For additional help, review Sections 2.2, 2.4, and 3.1.

In Exercises 1–6, solve the equation $f'(x) = 0$.

1. $f(x) = 4x^4 - 2x^2 + 1$

2. $f(x) = \frac{1}{3}x^3 - \frac{3}{2}x^2 - 10x$

3. $f(x) = 5x^{4/5} - 4x$

4. $f(x) = \frac{1}{2}x^2 - 3x^{5/3}$

5. $f(x) = \dfrac{x + 4}{x^2 + 1}$

6. $f(x) = \dfrac{x - 1}{x^2 + 4}$

In Exercises 7–10, use $g(x) = -x^5 - 2x^4 + 4x^3 + 2x - 1$ to determine the sign of the derivative.

7. $g'(-4)$

8. $g'(0)$

9. $g'(1)$

10. $g'(3)$

In Exercises 11 and 12, decide whether the function is increasing or decreasing on the given interval.

11. $f(x) = 2x^2 - 11x - 6$, $(3, 6)$

12. $f(x) = x^3 + 2x^2 - 4x - 8$, $(-2, 0)$

ENHANCED
WebAssign   Access end-of-section exercises online at **www.webassign.net**

3.3 Concavity and the Second-Derivative Test

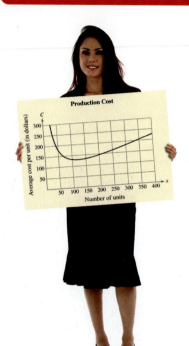

Production Cost

In Exercise 69, you will use the Second-Derivative Test to find the production level that will minimize the average cost per unit.

■ Determine the intervals on which the graphs of functions are concave upward or concave downward.

■ Find the points of inflection of the graphs of functions.

■ Use the Second-Derivative Test to find the relative extrema of functions.

■ Find the points of diminishing returns of input-output models.

Concavity

You already know that locating the intervals over which a function f increases or decreases helps to describe its graph. In this section, you will see that locating the intervals on which f' increases or decreases can determine where the graph of f is curving upward or curving downward. This property of curving upward or downward is defined formally as the **concavity** of the graph of the function.

Definition of Concavity

Let f be differentiable on an open interval I. The graph of f is

1. **concave upward** on I when f' is increasing on the interval.

2. **concave downward** on I when f' is decreasing on the interval.

In Figure 3.20, you can observe the following graphical interpretation of concavity.

1. A curve that is concave upward lies *above* its tangent line.

2. A curve that is concave downward lies *below* its tangent line.

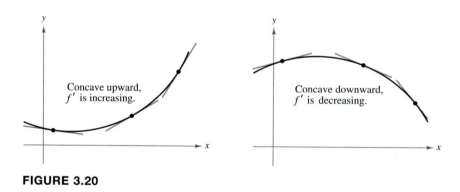

Concave upward, f' is increasing.

Concave downward, f' is decreasing.

FIGURE 3.20

To find the open intervals on which the graph of a function is concave upward or concave downward, you can use the second derivative of the function as follows.

Test for Concavity

Let f be a function whose second derivative exists on an open interval I.

1. If $f''(x) > 0$ for all x in I, then the graph of f is concave upward on I.

2. If $f''(x) < 0$ for all x in I, then the graph of f is concave downward on I.

| Example 1 | Determining Concavity |

a. The graph of the function

$$f(x) = x^2 \qquad \text{Original function}$$

is concave upward on the entire real number line because its second derivative

$$f''(x) = 2 \qquad \text{Second derivative}$$

is positive for all x. (See Figure 3.21.)

b. The graph of the function

$$f(x) = \sqrt{x} \qquad \text{Original function}$$

is concave downward for $x > 0$ because its second derivative

$$f''(x) = -\frac{1}{4}x^{-3/2} \qquad \text{Second derivative}$$

is negative for all $x > 0$. (See Figure 3.22.)

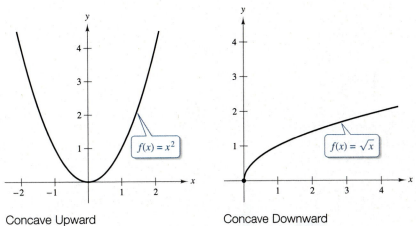

| Concave Upward | Concave Downward |
| **FIGURE 3.21** | **FIGURE 3.22** |

✔**Checkpoint 1**

Find the second derivative of f and discuss the concavity of its graph.

a. $f(x) = -2x^2$

b. $f(x) = -2\sqrt{x}$ ■

For a *continuous* function f, you can find the open intervals on which the graph of f is concave upward and concave downward as follows. [When there are x-values at which the function is not continuous, these values should be used, along with the points at which $f''(x) = 0$ or $f''(x)$ is undefined, to form the test intervals.]

Guidelines for Applying the Concavity Test

1. Locate the x-values at which $f''(x) = 0$ or $f''(x)$ is undefined.

2. Use these x-values to determine the test intervals.

3. Test the sign of $f''(x)$ in each test interval.

ALGEBRA TUTOR

xy

For help on the algebra in Example 2, see Example 1(a) in the *Chapter 3 Algebra Tutor*, on page 206.

Example 2 Applying the Test for Concavity

Determine the open intervals on which the graph of

$$f(x) = \frac{6}{x^2 + 3}$$

is concave upward or concave downward.

SOLUTION Begin by finding the second derivative of f.

$$f(x) = 6(x^2 + 3)^{-1} \qquad \text{Rewrite original function.}$$

$$f'(x) = 6(-1)(x^2 + 3)^{-2}(2x) \qquad \text{Chain Rule}$$

$$= \frac{-12x}{(x^2 + 3)^2} \qquad \text{Simplify.}$$

$$f''(x) = \frac{(x^2 + 3)^2(-12) - (-12x)(2)(x^2 + 3)(2x)}{(x^2 + 3)^4} \qquad \text{Quotient Rule}$$

$$= \frac{-12(x^2 + 3) + 48x^2}{(x^2 + 3)^3} \qquad \text{Simplify.}$$

$$= \frac{36(x^2 - 1)}{(x^2 + 3)^3} \qquad \text{Simplify.}$$

From this, you can see that $f''(x)$ is defined for all real numbers and $f''(x) = 0$ when $x = \pm 1$. So, you can test the concavity of f by testing the intervals

$$(-\infty, -1), \quad (-1, 1), \quad \text{and} \quad (1, \infty). \qquad \text{Test intervals}$$

The results are shown in the table and in Figure 3.23.

STUDY TIP

In Example 2, f' is increasing on the interval $(1, \infty)$ even though f is decreasing there. Be sure you see that the increasing or decreasing of f' does not necessarily correspond to the increasing or decreasing of f.

Interval	$-\infty < x < -1$	$-1 < x < 1$	$1 < x < \infty$
Test value	$x = -2$	$x = 0$	$x = 2$
Sign of $f''(x)$	$f''(-2) > 0$	$f''(0) < 0$	$f''(2) > 0$
Conclusion	Concave upward	Concave downward	Concave upward

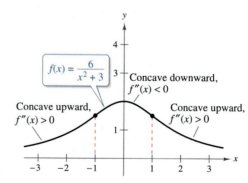

FIGURE 3.23

✓**Checkpoint 2**

Determine the intervals on which the graph of

$$f(x) = \frac{12}{x^2 + 4}$$

is concave upward or concave downward.

Points of Inflection

If the tangent line to a graph exists at a point at which the concavity changes, then the point is a **point of inflection.** Three examples of inflection points are shown in Figure 3.24. (Note that the third graph has a vertical tangent line at its point of inflection.)

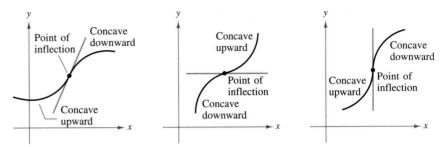

The graph *crosses* its tangent line at a point of inflection.
FIGURE 3.24

> ### STUDY TIP
>
> As shown in Figure 3.24, a graph crosses its tangent line at a point of inflection.

Definition of Point of Inflection

If the graph of a continuous function has a tangent line at a point where its concavity changes from upward to downward (or downward to upward), then the point is a **point of inflection.**

Because a point of inflection occurs where the concavity of a graph changes, it must be true that at such points the sign of f'' changes. So, to locate possible points of inflection, you need to determine only the values of x for which $f''(x) = 0$ or for which $f''(x)$ does not exist. This parallels the procedure for locating the relative extrema of f by determining the critical numbers of f.

Property of Points of Inflection

If $(c, f(c))$ is a point of inflection of the graph of f, then either $f''(c) = 0$ or $f''(c)$ is undefined.

Example 3   Finding a Point of Inflection

Discuss the concavity of the graph of $f(x) = 2x^3 + 1$ and find its point of inflection.

SOLUTION   Differentiating twice produces the following.

$$f(x) = 2x^3 + 1 \qquad \text{Write original function.}$$
$$f'(x) = 6x^2 \qquad \text{Find first derivative.}$$
$$f''(x) = 12x \qquad \text{Find second derivative.}$$

Setting $f''(x) = 0$, you can determine that the only possible point of inflection occurs at $x = 0$. After testing the intervals $(-\infty, 0)$ and $(0, \infty)$, you can determine that the graph is concave downward on $(-\infty, 0)$ and concave upward on $(0, \infty)$. Because the concavity changes at $x = 0$, you can conclude that the graph of f has a point of inflection at $(0, 1)$, as shown in Figure 3.25.

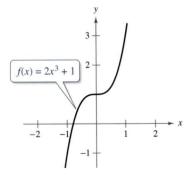

$f(x) = 2x^3 + 1$

FIGURE 3.25

✓ Checkpoint 3

Discuss the concavity of the graph of $f(x) = -x^3$ and find its point of inflection.   ■

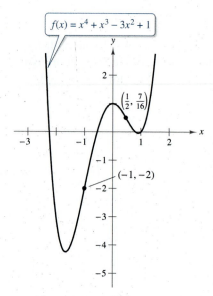

$f(x) = x^4 + x^3 - 3x^2 + 1$

$\left(\frac{1}{2}, \frac{7}{16}\right)$

$(-1, -2)$

Two Points of Inflection

FIGURE 3.26

Example 4   Finding Points of Inflection

Discuss the concavity of the graph of

$$f(x) = x^4 + x^3 - 3x^2 + 1$$

and find its points of inflection.

SOLUTION   Begin by finding the second derivative of f.

$f(x) = x^4 + x^3 - 3x^2 + 1$	Write original function.
$f'(x) = 4x^3 + 3x^2 - 6x$	Find first derivative.
$f''(x) = 12x^2 + 6x - 6$	Find second derivative.
$\quad\;\; = 6(2x - 1)(x + 1)$	Factor.

From this, you can see that the possible points of inflection occur at $x = \frac{1}{2}$ and $x = -1$. After testing the intervals $(-\infty, -1)$, $\left(-1, \frac{1}{2}\right)$, and $\left(\frac{1}{2}, \infty\right)$, you can determine that the graph is concave upward on $(-\infty, -1)$, concave downward on $\left(-1, \frac{1}{2}\right)$, and concave upward on $\left(\frac{1}{2}, \infty\right)$. Because the concavity changes at $x = -1$ and $x = \frac{1}{2}$, you can conclude that the graph of f has points of inflection at these x-values, as shown in Figure 3.26. The points of inflection are

$$(-1, -2) \quad \text{and} \quad \left(\frac{1}{2}, \frac{7}{16}\right).$$

✓ Checkpoint 4

Discuss the concavity of the graph of

$$f(x) = x^4 - 2x^3 + 1$$

and find its points of inflection.   ■

It is possible for the second derivative to be zero at a point that is *not* a point of inflection. For example, compare the graphs of

$$f(x) = x^3 \quad \text{and} \quad g(x) = x^4$$

as shown in Figure 3.27. Both second derivatives are zero when $x = 0$, but only the graph of f has a point of inflection at $x = 0$. This shows that before concluding that a point of inflection exists at a value of x for which $f''(x) = 0$, you must test to be certain that the concavity actually changes at that point.

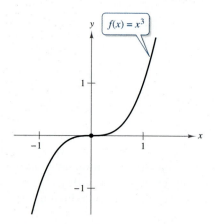

$f''(0) = 0$, and $(0, 0)$ is a point of inflection.

$g''(0) = 0$, but $(0, 0)$ is not a point of inflection.

FIGURE 3.27

The Second-Derivative Test

The second derivative can be used to perform a simple test for relative minima and relative maxima. If f is a function such that $f'(c) = 0$ and the graph of f is concave upward at $x = c$, then $f(c)$ is a relative minimum of f. Similarly, if f is a function such that $f'(c) = 0$ and the graph of f is concave downward at $x = c$, then $f(c)$ is a relative maximum of f, as shown in Figure 3.28.

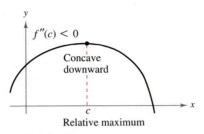
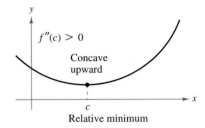

FIGURE 3.28

Second-Derivative Test

Let $f'(c) = 0$, and let f'' exist on an open interval containing c.

1. If $f''(c) > 0$, then $f(c)$ is a relative minimum.

2. If $f''(c) < 0$, then $f(c)$ is a relative maximum.

3. If $f''(c) = 0$, then the test fails. In such cases, you can use the First-Derivative Test to determine whether $f(c)$ is a relative minimum, a relative maximum, or neither.

Example 5   Using the Second-Derivative Test

Find the relative extrema of $f(x) = -3x^5 + 5x^3$.

SOLUTION   Begin by finding the first derivative of f.

$$f'(x) = -15x^4 + 15x^2 = 15x^2(1 - x^2)$$

From this derivative, you can see that $x = 0$, $x = -1$, and $x = 1$ are the only critical numbers of f. Using the second derivative

$$f''(x) = -60x^3 + 30x = 30x(1 - 2x^2)$$

you can apply the Second-Derivative Test, as shown.

Point	$(-1, -2)$	$(0, 0)$	$(1, 2)$
Sign of $f''(x)$	$f''(-1) > 0$	$f''(0) = 0$	$f''(1) < 0$
Conclusion	Relative minimum	Test fails.	Relative maximum

Because the Second-Derivative Test fails at $(0, 0)$, you can use the First-Derivative Test and observe that f is positive on both sides of $x = 0$. So, $(0, 0)$ is neither a relative minimum nor a relative maximum. A test for concavity would show that $(0, 0)$ is a point of inflection. The graph of f is shown in Figure 3.29.

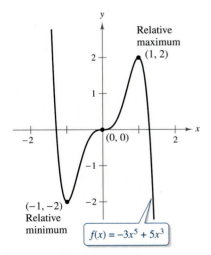

FIGURE 3.29

✓**Checkpoint 5**

Find all relative extrema of $f(x) = x^4 - 4x^3 + 1$.

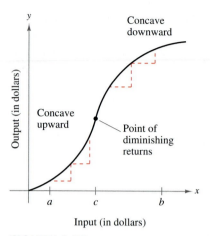

FIGURE 3.30

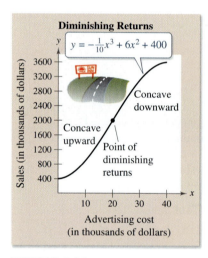

FIGURE 3.31

Extended Application: Diminishing Returns

In economics, the notion of concavity is related to the concept of **diminishing returns.** Consider a function

Output → ← Input

$$y = f(x)$$

where x measures input (in dollars) and y measures output (in dollars). In Figure 3.30, notice that the graph of this function is concave upward on the interval (a, c) and is concave downward on the interval (c, b). On the interval (a, c), each additional dollar of input returns more than the previous input dollar. By contrast, on the interval (c, b), each additional dollar of input returns less than the previous input dollar. The point $(c, f(c))$ is called the **point of diminishing returns.** An increased investment beyond this point is usually considered a poor use of capital.

Example 6   **Exploring Diminishing Returns**

By increasing its advertising cost x (in thousands of dollars) for a product, a company discovers that it can increase the sales y (in thousands of dollars) according to the model

$$y = -\frac{1}{10}x^3 + 6x^2 + 400, \quad 0 \le x \le 40.$$

Find the point of diminishing returns for this product.

SOLUTION   Begin by finding the first and second derivatives.

$$y' = 12x - \frac{3x^2}{10} \qquad \text{First derivative}$$

$$y'' = 12 - \frac{3x}{5} \qquad \text{Second derivative}$$

The second derivative is zero only when $x = 20$. By testing for concavity on the intervals $(0, 20)$ and $(20, 40)$, you can conclude that the graph has a point of diminishing returns when $x = 20$, as shown in Figure 3.31. So, the point of diminishing returns for this product occurs when $20,000 is spent on advertising.

✓ **Checkpoint 6**

Find the point of diminishing returns for the model below, where R is the revenue (in thousands of dollars) and x is the advertising cost (in thousands of dollars).

$$R = \frac{1}{20,000}(450x^2 - x^3), \quad 0 \le x \le 300 \qquad ■$$

SUMMARIZE   (Section 3.3)

1. State the test for concavity *(page 160)*. For examples of applying the test for concavity, see Examples 1 and 2.

2. State the definition of point of inflection *(page 163)*. For examples of finding points of inflection, see Examples 3 and 4.

3. State the Second-Derivative Test *(page 165)*. For an example of using the Second-Derivative Test, see Example 5.

4. Describe a real-life example of how the second derivative can be used to find the point of diminishing returns for a product *(page 166, Example 6)*.

SKILLS WARM UP 3.3

The following warm-up exercises involve skills that were covered in earlier sections. You will use these skills in the exercise set for this section. For additional help, review Sections 2.2, 2.4, 2.5, 2.6, and 3.1.

In Exercises 1–6, find the second derivative of the function.

1. $f(x) = 4x^4 - 9x^3 + 5x - 1$

2. $g(s) = (s^2 - 1)(s^2 - 3s + 2)$

3. $g(x) = (x^2 + 1)^4$

4. $f(x) = (x - 3)^{4/3}$

5. $h(x) = \dfrac{4x + 3}{5x - 1}$

6. $f(x) = \dfrac{2x - 1}{3x + 2}$

In Exercises 7–10, find the critical numbers of the function.

7. $f(x) = 5x^3 - 5x + 11$

8. $f(x) = x^4 - 4x^3 - 10$

9. $g(t) = \dfrac{16 + t^2}{t}$

10. $h(x) = \dfrac{x^4 - 50x^2}{8}$

3.4 Optimization Problems

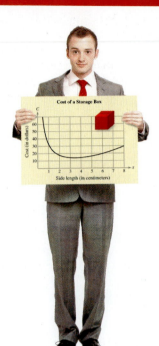

In Exercise 13, you will use primary equations, secondary equations, and derivatives to find the dimensions of a box that will minimize the cost of making the box.

■ Solve real-life optimization problems.

Solving Optimization Problems

One of the most common applications of calculus is the determination of optimum (minimum or maximum) values. Before learning a general method for solving optimization problems, consider the next example.

 Example 1 Finding the Maximum Volume

A manufacturer wants to design an open box that has a square base and a surface area S of 108 square inches, as shown in Figure 3.32. What dimensions will produce a box with a maximum volume?

SOLUTION Because the base of the box is square, the volume is

$$V = x^2 h. \qquad \text{Primary equation}$$

This equation is called the **primary equation** because it gives a formula for the quantity to be optimized. The surface area of the box is

$$S = (\text{area of base}) + (\text{area of four sides})$$
$$108 = x^2 + 4xh. \qquad \text{Secondary equation}$$

Open Box with Square Base:
$S = x^2 + 4xh = 108$
FIGURE 3.32

Because V is to be optimized, it helps to express V as a function of just one variable. To do this, solve the secondary equation for h in terms of x to obtain

$$h = \frac{108 - x^2}{4x}$$

and substitute for h in the primary equation.

$$V = x^2 h = x^2 \left(\frac{108 - x^2}{4x} \right) = 27x - \frac{1}{4}x^3 \qquad \text{Function of one variable}$$

Before finding which x-value yields a maximum value of V, you need to determine the *feasible domain* of the function. That is, what values of x make sense in this problem? Because x must be nonnegative and the area of the base ($A = x^2$) is at most 108, you can conclude that the feasible domain is

$$0 \le x \le \sqrt{108}. \qquad \text{Feasible domain}$$

Using the techniques described in the first three sections of this chapter, you can determine that $\left(\text{on the interval } 0 \le x \le \sqrt{108}\right)$ this function has an absolute maximum at $x = 6$ inches and $h = 3$ inches.

ALGEBRA TUTOR xy

For help on the algebra in Example 1, see Example 1(c) in the *Chapter 3 Algebra Tutor*, on page 206.

✓**Checkpoint 1**

Use a graphing utility to graph the volume function

$$V = 27x - \frac{1}{4}x^3$$

from Example 1 on $0 \le x \le \sqrt{108}$. Verify that the function has an absolute maximum at $x = 6$. What is the maximum volume?

In studying Example 1, be sure that you understand the basic question that it asks. Remember that you are not ready to begin solving an optimization problem until you have clearly identified the problem. Once you are sure you understand what is being asked, you are ready to begin considering a method for solving the problem.

For instance, in Example 1, you should realize that there are infinitely many open boxes having 108 square inches of surface area. To begin solving this problem, you might ask yourself which basic shape would seem to yield a maximum volume. Should the box be tall, squat, or nearly cubical? You might even try calculating a few volumes, as shown in Figure 3.33, to see if you can get a better feeling for what the optimum dimensions should be.

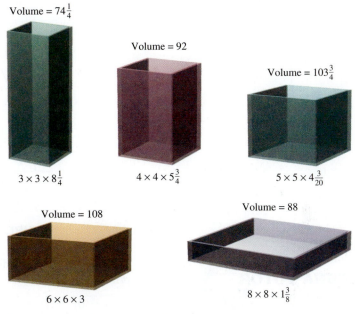

Volume = $74\frac{1}{4}$

$3 \times 3 \times 8\frac{1}{4}$

Volume = 92

$4 \times 4 \times 5\frac{3}{4}$

Volume = $103\frac{3}{4}$

$5 \times 5 \times 4\frac{3}{20}$

Volume = 108

$6 \times 6 \times 3$

Volume = 88

$8 \times 8 \times 1\frac{3}{8}$

Which box has the greatest volume?

FIGURE 3.33

There are several steps in the solution of Example 1. The first step is to sketch a diagram and identify all *known* quantities and all quantities *to be determined*. The second step is to write a primary equation for the quantity to be optimized. Then, a secondary equation is used to rewrite the primary equation as a function of one variable. Finally, calculus is used to determine the optimum value. These steps are summarized below.

Guidelines for Solving Optimization Problems

1. Identify all given quantities and all quantities to be determined. If possible, make a sketch.

2. Write a **primary equation** for the quantity that is to be maximized or minimized. (A summary of several common formulas is given in Appendix D.)

3. Reduce the primary equation to one having a single independent variable. This may involve the use of a **secondary equation** that relates the independent variables of the primary equation.

4. Determine the feasible domain of the primary equation. That is, determine the values for which the stated problem makes sense.

5. Determine the desired maximum or minimum value by the calculus techniques discussed in Sections 3.1 through 3.3.

> **Example 2**   **Finding a Minimum Distance**

Find the points on the graph of

$$y = 4 - x^2$$

that are closest to $(0, 2)$.

SOLUTION

1. Figure 3.34 shows that there are two points at a minimum distance from $(0, 2)$.

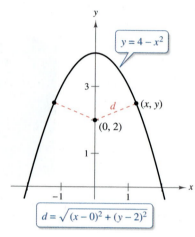

$$y = 4 - x^2$$

(x, y)

$(0, 2)$

$$d = \sqrt{(x - 0)^2 + (y - 2)^2}$$

FIGURE 3.34

2. You are asked to minimize the distance d. So, you can use the Distance Formula to obtain a primary equation.

$$d = \sqrt{(x - 0)^2 + (y - 2)^2} \qquad \text{Primary equation}$$

3. Using the secondary equation $y = 4 - x^2$, you can rewrite the primary equation as a function of a single variable.

$$d = \sqrt{x^2 + (4 - x^2 - 2)^2} \qquad \text{Substitute } 4 - x^2 \text{ for } y.$$
$$= \sqrt{x^2 + (2 - x^2)^2} \qquad \text{Simplify.}$$
$$= \sqrt{x^2 + 4 - 4x^2 + x^4} \qquad \text{Expand binomial.}$$
$$= \sqrt{x^4 - 3x^2 + 4} \qquad \text{Combine like terms.}$$

Because d is smallest when the expression under the radical is smallest, you simplify the problem by finding the minimum value of $f(x) = x^4 - 3x^2 + 4$.

4. The domain of f is the entire real number line.

5. To find the minimum value of $f(x)$, first find the critical numbers of f.

$$f'(x) = 4x^3 - 6x \qquad \text{Find derivative of } f.$$
$$0 = 4x^3 - 6x \qquad \text{Set derivative equal to 0.}$$
$$0 = 2x(2x^2 - 3) \qquad \text{Factor.}$$
$$x = 0, x = \sqrt{\tfrac{3}{2}}, x = -\sqrt{\tfrac{3}{2}} \qquad \text{Critical numbers}$$

The First-Derivative Test verifies that $x = 0$ yields a relative maximum, whereas both $\sqrt{3/2}$ and $-\sqrt{3/2}$ yield a minimum. So, the points closest to $(0, 2)$ are

$$\left(\sqrt{\tfrac{3}{2}}, \tfrac{5}{2}\right) \quad \text{and} \quad \left(-\sqrt{\tfrac{3}{2}}, \tfrac{5}{2}\right).$$

ALGEBRA TUTOR xy

For help on the algebra in Example 2, see Example 1(b) in the *Chapter 3 Algebra Tutor*, on page 206.

✓ **Checkpoint 2**

Find the points on the graph of $y = 4 - x^2$ that are closest to $(0, 3)$.

Example 3 Finding a Minimum Area

A rectangular page will contain 24 square inches of print. The margins at the top and bottom of the page are $1\frac{1}{2}$ inches wide. The margins on each side are 1 inch wide. What should the dimensions of the page be to minimize the amount of paper used?

SOLUTION

1. A diagram of the page is shown in Figure 3.35.

2. Letting A be the area to be minimized, the primary equation is

   $$A = (x + 3)(y + 2).$$ Primary equation

3. The printed area inside the margins is given by

   $$24 = xy.$$ Secondary equation

   Solving this equation for y produces

   $$y = \frac{24}{x}.$$

   By substituting this result into the primary equation, you obtain

   $$A = (x + 3)\left(\frac{24}{x} + 2\right)$$ Write as a function of one variable.

   $$= (x + 3)\left(\frac{24 + 2x}{x}\right)$$ Rewrite second factor as a single fraction.

   $$= \frac{2x^2}{x} + \frac{30x}{x} + \frac{72}{x}$$ Multiply and separate into terms.

   $$= 2x + 30 + \frac{72}{x}.$$ Simplify.

4. Because x must be positive, the feasible domain is $x > 0$.

5. To find the minimum area, begin by finding the critical numbers of A.

   $$\frac{dA}{dx} = 2 - \frac{72}{x^2}$$ Find derivative of A.

   $$0 = 2 - \frac{72}{x^2}$$ Set derivative equal to 0.

   $$-2 = -\frac{72}{x^2}$$ Subtract 2 from each side.

   $$x^2 = 36$$ Simplify.

   $$x = \pm 6$$ Critical numbers

   Because $x = -6$ is not in the feasible domain, you need to consider only the critical number $x = 6$. Using the First-Derivative Test, it follows that A is a minimum when $x = 6$. So, the dimensions of the page should be

   $$x + 3 = 6 + 3 = 9 \text{ inches}   \text{ by }   y + 2 = \frac{24}{6} + 2 = 6 \text{ inches.}$$

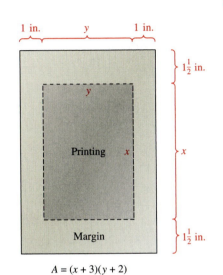

1 in. y 1 in.

$1\frac{1}{2}$ in.

y

Printing x x

Margin $1\frac{1}{2}$ in.

$A = (x + 3)(y + 2)$

FIGURE 3.35

✓ Checkpoint 3

A rectangular page will contain 54 square inches of print. The margins at the top and bottom of the page are $1\frac{1}{2}$ inches wide. The margins on each side are 1 inch wide. What should the dimensions of the page be to minimize the amount of paper used? ■

As applications go, the examples described in this section are fairly simple, and yet the resulting primary equations are quite complicated. Real-life applications often involve equations that are at least as complex as the ones in the examples. Remember that one of the main goals of this course is to enable you to use the power of calculus to analyze equations that at first glance seem formidable.

Also remember that once you have found the primary equation, you can use the graph of the equation to help solve the problem. For instance, the graphs of the primary equations in Examples 1 through 3 are shown in Figure 3.36.

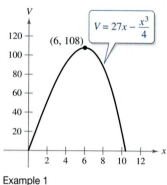

$$V = 27x - \frac{x^3}{4}$$

Example 1

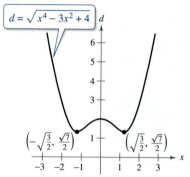

$$d = \sqrt{x^4 - 3x^2 + 4}$$

Example 2

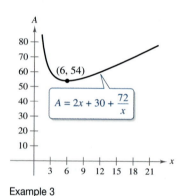

$$A = 2x + 30 + \frac{72}{x}$$

Example 3

FIGURE 3.36

SUMMARIZE   (Section 3.4)

1. State what is meant by the primary equation of an optimization problem *(page 168)*. For examples of primary equations in optimization problems, see Examples 1, 2, and 3.

2. State what is meant by the feasible domain of a function *(page 168)*. For examples of feasible domains, see Examples 1, 2, and 3.

3. State what is meant by the secondary equation of an optimization problem *(page 169)*. For examples of secondary equations in optimization problems, see Examples 1, 2, and 3.

4. State the guidelines for solving optimization problems *(page 169)*. For examples of solving optimization problems, see Examples 2 and 3.

5. Describe a real-life example of how solving an optimization problem can be used to determine the dimensions of a page so that the amount of paper used is minimized *(page 171, Example 3)*.

SKILLS WARM UP 3.4
The following warm-up exercises involve skills that were covered in earlier sections. You will use these skills in the exercise set for this section. For additional help, review Section 3.1.

In Exercises 1–4, write a formula for the written statement.

1. The sum of one number and half a second number is 12.

2. The product of one number and twice another is 24.

3. The area of a rectangle is 24 square units.

4. The distance between two points is 10 units.

In Exercises 5–10, find the critical numbers of the function.

5. $y = x^2 + 6x - 9$

6. $y = 2x^3 - x^2 - 4x$

7. $y = 5x + \dfrac{125}{x}$

8. $y = 3x + \dfrac{96}{x^2}$

9. $y = \dfrac{x^2 + 1}{x}$

10. $y = \dfrac{x}{x^2 + 9}$

ENHANCED WebAssign Access end-of-section exercises online at **www.webassign.net**

QUIZ YOURSELF

Take this quiz as you would take a quiz in class. When you are done, check your work against the answers given in the back of the book.

In Exercises 1–3, find the critical numbers and the open intervals on which the function is increasing or decreasing. Use a graphing utility to verify your results.

1. $f(x) = x^2 - 6x + 1$ **2.** $f(x) = 2x^3 + 12x^2$

3. $f(x) = \dfrac{x}{x^2 + 25}$

In Exercises 4–6, find all relative extrema of the function.

4. $f(x) = x^3 + 3x^2 - 5$ **5.** $f(x) = x^4 - 8x^2 + 3$

6. $f(x) = 2x^{2/3}$

In Exercises 7–9, find the absolute extrema of the function on the closed interval. Use a graphing utility to verify your results.

7. $f(x) = x^2 + 2x - 8, \ [-2, 1]$ **8.** $f(x) = x^3 - 27x, \ [-4, 4]$

9. $f(x) = \dfrac{x}{x^2 + 1}, \ [0, 2]$

In Exercises 10 and 11, discuss the concavity of the graph of the function and find the points of inflection.

10. $f(x) = x^3 - 6x^2 + 7x$ **11.** $f(x) = x^4 - 24x^2$

In Exercises 12 and 13, use the Second-Derivative Test to find all relative extrema of the function.

12. $f(x) = 2x^3 + 3x^2 - 12x + 16$ **13.** $f(x) = 2x + \dfrac{18}{x}$

14. By increasing its advertising cost x for a product, a company discovers that it can increase the sales S according to the model

$$S = \frac{1}{3600}(360x^2 - x^3), \quad 0 \le x \le 240$$

where x and S are in thousands of dollars. Find the point of diminishing returns for this product.

15. A gardener has 200 feet of fencing to enclose a rectangular garden adjacent to a river (see figure). No fencing is needed along the river. What dimensions should be used so that the area of the garden will be a maximum?

16. The resident population P (in thousands) of Maine from 2000 through 2009 can be modeled by

$$P = 0.001t^3 - 0.64t^2 + 10.3t + 1276, \quad 0 \le t \le 9$$

where t is the year, with $t = 0$ corresponding to 2000. *(Source: U.S. Census Bureau)*

(a) During which year(s) was the population increasing? decreasing?

(b) During which year, from 2000 through 2009, was the population the greatest? the least?

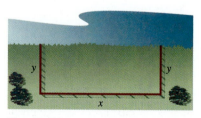

Figure for 15

3.5 Business and Economics Applications

■ Solve business and economics optimization problems.
■ Find the price elasticity of demand for demand functions.
■ Recognize basic business terms and formulas.

Optimization in Business and Economics

The problems in this section are primarily optimization problems. So, the five-step procedure used in Section 3.4 is an appropriate strategy to follow.

Example 1 Finding the Maximum Revenue

A company has determined that its total revenue (in dollars) for a product can be modeled by

$$R = -x^3 + 450x^2 + 52{,}500x$$

where x is the number of units produced (and sold). What production level will yield a maximum revenue?

SOLUTION

1. A sketch of the revenue function is shown in Figure 3.37.

2. The primary equation is the given revenue function.

$$R = -x^3 + 450x^2 + 52{,}500x$$

3. Because R is already given as a function of one variable, you do not need a secondary equation.

4. The feasible domain of the primary equation is

$$0 \le x \le 546. \qquad \text{\color{red}Feasible domain}$$

This is determined by finding the x-intercepts of the revenue function, as shown in Figure 3.37.

5. To maximize the revenue, find the critical numbers.

$$\frac{dR}{dx} = -3x^2 + 900x + 52{,}500 = 0 \qquad \text{\color{red}Set derivative equal to 0.}$$

$$-3(x - 350)(x + 50) = 0 \qquad \text{\color{red}Factor.}$$

$$x = 350, \, x = -50 \qquad \text{\color{red}Critical numbers}$$

The only critical number in the feasible domain is $x = 350$. From the graph of the function, you can see that the production level of 350 units corresponds to a maximum revenue.

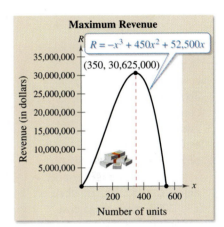

Maximum Revenue

$R = -x^3 + 450x^2 + 52{,}500x$

(350, 30,625,000)

Revenue (in dollars)

Number of units

Maximum revenue occurs when $dR/dx = 0$.

FIGURE 3.37

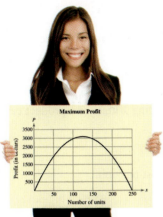

In Exercise 15, you will use derivatives to find the price that yields a maximum profit.

✓ Checkpoint 1

Find the number of units that must be produced to maximize the revenue function $R = -x^3 + 150x^2 + 9375x$, where R is the total revenue (in dollars) and x is the number of units produced (and sold). What is the maximum revenue?

To study the effects of production levels on cost, one method economists use is the **average cost function** $\overline{C}$, which is defined as

$$\overline{C} = \frac{C}{x}$$

Average cost function

where $C = f(x)$ is the total cost function and x is the number of units produced.

 Example 2   **Finding the Minimum Average Cost**

A company estimates that the cost (in dollars) of producing x units of a product can be modeled by

$$C = 800 + 0.04x + 0.0002x^2.$$

Find the production level that minimizes the average cost per unit.

SOLUTION

1. C represents the total cost, x represents the number of units produced, and $\overline{C}$ represents the average cost per unit.

2. The primary equation is

$$\overline{C} = \frac{C}{x}.$$

Primary equation

3. Substituting the given equation for C produces

$$\overline{C} = \frac{800 + 0.04x + 0.0002x^2}{x}$$

Substitute for C.

$$= \frac{800}{x} + 0.04 + 0.0002x.$$

Function of one variable

4. The feasible domain of this function is

$$x > 0$$

Feasible domain

because the company cannot produce a negative number of units.

5. You can find the critical numbers as shown.

$$\frac{d\overline{C}}{dx} = -\frac{800}{x^2} + 0.0002 = 0$$

Set derivative equal to 0.

$$0.0002 = \frac{800}{x^2}$$

$$x^2 = \frac{800}{0.0002}$$

Multiply each side by x^2 and divide each side by 0.0002.

$$x^2 = 4{,}000{,}000$$

$$x = \pm 2000$$

Critical numbers

By choosing the positive value of x and sketching the graph of $\overline{C}$, as shown in Figure 3.38, you can see that a production level of $x = 2000$ minimizes the average cost per unit.

STUDY TIP

To see that $x = 2000$ corresponds to a minimum average cost in Example 2, try evaluating $\overline{C}$ for several values of x. For instance, when $x = 400$, the average cost per unit is $\overline{C} = \$2.12$, but when $x = 2000$, the average cost per unit is $\overline{C} = \$0.84$.

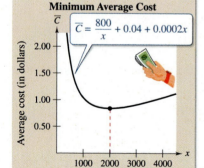

Minimum Average Cost

$$\overline{C} = \frac{800}{x} + 0.04 + 0.0002x$$

Minimum average cost occurs when $d\overline{C}/dx = 0$.

FIGURE 3.38

✓ **Checkpoint 2**

Find the production level that minimizes the average cost per unit for the cost function

$$C = 400 + 0.05x + 0.0025x^2$$

where C is the cost (in dollars) of producing x units of a product. ■

Example 3 Finding the Maximum Revenue

A business sells 2000 units of a product per month at a price of $10 each. It can sell 250 more items per month for each $0.25 reduction in price. What price per unit will maximize the monthly revenue?

SOLUTION

1. Let x represent the number of units sold in a month, let p represent the price per unit, and let R represent the monthly revenue.

2. Because the revenue is to be maximized, the primary equation is

$$R = xp. \qquad \text{Primary equation}$$

3. A price of $p = \$10$ corresponds to $x = 2000$, and a price of $p = \$9.75$ corresponds to $x = 2250$. Using this information, you can use the two-point form to write the demand equation.

$$p - 10 = \frac{10 - 9.75}{2000 - 2250}(x - 2000) \qquad \text{Two-point form}$$

$$p - 10 = -0.001(x - 2000) \qquad \text{Simplify.}$$

$$p = -0.001x + 12 \qquad \text{Secondary equation}$$

Substituting this value into the revenue equation produces

$$R = x(-0.001x + 12) \qquad \text{Substitute for } p.$$

$$= -0.001x^2 + 12x. \qquad \text{Function of one variable}$$

4. The feasible domain of the revenue function is

$$0 \le x \le 12,000. \qquad \text{Feasible domain}$$

This is determined by finding the x-intercepts of the revenue function.

5. To maximize the revenue, find the critical numbers.

$$\frac{dR}{dx} = 12 - 0.002x = 0 \qquad \text{Set derivative equal to 0.}$$

$$-0.002x = -12$$

$$x = 6000 \qquad \text{Critical number}$$

From the graph of R in Figure 3.39, you can see that this production level yields a maximum revenue. The price that corresponds to this production level is

$$p = 12 - 0.001x \qquad \text{Demand function}$$

$$= 12 - 0.001(6000) \qquad \text{Substitute 6000 for } x.$$

$$= \$6. \qquad \text{Price per unit}$$

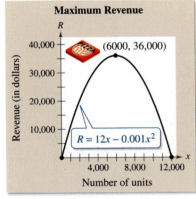

Maximum Revenue

$R = 12x - 0.001x^2$

FIGURE 3.39

✓ Checkpoint 3

Find the price per unit that will maximize the monthly revenue for the business in Example 3 when it can sell only 200 more items per month for each $0.25 reduction in price. ■

In Example 3, the revenue function was written as a function of x. It could also have been written as a function of p. That is,

$$R = 1000(12p - p^2).$$

By finding the critical numbers of this function, you can determine that the maximum revenue occurs at $p = 6$.

ALGEBRA TUTOR

For help on the algebra in Example 4, see Example 2(b) in the *Chapter 3 Algebra Tutor*, on page 207.

Example 4 **Finding the Maximum Profit**

The marketing department of a business has determined that the demand for a product can be modeled by $p = 50/\sqrt{x}$, where p is the price per unit (in dollars) and x is the number of units. The cost (in dollars) of producing x units is given by $C = 0.5x + 500$. What price will yield a maximum profit?

SOLUTION

1. Let R represent the revenue, P the profit, p the price per unit, x the number of units, and C the total cost of producing x units.

2. Because you are maximizing the profit, the primary equation is

$$P = R - C. \qquad \text{Primary equation}$$

3. Because the revenue is $R = xp$, you can write the profit function as

$$
\begin{aligned}
P &= R - C \\
&= xp - (0.5x + 500) & \text{Substitute for } R \text{ and } C. \\
&= x\left(\frac{50}{\sqrt{x}}\right) - 0.5x - 500 & \text{Substitute for } p. \\
&= 50\sqrt{x} - 0.5x - 500. & \text{Function of one variable}
\end{aligned}
$$

4. The feasible domain of the function is $127 < x \le 7872$. (When x is less than 127 or greater than or equal to 7872, the profit is negative.)

5. To maximize the profit, find the critical numbers.

$$
\begin{aligned}
\frac{dP}{dx} &= \frac{25}{\sqrt{x}} - 0.5 = 0 & \text{Set derivative equal to 0.} \\
\frac{25}{\sqrt{x}} &= 0.5 & \text{Add 0.5 to each side.} \\
50 &= \sqrt{x} & \text{Isolate } x\text{-term on one side.} \\
2500 &= x & \text{Critical number}
\end{aligned}
$$

From the graph of the profit function shown in Figure 3.40, you can see that a maximum profit occurs at $x = 2500$. The price that corresponds to $x = 2500$ is

$$p = \frac{50}{\sqrt{x}} = \frac{50}{\sqrt{2500}} = \frac{50}{50} = \$1.00. \qquad \text{Price per unit}$$

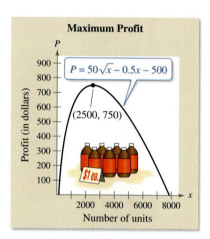

FIGURE 3.40

✓ **Checkpoint 4**

Find the price that will maximize profit for the demand and cost functions

$$p = \frac{40}{\sqrt{x}} \quad \text{and} \quad C = 2x + 50$$

where p is the price per unit (in dollars), x is the number of units, and C is the cost (in dollars). ■

To find the maximum profit in Example 4, the equation $P = R - C$ was differentiated and set equal to zero. From the equation

$$\frac{dP}{dx} = \frac{dR}{dx} - \frac{dC}{dx} = 0$$

it follows that the maximum profit occurs when the marginal revenue is equal to the marginal cost, as shown in Figure 3.41.

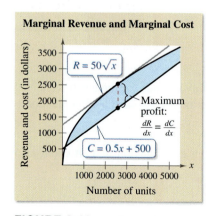

FIGURE 3.41

Price Elasticity of Demand

One way in which economists measure the responsiveness of consumers to a change in the price of a product is with **price elasticity of demand.** For example, a drop in the price of fresh tomatoes might result in a much greater demand for fresh tomatoes; such a demand is called **elastic.** On the other hand, the demand for items such as coffee and gasoline is relatively unresponsive to changes in price; the demand for such items is called **inelastic.**

More formally, the elasticity of demand is the percent change of a quantity demanded x, divided by the percent change in its price p. You can develop a formula for price elasticity of demand using the approximation

$$\frac{\Delta p}{\Delta x} \approx \frac{dp}{dx}$$

which is based on the definition of the derivative. Using this approximation, you can write

$$
\begin{aligned}
\text{Price elasticity of demand} &= \frac{\text{rate of change in demand}}{\text{rate of change in price}} \\
&= \frac{\Delta x/x}{\Delta p/p} \\
&= \frac{p/x}{\Delta p/\Delta x} \\
&\approx \frac{p/x}{dp/dx}.
\end{aligned}
$$

Definition of Price Elasticity of Demand

If $p = f(x)$ is a differentiable function, then the **price elasticity of demand** is given by

$$\eta = \frac{p/x}{dp/dx}$$

where η is the lowercase Greek letter eta. For a given price, the demand is **elastic** when $|\eta| > 1$, the demand is **inelastic** when $|\eta| < 1$, and the demand has **unit elasticity** when $|\eta| = 1$.

Price elasticity of demand is related to the total revenue function, as indicated in Figure 3.42 and the list below.

1. If the demand is *elastic*, then a decrease in price is accompanied by an increase in unit sales sufficient to increase the total revenue.

2. If the demand is *inelastic*, then a decrease in price is not accompanied by an increase in unit sales sufficient to increase the total revenue.

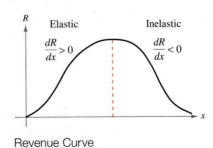

Revenue Curve
FIGURE 3.42

Demand Function of a Product

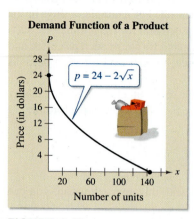

FIGURE 3.43

Example 5 Comparing Elasticity and Revenue

The demand function for a product is modeled by

$$p = 24 - 2\sqrt{x}, \quad 0 \le x \le 144$$

where p is the price per unit (in dollars) and x is the number of units. (See Figure 3.43.)

a. Determine when the demand is elastic, inelastic, and of unit elasticity.

b. Use the result of part (a) to describe the behavior of the revenue function.

SOLUTION

a. The price elasticity of demand is given by

$$\eta = \frac{p/x}{dp/dx} \qquad \text{Formula for price elasticity of demand}$$

$$= \frac{\dfrac{24 - 2\sqrt{x}}{x}}{\dfrac{-1}{\sqrt{x}}} \qquad \text{Substitute for } p/x \text{ and } dp/dx.$$

$$= \frac{\left(\dfrac{24 - 2\sqrt{x}}{x}\right)(-\sqrt{x})}{\left(\dfrac{-1}{\sqrt{x}}\right)(-\sqrt{x})} \qquad \text{Multiply numerator and denominator by } -\sqrt{x}.$$

$$= \frac{-24\sqrt{x} + 2x}{x} \qquad \text{Simplify.}$$

$$= -\frac{24\sqrt{x}}{x} + 2. \qquad \text{Rewrite as two fractions and simplify.}$$

The demand is of unit elasticity when $|\eta| = 1$. In the interval $[0, 144]$, the only solution of the equation

$$|\eta| = \left| -\frac{24\sqrt{x}}{x} + 2 \right| = 1 \qquad \text{Unit elasticity}$$

is $x = 64$. So, the demand is of unit elasticity when $x = 64$. For x-values in the interval $(0, 64)$,

$$|\eta| = \left| -\frac{24\sqrt{x}}{x} + 2 \right| > 1, \quad 0 < x < 64 \qquad \text{Elastic}$$

which implies that the demand is elastic when $0 < x < 64$. For x-values in the interval $(64, 144)$,

$$|\eta| = \left| -\frac{24\sqrt{x}}{x} + 2 \right| < 1, \quad 64 < x < 144 \qquad \text{Inelastic}$$

which implies that the demand is inelastic when $64 < x < 144$.

Revenue Function of a Product

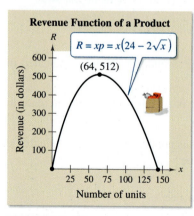

FIGURE 3.44

b. From part (a), you can conclude that the revenue function R is increasing on the open interval $(0, 64)$, is decreasing on the open interval $(64, 144)$, and is a maximum when $x = 64$, as indicated in Figure 3.44.

✓ **Checkpoint 5**

The demand function for a product is modeled by $p = 36 - 2\sqrt{x}, 0 \le x \le 324$, where p is the price per unit (in dollars) and x is the number of units. Determine when the demand is elastic, inelastic, and of unit elasticity.

Business Terms and Formulas

This section concludes with a summary of the basic business terms and formulas used in this section. A summary of the graphs of the demand, revenue, cost, and profit functions is shown in Figure 3.45.

Summary of Business Terms and Formulas

x = number of units produced (or sold)	η = price elasticity of demand
p = price per unit	$= \dfrac{p/x}{dp/dx}$
R = total revenue from selling x units = xp	$\dfrac{dR}{dx}$ = marginal revenue
C = total cost of producing x units	$\dfrac{dC}{dx}$ = marginal cost
P = total profit from selling x units = $R - C$	$\dfrac{dP}{dx}$ = marginal profit
$\overline{C}$ = average cost per unit = $\dfrac{C}{x}$	

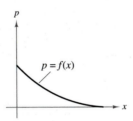

Demand Function
Quantity demanded increases as price decreases.

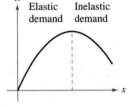

Revenue Function
The low prices required to sell more units eventually result in a decreasing revenue.

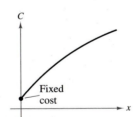

Cost Function
The total cost to produce x units includes the fixed cost.

Profit Function
The break-even point occurs when $R = C$.

FIGURE 3.45

SUMMARIZE   (Section 3.5)

1. Describe a real-life example of how optimization can be used to find the maximum revenue for a product *(page 175, Example 1)*.

2. State the definition of the average cost function *(page 176)*. For an example of an average cost function, see Example 2.

3. State the definition of price elasticity of demand *(page 179)*. For an example of price elasticity of demand, see Example 5.

SKILLS WARM UP 3.5   The following warm-up exercises involve skills that were covered in a previous course or earlier sections. You will use these skills in the exercise set for this section. For additional help, review Appendix Sections A.2 and A.3, and Section 2.3.

In Exercises 1–4, evaluate the expression for $x = 150$.

1. $\left| -\dfrac{300}{x} + 3 \right|$

2. $\left| -\dfrac{600}{5x} + 2 \right|$

3. $\left| \dfrac{(20x^{-1/2})/x}{-10x^{-3/2}} \right|$

4. $\left| \dfrac{(4000/x^2)/x}{-8000x^{-3}} \right|$

In Exercises 5–10, find the marginal revenue, marginal cost, or marginal profit.

5. $C = 650 + 1.2x + 0.003x^2$

6. $P = 0.01x^2 + 11x$

7. $P = -0.7x^2 + 7x - 50$

8. $C = 1700 + 4.2x + 0.001x^3$

9. $R = 14x - \dfrac{x^2}{2000}$

10. $R = 3.4x - \dfrac{x^2}{1500}$

ENHANCED WebAssign   Access end-of-section exercises online at **www.webassign.net**

3.6 Asymptotes

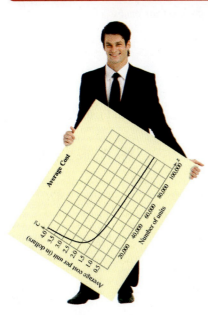

In Exercise 61, you will use limits at infinity to find the limit of an average cost function as the number of units produced increases.

■ Find the vertical asymptotes of functions and find infinite limits.
■ Find the horizontal asymptotes of functions and find limits at infinity.
■ Use asymptotes to answer questions about real-life situations.

Vertical Asymptotes and Infinite Limits

In the first three sections of this chapter, you studied ways in which you can use calculus to help analyze the graph of a function. In this section, you will study another valuable aid to curve sketching: the determination of vertical and horizontal asymptotes.

Recall from Section 1.5, Example 11, that the function

$$f(x) = \frac{3}{x-2}$$

is unbounded as x approaches 2 (see Figure 3.46).

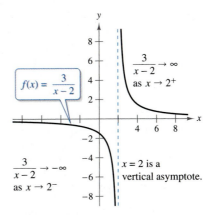

$$f(x) = \frac{3}{x-2}$$

$\dfrac{3}{x-2} \to \infty$ as $x \to 2^{+}$

$\dfrac{3}{x-2} \to -\infty$ as $x \to 2^{-}$

$x = 2$ is a vertical asymptote.

FIGURE 3.46

This type of behavior is described by saying that the line

$$x = 2 \qquad \text{Vertical asymptote}$$

is a **vertical asymptote** of the graph of f. The type of limit in which $f(x)$ approaches infinity (or negative infinity) as x approaches c from the left or from the right is an **infinite limit.** The infinite limits for the function $f(x) = 3/(x-2)$ can be written as

$$\lim_{x \to 2^{-}} \frac{3}{x-2} = -\infty$$

and

$$\lim_{x \to 2^{+}} \frac{3}{x-2} = \infty.$$

Definition of Vertical Asymptote

If $f(x)$ approaches infinity (or negative infinity) as x approaches c from the right or from the left, then the line

$$x = c$$

is a **vertical asymptote** of the graph of f.

One of the most common instances of a vertical asymptote is the graph of a *rational function*—that is, a function of the form $f(x) = p(x)/q(x)$, where $p(x)$ and $q(x)$ are polynomials. If c is a real number such that $q(c) = 0$ and $p(c) \neq 0$, then the graph of f has a vertical asymptote at $x = c$. Example 1 shows four cases.

Example 1   Finding Infinite Limits

Limit from the left   *Limit from the right*

a. $\lim\limits_{x \to 1^-} \dfrac{1}{x - 1} = -\infty$ $\lim\limits_{x \to 1^+} \dfrac{1}{x - 1} = \infty$ See Figure 3.47(a).

b. $\lim\limits_{x \to 1^-} \dfrac{-1}{x - 1} = \infty$ $\lim\limits_{x \to 1^+} \dfrac{-1}{x - 1} = -\infty$ See Figure 3.47(b).

c. $\lim\limits_{x \to 1^-} \dfrac{-1}{(x - 1)^2} = -\infty$ $\lim\limits_{x \to 1^+} \dfrac{-1}{(x - 1)^2} = -\infty$ See Figure 3.47(c).

d. $\lim\limits_{x \to 1^-} \dfrac{1}{(x - 1)^2} = \infty$ $\lim\limits_{x \to 1^+} \dfrac{1}{(x - 1)^2} = \infty$ See Figure 3.47(d).

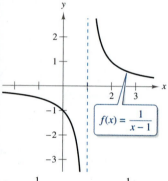

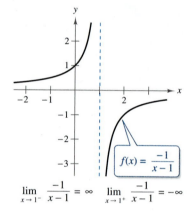

$\lim\limits_{x \to 1^-} \dfrac{1}{x - 1} = -\infty$ $\lim\limits_{x \to 1^+} \dfrac{1}{x - 1} = \infty$ $\lim\limits_{x \to 1^-} \dfrac{-1}{x - 1} = \infty$ $\lim\limits_{x \to 1^+} \dfrac{-1}{x - 1} = -\infty$

(a)   (b)

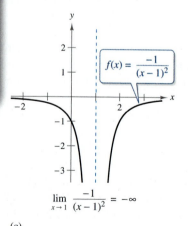

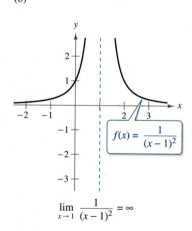

$\lim\limits_{x \to 1} \dfrac{-1}{(x - 1)^2} = -\infty$ $\lim\limits_{x \to 1} \dfrac{1}{(x - 1)^2} = \infty$

(c)   (d)

FIGURE 3.47

✓ **Checkpoint 1**

Find each limit.

a. $\lim\limits_{x \to 2^-} \dfrac{1}{x - 2}$ **b.** $\lim\limits_{x \to 2^+} \dfrac{1}{x - 2}$ **c.** $\lim\limits_{x \to -3^-} \dfrac{1}{x + 3}$ **d.** $\lim\limits_{x \to -3^+} \dfrac{1}{x + 3}$

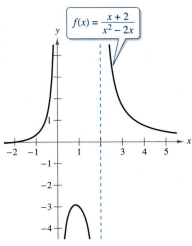

$f(x) = \dfrac{x + 2}{x^2 - 2x}$

Vertical Asymptotes at $x = 0$ and $x = 2$

FIGURE 3.48

Each of the graphs in Example 1 has only one vertical asymptote. As shown in the next example, the graph of a rational function can have more than one vertical asymptote.

Example 2 **Finding Vertical Asymptotes**

Determine all vertical asymptotes of the graph of

$$f(x) = \frac{x + 2}{x^2 - 2x}.$$

SOLUTION The possible vertical asymptotes correspond to the x-values for which the denominator is zero.

$x^2 - 2x = 0$	Set denominator equal to 0.
$x(x - 2) = 0$	Factor.
$x = 0, x = 2$	Zeros of denominator

Because the numerator of f is not zero at either of these x-values, you can conclude that the graph of f has two vertical asymptotes—one at $x = 0$ and one at $x = 2$, as shown in Figure 3.48.

✓ **Checkpoint 2**

Determine all vertical asymptotes of the graph of

$$f(x) = \frac{x + 4}{x^2 - 4x}.$$

Example 3 **Finding Vertical Asymptotes**

Determine all vertical asymptotes of the graph of

$$f(x) = \frac{x^2 + 2x - 8}{x^2 - 4}.$$

SOLUTION First factor the numerator and denominator. Then divide out common factors.

$f(x) = \dfrac{x^2 + 2x - 8}{x^2 - 4}$	Write original function.
$= \dfrac{(x + 4)(x - 2)}{(x + 2)(x - 2)}$	Factor numerator and denominator.
$= \dfrac{(x + 4)\cancel{(x - 2)}}{(x + 2)\cancel{(x - 2)}}$	Divide out common factors.
$= \dfrac{x + 4}{x + 2}, \quad x \neq 2$	Simplify.

For all values of x other than $x = 2$, the graph of this simplified function is the same as the graph of f. So, you can conclude that the graph of f has only one vertical asymptote. This occurs at $x = -2$, as shown in Figure 3.49.

✓ **Checkpoint 3**

Determine all vertical asymptotes of the graph of

$$f(x) = \frac{x^2 + 4x + 3}{x^2 - 9}.$$

Undefined when $x = 2$

$f(x) = \dfrac{x^2 + 2x - 8}{x^2 - 4}$

Vertical Asymptote at $x = -2$

FIGURE 3.49

When you use a graphing utility to graph a function that has a vertical asymptote, the utility may try to connect separate branches of the graph. For instance, the figure below shows the graph of

$$f(x) = \frac{3}{x - 2}$$

on a graphing calculator.

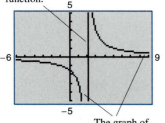

This line is not part of the graph of the function.

The graph of the function has two branches.

From Example 3, you know that the graph of

$$f(x) = \frac{x^2 + 2x - 8}{x^2 - 4}$$

has a vertical asymptote at $x = -2$. This implies that the limit of $f(x)$ as $x \to -2$ from the right (or from the left) is either ∞ or $-\infty$. But without looking at the graph, how can you determine that the limit from the left is *negative* infinity and the limit from the right is *positive* infinity? That is, why is the limit from the left

$$\lim_{x \to -2^-} \frac{x^2 + 2x - 8}{x^2 - 4} = -\infty \qquad \text{Limit from the left}$$

and why is the limit from the right

$$\lim_{x \to -2^+} \frac{x^2 + 2x - 8}{x^2 - 4} = \infty? \qquad \text{Limit from the right}$$

It is cumbersome to determine these limits analytically, and you may find the graphical method shown in Example 4 to be more efficient.

Example 4   Determining Infinite Limits

Find the limits.

$$\lim_{x \to 1^-} \frac{x^2 - 3x}{x - 1} \quad \text{and} \quad \lim_{x \to 1^+} \frac{x^2 - 3x}{x - 1}$$

SOLUTION   Begin by considering the function

$$f(x) = \frac{x^2 - 3x}{x - 1}.$$

Because the denominator is zero when $x = 1$ and the numerator is not zero when $x = 1$, it follows that the graph of the function has a vertical asymptote at $x = 1$. This implies that each of the given limits is either ∞ or $-\infty$. To determine which, use a graphing utility to graph the function, as shown in Figure 3.50. From the graph, you can see that the limit from the left is positive infinity and the limit from the right is negative infinity. That is,

$$\lim_{x \to 1^-} \frac{x^2 - 3x}{x - 1} = \infty \qquad \text{Limit from the left}$$

and

$$\lim_{x \to 1^+} \frac{x^2 - 3x}{x - 1} = -\infty. \qquad \text{Limit from the right}$$

From the left, $f(x)$ approaches positive infinity.

From the right, $f(x)$ approaches negative infinity.

FIGURE 3.50

✓Checkpoint 4

Find the limits.

$$\lim_{x \to 2^-} \frac{x^2 - 4x}{x - 2} \quad \text{and} \quad \lim_{x \to 2^+} \frac{x^2 - 4x}{x - 2}$$

In Example 4, try evaluating $f(x)$ at x-values that are just barely to the left of 1. You will find that you can make the values of $f(x)$ arbitrarily large by choosing x sufficiently close to 1. For instance, $f(0.99999) \approx 199,999$.

Horizontal Asymptotes and Limits at Infinity

Another type of limit, called a **limit at infinity,** specifies a finite value approached by a function as x increases (or decreases) without bound.

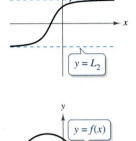

> ### Definition of Horizontal Asymptote
>
> If f is a function and L_1 and L_2 are real numbers, then the statements
>
> $$\lim_{x\to\infty} f(x) = L_1 \quad \text{and} \quad \lim_{x\to-\infty} f(x) = L_2$$
>
> denote **limits at infinity.** The lines $y = L_1$ and $y = L_2$ are **horizontal asymptotes** of the graph of f.

Figure 3.51 shows two ways in which the graph of a function can approach one or more horizontal asymptotes. Note that it is possible for the graph of a function to cross its horizontal asymptote.

Limits at infinity share many of the properties of limits discussed in Section 1.5. When finding horizontal asymptotes, you can use the property that

$$\lim_{x\to\infty} \frac{1}{x^r} = 0, \quad r > 0 \quad \text{and} \quad \lim_{x\to-\infty} \frac{1}{x^r} = 0, \quad r > 0.$$

FIGURE 3.51

(The second limit assumes that x^r is defined when $x < 0$.)

Example 5 Finding Limits at Infinity

Find the limit: $\displaystyle\lim_{x\to\infty} \left(5 - \frac{2}{x^2}\right)$.

SOLUTION

$$\lim_{x\to\infty} \left(5 - \frac{2}{x^2}\right) = \lim_{x\to\infty} 5 - \lim_{x\to\infty} \frac{2}{x^2} \qquad \color{red}{\lim_{x\to\infty} [f(x) - g(x)] = \lim_{x\to\infty} f(x) - \lim_{x\to\infty} g(x)}$$

$$= \lim_{x\to\infty} 5 - 2\left(\lim_{x\to\infty} \frac{1}{x^2}\right) \qquad \color{red}{\lim_{x\to\infty} cf(x) = c \lim_{x\to\infty} f(x)}$$

$$= 5 - 2(0)$$

$$= 5$$

You can verify this limit by sketching the graph of

$$f(x) = 5 - \frac{2}{x^2}$$

as shown in Figure 3.52. Note that the graph has $y = 5$ as a horizontal asymptote to the right. By evaluating the limit

$$\lim_{x\to-\infty} \left(5 - \frac{2}{x^2}\right)$$

you can show that $y = 5$ is also a horizontal asymptote to the left.

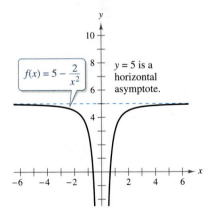

$f(x) = 5 - \dfrac{2}{x^2}$

$y = 5$ is a horizontal asymptote.

FIGURE 3.52

✓ Checkpoint 5

Find the limit: $\displaystyle\lim_{x\to\infty} \left(2 + \frac{5}{x^2}\right)$.

There is an easy way to determine whether the graph of a *rational* function has a horizontal asymptote. This shortcut is based on a comparison of the degrees of the numerator and denominator of the rational function.

Horizontal Asymptotes of Rational Functions

Let $f(x) = p(x)/q(x)$ be a rational function.

1. If the degree of the numerator is less than the degree of the denominator, then $y = 0$ is a horizontal asymptote of the graph of f (to the left and to the right).

2. If the degree of the numerator is equal to the degree of the denominator, then $y = a/b$ is a horizontal asymptote of the graph of f (to the left and to the right), where a and b are the leading coefficients of $p(x)$ and $q(x)$, respectively.

3. If the degree of the numerator is greater than the degree of the denominator, then the graph of f has no horizontal asymptote.

Example 6   Finding Horizontal Asymptotes

Find the horizontal asymptote of the graph of each function.

a. $y = \dfrac{-2x + 3}{3x^2 + 1}$ **b.** $y = \dfrac{-2x^2 + 3}{3x^2 + 1}$ **c.** $y = \dfrac{-2x^3 + 3}{3x^2 + 1}$

SOLUTION

a. Because the degree of the numerator is less than the degree of the denominator, $y = 0$ is a horizontal asymptote. [See Figure 3.53(a).]

b. Because the degree of the numerator is equal to the degree of the denominator, the line $y = -\frac{2}{3}$ is a horizontal asymptote. [See Figure 3.53(b).]

c. Because the degree of the numerator is greater than the degree of the denominator, the graph has no horizontal asymptote. [See Figure 3.53(c).]

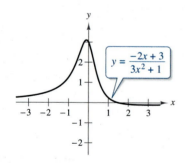
(a) $y = 0$ is a horizontal asymptote.

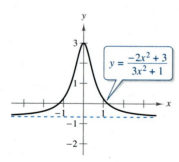
(b) $y = -\frac{2}{3}$ is a horizontal asymptote.

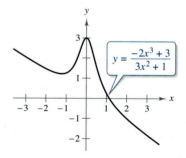
(c) No horizontal asymptote

FIGURE 3.53

✓ **Checkpoint 6**

Find the horizontal asymptote of the graph of each function.

a. $y = \dfrac{2x + 1}{4x^2 + 5}$ **b.** $y = \dfrac{2x^2 + 1}{4x^2 + 5}$ **c.** $y = \dfrac{2x^3 + 1}{4x^2 + 5}$

Some functions have two horizontal asymptotes: one to the right and one to the left (see Exercises 59 and 60).

Applications of Asymptotes

There are many examples of asymptotic behavior in real life. For instance, Example 7 describes the asymptotic behavior of an average cost function.

Example 7   **Modeling Average Cost**

A small business invests $5000 in a new product. In addition to this initial investment, the product will cost $0.50 per unit to produce.

a. Find the average cost per unit when 1000 units are produced.

b. Find the average cost per unit when 10,000 units are produced.

c. Find the average cost per unit when 100,000 units are produced.

d. What is the limit of the average cost as the number of units produced increases?

SOLUTION   From the given information, you can model the total cost C (in dollars) by

$$C = 0.5x + 5000 \qquad \text{Total cost function}$$

where x is the number of units produced. This implies that the average cost function is

$$\overline{C} = \frac{C}{x} = 0.5 + \frac{5000}{x}. \qquad \text{Average cost function}$$

a. When only 1000 units are produced, the average cost per unit is

$$\overline{C} = 0.5 + \frac{5000}{1000} \qquad \text{Substitute 1000 for } x.$$

$$= \$5.50. \qquad \text{Average cost for 1000 units}$$

b. When 10,000 units are produced, the average cost per unit is

$$\overline{C} = 0.5 + \frac{5000}{10,000} \qquad \text{Substitute 10,000 for } x.$$

$$= \$1.00. \qquad \text{Average cost for 10,000 units}$$

c. When 100,000 units are produced, the average cost per unit is

$$\overline{C} = 0.5 + \frac{5000}{100,000} \qquad \text{Substitute 100,000 for } x.$$

$$= \$0.55. \qquad \text{Average cost for 100,000 units}$$

d. As x approaches infinity, the limiting average cost per unit is

$$\lim_{x \to \infty} \left(0.5 + \frac{5000}{x} \right) = \$0.50.$$

As shown in Figure 3.54, this example points out one of the major problems faced by small businesses. That is, it is difficult to have competitively low prices when the production level is low.

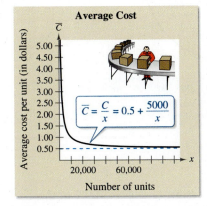

Average Cost

Average cost per unit (in dollars)

$$\overline{C} = \frac{C}{x} = 0.5 + \frac{5000}{x}$$

5.00, 4.50, 4.00, 3.50, 3.00, 2.50, 2.00, 1.50, 1.00, 0.50

20,000   60,000

Number of units

As $x \to \infty$, the average cost per unit approaches $0.50.

FIGURE 3.54

✓ **Checkpoint 7**

A small business invests $25,000 in a new product. In addition, the product will cost $0.75 per unit to produce. Find the cost function and the average cost function. What is the limit of the average cost function as the number of units produced increases? ■

In Example 7, suppose that the small business had made an initial investment of $50,000. How would this change the answers to the questions? Would it change the average cost of producing x units? Would it change the limiting average cost per unit?

Example 8   **Modeling Smokestack Emission**

A manufacturing plant has determined that the cost C (in dollars) of removing $p\%$ of the smokestack pollutants of its main smokestack is modeled by

$$C = \frac{80,000p}{100 - p}, \quad 0 \le p < 100.$$

What is the vertical asymptote of this function? What does the vertical asymptote mean to the plant owners?

SOLUTION   The graph of the cost function is shown in Figure 3.55. From the graph, you can see that $p = 100$ is the vertical asymptote. This means that as the plant attempts to remove higher and higher percents of the pollutants, the cost increases dramatically. For instance, the cost of removing 85% of the pollutants is

$$C = \frac{80,000(85)}{100 - 85} \approx \$453,333 \qquad \text{Cost for 85\% removal}$$

but the cost of removing 90% is

$$C = \frac{80,000(90)}{100 - 90} = \$720,000. \qquad \text{Cost for 90\% removal}$$

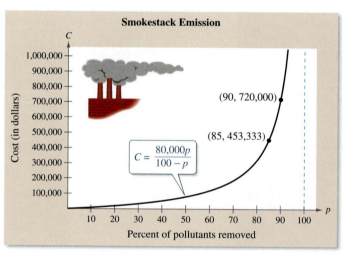

Smokestack Emission

(90, 720,000)

(85, 453,333)

$C = \dfrac{80,000p}{100 - p}$

Cost (in dollars)

Percent of pollutants removed

FIGURE 3.55

 Checkpoint 8

According to the cost function in Example 8, is it possible to remove 100% of the smokestack pollutants? Why or why not?

SUMMARIZE   (Section 3.6)

1. State the definition of vertical asymptote *(page 183)*. For examples of vertical asymptotes, see Examples 1, 2, and 3.

2. State the definition of horizontal asymptote *(page 187)*. For examples of horizontal asymptotes, see Example 6.

3. Describe a real-life example of how asymptotic behavior can be used to analyze the average cost for a new product *(page 189, Example 7)*.

SKILLS WARM UP 3.6   The following warm-up exercises involve skills that were covered in earlier sections. You will use these skills in the exercise set for this section. For additional help, review Sections 1.5, 2.3, and 3.5.

In Exercises 1–8, find the limit.

1. $\lim\limits_{x \to 2} (x + 1)$

2. $\lim\limits_{x \to -1} (3x + 4)$

3. $\lim\limits_{x \to -3} \dfrac{2x^2 + x - 15}{x + 3}$

4. $\lim\limits_{x \to 2} \dfrac{3x^2 - 8x + 4}{x - 2}$

5. $\lim\limits_{x \to 2^+} \dfrac{x^2 - 5x + 6}{x^2 - 4}$

6. $\lim\limits_{x \to 1^-} \dfrac{x^2 - 6x + 5}{x^2 - 1}$

7. $\lim\limits_{x \to 0^+} \sqrt{x}$

8. $\lim\limits_{x \to 1^+} \left(x + \sqrt{x - 1}\right)$

In Exercises 9–12, find the average cost and the marginal cost.

9. $C = 150 + 3x$

10. $C = 1900 + 1.7x + 0.002x^2$

11. $C = 0.005x^2 + 0.5x + 1375$

12. $C = 760 + 0.05x$

3.7 Curve Sketching: A Summary

■ Analyze the graphs of functions.

■ Recognize the graphs of simple polynomial functions.

Summary of Curve-Sketching Techniques

It would be difficult to overstate the importance of using graphs in mathematics. Descartes's introduction of analytic geometry contributed significantly to the rapid advances in calculus that began during the mid-seventeenth century.

So far, you have studied several concepts that are useful in analyzing the graph of a function.

- x-intercepts and y-intercepts   (Section 1.2)
- Domain and range   (Section 1.4)
- Continuity   (Section 1.6)
- Differentiability   (Section 2.1)
- Relative extrema   (Section 3.2)
- Concavity   (Section 3.3)
- Points of inflection   (Section 3.3)
- Vertical asymptotes   (Section 3.6)
- Horizontal asymptotes   (Section 3.6)

In Exercise 45, you will analyze the graph of the Social Security average monthly benefits to determine whether the model is a good fit for the data.

When you are sketching the graph of a function, either by hand or with a graphing utility, remember that you cannot normally show the *entire* graph. The decision as to which part of the graph to show is crucial. For instance, which of the viewing windows in Figure 3.56 better represents the graph of

$$f(x) = x^3 - 25x^2 + 74x - 20?$$

Figure 3.56(a) gives a more complete view of the graph, but the context of the problem might indicate that Figure 3.56(b) is better.

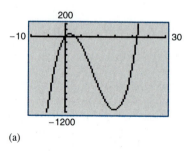

(a)

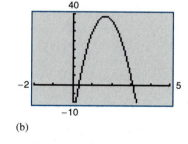

(b)

FIGURE 3.56

Guidelines for Analyzing the Graph of a Function

1. Determine the domain and range of the function. When the function models a real-life situation, consider the context.

2. Determine the intercepts and asymptotes of the graph.

3. Locate the x-values at which $f'(x)$ and $f''(x)$ are zero or undefined. Use the results to determine where the relative extrema and the points of inflection occur.

Example 1   Analyzing a Graph

Analyze the graph of

$$f(x) = x^3 + 3x^2 - 9x + 5.$$

SOLUTION   The y-intercept occurs at $(0, 5)$. Because this function factors as

$$f(x) = (x - 1)^2(x + 5) \qquad \text{Factored form}$$

the x-intercepts occur at $(-5, 0)$ and $(1, 0)$. The first derivative is

$$f'(x) = 3x^2 + 6x - 9 \qquad \text{First derivative}$$
$$= 3(x - 1)(x + 3). \qquad \text{Factored form}$$

So, the critical numbers of f are $x = 1$ and $x = -3$. The second derivative of f is

$$f''(x) = 6x + 6 \qquad \text{Second derivative}$$
$$= 6(x + 1) \qquad \text{Factored form}$$

which implies that the second derivative is zero when $x = -1$. By testing the values of $f'(x)$ and $f''(x)$, as shown in the table, you can see that f has one relative minimum, one relative maximum, and one point of inflection. The graph of f is shown in Figure 3.57.

	$f(x)$	$f'(x)$	$f''(x)$	Characteristics of graph
x in $(-\infty, -3)$		+	−	Increasing, concave downward
$x = -3$	32	0	−	Relative maximum
x in $(-3, -1)$		−	−	Decreasing, concave downward
$x = -1$	16	−	0	Point of inflection
x in $(-1, 1)$		−	+	Decreasing, concave upward
$x = 1$	0	0	+	Relative minimum
x in $(1, \infty)$		+	+	Increasing, concave upward

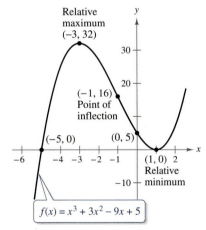

Relative maximum $(-3, 32)$

$(-1, 16)$ Point of inflection

$(-5, 0)$ $(0, 5)$

$(1, 0)$ Relative minimum

$f(x) = x^3 + 3x^2 - 9x + 5$

FIGURE 3.57

✓ Checkpoint 1

Analyze the graph of

$$f(x) = -x^3 + 3x^2 + 9x - 27.$$

Example 2   Analyzing a Graph

Analyze the graph of

$$f(x) = x^4 - 12x^3 + 48x^2 - 64x.$$

SOLUTION   One of the intercepts occurs at $(0, 0)$. Because this function factors as

$$f(x) = x(x^3 - 12x^2 + 48x - 64)$$
$$= x(x - 4)^3 \qquad \text{Factored form}$$

a second x-intercept occurs at $(4, 0)$. The first derivative is

$$f'(x) = 4x^3 - 36x^2 + 96x - 64 \qquad \text{First derivative}$$
$$= 4(x - 1)(x - 4)^2. \qquad \text{Factored form}$$

So, the critical numbers of f are $x = 1$ and $x = 4$. The second derivative of f is

$$f''(x) = 12x^2 - 72x + 96 \qquad \text{Second derivative}$$
$$= 12(x - 4)(x - 2) \qquad \text{Factored form}$$

which implies that the second derivative is zero when $x = 2$ and $x = 4$. By testing the values of $f'(x)$ and $f''(x)$, as shown in the table, you can see that f has one relative minimum and two points of inflection. The graph is shown in Figure 3.58.

	$f(x)$	$f'(x)$	$f''(x)$	Characteristics of graph
x in $(-\infty, 1)$		$-$	$+$	Decreasing, concave upward
$x = 1$	-27	0	$+$	Relative minimum
x in $(1, 2)$		$+$	$+$	Increasing, concave upward
$x = 2$	-16	$+$	0	Point of inflection
x in $(2, 4)$		$+$	$-$	Increasing, concave downward
$x = 4$	0	0	0	Point of inflection
x in $(4, \infty)$		$+$	$+$	Increasing, concave upward

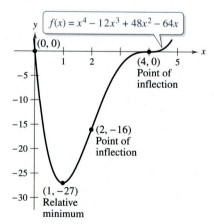

FIGURE 3.58

✓ Checkpoint 2

Analyze the graph of

$$f(x) = x^4 - 4x^3 + 5.$$

The fourth-degree polynomial function in Example 2 has one relative minimum and no relative maxima. In general, a polynomial function of degree n can have *at most* $n - 1$ relative extrema, and *at most* $n - 2$ points of inflection. Moreover, polynomial functions of even degree must have at least one relative extremum.

Example 3   Analyzing a Graph

Analyze the graph of

$$f(x) = \frac{x^2 - 2x + 4}{x - 2}.$$

SOLUTION   The y-intercept occurs at $(0, -2)$. Using the Quadratic Formula on the numerator, you can see that there are no x-intercepts. Because the denominator is zero when $x = 2$ (and the numerator is not zero when $x = 2$), it follows that $x = 2$ is a vertical asymptote of the graph. There are no horizontal asymptotes because the degree of the numerator is greater than the degree of the denominator. The first derivative is

$$f'(x) = \frac{(x - 2)(2x - 2) - (x^2 - 2x + 4)}{(x - 2)^2} \qquad \text{First derivative}$$

$$= \frac{x(x - 4)}{(x - 2)^2}. \qquad \text{Factored form}$$

So, the critical numbers of f are $x = 0$ and $x = 4$. The second derivative is

$$f''(x) = \frac{(x - 2)^2(2x - 4) - (x^2 - 4x)(2)(x - 2)}{(x - 2)^4} \qquad \text{Second derivative}$$

$$= \frac{(x - 2)(2x^2 - 8x + 8 - 2x^2 + 8x)}{(x - 2)^4}$$

$$= \frac{8}{(x - 2)^3}. \qquad \text{Factored form}$$

Because the second derivative has no zeros and because $x = 2$ is not in the domain of the function, you can conclude that the graph has no points of inflection. By testing the values of $f'(x)$ and $f''(x)$, as shown in the table, you can see that f has one relative minimum and one relative maximum. The graph of f is shown in Figure 3.59.

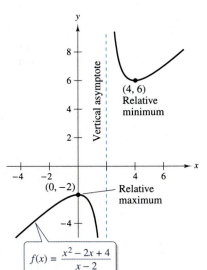

$$f(x) = \frac{x^2 - 2x + 4}{x - 2}$$

FIGURE 3.59

	$f(x)$	$f'(x)$	$f''(x)$	Characteristics of graph
x in $(-\infty, 0)$		$+$	$-$	Increasing, concave downward
$x = 0$	-2	0	$-$	Relative maximum
x in $(0, 2)$		$-$	$-$	Decreasing, concave downward
$x = 2$	Undef.	Undef.	Undef.	Vertical asymptote
x in $(2, 4)$		$-$	$+$	Decreasing, concave upward
$x = 4$	6	0	$+$	Relative minimum
x in $(4, \infty)$		$+$	$+$	Increasing, concave upward

✓ Checkpoint 3

Analyze the graph of

$$f(x) = \frac{x^2}{x - 1}.$$

Example 4 Analyzing a Graph

Analyze the graph of

$$f(x) = \frac{2(x^2 - 9)}{x^2 - 4}.$$

SOLUTION Begin by writing the function in factored form.

$$f(x) = \frac{2(x - 3)(x + 3)}{(x - 2)(x + 2)} \qquad \text{Factored form}$$

The y-intercept is $\left(0, \frac{9}{2}\right)$, and the x-intercepts are $(-3, 0)$ and $(3, 0)$. The graph of f has vertical asymptotes at $x = \pm 2$ and a horizontal asymptote at $y = 2$. The first derivative is

$$f'(x) = \frac{2[(x^2 - 4)(2x) - (x^2 - 9)(2x)]}{(x^2 - 4)^2} \qquad \text{First derivative}$$

$$= \frac{2(2x^3 - 8x - 2x^3 + 18x)}{(x^2 - 4)^2} \qquad \text{Multiply.}$$

$$= \frac{20x}{(x^2 - 4)^2}. \qquad \text{Factored form}$$

So, the critical number of f is $x = 0$. The second derivative of f is

$$f''(x) = \frac{(x^2 - 4)^2(20) - (20x)(2)(x^2 - 4)(2x)}{(x^2 - 4)^4} \qquad \text{Second derivative}$$

$$= \frac{20(x^2 - 4)(x^2 - 4 - 4x^2)}{(x^2 - 4)^4}$$

$$= -\frac{20(3x^2 + 4)}{(x^2 - 4)^3}. \qquad \text{Factored form}$$

Because the second derivative has no zeros and $x = \pm 2$ are not in the domain of the function, you can conclude that the graph has no points of inflection. By testing the values of $f'(x)$ and $f''(x)$, as shown in the table, you can see that f has one relative minimum. The graph of f is shown in Figure 3.60.

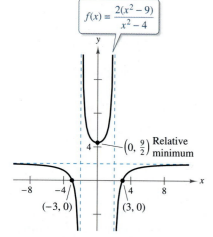

FIGURE 3.60

	$f(x)$	$f'(x)$	$f''(x)$	Characteristics of graph
x in $(-\infty, -2)$		$-$	$-$	Decreasing, concave downward
$x = -2$	Undef.	Undef.	Undef.	Vertical asymptote
x in $(-2, 0)$		$-$	$+$	Decreasing, concave upward
$x = 0$	$\frac{9}{2}$	0	$+$	Relative minimum
x in $(0, 2)$		$+$	$+$	Increasing, concave upward
$x = 2$	Undef.	Undef.	Undef.	Vertical asymptote
x in $(2, \infty)$		$+$	$-$	Increasing, concave downward

✔ **Checkpoint 4**

Analyze the graph of

$$f(x) = \frac{x^2 + 1}{x^2 - 1}.$$

TECH TUTOR

Some graphing utilities will not graph the function in Example 5 properly when the function is entered as

$$f(x) = 2x\text{^}(5/3) - 5x\text{^}(4/3).$$

To correct for this, you can enter the function as

$$f(x) = 2\left(\sqrt[3]{x}\right)\text{^}5 - 5\left(\sqrt[3]{x}\right)\text{^}4.$$

Try entering both functions into a graphing utility to see whether both functions produce correct graphs.

ALGEBRA TUTOR xy

For help on the algebra in Example 5, see Example 2(a) in the *Chapter 3 Algebra Tutor*, on page 207.

Example 5   **Analyzing a Graph**

Analyze the graph of

$$f(x) = 2x^{5/3} - 5x^{4/3}.$$

SOLUTION   Begin by writing the function in factored form.

$$f(x) = x^{4/3}(2x^{1/3} - 5) \qquad \text{Factored form}$$

One of the intercepts is $(0, 0)$. A second x-intercept occurs when $2x^{1/3} - 5 = 0$.

$$2x^{1/3} - 5 = 0$$
$$2x^{1/3} = 5$$
$$x^{1/3} = \frac{5}{2}$$
$$x = \left(\frac{5}{2}\right)^3$$
$$x = \frac{125}{8}$$

The first derivative is

$$f'(x) = \frac{10}{3}x^{2/3} - \frac{20}{3}x^{1/3} \qquad \text{First derivative}$$
$$= \frac{10}{3}x^{1/3}(x^{1/3} - 2). \qquad \text{Factored form}$$

So, the critical numbers of f are $x = 0$ and $x = 8$. The second derivative is

$$f''(x) = \frac{20}{9}x^{-1/3} - \frac{20}{9}x^{-2/3} \qquad \text{Second derivative}$$
$$= \frac{20}{9}x^{-2/3}(x^{1/3} - 1)$$
$$= \frac{20(x^{1/3} - 1)}{9x^{2/3}}. \qquad \text{Factored form}$$

So, possible points of inflection occur at $x = 1$ and when $x = 0$. By testing the values of $f'(x)$ and $f''(x)$, as shown in the table, you can see that f has one relative maximum, one relative minimum, and one point of inflection. The graph of f is shown in Figure 3.61.

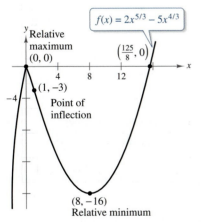

$f(x) = 2x^{5/3} - 5x^{4/3}$

Relative maximum $(0, 0)$

$\left(\frac{125}{8}, 0\right)$

$(1, -3)$ Point of inflection

$(8, -16)$ Relative minimum

FIGURE 3.61

	$f(x)$	$f'(x)$	$f''(x)$	Characteristics of graph
x in $(-\infty, 0)$		$+$	$-$	Increasing, concave downward
$x = 0$	0	0	Undef.	Relative maximum
x in $(0, 1)$		$-$	$-$	Decreasing, concave downward
$x = 1$	-3	$-$	0	Point of inflection
x in $(1, 8)$		$-$	$+$	Decreasing, concave upward
$x = 8$	-16	0	$+$	Relative minimum
x in $(8, \infty)$		$+$	$+$	Increasing, concave upward

✓**Checkpoint 5**

Analyze the graph of $f(x) = 2x^{3/2} - 6x^{1/2}$.

Summary of Simple Polynomial Graphs

A summary of the graphs of polynomial functions of degrees 0, 1, 2, and 3 is shown in Figure 3.62. Because of their simplicity, lower-degree polynomial functions are commonly used as mathematical models.

Constant function (degree 0): $y = a$

Horizontal line

Linear function (degree 1): $y = ax + b$

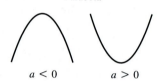

Line of slope a

$a < 0$ $a > 0$

Quadratic function (degree 2): $y = ax^2 + bx + c$

Parabola

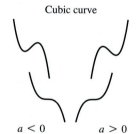

$a < 0$ $a > 0$

Cubic function (degree 3): $y = ax^3 + bx^2 + cx + d$

Cubic curve

$a < 0$ $a > 0$

FIGURE 3.62

SUMMARIZE   (Section 3.7)

1. List the concepts you have learned that are useful in analyzing the graph of a function *(page 192)*. For an example that uses some of these concepts to analyze the graph of a function, see Example 1.

2. State the guidelines for analyzing the graph of a function *(page 192)*. For examples that use these guidelines, see Examples 3, 4, and 5.

3. State a general rule relating the degree n of a polynomial function with (a) the number of relative extrema and (b) the number of points of inflection *(page 195)*. For an example where this rule can be used to analyze the graph of a polynomial function, see Example 2.

SKILLS WARM UP 3.7

The following warm-up exercises involve skills that were covered in earlier sections. You will use these skills in the exercise set for this section. For additional help, review Sections 3.1 and 3.6.

In Exercises 1–4, find the vertical and horizontal asymptotes of the graph.

1. $f(x) = \dfrac{1}{x^2}$

2. $f(x) = \dfrac{8}{(x - 2)^2}$

3. $f(x) = \dfrac{40x}{x + 3}$

4. $f(x) = \dfrac{x^2 - 3}{x^2 - 4x + 3}$

In Exercises 5–10, determine the open intervals on which the function is increasing or decreasing.

5. $f(x) = x^2 + 4x + 2$

6. $f(x) = -x^2 - 8x + 1$

7. $f(x) = x^3 - 3x + 1$

8. $f(x) = \dfrac{-x^3 + x^2 - 1}{x^2}$

9. $f(x) = \dfrac{x - 2}{x - 1}$

10. $f(x) = -x^3 - 4x^2 + 3x + 2$

ENHANCED WebAssign   Access end-of-section exercises online at **www.webassign.net**

3.8 Differentials and Marginal Analysis

■ Find the differentials of functions.
■ Use differentials in economics to approximate changes in revenue, cost, and profit.
■ Find the differential of a function using differentiation formulas.

Differentials

When the derivative was defined in Section 2.1 as the limit of the ratio $\Delta y/\Delta x$, it seemed natural to retain the quotient symbolism for the limit itself. So, the derivative of y with respect to x was denoted by

$$\frac{dy}{dx} = \lim_{\Delta x \to 0} \frac{\Delta y}{\Delta x}$$

even though dy/dx was not interpreted as the quotient of two separate quantities. In this section, you will see that the quantities dy and dx can be assigned meanings in such a way that their quotient, when $dx \neq 0$, is equal to the derivative of y with respect to x.

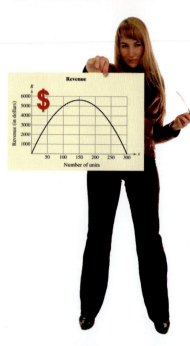

In Exercise 35, you will use differentials to approximate the change in revenue for a one-unit increase in sales of a product.

Definition of Differentials

Let $y = f(x)$ represent a differentiable function. The **differential of x** (denoted by dx) is any nonzero real number. The **differential of y** (denoted by dy) is
$dy = f'(x)\, dx$.

In the definition of differentials, dx can have any nonzero value. In most applications, however, dx is chosen to be small, and this choice is denoted by $dx = \Delta x$.

One use of differentials is in approximating the change in $f(x)$ that corresponds to a change in x, as shown in Figure 3.63. This change is denoted by

$$\Delta y = f(x + \Delta x) - f(x). \qquad \text{Change in } y$$

In Figure 3.63, notice that as Δx gets smaller and smaller, the values of dy and Δy get closer and closer. That is, when Δx is small, $dy \approx \Delta y$. This **tangent line approximation** is the basis for most applications of differentials.

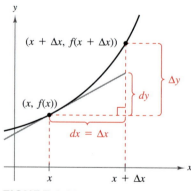

FIGURE 3.63

Note in Figure 3.63 that near the point of tangency, the graph of f is very close to the tangent line. This is the essence of the approximations used in this section. In other words, near the point of tangency, $dy \approx \Delta y$.

> **Example 1**   Comparing Δy and dy

Consider the function given by

$$f(x) = x^2.$$

Find the value of dy when $x = 1$ and $dx = 0.01$. Compare this with the value of Δy when $x = 1$ and $\Delta x = 0.01$.

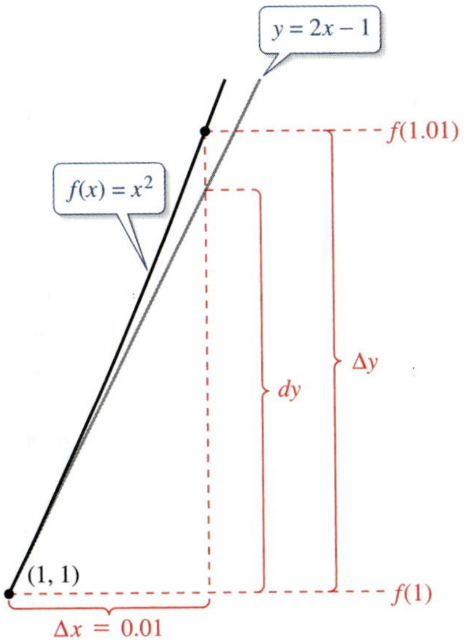

SOLUTION   Begin by finding the derivative of f.

$$f'(x) = 2x \qquad \text{Derivative of } f$$

When $x = 1$ and $dx = 0.01$, the value of the differential dy is

$$\begin{aligned} dy &= f'(x)\,dx & &\text{Differential of } y \\ &= f'(1)(0.01) & &\text{Substitute 1 for } x \text{ and 0.01 for } dx. \\ &= 2(1)(0.01) & &\text{Use } f'(x) = 2x. \\ &= 0.02. & &\text{Simplify.} \end{aligned}$$

When $x = 1$ and $\Delta x = 0.01$, the value of Δy is

$$\begin{aligned} \Delta y &= f(x + \Delta x) - f(x) & &\text{Change in } y \\ &= f(1.01) - f(1) & &\text{Substitute 1 for } x \text{ and 0.01 for } \Delta x. \\ &= (1.01)^2 - (1)^2 \\ &= 1.0201 - 1 \\ &= 0.0201. & &\text{Simplify.} \end{aligned}$$

Note that $dy \approx \Delta y$, as shown in Figure 3.64.

FIGURE 3.64

> ✓ **Checkpoint 1**

Find the value of dy when $x = 2$ and $dx = 0.01$ for $f(x) = x^4$. Compare this with the value of Δy when $x = 2$ and $\Delta x = 0.01$. ■

In Example 1, the tangent line to the graph of $f(x) = x^2$ at $x = 1$ is

$$y = 2x - 1 \quad \text{or} \quad g(x) = 2x - 1. \qquad \text{Tangent line to the graph of } f \text{ at } x = 1$$

For x-values near 1, this line is close to the graph of f, as shown in Figure 3.64. For instance,

$$f(1.01) = 1.01^2 = 1.0201 \quad \text{and} \quad g(1.01) = 2(1.01) - 1 = 1.02.$$

The validity of the approximation

$$dy \approx \Delta y, \quad dx \neq 0$$

stems from the definition of the derivative. That is, the existence of the limit

$$f'(x) = \lim_{\Delta x \to 0} \frac{f(x + \Delta x) - f(x)}{\Delta x}$$

implies that when Δx is close to zero, then $f'(x)$ is close to the difference quotient. So, you can write

$$\frac{f(x + \Delta x) - f(x)}{\Delta x} \approx f'(x)$$

$$f(x + \Delta x) - f(x) \approx f'(x)\,\Delta x$$

$$\Delta y \approx f'(x)\,\Delta x.$$

Substituting dx for Δx and dy for $f'(x)\,dx$ produces $\Delta y \approx dy$.

Marginal Analysis

Differentials are used in economics to approximate changes in revenue, cost, and profit. Let $R = f(x)$ be the total revenue for selling x units of a product. When the number of units increases by 1, the change in x is $\Delta x = 1$, and the change in R is

$$\Delta R = f(x + \Delta x) - f(x) \approx dR = \frac{dR}{dx}\, dx.$$

In other words, you can use the differential dR to approximate the change in the revenue that accompanies the sale of one additional unit. Similarly, the differentials dC and dP can be used to approximate the changes in cost and profit that accompany the sale (or production) of one additional unit.

 Example 2   **Using Marginal Analysis**

The demand function for a product is modeled by

$$p = 400 - x, \quad 0 \le x \le 400$$

where p is the price per unit (in dollars) and x is the number of units. Use differentials to approximate the change in revenue as sales increase from 149 units to 150 units. Compare this with the actual change in revenue.

SOLUTION   Begin by finding the revenue function. Because the demand is given by $p = 400 - x$, the revenue is

$$
\begin{aligned}
R &= xp & &\text{Formula for revenue} \\
&= x(400 - x) & &\text{Use } p = 400 - x. \\
&= 400x - x^2. & &\text{Multiply.}
\end{aligned}
$$

Next, find the marginal revenue, dR/dx.

$$\frac{dR}{dx} = 400 - 2x \qquad \text{Power Rule}$$

When $x = 149$ and $dx = \Delta x = 1$, the approximate change in the revenue is

$$
\begin{aligned}
\Delta R &\approx dR \\
&= \frac{dR}{dx}\, dx \\
&= (400 - 2x)\, dx \\
&= [400 - 2(149)](1) \\
&= \$102.
\end{aligned}
$$

When x increases from 149 to 150 and $R = f(x) = 400x - x^2$, the actual change in revenue is

$$
\begin{aligned}
\Delta R &= f(x + \Delta x) - f(x) \\
&= \left[400(150) - 150^2\right] - \left[400(149) - 149^2\right] \\
&= 37{,}500 - 37{,}399 \\
&= \$101.
\end{aligned}
$$

✓ Checkpoint 2

The demand function for a product is modeled by $p = 200 - x$, $0 \le x \le 200$, where p is the price per unit (in dollars) and x is the number of units. Use differentials to approximate the change in revenue as sales increase from 89 to 90 units. Compare this with the actual change in revenue.

Example 3 Using Marginal Analysis

The profit (in dollars) derived from selling x units of an item is modeled by

$$P = 0.0002x^3 + 10x.$$

Use the differential dP to approximate the change in profit when the production level changes from 50 to 51 units. Compare this with the actual gain in profit obtained by increasing the production level from 50 to 51 units.

SOLUTION   The marginal profit is

$$\frac{dP}{dx} = 0.0006x^2 + 10.$$

When $x = 50$ and $dx = \Delta x = 1$, the approximate change in profit is

$$\Delta P \approx dP$$
$$= \frac{dP}{dx}\, dx$$
$$= (0.0006x^2 + 10)\, dx$$
$$= [0.0006(50)^2 + 10](1)$$
$$= \$11.50.$$

When x changes from 50 to 51 units and $P = f(x) = 0.0002x^3 + 10x$, the actual change in profit is

$$\Delta P = f(x + \Delta x) - f(x)$$
$$= [(0.0002)(51)^3 + 10(51)] - [(0.0002)(50)^3 + 10(50)]$$
$$\approx 536.53 - 525.00$$
$$= \$11.53.$$

These values are shown graphically in Figure 3.65.

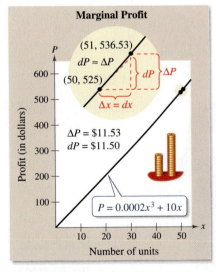

Marginal Profit

FIGURE 3.65

✓ **Checkpoint 3**

Use the differential dP to approximate the change in profit for the profit function in Example 3 when the production level changes from 40 to 41 units. Compare this with the actual gain in profit obtained by increasing the production level from 40 to 41 units.

Formulas for Differentials

You can use the definition of differentials to rewrite each differentiation rule in **differential form.**

Differential Forms of Differentiation Rules

Constant Multiple Rule: $d[cu] = c\,du$

Sum or Difference Rule: $d[u \pm v] = du \pm dv$

Product Rule: $d[uv] = u\,dv + v\,du$

Quotient Rule: $d\left[\dfrac{u}{v}\right] = \dfrac{v\,du - u\,dv}{v^2}$

Constant Rule: $d[c] = 0$

Power Rule: $d[x^n] = nx^{n-1}\,dx$

The next example compares the derivatives and differentials of several simple functions.

Example 4   **Finding Differentials**

Function	Derivative	Differential
a. $y = x^2$	$\dfrac{dy}{dx} = 2x$	$dy = 2x\,dx$
b. $y = \dfrac{3x + 2}{5}$	$\dfrac{dy}{dx} = \dfrac{3}{5}$	$dy = \dfrac{3}{5}\,dx$
c. $y = 2x^2 - 3x$	$\dfrac{dy}{dx} = 4x - 3$	$dy = (4x - 3)\,dx$
d. $y = \dfrac{1}{x}$	$\dfrac{dy}{dx} = -\dfrac{1}{x^2}$	$dy = -\dfrac{1}{x^2}\,dx$

 Checkpoint 4

Find the differential dy of each function.

a. $y = 4x^3$ **b.** $y = \dfrac{2x + 1}{3}$

c. $y = 3x^2 - 2x$ **d.** $y = \dfrac{1}{x^2}$

SUMMARIZE   (Section 3.8)

1. State the definition of differentials *(page 200)*. For an example of a differential, see Example 1.

2. Explain what is meant by marginal analysis *(page 202)*. For examples of marginal analysis, see Examples 2 and 3.

3. State the differential forms of the differential rules *(page 204)*. For an example that uses the differential forms of the differential rules, see Example 4.

SKILLS WARM UP 3.8

The following warm-up exercises involve skills that were covered in earlier sections. You will use these skills in the exercise set for this section. For additional help, review Sections 2.2 and 2.4.

In Exercises 1–12, find the derivative.

1. $C = 44 + 0.09x^2$

2. $C = 250 + 0.15x$

3. $R = x(1.25 + 0.02\sqrt{x})$

4. $R = x(15.5 - 1.55x)$

5. $P = -0.03x^{1/3} + 1.4x - 2250$

6. $P = -0.02x^2 + 25x - 1000$

7. $A = \frac{1}{4}\sqrt{3}x^2$

8. $A = 6x^2$

9. $C = 2\pi r$

10. $P = 4w$

11. $S = 4\pi r^2$

12. $P = 2x + \sqrt{2}x$

In Exercises 13–16, write a formula for the quantity.

13. Area A of a circle of radius r

14. Area A of a square of side x

15. Volume V of a cube of edge x

16. Volume V of a sphere of radius r

ENHANCED
WebAssign Access end-of-section exercises online at **www.webassign.net**

ALGEBRA TUTOR

xy

Solving Equations

Much of the algebra in Chapter 3 involves simplifying algebraic expressions (see pages 133 and 134) and solving algebraic equations (see page 58). The Algebra Tutor on page 58 illustrates some of the basic techniques for solving equations. On these two pages, you can review some of the more complicated techniques for solving equations.

When solving an equation, remember that your basic goal is to isolate the variable on one side of the equation. To do this, you use inverse operations. For instance, to isolate x in

$$x - 2 = 0$$

you add 2 to each side of the equation, because *addition* is the inverse operation of *subtraction*. To isolate x in

$$\sqrt{x} = 2$$

you square each side of the equation, because *squaring* is the inverse operation of *taking the square root*.

Example 1   Solving Equations

Solve each equation.

a. $\dfrac{36(x^2 - 1)}{(x^2 + 3)^3} = 0$ **b.** $0 = 2x(2x^2 - 3)$ **c.** $\dfrac{dV}{dx} = 0$, where $V = 27x - \dfrac{1}{4}x^3$

SOLUTION

a. $\dfrac{36(x^2 - 1)}{(x^2 + 3)^3} = 0$ —— Example 2, page 162

$36(x^2 - 1) = 0$ —— A fraction is equal to zero only when its numerator is zero.

$x^2 - 1 = 0$ —— Divide each side by 36.

$x^2 = 1$ —— Add 1 to each side.

$x = \pm 1$ —— Take the square root of each side.

b. $0 = 2x(2x^2 - 3)$ —— Example 2, page 170

$2x = 0 \implies x = 0$ —— Set first factor equal to zero.

$2x^2 - 3 = 0 \implies x = \pm\sqrt{\dfrac{3}{2}}$ —— Set second factor equal to zero.

c. $V = 27x - \dfrac{1}{4}x^3$ —— Example 1, page 168

$\dfrac{dV}{dx} = 27 - \dfrac{3}{4}x^2$ —— Find derivative of V.

$0 = 27 - \dfrac{3}{4}x^2$ —— Set derivative equal to 0.

$\dfrac{3}{4}x^2 = 27$ —— Add $\frac{3}{4}x^2$ to each side.

$x^2 = 36$ —— Multiply each side by $\frac{4}{3}$.

$x = \pm 6$ —— Take the square root of each side.

Example 2 Solving Equations

Solve each equation.

a. $\dfrac{20(x^{1/3} - 1)}{9x^{2/3}} = 0$ **b.** $\dfrac{25}{\sqrt{x}} - 0.5 = 0$

c. $x^2(4x - 3) = 0$ **d.** $\dfrac{4x}{3(x^2 - 4)^{1/3}} = 0$

e. $g'(x) = 0$, where $g(x) = (x - 2)(x + 1)^2$

SOLUTION

a. $\dfrac{20(x^{1/3} - 1)}{9x^{2/3}} = 0$ Example 5, page 197

$20(x^{1/3} - 1) = 0$ A fraction is equal to zero only when its numerator is zero.

$x^{1/3} - 1 = 0$ Divide each side by 20.

$x^{1/3} = 1$ Add 1 to each side.

$x = 1$ Cube each side.

b. $\dfrac{25}{\sqrt{x}} - 0.5 = 0$ Example 4, page 178

$\dfrac{25}{\sqrt{x}} = 0.5$ Add 0.5 to each side.

$25 = 0.5\sqrt{x}$ Multiply each side by $\sqrt{x}$.

$50 = \sqrt{x}$ Divide each side by 0.5.

$2500 = x$ Square both sides.

c. $x^2(4x - 3) = 0$ Example 2, page 155

$x^2 = 0$ ⟹ $x = 0$ Set first factor equal to zero.

$4x - 3 = 0$ ⟹ $x = \frac{3}{4}$ Set second factor equal to zero.

d. $\dfrac{4x}{3(x^2 - 4)^{1/3}} = 0$ Example 5, page 148

$4x = 0$ A fraction is equal to zero only when its numerator is zero.

$x = 0$ Divide each side by 4.

e. $g(x) = (x - 2)(x + 1)^2$ Exercise 45

$g'(x) = (x - 2)(2)(x + 1) + (x + 1)^2(1)$ Find derivative of g.

$(x - 2)(2)(x + 1) + (x + 1)^2(1) = 0$ Set derivative equal to zero.

$(x + 1)[2(x - 2) + (x + 1)] = 0$ Factor.

$(x + 1)(2x - 4 + x + 1) = 0$ Multiply factors.

$(x + 1)(3x - 3) = 0$ Combine like terms.

$x + 1 = 0$ ⟹ $x = -1$ Set first factor equal to zero.

$3x - 3 = 0$ ⟹ $x = 1$ Set second factor equal to zero.

SUMMARY AND STUDY STRATEGIES

After studying this chapter, you should have acquired the following skills. The exercise numbers are keyed to the Review Exercises that begin on page 210. Answers to odd-numbered Review Exercises are given in the back of the text.*

Section 3.1

Review Exercises

- Find the critical numbers of a function.   *1–6*

  c is a critical number of f when $f'(c) = 0$ or $f'(c)$ is undefined.

- Find the open intervals on which a function is increasing or decreasing.   *7–12*

  f is increasing when $f'(x) > 0$.

  f is decreasing when $f'(x) < 0$.

- Find intervals on which a real-life model is increasing or decreasing.   *13, 14*

Section 3.2

- Use the First-Derivative Test to find the relative extrema of a function.   *15–24*
- Find the absolute extrema of a continuous function on a closed interval.   *25–32*
- Find minimum and maximum values of a real-life model and interpret the results in context.   *33, 34*

Section 3.3

- Find the open intervals on which the graph of a function is concave upward or concave downward.   *35–38*

  f is concave upward when $f''(x) > 0$.

  f is concave downward when $f''(x) < 0$.

- Find the points of inflection of the graph of a function.   *39–42*
- Use the Second-Derivative Test to find the relative extrema of a function.   *43–48*
- Find the point of diminishing returns of an input-output model.   *49, 50*

Section 3.4

- Solve real-life optimization problems.   *51–54*

Section 3.5

- Solve business and economics optimization problems.   *55–60*
- Find the price elasticity of demand for a demand function.   *61, 62*

---

* A wide range of valuable study aids are available to help you master the material in this chapter. The *Student Solutions Manual* includes step-by-step solutions to all odd-numbered exercises to help you review and prepare. The student website at *www.cengagebrain.com* offers algebra help and a *Graphing Technology Guide*, which contains step-by-step commands and instructions for a wide variety of graphing calculators.

Section 3.6

Section 3.7

Section 3.8

Study Strategies

■ **Solve Problems Graphically, Analytically, and Numerically**   When analyzing the graph of a function, use a variety of problem-solving strategies. For instance, if you were asked to analyze the graph of

$$f(x) = x^3 - 4x^2 + 5x - 4$$

you could begin *graphically*. That is, you could use a graphing utility to find a viewing window that appears to show the important characteristics of the graph. From the graph shown below, the function appears to have one relative minimum, one relative maximum, and one point of inflection.

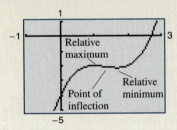

Next, you could use calculus to *analyze* the graph. Because the derivative of f is

$$f'(x) = 3x^2 - 8x + 5 = (3x - 5)(x - 1)$$

the critical numbers of f are $x = \frac{5}{3}$ and $x = 1$. By the First-Derivative Test, you can conclude that $x = \frac{5}{3}$ yields a relative minimum and $x = 1$ yields a relative maximum. Because

$$f''(x) = 6x - 8$$

you can conclude that $x = \frac{4}{3}$ yields a point of inflection. Finally, you could analyze the graph *numerically*. For instance, you could construct a table of values and observe that f is increasing on the interval $(-\infty, 1)$, decreasing on the interval $\left(1, \frac{5}{3}\right)$, and increasing on the interval $\left(\frac{5}{3}, \infty\right)$.

■ **Problem-Solving Strategies**   When you get stuck while trying to solve an optimization problem, consider the strategies below.

1. *Draw a Diagram.*   If feasible, draw a diagram that represents the problem. Label all known values and unknown values on the diagram.

2. *Solve a Simpler Problem.*   Simplify the problem, or write several simple examples of the problem. For instance, if you are asked to find the dimensions that will produce a maximum area, try calculating the areas of several examples.

3. *Rewrite the Problem in Your Own Words.*   Rewriting a problem can help you understand it better.

4. *Guess and Check.*   Try guessing the answer, then check your guess in the statement of the original problem. By refining your guesses, you may be able to think of a general strategy for solving the problem.

Review Exercises

Finding Critical Numbers In Exercises 1–6, find the critical numbers of the function.

1. $f(x) = -x^2 + 2x + 4$

2. $y = 3x^2 + 18x$

3. $y = 4x^3 - 108x$

4. $f(x) = x^4 - 8x^2 + 13$

5. $g(x) = (x - 1)^2(x - 3)$

6. $h(x) = \sqrt{x}(x - 3)$

Intervals on Which f Is Increasing or Decreasing In Exercises 7–12, find the critical numbers and the open intervals on which the function is increasing or decreasing. Use a graphing utility to verify your results.

7. $f(x) = x^2 + x - 2$

8. $g(x) = (x + 2)^3$

9. $f(x) = -x^3 + 6x^2 - 2$

10. $y = x^3 - 12x^2$

11. $y = (x - 1)^{2/3}$

12. $y = 2x^{1/3} - 3$

13. Revenue The revenue R of Chipotle Mexican Grill (in millions of dollars) from 2004 through 2009 can be modeled by

$$R = 6.268t^2 + 136.07t - 191.3, \quad 4 \le t \le 9$$

where t is the time in years, with $t = 4$ corresponding to 2004. Show that the sales were increasing from 2004 through 2009. *(Source: Chipotle Mexican Grill, Inc.)*

14. Revenue The revenue R of Cintas (in millions of dollars) from 2000 through 2010 can be modeled by

$$R = -5.5778t^3 + 67.524t^2 + 45.22t + 1969.2$$

for $0 \le t \le 10$, where t is the time in years, with $t = 0$ corresponding to 2000. *(Source: Cintas Corporation)*

(a) Use a graphing utility to graph the model. Then graphically estimate the years during which the revenue was increasing and the years during which the revenue was decreasing.

(b) Use the test for increasing and decreasing functions to verify the result of part (a).

Finding Relative Extrema In Exercises 15–24, use the First-Derivative Test to find all relative extrema of the function. Use a graphing utility to verify your result.

15. $f(x) = 4x^3 - 6x^2 - 2$

16. $f(x) = \frac{1}{4}x^4 - 8x$

17. $g(x) = x^2 - 16x + 12$

18. $h(x) = 4 + 10x - x^2$

19. $h(x) = 2x^2 - x^4$

20. $s(x) = x^4 - 8x^2 + 3$

21. $f(x) = \dfrac{6}{x^2 + 1}$

22. $f(x) = \dfrac{2}{x^2 - 1}$

23. $h(x) = \dfrac{x^2}{x - 2}$

24. $g(x) = x - 6\sqrt{x}, \quad x > 0$

Finding Extrema on a Closed Interval In Exercises 25–32, find the absolute extrema of the function on the closed interval. Use a graphing utility to verify your result.

25. $f(x) = x^2 + 5x + 6; \quad [-3, 0]$

26. $f(x) = x^4 - 2x^3; \quad [0, 2]$

27. $f(x) = x^3 - 12x + 1; \quad [-4, 4]$

28. $f(x) = x^3 + 2x^2 - 3x + 4; \quad [-3, 2]$

29. $f(x) = 2\sqrt{x} - x; \quad [0, 9]$

30. $f(x) = \dfrac{x}{\sqrt{x^2 + 1}}; \quad [0, 2]$

31. $f(x) = \dfrac{2x}{x^2 + 1}; \quad [-1, 2]$

32. $f(x) = \dfrac{8}{x} + x; \quad [1, 4]$

33. Surface Area A right circular cylinder of radius r and height h has a volume of 25 cubic inches (see figure). The total surface area of the cylinder in terms of r is given by

$$S = 2\pi r\left(r + \frac{25}{\pi r^2}\right).$$

Find the radius that will minimize the surface area. Use a graphing utility to verify your result.

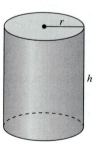

34. Profit   The profit P (in dollars) made by a company from selling x tablet computers can be modeled by

$$P = 1.64x - \frac{x^2}{15,000} - 2500.$$

Find the number of units sold that will yield a maximum profit. What is the maximum profit?

Applying the Test for Concavity   In Exercises 35–38, determine the open intervals on which the graph of the function is concave upward or concave downward.

35. $f(x) = (x - 2)^3$

36. $h(x) = x^5 - 10x^2$

37. $g(x) = \frac{1}{4}(-x^4 + 8x^2 - 12)$

38. $h(x) = x^3 - 6x$

Finding Points of Inflection   In Exercises 39–42, discuss the concavity of the graph of the function and find the points of inflection.

39. $f(x) = \frac{1}{2}x^4 - 4x^3$

40. $f(x) = \frac{1}{4}x^4 - 2x^2 - x$

41. $f(x) = x^3(x - 3)^2$

42. $f(x) = (x - 1)^2(x - 3)$

Using the Second-Derivative Test   In Exercises 43–48, use the Second-Derivative Test to find all relative extrema of the function.

43. $f(x) = x^3 - 6x^2 + 12x$

44. $f(x) = x^4 - 32x^2 + 12$

45. $f(x) = x^5 - 5x^3$

46. $f(x) = x(x^2 - 3x - 9)$

47. $f(x) = 2x^2(1 - x^2)$

48. $f(x) = x - 4\sqrt{x + 1}$

Point of Diminishing Returns   In Exercises 49 and 50, find the point of diminishing returns for the function. For each function, R is the revenue (in thousands of dollars) and x is the amount spent (in thousands of dollars) on advertising. Use a graphing utility to verify your result.

49. $R = \frac{1}{1500}(150x^2 - x^3), \quad 0 \le x \le 100$

50. $R = -\frac{2}{3}(x^3 - 12x^2 - 6), \quad 0 \le x \le 8$

51. Minimum Perimeter   Find the length and width of a rectangle that has an area of 225 square meters and a minimum perimeter.

52. Maximum Volume   A rectangular solid with a square base has a surface area of 432 square centimeters.

(a) Determine the dimensions that yield the maximum volume.

(b) Find the maximum volume.

53. Maximum Volume   An open box is to be made from a 10-inch by 16-inch rectangular piece of material by cutting equal squares from the corners and turning up the sides (see figure). Find the volume of the largest box that can be made.

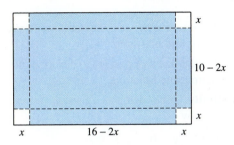

54. Minimum Area   A rectangular page is to contain 108 square inches of print. The margins at the top and bottom of the page are to be $\frac{3}{4}$ inch wide. The margins on each side are to be 1 inch wide. Find the dimensions of the page that will minimize the amount of paper used.

Finding the Maximum Revenue   In Exercises 55 and 56, find the number of units x that produces a maximum revenue R.

55. $R = 450x - 0.25x^2$

56. $R = 36x^2 - 0.05x^3$

Finding the Minimum Average Cost   In Exercises 57 and 58, find the number of units x that produces the minimum average cost per unit $\overline{C}$.

57. $C = 0.2x^2 + 10x + 4500$

58. $C = 0.03x^3 + 30x + 3840$

59. Maximum Profit   A commodity has a demand function modeled by

$$p = 36 - 4x$$

and a total cost function modeled by

$$C = 2x^2 + 6$$

where x is the number of units.

(a) What price yields a maximum profit?

(b) When the profit is maximized, what is the average cost per unit?

60. Maximum Profit   The profit P (in thousands of dollars) for a company in terms of the amount s spent on advertising (in thousands of dollars) can be modeled by

$$P = -4s^3 + 72s^2 - 240s + 500.$$

Find the amount of advertising that maximizes the profit. Find the point of diminishing returns.

61. Elasticity The demand function for a product is modeled by

$$p = 60 - 0.04x, \quad 0 \le x \le 1500$$

where p is the price (in dollars) and x is the number of units.

(a) Determine when the demand is elastic, inelastic, and of unit elasticity.

(b) Use the result of part (a) to describe the behavior of the revenue function.

62. Elasticity The demand function for a product is modeled by

$$p = 960 - x, \quad 0 \le x \le 960$$

where p is the price (in dollars) and x is the number of units.

(a) Determine when the demand is elastic, inelastic, and of unit elasticity.

(b) Use the result of part (a) to describe the behavior of the revenue function.

Finding Vertical Asymptotes In Exercises 63–66, determine all vertical asymptotes of the graph of the function.

63. $f(x) = \dfrac{x + 4}{x^2 + 7x}$

64. $f(x) = \dfrac{x - 1}{x^2 - 4}$

65. $f(x) = \dfrac{x^2 - 16}{2x^2 + 9x + 4}$

66. $f(x) = \dfrac{x^2 + 6x + 9}{x^2 - 5x - 24}$

 Determining Infinite Limits In Exercises 67–70, use a graphing utility to find the limit.

67. $\displaystyle\lim_{x \to 0^+} \left(x - \frac{1}{x^3} \right)$

68. $\displaystyle\lim_{x \to 0^-} \left(3 + \frac{1}{x} \right)$

69. $\displaystyle\lim_{x \to -1^+} \frac{x^2 - 2x + 1}{x + 1}$

70. $\displaystyle\lim_{x \to 3^-} \frac{3x^2 + 1}{x^2 - 9}$

Finding Horizontal Asymptotes In Exercises 71–74, find the horizontal asymptote of the graph of the function.

71. $f(x) = \dfrac{2x^2}{3x^2 + 5}$

72. $f(x) = \dfrac{3x^2 - 2x + 3}{x + 1}$

73. $f(x) = \dfrac{3x}{x^2 + 1}$

74. $f(x) = \dfrac{x}{x - 2} + \dfrac{2x}{x + 2}$

75. Average Cost The cost C (in dollars) of producing x units of a product is

$$C = 0.75x + 4000.$$

(a) Find the average cost function $\overline{C}$.

(b) Find $\overline{C}$ when $x = 100$ and when $x = 1000$.

(c) Determine the limit of the average cost function as x approaches infinity. Interpret the limit in the context of the problem.

76. Average Cost The cost C (in dollars) of producing x units of a product is

$$C = 1.50x + 8000.$$

(a) Find the average cost function $\overline{C}$.

(b) Find $\overline{C}$ when $x = 1000$ and when $x = 10,000$.

(c) Determine the limit of the average cost function as x approaches infinity. Interpret the limit in the context of the problem.

77. Seizing Drugs The cost C (in millions of dollars) for the federal government to seize $p\%$ of an illegal drug as it enters the country is modeled by

$$C = \frac{250p}{100 - p}, \quad 0 \le p < 100.$$

(a) Find the costs of seizing 20%, 50%, and 90%.

(b) Find the limit of C as $p \to 100^-$. Interpret the limit in the context of the problem. Use a graphing utility to verify your result.

78. Removing Pollutants The cost C (in dollars) of removing $p\%$ of the air pollutants in the stack emission of a utility company that burns coal is modeled by

$$C = \frac{160,000p}{100 - p}, \quad 0 \le p < 100.$$

(a) Find the costs of seizing 25%, 50%, and 75%.

(b) Find the limit of C as $p \to 100^-$. Interpret the limit in the context of the problem. Use a graphing utility to verify your result.

Analyzing a Graph In Exercises 79–90, analyze and sketch the graph of the function. Label any intercepts, relative extrema, points of inflection, and asymptotes.

79. $f(x) = 4x - x^2$

80. $f(x) = 4x^3 - x^4$

81. $f(x) = x^3 - 6x^2 + 3x + 10$

82. $f(x) = -x^3 + 3x^2 + 9x - 2$

83. $f(x) = x^4 - 4x^3 + 16x - 16$

84. $f(x) = x^5 + 1$

85. $f(x) = x\sqrt{16 - x^2}$

86. $f(x) = x^2\sqrt{9 - x^2}$

87. $f(x) = \dfrac{x + 1}{x - 1}$

88. $f(x) = \dfrac{x - 1}{3x^2 + 1}$

89. $f(x) = 3x^{2/3} - 2x$

90. $f(x) = x^{4/5}$

91. **Bacteria** The data in the table show the number N of bacteria in a culture at time t, where t is measured in days.

t	1	2	3	4	5	6	7	8
N	25	200	804	1756	2296	2434	2467	2473

A model for these data is

$$N = \dfrac{24{,}670 - 35{,}153t + 13{,}250t^2}{100 - 39t + 7t^2}, \quad 1 \le t \le 8.$$

(a) Use a graphing utility to create a scatter plot of the data and graph the model in the same viewing window. How well does the model fit the data?

(b) Use the model to predict the number of bacteria in the culture after 10 days.

(c) Should this model be used to predict the number of bacteria in the culture after a few months? Why or why not?

92. **Meteorology** The monthly average high temperatures T (in degrees Fahrenheit) in New York City can be modeled by

$$T = \dfrac{31.6 - 1.822t + 0.0984t^2}{1 - 0.194t + 0.0131t^2}, \quad 1 \le t \le 12$$

where t is the month, with $t = 1$ corresponding to January. Use a graphing utility to graph the model and find all absolute extrema. Interpret the meaning of these values in the context of the problem. *(Source: National Climatic Data Center)*

Comparing Δy and dy In Exercises 93–96, compare the values of dy and Δy for the function.

Function	x-Value	Differential of x
93. $f(x) = 2x^2$	$x = 2$	$\Delta x = dx = 0.01$
94. $f(x) = x^4 + 3$	$x = 1$	$\Delta x = dx = 0.1$
95. $f(x) = 6x - x^3$	$x = 3$	$\Delta x = dx = 0.1$
96. $f(x) = 5x^{3/2}$	$x = 9$	$\Delta x = dx = 0.01$

Marginal Analysis In Exercises 97–102, use differentials to approximate the change in cost, revenue, or profit corresponding to an increase in sales of one unit. For instance, in Exercise 97, approximate the change in cost as x increases from 10 to 11.

Function	x-Value
97. $C = 40x^2 + 1225$	$x = 10$
98. $C = 1.5\sqrt[3]{x} + 500$	$x = 125$
99. $R = 6.25x + 0.4x^{3/2}$	$x = 225$
100. $R = 80x - 0.35x^2$	$x = 80$
101. $P = 0.003x^2 + 0.019x - 1200$	$x = 750$
102. $P = -0.2x^3 + 3000x - 7500$	$x = 50$

Finding Differentials In Exercises 103–108, find the differential dy.

103. $y = 0.5x^3$

104. $y = 7x^4 + 2x^2$

105. $y = (3x^2 - 2)^3$

106. $y = \sqrt{36 - x^2}$

107. $y = \dfrac{2 - x}{x + 5}$

108. $y = \dfrac{3x^2}{x - 4}$

109. **Profit** The profit P (in dollars) for a company producing x units is

$$P = -0.8x^2 + 324x - 2000.$$

(a) Use differentials to approximate the change in profit when the production level changes from 100 to 101 units.

(b) Compare this with the actual change in profit.

110. **Demand** The demand function for a product is

$$p = 108 - 0.2x$$

where p is the price per unit (in dollars) and x is the number of units.

(a) Use differentials to approximate the change in revenue as sales increase from 20 units to 21 units.

(b) Repeat part (a) when sales increase from 40 units to 41 units.

111. **Physiology: Body Surface Area** The body surface area (BSA) of a 180-centimeter-tall (about six-foot-tall) person is modeled by

$$B = 0.1\sqrt{5w}$$

where B is the BSA (in square meters) and w is the weight (in kilograms). Use differentials to approximate the change in the person's BSA when the person's weight changes from 90 kilograms to 95 kilograms.

TEST YOURSELF

Take this test as you would take a test in class. When you are done, check your work against the answers given in the back of the book.

In Exercises 1–3, find the critical numbers and the open intervals on which the function is increasing or decreasing .

1. $f(x) = 3x^2 - 4$

2. $f(x) = x^3 - 12x$

3. $f(x) = (x - 5)^4$

In Exercises 4–6, use the First-Derivative Test to find all relative extrema of the function.

4. $f(x) = \frac{1}{3}x^3 - 9x + 4$

5. $f(x) = 2x^4 - 4x^2 - 5$

6. $f(x) = \frac{5}{x^2 + 2}$

In Exercises 7–9, find the absolute extrema of the function on the closed interval. Use a graphing utility to verify your result.

7. $f(x) = x^2 + 6x + 8, \ [-4, 0]$

8. $f(x) = 12\sqrt{x} - 4x, \ [0, 5]$

9. $f(x) = \frac{6}{x} + \frac{x}{2}, \ [1, 6]$

In Exercises 10 and 11, determine the open intervals on which the graph of the function is concave upward or concave downward.

10. $f(x) = x^5 - 80x^2$

11. $f(x) = \frac{20}{3x^2 + 8}$

In Exercises 12 and 13, discuss the concavity of the graph of the function and find the points of inflection.

12. $f(x) = x^4 + 6$

13. $f(x) = x^4 - 54x^2 + 230$

In Exercises 14 and 15, use the Second-Derivative Test to find all relative extrema of the function.

14. $f(x) = x^3 - 6x^2 - 36x + 50$

15. $f(x) = \frac{3}{5}x^5 - 9x^3$

In Exercises 16–18, find the vertical and horizontal asymptotes of the graph of the function.

16. $f(x) = \frac{3x + 2}{x - 5}$

17. $f(x) = \frac{2x^2}{x^2 + 3}$

18. $f(x) = \frac{2x^2 - 5}{x - 1}$

In Exercises 19–21, analyze and sketch the graph of the function. Label any intercepts, relative extrema, points of inflection, and asymptotes.

19. $y = -x^3 + 3x^2 + 9x - 2$

20. $y = x^5 - 5x$

21. $y = \frac{x}{x^2 - 4}$

In Exercises 22–24, find the differential dy.

22. $y = 5x^2 - 3$

23. $y = \frac{1 - x}{x + 3}$

24. $y = (x + 4)^3$

25. The demand function for a product is modeled by

$$p = 280 - 0.4x, \quad 0 \le x \le 700$$

where p is the price per unit (in dollars) and x is the number of units. Determine when the demand is elastic, inelastic, and of unit elasticity.

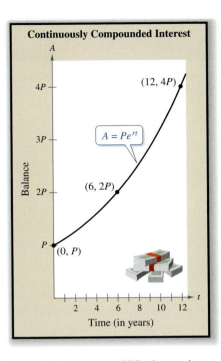

Continuously Compounded Interest

$$A = Pe^{rt}$$

(12, 4P)

(6, 2P)

(0, P)

Balance

Time (in years)

Example 3 on page 253 shows how an exponential growth model can be used to find the annual interest rate of an account.

4 Exponential and Logarithmic Functions

215

4.1 Exponential Functions

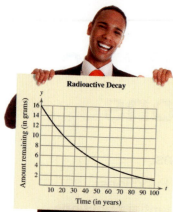

In Exercise 5, you will evaluate an exponential function to find the remaining amount of a radioactive material.

■ Use the properties of exponents to evaluate and simplify exponential expressions.
■ Sketch the graphs of exponential functions.

Exponential Functions

You are already familiar with the behavior of algebraic functions such as

$$f(x) = x^2$$
$$g(x) = \sqrt{x} = x^{1/2}$$

and

$$h(x) = \frac{1}{x} = x^{-1}$$

each of which involves a variable raised to a constant power. By interchanging roles and raising a constant to a variable power, you obtain another important class of functions called **exponential functions.** Some simple examples are

$$f(x) = 2^x$$
$$g(x) = \left(\frac{1}{10}\right)^x = \frac{1}{10^x}$$

and

$$h(x) = 3^{2x} = 9^x.$$

In general, you can use any positive number $a \neq 1$ as the base of an exponential function.

Definition of Exponential Function

If $a > 0$ and $a \neq 1$, then the **exponential function** with base a is given by

$$f(x) = a^x.$$

In the definition of an exponential function, the base $a = 1$ is excluded because it yields

$$f(x) = 1^x = 1.$$

This is a constant function, not an exponential function.

When working with exponential functions, the properties of exponents, shown below, are useful.

Properties of Exponents

Let a and b be positive real numbers.

1. $a^0 = 1$ **2.** $a^x a^y = a^{x+y}$ **3.** $\dfrac{a^x}{a^y} = a^{x-y}$

4. $(a^x)^y = a^{xy}$ **5.** $(ab)^x = a^x b^x$ **6.** $\left(\dfrac{a}{b}\right)^x = \dfrac{a^x}{b^x}$

7. $a^{-x} = \dfrac{1}{a^x}$

Example 1 Applying Properties of Exponents

a. $(2^2)(2^3) = 2^{2+3} = 2^5 = 32$ Apply Property 2.

b. $(2^2)(2^{-3}) = 2^{2-3} = 2^{-1} = \dfrac{1}{2}$ Apply Properties 2 and 7.

c. $(3^2)^3 = 3^{2(3)} = 3^6 = 729$ Apply Property 4.

d. $\left(\dfrac{1}{3}\right)^{-2} = \left(\dfrac{3}{1}\right)^2 = \dfrac{3^2}{1^2} = 9$ Apply Properties 7 and 6.

e. $\dfrac{3^2}{3^3} = 3^{2-3} = 3^{-1} = \dfrac{1}{3}$ Apply Properties 3 and 7.

f. $(2^{1/2})(3^{1/2}) = [(2)(3)]^{1/2} = 6^{1/2} = \sqrt{6}$ Apply Property 5.

✓ Checkpoint 1

Simplify each expression using the properties of exponents.

a. $(3^2)(3^3)$ **b.** $(3^2)(3^{-1})$ **c.** $(2^3)^2$

d. $(1/2)^{-3}$ **e.** $2^2/2^3$ **f.** $(2^{1/2})(5^{1/2})$ ■

Although Example 1 demonstrates the properties of exponents with integer and rational exponents, it is important to realize that the properties hold for *all* real exponents. With a calculator, you can obtain approximations of a^x for any positive number a and any real number x. Here are some examples.

$$2^{-0.6} \approx 0.660, \qquad \pi^{0.75} \approx 2.360, \qquad (1.56)^{\sqrt{2}} \approx 1.876$$

Example 2 Dating Organic Material

In living organic material, the ratio of radioactive carbon isotopes to the total number of carbon atoms is about 1 to 10^{12}. When organic material dies, its radioactive carbon isotopes begin to decay, with a half-life of about 5715 years. This means that after 5715 years, the ratio of isotopes to atoms will have decreased to one-half the original ratio; after a second 5715 years, the ratio will have decreased to one-fourth of the original; and so on. Figure 4.1 shows this decreasing ratio. The formula for the ratio R of carbon isotopes to carbon atoms is $R = (1/10^{12})(1/2)^{t/5715}$, where t is the time in years. Find the value of R for each period of time.

a. 10,000 years   **b.** 20,000 years   **c.** 25,000 years

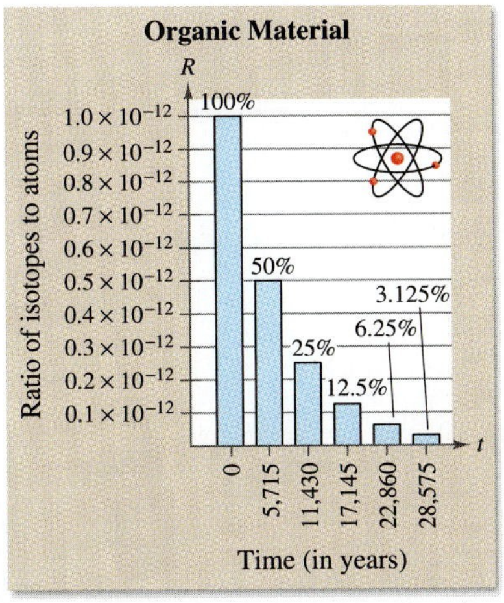

SOLUTION

a. $R = \left(\dfrac{1}{10^{12}}\right)\left(\dfrac{1}{2}\right)^{10,000/5715} \approx 2.973 \times 10^{-13}$ Ratio for 10,000 years

b. $R = \left(\dfrac{1}{10^{12}}\right)\left(\dfrac{1}{2}\right)^{20,000/5715} \approx 8.842 \times 10^{-14}$ Ratio for 20,000 years

c. $R = \left(\dfrac{1}{10^{12}}\right)\left(\dfrac{1}{2}\right)^{25,000/5715} \approx 4.821 \times 10^{-14}$ Ratio for 25,000 years

Organic Material

Ratio of isotopes to atoms

1.0×10^{-12} — 100%
0.9×10^{-12}
0.8×10^{-12}
0.7×10^{-12}
0.6×10^{-12}
0.5×10^{-12} — 50%
0.4×10^{-12} — 3.125%
0.3×10^{-12} — 25% — 6.25%
0.2×10^{-12} — 12.5%
0.1×10^{-12}

0   5,715   11,430   17,145   22,860   28,575

Time (in years)

FIGURE 4.1

✓ Checkpoint 2

Use the formula for the ratio of carbon isotopes to carbon atoms in Example 2 to find the value of R for each period of time.

a. 5000 years   **b.** 15,000 years   **c.** 30,000 years   ■

Graphs of Exponential Functions

The basic nature of the graph of an exponential function can be determined by the point-plotting method or by using a graphing utility.

Example 3   Graphing Exponential Functions

Sketch the graph of each exponential function.

a. $f(x) = 2^x$ **b.** $g(x) = \left(\frac{1}{2}\right)^x = 2^{-x}$ **c.** $h(x) = 3^x$

SOLUTION   To sketch these functions by hand, you can begin by constructing a table of values, as shown below.

x	-3	-2	-1	0	1	2	3	4
$f(x) = 2^x$	$\frac{1}{8}$	$\frac{1}{4}$	$\frac{1}{2}$	1	2	4	8	16
$g(x) = 2^{-x}$	8	4	2	1	$\frac{1}{2}$	$\frac{1}{4}$	$\frac{1}{8}$	$\frac{1}{16}$
$h(x) = 3^x$	$\frac{1}{27}$	$\frac{1}{9}$	$\frac{1}{3}$	1	3	9	27	81

The graphs of the three functions are shown in Figure 4.2. Note that the graphs of $f(x) = 2^x$ and $h(x) = 3^x$ are increasing, whereas the graph of $g(x) = 2^{-x}$ is decreasing.

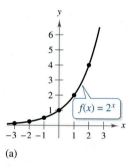

(a)

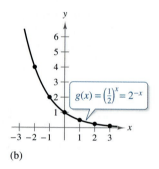

(b)

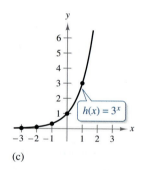

(c)

FIGURE 4.2

✓**Checkpoint 3**

Sketch the graph of

$f(x) = 5^x$.   ■

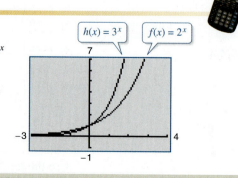

The forms of the graphs in Figure 4.2 are typical of the graphs of the exponential functions $y = a^{-x}$ and $y = a^x$, where $a > 1$. The basic characteristics of such graphs are summarized in Figure 4.3.

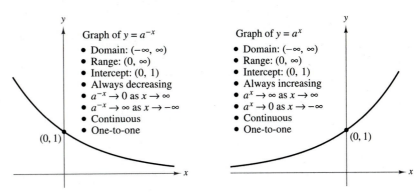

Graph of $y = a^{-x}$
- Domain: $(-\infty, \infty)$
- Range: $(0, \infty)$
- Intercept: $(0, 1)$
- Always decreasing
- $a^{-x} \to 0$ as $x \to \infty$
- $a^{-x} \to \infty$ as $x \to -\infty$
- Continuous
- One-to-one

Graph of $y = a^x$
- Domain: $(-\infty, \infty)$
- Range: $(0, \infty)$
- Intercept: $(0, 1)$
- Always increasing
- $a^x \to \infty$ as $x \to \infty$
- $a^x \to 0$ as $x \to -\infty$
- Continuous
- One-to-one

Characteristics of the Exponential Functions $y = a^{-x}$ and $y = a^x$ $(a > 1)$
FIGURE 4.3

Example 4   Graphing an Exponential Function

Sketch the graph of $f(x) = 3^{-x} - 1$.

SOLUTION   Begin by creating a table of values, as shown below.

x	-2	-1	0	1	2
$f(x)$	$3^2 - 1 = 8$	$3^1 - 1 = 2$	$3^0 - 1 = 0$	$3^{-1} - 1 = -\frac{2}{3}$	$3^{-2} - 1 = -\frac{8}{9}$

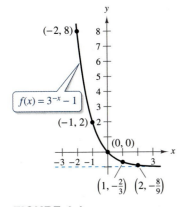

$(-2, 8)$

$f(x) = 3^{-x} - 1$

$(-1, 2)$

$(0, 0)$

$\left(1, -\frac{2}{3}\right)$ $\left(2, -\frac{8}{9}\right)$

FIGURE 4.4

From the limit

$$\lim_{x \to \infty} (3^{-x} - 1) = \lim_{x \to \infty} 3^{-x} - \lim_{x \to \infty} 1$$
$$= \lim_{x \to \infty} \frac{1}{3^x} - \lim_{x \to \infty} 1$$
$$= 0 - 1$$
$$= -1$$

you can see that $y = -1$ is a horizontal asymptote of the graph. The graph is shown in Figure 4.4.

✓ Checkpoint 4

Sketch the graph of $f(x) = 2^{-x} + 1$.   ■

SUMMARIZE   *(Section 4.1)*

1. State the definition of an exponential function *(page 216)*. For examples of exponential functions, see Example 3.

2. State the properties of exponents *(page 216)*. For examples of the properties of exponents, see Examples 1 and 2.

3. State the basic characteristics of the graphs of the exponential functions $y = a^{-x}$ and $y = a^x$ *(page 219)*. For an example of the graph of an exponential function, see Example 4.

SKILLS WARM UP 4.1   The following warm-up exercises involve skills that were covered in earlier sections. You will use these skills in the exercise set for this section. For additional help, review Section 1.4.

In Exercises 1–6, describe how the graph of g is related to the graph of f.

1. $g(x) = f(x + 2)$

2. $g(x) = -f(x)$

3. $g(x) = -1 + f(x)$

4. $g(x) = f(-x)$

5. $g(x) = f(x - 1)$

6. $g(x) = f(x) + 2$

In Exercises 7–12, evaluate each expression.

7. $25^{3/2}$

8. $64^{3/4}$

9. $27^{2/3}$

10. $\left(\dfrac{1}{5}\right)^3$

11. $\left(\dfrac{1}{8}\right)^{1/3}$

12. $\left(\dfrac{5}{8}\right)^2$

In Exercises 13–18, solve for x.

13. $2x - 6 = 4$

14. $3x + 1 = 5$

15. $(x + 4)^2 = 25$

16. $(x - 2)^2 = 8$

17. $x^2 + 4x - 5 = 0$

18. $2x^2 - 3x + 1 = 0$

ENHANCED
WebAssign   Access end-of-section exercises online at **www.webassign.net**

4.2 Natural Exponential Functions

■ Evaluate and graph functions involving the natural exponential function.
■ Solve compound interest problems.
■ Solve present value problems.

Natural Exponential Functions

In Section 4.1, exponential functions were introduced using an unspecified base a. In calculus, the most convenient (or natural) choice for a base is the irrational number e, whose decimal approximation is

$$e \approx 2.71828182846.$$

Although this choice of base may seem unusual, its convenience will become apparent as the rules for differentiating exponential functions are developed in Section 4.3. In that development, you will encounter the limit used in the definition of e.

Limit Definition of e

The irrational number e is defined to be the limit of $(1 + x)^{1/x}$ as $x \to 0$. That is,

$$\lim_{x \to 0} (1 + x)^{1/x} = e.$$

In Exercise 46, you will evaluate a natural exponential function to find the population of Las Vegas, Nevada for several years.

Example 1 Graphing the Natural Exponential Function

Complete the table of values for $f(x) = e^x$. Then sketch the graph of f.

x	-2	-1	0	1	2
$f(x)$					

SOLUTION Begin by completing the table as shown.

x	-2	-1	0	1	2
$f(x)$	$e^{-2} \approx 0.135$	$e^{-1} \approx 0.368$	$e^0 = 1$	$e^1 \approx 2.718$	$e^2 \approx 7.389$

Then use the point-plotting method to sketch the graph of f, as shown in Figure 4.5. Note that e^x is positive for all values of x. Moreover, the graph has the x-axis as a horizontal asymptote to the left. That is,

$$\lim_{x \to -\infty} e^x = 0.$$

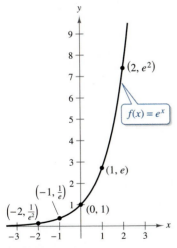

FIGURE 4.5

✓ **Checkpoint 1**

Complete the table of values for $g(x) = e^{-x}$. Then sketch the graph of g.

x	-2	-1	0	1	2
$g(x)$					

Exponential functions are often used to model the growth of a quantity or a population. When the quantity's growth *is not* restricted, an exponential model is often used. When the quantity's growth *is* restricted, the best model is often a **logistic growth model** of the form

$$f(t) = \frac{a}{1 + be^{-kt}}.$$

Graphs of both types of population growth models are shown in Figure 4.6. The graph of a logistic growth model is called a *logistic curve*.

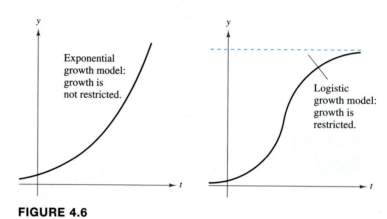

Exponential growth model: growth is not restricted.

Logistic growth model: growth is restricted.

FIGURE 4.6

Example 2   Modeling a Population

A bacterial culture is growing according to the logistic growth model

$$y = \frac{1.25}{1 + 0.25e^{-0.4t}}, \quad t \geq 0$$

where y is the culture weight (in grams) and t is the time (in hours). Find the weight of the culture after 0 hours, 1 hour, and 10 hours. What is the limit of the model as t increases without bound?

SOLUTION   The graph of the model is shown in Figure 4.7.

$$y = \frac{1.25}{1 + 0.25e^{-0.4(0)}} = 1 \text{ gram} \qquad \text{Weight when } t = 0$$

$$y = \frac{1.25}{1 + 0.25e^{-0.4(1)}} \approx 1.071 \text{ grams} \qquad \text{Weight when } t = 1$$

$$y = \frac{1.25}{1 + 0.25e^{-0.4(10)}} \approx 1.244 \text{ grams} \qquad \text{Weight when } t = 10$$

As t approaches infinity, the limit of y is

$$\lim_{t \to \infty} \frac{1.25}{1 + 0.25e^{-0.4t}} = \lim_{t \to \infty} \frac{1.25}{1 + (0.25/e^{0.4t})} = \frac{1.25}{1 + 0} = 1.25.$$

So, as t increases without bound, the weight of the culture approaches 1.25 grams.

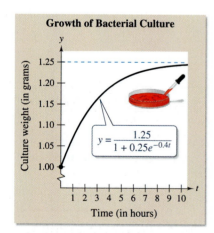

Growth of Bacterial Culture

Culture weight (in grams)

$$y = \frac{1.25}{1 + 0.25e^{-0.4t}}$$

Time (in hours)

When a culture is grown in a dish, the size of the dish and the available food limit the culture's growth.

FIGURE 4.7

✓ Checkpoint 2

A bacterial culture is growing according to the model $y = 1.50/(1 + 0.2e^{-0.5t})$, $t \geq 0$, where y is the culture weight (in grams) and t is the time (in hours). Find the weight of the culture after 0 hours, 1 hour, and 10 hours. What is the limit of the model as t increases without bound?

Extended Application: Compound Interest

An amount of P dollars is deposited in an account at an annual interest rate of r (in decimal form). What is the balance after 1 year? The answer depends on the number of times the interest is compounded, according to the formula

$$A = P\left(1 + \frac{r}{n}\right)^n$$

where n is the number of compoundings per year. The balances for a deposit of $1000 at 8%, for various compounding periods, are shown in the table.

Number of times compounded per year, n	Balance (in dollars), A
Annually, $n = 1$	$A = 1000\left(1 + \frac{0.08}{1}\right)^1 = \$1080.00$
Semiannually, $n = 2$	$A = 1000\left(1 + \frac{0.08}{2}\right)^2 = \$1081.60$
Quarterly, $n = 4$	$A = 1000\left(1 + \frac{0.08}{4}\right)^4 \approx \$1082.43$
Monthly, $n = 12$	$A = 1000\left(1 + \frac{0.08}{12}\right)^{12} \approx \$1083.00$
Daily, $n = 365$	$A = 1000\left(1 + \frac{0.08}{365}\right)^{365} \approx \$1083.28$

You may be surprised to discover that as n increases, the balance A approaches a limit, as indicated in the following development. In this development, let

$$x = \frac{r}{n}.$$

Then $x \to 0$ as $n \to \infty$, and you have

$$A = \lim_{n \to \infty} P\left(1 + \frac{r}{n}\right)^n$$

$$= P \lim_{n \to \infty} \left(1 + \frac{r}{n}\right)^n$$

$$= P \lim_{n \to \infty} \left[\left(1 + \frac{r}{n}\right)^{n/r}\right]^r$$

$$= P\left[\lim_{x \to 0} (1 + x)^{1/x}\right]^r \qquad \text{Substitute } x \text{ for } r/n.$$

$$= Pe^r.$$

This limit is the balance after 1 year of **continuous compounding.** So, for a deposit of $1000 at 8%, compounded continuously, the balance at the end of the year would be

$$A = 1000e^{0.08}$$

$$\approx \$1083.29.$$

Summary of Compound Interest Formulas

Let P be the amount deposited, t the number of years, A the balance, and r the annual interest rate (in decimal form).

1. Compounded n times per year: $A = P\left(1 + \frac{r}{n}\right)^{nt}$

2. Compounded continuously: $A = Pe^{rt}$

The average interest rates paid by banks on savings accounts have varied greatly during the past 30 years. At times savings accounts have earned as much as 12% annual interest, and at times they have earned less than 1%. The next example shows how the annual interest rate can affect the balance of an account.

Example 3   Finding Account Balances

You are creating a trust fund for your newborn nephew. You plan to deposit $12,000 in an account, with instructions that the account be turned over to your nephew on his 25th birthday. Compare the balances in the account for each situation. Which account should you choose?

a. 7%, compounded continuously

b. 7%, compounded quarterly

c. 11%, compounded continuously

d. 11%, compounded quarterly

SOLUTION

a. $12{,}000e^{0.07(25)} \approx 69{,}055.23$ 7%, compounded continuously

b. $12{,}000\left(1 + \dfrac{0.07}{4}\right)^{4(25)} \approx 68{,}017.87$ 7%, compounded quarterly

c. $12{,}000e^{0.11(25)} \approx 187{,}711.58$ 11%, compounded continuously

d. $12{,}000\left(1 + \dfrac{0.11}{4}\right)^{4(25)} \approx 180{,}869.07$ 11%, compounded quarterly

The balances in the accounts for parts (a) and (c) are shown in Figure 4.8. Notice the dramatic difference between the balances at 7% and 11%. You should choose the account described in part (c) because it earns more money than the other accounts.

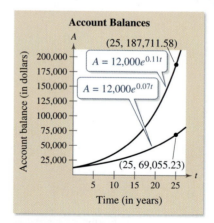

FIGURE 4.8

✓ Checkpoint 3

Find the balance in an account when $2000 is deposited for 10 years at an interest rate of 9%, compounded as follows. Compare the results and make a general statement about compounding.

a. quarterly   **b.** monthly

c. daily   **d.** continuously

In Example 3, note that the interest earned depends on the frequency with which the interest is compounded. The annual percentage rate is called the **stated rate** or **nominal rate.** However, the nominal rate does not reflect the actual rate at which interest is earned, which means that the compounding produced an **effective rate** that is larger than the nominal rate. In general, the effective rate corresponding to a nominal rate of r that is compounded n times per year is

$$\text{Effective rate} = r_{eff} = \left(1 + \frac{r}{n}\right)^n - 1.$$

 Example 4   **Finding the Effective Rate of Interest**

Find the effective rate of interest corresponding to a nominal rate of 6% per year compounded (a) annually, (b) semiannually, (c) quarterly, and (d) monthly.

SOLUTION

a. $r_{eff} = \left(1 + \dfrac{r}{n}\right)^n - 1$ Formula for effective rate of interest

$\phantom{r_{eff}} = \left(1 + \dfrac{0.06}{1}\right)^1 - 1$ Substitute for r and n.

$\phantom{r_{eff}} = 1.06 - 1$ Simplify.

$\phantom{r_{eff}} = 0.06$

So, the effective rate is 6% per year.

b. $r_{eff} = \left(1 + \dfrac{r}{n}\right)^n - 1$ Formula for effective rate of interest

$\phantom{r_{eff}} = \left(1 + \dfrac{0.06}{2}\right)^2 - 1$ Substitute for r and n.

$\phantom{r_{eff}} = (1.03)^2 - 1$ Simplify.

$\phantom{r_{eff}} = 0.0609$

So, the effective rate is 6.09% per year.

c. $r_{eff} = \left(1 + \dfrac{r}{n}\right)^n - 1$ Formula for effective rate of interest

$\phantom{r_{eff}} = \left(1 + \dfrac{0.06}{4}\right)^4 - 1$ Substitute for r and n.

$\phantom{r_{eff}} = (1.015)^4 - 1$ Simplify.

$\phantom{r_{eff}} \approx 0.0614$

So, the effective rate is about 6.14% per year.

d. $r_{eff} = \left(1 + \dfrac{r}{n}\right)^n - 1$ Formula for effective rate of interest

$\phantom{r_{eff}} = \left(1 + \dfrac{0.06}{12}\right)^{12} - 1$ Substitute for r and n.

$\phantom{r_{eff}} = (1.005)^{12} - 1$ Simplify.

$\phantom{r_{eff}} \approx 0.0617$

So, the effective rate is about 6.17% per year.

✓ **Checkpoint 4**

Repeat Example 4 using a nominal rate of 7%.

Present Value

In planning for the future, this problem often arises: "How much money P should be deposited now, at a fixed rate of interest r, in order to have a balance of A, t years from now?" The answer to this question is given by the **present value** of A.

To find the present value of a future investment, use the formula for compound interest as shown.

$$A = P\left(1 + \frac{r}{n}\right)^{nt} \qquad \text{\textcolor{red}{Formula for compound interest}}$$

Solving for P gives a present value of

$$P = \frac{A}{\left(1 + \dfrac{r}{n}\right)^{nt}} \quad \text{or} \quad P = \frac{A}{(1 + i)^N}$$

where $i = r/n$ is the interest rate per compounding period and $N = nt$ is the total number of compounding periods. You will learn another way to find the present value of a future investment in Section 6.1.

 **Example 5**   **Finding Present Value**

An investor is purchasing a 10-year certificate of deposit that pays an annual percentage rate of 8%, compounded monthly. How much should the person invest in order to obtain a balance of $15,000 at maturity?

SOLUTION   Here, $A = 15{,}000$, $r = 0.08$, $n = 12$, and $t = 10$. Using the formula for present value, you obtain

$$P = \frac{15{,}000}{\left(1 + \dfrac{0.08}{12}\right)^{12(10)}} \qquad \text{\textcolor{red}{Substitute for A, r, n, and t.}}$$

$$\approx 6757.85. \qquad \text{\textcolor{red}{Simplify.}}$$

So, the person should invest $6757.85 in the certificate of deposit.

✓**Checkpoint 5**

How much money should be deposited in an account paying 6% interest compounded monthly in order to have a balance of $20,000 after 3 years?

SUMMARIZE   (Section 4.2)

1. State the limit definition of e *(page 221)*. For an example of a graph of a natural exponential function, see Example 1.

2. Describe a real-life example of how an exponential function can be used to model a population *(page 222, Example 2)*.

3. State the compound interest formulas for n compoundings per year and for continuous compounding *(page 223)*. For applications of these formulas, see Example 3.

4. State the formula for finding the effective rate of interest *(page 225)*. For an application of this formula, see Example 4.

5. State the formula for present value *(page 226)*. For an application of this formula, see Example 5.

SKILLS WARM UP 4.2

The following warm-up exercises involve skills that were covered in earlier sections. You will use these skills in the exercise set for this section. For additional help, review Sections 1.6 and 3.6.

In Exercises 1–4, discuss the continuity of the function.

1. $f(x) = \dfrac{3x^2 + 2x + 1}{x^2 + 1}$

2. $f(x) = \dfrac{x + 1}{x^2 - 4}$

3. $f(x) = \dfrac{x^2 - 6x + 5}{x^2 - 3}$

4. $g(x) = \dfrac{x^2 - 9x + 20}{x - 4}$

In Exercises 5–12, find the horizontal asymptote of the graph of the function.

5. $f(x) = \dfrac{25}{1 + 4x}$

6. $f(x) = \dfrac{16x}{3 + x^2}$

7. $f(x) = \dfrac{8x^3 + 2}{2x^3 + x}$

8. $f(x) = \dfrac{x}{2x}$

9. $f(x) = \dfrac{3}{2 + (1/x)}$

10. $f(x) = \dfrac{6}{1 + x^{-2}}$

11. $f(x) = 2^{-x}$

12. $f(x) = \dfrac{7}{1 + 5x}$

ENHANCED WebAssign Access end-of-section exercises online at **www.webassign.net**

4.3 Derivatives of Exponential Functions

■ Find the derivatives of natural exponential functions.
■ Use calculus to analyze the graphs of real-life functions that involve the natural exponential function.
■ Explore the normal probability density function.

Derivatives of Exponential Functions

In Section 4.2, it was stated that the most convenient base for exponential functions is the irrational number e. The convenience of this base stems primarily from the fact that the function

$$f(x) = e^x$$

is its own derivative. You will see that this is not true of other exponential functions of the form

$$y = a^x$$

where $a \neq e$. To verify that $f(x) = e^x$ is its own derivative, notice that the limit

$$\lim_{\Delta x \to 0} (1 + \Delta x)^{1/\Delta x} = e$$

implies that for small values of Δx,

$$e \approx (1 + \Delta x)^{1/\Delta x}$$

or

$$e^{\Delta x} \approx 1 + \Delta x.$$

This approximation is used in the following derivation.

$$f'(x) = \lim_{\Delta x \to 0} \frac{f(x + \Delta x) - f(x)}{\Delta x} \qquad \text{Definition of derivative}$$

$$= \lim_{\Delta x \to 0} \frac{e^{x + \Delta x} - e^x}{\Delta x} \qquad \text{Use } f(x) = e^x.$$

$$= \lim_{\Delta x \to 0} \frac{e^x(e^{\Delta x} - 1)}{\Delta x} \qquad \text{Factor numerator.}$$

$$= \lim_{\Delta x \to 0} \frac{e^x[(1 + \Delta x) - 1]}{\Delta x} \qquad \text{Substitute } 1 + \Delta x \text{ for } e^{\Delta x}.$$

$$= \lim_{\Delta x \to 0} \frac{e^x(\Delta x)}{\Delta x} \qquad \text{Divide out common factor.}$$

$$= \lim_{\Delta x \to 0} e^x \qquad \text{Simplify.}$$

$$= e^x \qquad \text{Evaluate limit.}$$

When u is a function of x, you can apply the Chain Rule to obtain the derivative of e^u with respect to x. Both formulas are summarized below.

Derivative of the Natural Exponential Function

Let u be a differentiable function of x.

1. $\dfrac{d}{dx}[e^x] = e^x$ **2.** $\dfrac{d}{dx}[e^u] = e^u \dfrac{du}{dx}$

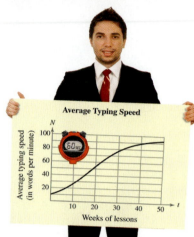

Average Typing Speed

In Exercise 48, you will use the derivative of an exponential function to find the rate of change of the average typing speed after 5, 10, and 30 weeks of lessons.

Example 1 Finding Slopes of Tangent Lines

Find the slopes of the tangent lines to

$$f(x) = e^x$$

at the points $(0, 1)$ and $(1, e)$. What conclusion can you make?

SOLUTION Because the derivative of f is

$$f'(x) = e^x \qquad \text{Derivative}$$

it follows that the slope of the tangent line to the graph of f is

$$f'(0) = e^0 = 1 \qquad \text{Slope at point } (0, 1)$$

at the point $(0, 1)$ and

$$f'(1) = e^1 = e \qquad \text{Slope at point } (1, e)$$

at the point $(1, e)$, as shown in Figure 4.9. From this pattern, you can see that the slope of the tangent line to the graph of $f(x) = e^x$ at any point (x, e^x) is equal to the y-coordinate of the point.

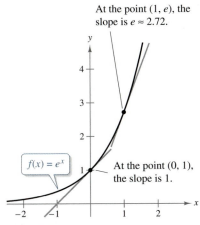

At the point $(1, e)$, the slope is $e \approx 2.72$.

$f(x) = e^x$

At the point $(0, 1)$, the slope is 1.

FIGURE 4.9

✔ Checkpoint 1

Find the slopes of the tangent lines to $f(x) = 2e^x$ at the points $(0, 2)$ and $(1, 2e)$. ■

Example 2 Differentiating Exponential Functions

Differentiate each function.

a. $f(x) = e^{2x}$ **b.** $f(x) = e^{-3x^2}$

c. $f(x) = 6e^{x^3}$ **d.** $f(x) = e^{-x}$

SOLUTION

a. Let $u = 2x$. Then $du/dx = 2$, and you can apply the Chain Rule.

$$f'(x) = e^u \frac{du}{dx} = e^{2x}(2) = 2e^{2x}$$

b. Let $u = -3x^2$. Then $du/dx = -6x$, and you can apply the Chain Rule.

$$f'(x) = e^u \frac{du}{dx} = e^{-3x^2}(-6x) = -6xe^{-3x^2}$$

c. Let $u = x^3$. Then $du/dx = 3x^2$, and you can apply the Chain Rule.

$$f'(x) = 6e^u \frac{du}{dx} = 6e^{x^3}(3x^2) = 18x^2e^{x^3}$$

d. Let $u = -x$. Then $du/dx = -1$, and you can apply the Chain Rule.

$$f'(x) = e^u \frac{du}{dx} = e^{-x}(-1) = -e^{-x}$$

✔ Checkpoint 2

Differentiate each function.

a. $f(x) = e^{3x}$ **b.** $f(x) = e^{-2x^3}$

c. $f(x) = 4e^{x^2}$ **d.** $f(x) = e^{-2x}$ ■

STUDY TIP

In Example 2, notice that when you differentiate an exponential function, the exponent does not change. For instance, the derivative of $f(x) = e^{3x}$ is $f'(x) = 3e^{3x}$. In both f and f', the exponent is $3x$.

The differentiation rules that you studied in Chapter 2 can be used with exponential functions, as shown in Example 3.

Example 3 Differentiating Exponential Functions

Differentiate each function.

a. $f(x) = 4e$ **b.** $f(x) = e^{2x-1}$

c. $f(x) = xe^x$ **d.** $f(x) = \dfrac{e^x - e^{-x}}{2}$

e. $f(x) = \dfrac{e^x}{x}$ **f.** $f(x) = xe^x - e^x$

SOLUTION

a. $f(x) = 4e$ Write original function.

   $f'(x) = 0$ Constant Rule

b. $f(x) = e^{2x-1}$ Write original function.

   $f'(x) = (e^{2x-1})(2)$ Chain Rule

   $= 2e^{2x-1}$ Simplify.

c. $f(x) = xe^x$ Write original function.

   $f'(x) = xe^x + e^x(1)$ Product Rule

   $= xe^x + e^x$ Simplify.

d. $f(x) = \dfrac{e^x - e^{-x}}{2}$ Write original function.

   $= \dfrac{1}{2}(e^x - e^{-x})$ Rewrite.

   $f'(x) = \dfrac{1}{2}[e^x - e^{-x}(-1)]$ Constant Multiple and Chain Rules

   $= \dfrac{1}{2}(e^x + e^{-x})$ Simplify.

e. $f(x) = \dfrac{e^x}{x}$ Write original function.

   $f'(x) = \dfrac{xe^x - e^x(1)}{x^2}$ Quotient Rule

   $= \dfrac{e^x(x-1)}{x^2}$ Simplify.

f. $f(x) = xe^x - e^x$ Write original function.

   $f'(x) = [xe^x + e^x(1)] - e^x$ Product and Difference Rules

   $= xe^x + e^x - e^x$

   $= xe^x$ Simplify.

✓ Checkpoint 3

Differentiate each function.

a. $f(x) = 9e$ **b.** $f(x) = e^{3x+1}$ **c.** $f(x) = x^2e^x$

d. $f(x) = \dfrac{e^x + e^{-x}}{2}$ **e.** $f(x) = \dfrac{e^x}{x^2}$ **f.** $f(x) = x^2e^x - e^x$

Applications

In Chapter 3, you learned how to use derivatives to analyze the graphs of functions. The next example applies those techniques to a function composed of exponential functions. In the example, notice that

$$e^a = e^b$$

implies that $a = b$.

 Example 4   **Analyzing a Catenary**

When a telephone wire is hung between two poles, the wire forms a U-shaped curve called a **catenary.** For instance, the function

$$y = 30(e^{x/60} + e^{-x/60}), \quad -30 \le x \le 30$$

models the shape of a telephone wire strung between two poles that are 60 feet apart (x and y are measured in feet). Show that the lowest point on the wire is midway between the two poles. How much does the wire sag between the two poles?

SOLUTION   First, find the derivative of the function.

$y = 30(e^{x/60} + e^{-x/60})$	Write original function.
$y' = 30\left[e^{x/60}\left(\tfrac{1}{60}\right) + e^{-x/60}\left(-\tfrac{1}{60}\right)\right]$	Derivative
$\quad = 30\left(\tfrac{1}{60}\right)(e^{x/60} - e^{-x/60})$	Factor out $\tfrac{1}{60}$.
$\quad = \tfrac{1}{2}(e^{x/60} - e^{-x/60})$	Simplify.

To find the critical numbers, set the derivative equal to zero.

$\tfrac{1}{2}(e^{x/60} - e^{-x/60}) = 0$	Set derivative equal to 0.
$e^{x/60} - e^{-x/60} = 0$	Multiply each side by 2.
$e^{x/60} = e^{-x/60}$	Add $e^{-x/60}$ to each side.
$\dfrac{x}{60} = -\dfrac{x}{60}$	If $e^a = e^b$, then $a = b$.
$x = -x$	Multiply each side by 60.
$2x = 0$	Add x to each side.
$x = 0$	Divide each side by 2.

Using the First-Derivative Test, you can determine that the critical number $x = 0$ yields a relative minimum of the function. From the graph in Figure 4.10, you can see that this relative minimum is actually a minimum on the interval

$$[-30, 30].$$

So, you can conclude that the lowest point on the wire lies midway between the two poles. To find how much the wire sags between the two poles, you can compare its height at each pole with its height at the relative minimum.

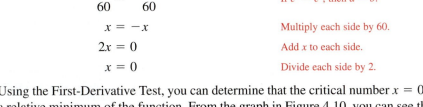

$y = 30(e^{-30/60} + e^{-(-30)/60}) \approx 67.7$ feet	Height at left pole
$y = 30(e^{0/60} + e^{-(0)/60}) = 60$ feet	Height at relative minimum
$y = 30(e^{30/60} + e^{-(30)/60}) \approx 67.7$ feet	Height at right pole

From this, you can see that the wire sags about 7.7 feet.

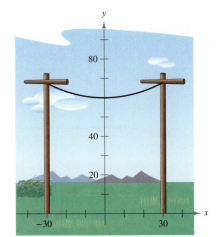

FIGURE 4.10

✓ Checkpoint 4

Use a graphing utility to graph the function in Example 4. Verify the minimum value. Use the information in the example to choose an appropriate viewing window. ■

 **Example 5**   **Finding a Maximum Revenue**

The demand function for a product is modeled by

$$p = 56e^{-0.000012x} \qquad \text{Demand function}$$

where p is the price per unit (in dollars) and x is the number of units. What price will yield a maximum revenue? What is the maximum revenue at this price?

SOLUTION   The revenue function is

$$R = xp = 56xe^{-0.000012x}. \qquad \text{Revenue function}$$

To find the maximum revenue *analytically*, you would first find the marginal revenue

$$\frac{dR}{dx} = 56xe^{-0.000012x}(-0.000012) + e^{-0.000012x}(56).$$

You would then set dR/dx equal to zero

$$56xe^{-0.000012x}(-0.000012) + e^{-0.000012x}(56) = 0$$

and solve for x. At this point, you can see that the analytical approach is rather cumbersome. In this problem, it is easier to use a *graphical* approach. After experimenting to find a reasonable viewing window, you can obtain a graph of R that is similar to that shown in Figure 4.11. Using the *maximum* feature, you can conclude that the maximum revenue occurs when x is about 83,333 units. To find the price that corresponds to this production level, substitute $x \approx 83,333$ into the demand function.

$$p \approx 56e^{-0.000012(83,333)} \approx \$20.60.$$

So, a price of about \$20.60 will yield a maximum revenue of

$$R \approx 56(83,333)e^{-0.000012(83,333)}$$
$$\approx \$1,716,771. \qquad \text{Maximum revenue}$$

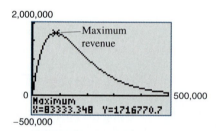

Use the *maximum* feature to approximate the x-value that corresponds to the maximum revenue.

FIGURE 4.11

 **Checkpoint 5**

The demand function for a product is modeled by

$$p = 50e^{-0.0000125x}$$

where p is the price per unit (in dollars) and x is the number of units. What price will yield a maximum revenue? What is the maximum revenue at this price?   ■

Try solving Example 5 analytically. When you do this, you must solve the equation

$$56xe^{-0.000012x}(-0.000012) + e^{-0.000012x}(56) = 0.$$

Explain how you would solve this equation. What is the solution?

The Normal Probability Density Function

If you take a course in statistics or quantitative business analysis, you will spend quite a bit of time studying the characteristics and use of the **normal probability density function** given by

$$f(x) = \frac{1}{\sigma\sqrt{2\pi}}e^{-(x-\mu)^2/(2\sigma^2)}$$

where σ is the lowercase Greek letter sigma, and μ is the lowercase Greek letter mu. In this formula, σ represents the *standard deviation* of the probability distribution, and μ represents the *mean* of the probability distribution.

Example 6   Exploring a Probability Density Function

Show that the graph of the normal probability density function

$$f(x) = \frac{1}{\sqrt{2\pi}}e^{-x^2/2}$$

has points of inflection at $x = \pm 1$.

SOLUTION   Begin by finding the second derivative of the function.

$$f'(x) = \frac{1}{\sqrt{2\pi}}(-x)e^{-x^2/2} \qquad \text{First derivative}$$

$$f''(x) = \frac{1}{\sqrt{2\pi}}[(-x)(-x)e^{-x^2/2} + (-1)e^{-x^2/2}] \qquad \text{Second derivative}$$

$$= \frac{1}{\sqrt{2\pi}}(e^{-x^2/2})(x^2 - 1) \qquad \text{Simplify.}$$

By setting the second derivative equal to 0, you can determine that $x = \pm 1$. By testing the concavity of the graph, you can then conclude that these x-values yield points of inflection, as shown in Figure 4.12.

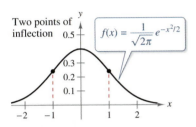

Two points of inflection

$$f(x) = \frac{1}{\sqrt{2\pi}}e^{-x^2/2}$$

The graph of the normal probability density function is bell-shaped.

FIGURE 4.12

✓ Checkpoint 6

Graph the normal probability density function

$$f(x) = \frac{1}{4\sqrt{2\pi}}e^{-x^2/32}$$

and approximate the points of inflection.

SUMMARIZE   (Section 4.3)

1. State the derivative of the natural exponential function (*page 228*). For examples of the derivative of the natural exponential function, see Examples 2 and 3.

2. Describe a real-life example of how a natural exponential function can be used to analyze the graph of a catenary (*page 231, Example 4*).

3. Describe a real-life example of how a natural exponential function can be used to analyze a company's maximum revenue (*page 232, Example 5*).

4. Describe a use of the natural exponential function in statistics (*page 233*). For an example of the natural exponential function in statistics, see Example 6.

SKILLS WARM UP 4.3   The following warm-up exercises involve skills that were covered in a previous course or in earlier sections. You will use these skills in the exercise set for this section. For additional help, review Appendix Section A.4, and Sections 2.2, 2.4, and 3.2.

In Exercises 1–4, factor the expression.

1. $x^2 e^x - \frac{1}{2} e^x$

2. $(xe^{-x})^{-1} + e^x$

3. $xe^x - e^{2x}$

4. $e^x - xe^{-x}$

In Exercises 5–8, find the derivative of the function.

5. $f(x) = \dfrac{3}{7x^2}$

6. $g(x) = 3x^2 - \dfrac{x}{6}$

7. $f(x) = (4x - 3)(x^2 + 9)$

8. $f(t) = \dfrac{t - 2}{\sqrt{t}}$

In Exercises 9 and 10, find the relative extrema of the function.

9. $f(x) = \frac{1}{8} x^3 - 2x$

10. $f(x) = x^4 - 2x^2 + 5$

ENHANCED
WebAssign   Access end-of-section exercises online at **www.webassign.net**

QUIZ YOURSELF

Take this quiz as you would take a quiz in class. When you are done, check your work against the answers given in the back of the book.

In Exercises 1–8, use properties of exponents to simplify the expression.

1. $4^3(4^2)$

2. $\left(\dfrac{1}{6}\right)^{-3}$

3. $\dfrac{3^8}{3^5}$

4. $(5^{1/2})(3^{1/2})$

5. $(e^2)(e^5)$

6. $(e^{2/3})(e^3)$

7. $\dfrac{e^2}{e^{-4}}$

8. $(e^{-1})^{-3}$

In Exercises 9–14, sketch the graph of the function.

9. $f(x) = 3^x - 2$

10. $f(x) = 5^{-x} + 2$

11. $f(x) = 6^{x-3}$

12. $f(x) = e^{x+2}$

13. $f(x) = e^x + 3$

14. $f(x) = e^{-2x} + 1$

15. After t years, the remaining mass y (in grams) of an initial mass of 35 grams of a radioactive element whose half-life is 80 years is given by

$$y = 35\left(\frac{1}{2}\right)^{t/80}, \quad t \geq 0.$$

How much of the initial mass remains after 50 years?

16. With an annual rate of inflation of 4.5% over the next 10 years, the approximate cost C of goods or services during any year in the decade is given by

$$C(t) = P(1.045)^t, \quad 0 \leq t \leq 10$$

where t is the time (in years) and P is the present cost. The price of a baseball game ticket is presently $20. Estimate the price 10 years from now.

17. For $P = \$3000$, $r = 3.5\%$, and $t = 5$ years, find the balance in an account when interest is compounded (a) quarterly, (b) monthly, and (c) continuously.

18. How much should be deposited in an account paying 6% interest compounded monthly in order to have a balance of $14,000 after 5 years?

In Exercises 19–22, find the derivative of the function.

19. $y = e^{5x}$

20. $y = e^{x-4}$

21. $y = 5e^{x+2}$

22. $y = 3e^x - xe^x$

23. Determine an equation of the tangent line to

$$y = e^{-2x}$$

at the point $(0, 1)$.

24. Analyze and sketch the graph of

$$f(x) = 0.5x^2 e^{-0.5x}.$$

Label any relative extrema, points of inflection, and asymptotes.

4.4 Logarithmic Functions

■ Sketch the graphs of natural logarithmic functions.

■ Use properties of logarithms to simplify, expand, and condense logarithmic expressions.

■ Use inverse properties of exponential and logarithmic functions to solve exponential and logarithmic equations.

■ Use properties of natural logarithms to answer questions about real-life situations.

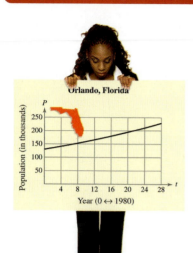

In Exercise 77, you will solve a natural exponential equation to predict when the population of Orlando, Florida will reach 300,000.

The Natural Logarithmic Function

From your previous algebra courses, you should be somewhat familiar with logarithms. For instance, the **common logarithm** $\log_{10} x$ is defined as

$$\log_{10} x = b \quad \text{if and only if} \quad 10^b = x.$$

The base of common logarithms is 10. In calculus, the most useful base for logarithms is the number e.

Definition of the Natural Logarithmic Function

The **natural logarithmic function,** denoted by $\ln x$, is defined as

$$\ln x = b \quad \text{if and only if} \quad e^b = x.$$

$\ln x$ is read as "el en of x" or as "the natural log of x."

This definition implies that the natural logarithmic function and the natural exponential function are inverse functions. So, every logarithmic equation can be written in an equivalent exponential form, and every exponential equation can be written in logarithmic form. Here are some examples.

Logarithmic form:	*Exponential form:*
$\ln 1 = 0$	$e^0 = 1$
$\ln e = 1$	$e^1 = e$
$\ln \dfrac{1}{e} = -1$	$e^{-1} = \dfrac{1}{e}$
$\ln 2 \approx 0.693$	$e^{0.693} \approx 2$
$\ln 0.1 \approx -2.303$	$e^{-2.303} \approx 0.1$

Because the functions $f(x) = e^x$ and $g(x) = \ln x$ are inverse functions, their graphs are reflections of each other in the line

$$y = x.$$

This reflective property is illustrated in Figure 4.13. The figure also contains a summary of several properties of the graph of the natural logarithmic function.

Notice that the domain of the natural logarithmic function is the set of *positive real numbers*—be sure you see that $\ln x$ is not defined for zero or for negative numbers. You can test this on your calculator. When you try evaluating

$$\ln(-1) \quad \text{or} \quad \ln 0$$

your calculator should indicate that the value is not a real number.

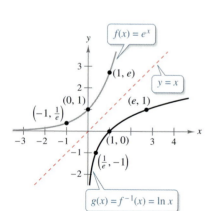

$g(x) = \ln x$

• Domain: $(0, \infty)$
• Range: $(-\infty, \infty)$
• Intercept: $(1, 0)$
• Always increasing
• $\ln x \to \infty$ as $x \to \infty$
• $\ln x \to -\infty$ as $x \to 0^+$
• Continuous
• One-to-one

FIGURE 4.13

Example 1   **Graphing Logarithmic Functions**

Sketch the graph of each function.

a. $f(x) = \ln(x + 1)$ **b.** $f(x) = 2 \ln(x - 2)$

SOLUTION

a. Because the natural logarithmic function is defined only for positive values, the domain of the function is $x + 1 > 0$, or

$x > -1$.   Domain

To sketch the graph, begin by constructing a table of values, as shown below. Then plot the points in the table and connect them with a smooth curve, as shown in Figure 4.14(a).

x	-0.5	0	0.5	1	1.5	2
$\ln(x + 1)$	-0.693	0	0.405	0.693	0.916	1.099

b. The domain of this function is $x - 2 > 0$, or

$x > 2$.   Domain

A table of values for the function is shown below, and its graph is shown in Figure 4.14(b).

x	2.5	3	3.5	4	4.5	5
$2 \ln(x - 2)$	-1.386	0	0.811	1.386	1.833	2.197

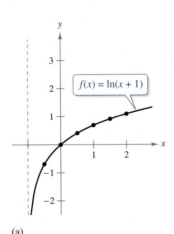

(a)

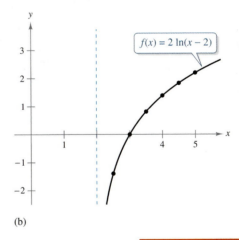

(b)

FIGURE 4.14

✓ Checkpoint 1

Complete the table and sketch the graph of

$f(x) = \ln(x + 2)$.

x	-1.5	-1	-0.5	0	0.5	1
$f(x)$						

TECH TUTOR

What happens when you take the logarithm of a negative number? Some graphing utilities do not give an error message for $\ln(-1)$. Instead, the graphing utility displays a complex number. For the purpose of this text, however, it is assumed that the domain of the logarithmic function is the set of positive real numbers.

STUDY TIP

How does the graph of $f(x) = \ln(x + 1)$ relate to the graph of $y = \ln x$? The graph of f is a translation of the graph of $y = \ln x$ one unit to the left.

Properties of Logarithmic Functions

Recall from Section 1.4 that inverse functions have the property that

$$f(f^{-1}(x)) = x \quad \text{and} \quad f^{-1}(f(x)) = x.$$

The properties listed below follow from the fact that the natural logarithmic function and the natural exponential function are inverse functions.

Inverse Properties of Logarithms and Exponents

1. $\ln e^x = x$ 2. $e^{\ln x} = x$

Example 2 Applying Inverse Properties

Simplify each expression.

a. $\ln e^{\sqrt{2}}$ **b.** $e^{\ln 3x}$

SOLUTION

a. Because $\ln e^x = x$, it follows that

$$\ln e^{\sqrt{2}} = \sqrt{2}.$$

b. Because $e^{\ln x} = x$, it follows that

$$e^{\ln 3x} = 3x.$$

✓Checkpoint 2

Simplify each expression.

a. $\ln e^3$ **b.** $e^{\ln(x+1)}$ ■

Most of the properties of exponential functions can be rewritten in terms of logarithmic functions. For instance, the property

$$e^x e^y = e^{x+y}$$

states that you can multiply two exponential expressions by adding their exponents. In terms of logarithms, this property becomes

$$\ln xy = \ln x + \ln y.$$

This property and two other properties of logarithms are summarized below.

Properties of Logarithms

1. $\ln xy = \ln x + \ln y$ 2. $\ln \dfrac{x}{y} = \ln x - \ln y$ 3. $\ln x^n = n \ln x$

STUDY TIP

There is no general property that can be used to rewrite $\ln(x + y)$. Specifically, $\ln(x + y)$ is not equal to $\ln x + \ln y$.

Rewriting a logarithm of a single quantity as the sum, difference, or multiple of logarithms is called *expanding* the logarithmic expression. The reverse procedure is called *condensing* a logarithmic expression.

Example 3   Expanding Logarithmic Expressions

Use the properties of logarithms to rewrite each expression as a sum, difference, or multiple of logarithms. (Assume $x > 0$ and $y > 0$.)

a. $\ln \dfrac{10}{9}$ **b.** $\ln \sqrt{x^2 + 1}$ **c.** $\ln \dfrac{xy}{5}$ **d.** $\ln[x^2(x + 1)]$

SOLUTION

a. $\ln \dfrac{10}{9} = \ln 10 - \ln 9$ Property 2

b. $\ln \sqrt{x^2 + 1} = \ln(x^2 + 1)^{1/2}$ Rewrite with rational exponent.

$\qquad\qquad\quad = \dfrac{1}{2} \ln(x^2 + 1)$ Property 3

c. $\ln \dfrac{xy}{5} = \ln(xy) - \ln 5$ Property 2

$\qquad\quad = \ln x + \ln y - \ln 5$ Property 1

d. $\ln[x^2(x + 1)] = \ln x^2 + \ln(x + 1)$ Property 1

$\qquad\qquad\quad = 2 \ln x + \ln(x + 1)$ Property 3

✓Checkpoint 3

Use the properties of logarithms to rewrite each expression as a sum, difference, or multiple of logarithms. (Assume $x > 0$ and $y > 0$.)

a. $\ln \dfrac{2}{5}$ **b.** $\ln \sqrt[3]{x + 2}$ **c.** $\ln \dfrac{x}{5y}$ **d.** $\ln x(x + 1)^2$ ■

Example 4   Condensing Logarithmic Expressions

Use the properties of logarithms to rewrite each expression as the logarithm of a single quantity. (Assume $x > 0$ and $y > 0$.)

a. $\ln x + 2 \ln y$ **b.** $2 \ln(x + 2) - 3 \ln x$

SOLUTION

a. $\ln x + 2 \ln y = \ln x + \ln y^2$ Property 3

$\qquad\qquad\quad = \ln xy^2$ Property 1

b. $2 \ln(x + 2) - 3 \ln x = \ln(x + 2)^2 - \ln x^3$ Property 3

$\qquad\qquad\qquad\qquad = \ln \dfrac{(x + 2)^2}{x^3}$ Property 2

✓Checkpoint 4

Use the properties of logarithms to rewrite each expression as the logarithm of a single quantity. (Assume $x > 0$ and $y > 0$.)

a. $4 \ln x + 3 \ln y$ **b.** $\ln(x + 1) - 2 \ln(x + 3)$ ■

TECH TUTOR

Try using a graphing utility to verify the results of Example 3(b). That is, try graphing the functions

$$y = \ln \sqrt{x^2 + 1}$$

and

$$y = \frac{1}{2} \ln(x^2 + 1).$$

Because these two functions are equivalent, their graphs should coincide.

Solving Exponential and Logarithmic Equations

To solve an exponential equation, first isolate the exponential expression. Then take the logarithm of each side of the equation and solve for the variable.

Example 5   Solving Exponential Equations

Solve each equation.

a. $e^x = 5$ **b.** $10 + e^{0.1t} = 14$

SOLUTION

a.
$e^x = 5$	Write original equation.
$\ln e^x = \ln 5$	Take natural log of each side.
$x = \ln 5$	Inverse property: $\ln e^x = x$

b.
$10 + e^{0.1t} = 14$	Write original equation.
$e^{0.1t} = 4$	Subtract 10 from each side.
$\ln e^{0.1t} = \ln 4$	Take natural log of each side.
$0.1t = \ln 4$	Inverse property: $\ln e^{0.1t} = 0.1t$
$t = 10 \ln 4$	Multiply each side by 10.

✓ Checkpoint 5

Solve each equation.

a. $e^x = 6$ **b.** $5 + e^{0.2t} = 10$ ■

To solve a logarithmic equation, first isolate the logarithmic expression. Then exponentiate each side of the equation and solve for the variable.

Example 6   Solving Logarithmic Equations

Solve each equation.

a. $\ln x = 5$

b. $3 + 2 \ln x = 7$

SOLUTION

a.
$\ln x = 5$	Write original equation.
$e^{\ln x} = e^5$	Exponentiate each side.
$x = e^5$	Inverse property: $e^{\ln x} = x$

b.
$3 + 2 \ln x = 7$	Write original equation.
$2 \ln x = 4$	Subtract 3 from each side.
$\ln x = 2$	Divide each side by 2.
$e^{\ln x} = e^2$	Exponentiate each side.
$x = e^2$	Inverse property: $e^{\ln x} = x$

✓ Checkpoint 6

Solve each equation.

a. $\ln x = 4$

b. $4 + 5 \ln x = 19$ ■

Application

Example 7 **Finding Doubling Time**

You deposit P dollars in an account whose annual interest rate is r, compounded continuously. How long will it take for your balance to double?

SOLUTION The balance in the account after t years is $A = Pe^{rt}$. So, the balance will have doubled when $Pe^{rt} = 2P$. To find the "doubling time," solve this equation for t.

$Pe^{rt} = 2P$	Balance in account has doubled.
$e^{rt} = 2$	Divide each side by P.
$\ln e^{rt} = \ln 2$	Take natural log of each side.
$rt = \ln 2$	Inverse property: $\ln e^{rt} = rt$
$t = \dfrac{1}{r} \ln 2$	Divide each side by r.

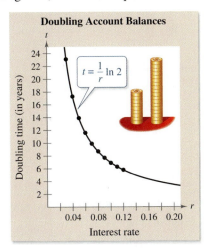

Doubling Account Balances

$t = \dfrac{1}{r} \ln 2$

FIGURE 4.15

From this result, you can see that the time it takes for the balance to double is inversely proportional to the interest rate r. The table shows the doubling times for several interest rates. Notice that the doubling time decreases as the rate increases. The relationship between doubling time and the interest rate is shown graphically in Figure 4.15.

r	3%	4%	5%	6%	7%	8%	9%	10%	11%	12%
t	23.1	17.3	13.9	11.6	9.9	8.7	7.7	6.9	6.3	5.8

✓ Checkpoint 7

Use the equation found in Example 7 to determine the amount of time it would take for your balance to double at an interest rate of 8.75%. ■

SUMMARIZE (Section 4.4)

1. State the definition of the natural logarithmic function *(page 236)*. For an example of graphing logarithmic functions, see Example 1.

2. State the inverse properties of logarithms and exponents *(page 238)*. For an example of applying these properties, see Example 2.

3. State the properties of logarithms *(page 238)*. For examples of using these properties to expand and condense logarithmic expressions, see Examples 3 and 4.

4. Identify the properties of logarithms and exponents used to solve the exponential and logarithmic equations in Examples 5 and 6 *(page 240)*.

5. Describe a real-life example of how a logarithm is used to determine how long it will take for an account balance to double *(page 241, Example 7)*.

SKILLS WARM UP 4.4

The following warm-up exercises involve skills that were covered in a previous course or in earlier sections. You will use these skills in the exercise set for this section. For additional help, review Appendix Section A.1, and Sections 1.4 and 4.2.

In Exercises 1–4, find the inverse function of f.

1. $f(x) = 5x$

2. $f(x) = x - 6$

3. $f(x) = 3x + 2$

4. $f(x) = \dfrac{3}{4}x - 9$

In Exercises 5–8, solve for x.

5. $0 < x + 4$

6. $0 < x^2 + 1$

7. $0 < \sqrt{x^2 - 1}$

8. $0 < x - 5$

In Exercises 9 and 10, find the balance in the account after 10 years.

9. $P = \$1900$, $r = 6\%$, compounded continuously

10. $P = \$2500$, $r = 3\%$, compounded continuously

ENHANCED
WebAssign   Access end-of-section exercises online at **www.webassign.net**

4.5 Derivatives of Logarithmic Functions

■ Find the derivatives of natural logarithmic functions.
■ Find the derivatives of exponential and logarithmic functions involving other bases.

Derivatives of Logarithmic Functions

Implicit differentiation can be used to develop the derivative of the natural logarithmic function.

$$y = \ln x \qquad \text{Natural logarithmic function}$$
$$e^y = x \qquad \text{Write in exponential form.}$$
$$\frac{d}{dx}[e^y] = \frac{d}{dx}[x] \qquad \text{Differentiate with respect to } x.$$
$$e^y \frac{dy}{dx} = 1 \qquad \text{Chain Rule}$$
$$\frac{dy}{dx} = \frac{1}{e^y} \qquad \text{Divide each side by } e^y.$$
$$\frac{dy}{dx} = \frac{1}{x} \qquad \text{Substitute } x \text{ for } e^y.$$

This result and its Chain Rule version are summarized below.

Derivative of the Natural Logarithmic Function

Let u be a differentiable function of x.

1. $\dfrac{d}{dx}[\ln x] = \dfrac{1}{x}$ **2.** $\dfrac{d}{dx}[\ln u] = \dfrac{1}{u}\dfrac{du}{dx}$

In Exercise 73, you will use the derivative of a logarithmic function to find the rate of change of a demand function.

Example 1   Differentiating a Logarithmic Function

Find the derivative of

$$f(x) = \ln 2x.$$

SOLUTION Let $u = 2x$. Then $du/dx = 2$, and you can apply the Chain Rule as shown.

$$f'(x) = \frac{1}{u}\frac{du}{dx} \qquad \text{Chain Rule}$$
$$= \frac{1}{2x}(2)$$
$$= \frac{1}{x} \qquad \text{Simplify.}$$

✓ **Checkpoint 1**

Find the derivative of

$$f(x) = \ln 5x.$$

Example 2 **Differentiating Logarithmic Functions**

Find the derivative of each function.

a. $f(x) = \ln(2x^2 + 4)$ **b.** $f(x) = x \ln x$ **c.** $f(x) = \dfrac{\ln x}{x}$

SOLUTION

a. $f'(x) = \dfrac{1}{u}\dfrac{du}{dx}$ Chain Rule

$\qquad = \dfrac{1}{2x^2 + 4}(4x)$ $u = 2x^2 + 4,\ du/dx = 4x$

$\qquad = \dfrac{2x}{x^2 + 2}$ Simplify.

b. $f'(x) = x\dfrac{d}{dx}[\ln x] + (\ln x)\dfrac{d}{dx}[x]$ Product Rule

$\qquad = x\left(\dfrac{1}{x}\right) + (\ln x)(1)$

$\qquad = 1 + \ln x$ Simplify.

c. $f'(x) = \dfrac{x\dfrac{d}{dx}[\ln x] - (\ln x)\dfrac{d}{dx}[x]}{x^2}$ Quotient Rule

$\qquad = \dfrac{x\left(\dfrac{1}{x}\right) - \ln x}{x^2}$

$\qquad = \dfrac{1 - \ln x}{x^2}$ Simplify.

✓**Checkpoint 2**

Find the derivative of each function.

a. $f(x) = \ln(x^2 - 4)$ **b.** $f(x) = x^2 \ln x$ **c.** $f(x) = -\dfrac{\ln x}{x^2}$ ■

STUDY TIP

When you are differentiating logarithmic functions, it is often helpful to use the properties of logarithms to rewrite the function *before* differentiating. To see the advantage of rewriting before differentiating, try using the Chain Rule to differentiate $f(x) = \ln\sqrt{x + 1}$ and compare your work with that shown in Example 3.

Example 3 **Rewriting Before Differentiating**

$f(x) = \ln\sqrt{x + 1}$ Original function

$\quad = \ln(x + 1)^{1/2}$ Rewrite with rational exponent.

$\quad = \dfrac{1}{2}\ln(x + 1)$ Property of logarithms

$f'(x) = \dfrac{1}{2}\left(\dfrac{1}{x + 1}\right)$ Differentiate.

$\quad = \dfrac{1}{2(x + 1)}$ Simplify.

✓**Checkpoint 3**

Find the derivative of

$f(x) = \ln\sqrt[3]{x + 1}.$ ■

Example 4 Rewriting Before Differentiating

Find the derivative of $f(x) = \ln[x(x^2 + 1)^2]$.

SOLUTION

$$
\begin{aligned}
f(x) &= \ln[x(x^2 + 1)^2] && \text{Write original function.} \\
&= \ln x + \ln(x^2 + 1)^2 && \text{Logarithmic properties} \\
&= \ln x + 2\ln(x^2 + 1) && \text{Logarithmic properties} \\
f'(x) &= \frac{1}{x} + 2\left(\frac{2x}{x^2 + 1}\right) && \text{Differentiate.} \\
&= \frac{1}{x} + \frac{4x}{x^2 + 1} && \text{Simplify.}
\end{aligned}
$$

✓ **Checkpoint 4**

Find the derivative of $f(x) = \ln[x^2\sqrt{x^2 + 1}]$. ■

Finding the derivative of the function in Example 4 without first rewriting would be a formidable task.

$$
f'(x) = \frac{1}{x(x^2 + 1)^2} \frac{d}{dx}[x(x^2 + 1)^2]
$$

You might try showing that this yields the same result obtained in Example 4, but be careful—the algebra is messy.

Example 5 Finding an Equation of a Tangent Line

Find an equation of the tangent line to the graph of $f(x) = 2 + 3x \ln x$ at the point $(1, 2)$.

SOLUTION   Begin by finding the derivative of f.

$$
\begin{aligned}
f(x) &= 2 + 3x \ln x && \text{Write original function.} \\
f'(x) &= 3x\left(\frac{1}{x}\right) + (\ln x)(3) && \text{Differentiate.} \\
&= 3 + 3 \ln x && \text{Simplify.}
\end{aligned}
$$

The slope of the line tangent to the graph of f at $(1, 2)$ is

$$
f'(1) = 3 + 3 \ln 1 = 3 + 3(0) = 3.
$$

Using the point-slope form of a line, you can find the equation of the tangent line to be

$$
y = 3x - 1.
$$

The graph of the function and the tangent line are shown in Figure 4.16.

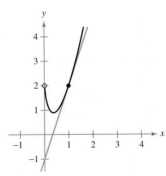

FIGURE 4.16

✓ **Checkpoint 5**

Find an equation of the tangent line to the graph of $f(x) = 4 \ln x$ at the point $(1, 0)$. ■

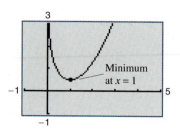

FIGURE 4.17

Example 6 — Analyzing a Graph

Analyze the graph of the function $f(x) = \dfrac{x^2}{2} - \ln x$.

SOLUTION From Figure 4.17, it appears that the function has a minimum at $x = 1$. To find the minimum analytically, find the critical numbers by setting the derivative of f equal to zero and solving for x.

$$f(x) = \frac{x^2}{2} - \ln x \qquad \text{Write original function.}$$

$$f'(x) = x - \frac{1}{x} \qquad \text{Differentiate.}$$

$$x - \frac{1}{x} = 0 \qquad \text{Set derivative equal to 0.}$$

$$x = \frac{1}{x} \qquad \text{Add } 1/x \text{ to each side.}$$

$$x^2 = 1 \qquad \text{Multiply each side by } x.$$

$$x = \pm 1 \qquad \text{Take square root of each side.}$$

Of these two possible critical numbers, only the positive one lies in the domain of f. By applying the First-Derivative Test, you can confirm that the function has a relative minimum at $x = 1$.

✓Checkpoint 6

Determine the relative extrema of the function $f(x) = x - 2 \ln x$. ■

 ### Example 7 — Finding a Rate of Change

A group of 200 college students was tested every 6 months over a four-year period. The group was composed of students who took Spanish during the fall semester of their freshman year and did not take subsequent Spanish courses. The average test score p (in percent) is modeled by

$$p = 91.6 - 15.6 \ln(t + 1), \quad 0 \le t \le 48$$

where t is the time in months, as shown in Figure 4.18. At what rate was the average score changing after 1 year?

SOLUTION The rate of change is

$$\frac{dp}{dt} = -\frac{15.6}{t + 1}.$$

The rate of change when $t = 12$ is

$$\frac{dp}{dt} = -\frac{15.6}{12 + 1} = -\frac{15.6}{13} = -1.2.$$

This means that the average score was decreasing at the rate of 1.2% per month.

Human Memory Model

Average test score (in percent) vs. Time (in months)

FIGURE 4.18

✓Checkpoint 7

Suppose the average test score in Example 7 was modeled by

$$p = 92.3 - 16.9 \ln(t + 1), \quad 0 \le t \le 48$$

where t is the time in months. How would the rate at which the average test score was changing after 1 year compare with that of the model in Example 7? ■

Other Bases

This chapter began with a definition of a general exponential function

$$f(x) = a^x$$

where a is a positive number such that $a \neq 1$. The corresponding **logarithm to the base a** is defined by

$$\log_a x = b \quad \text{if and only if} \quad a^b = x.$$

As with the natural logarithmic function, the domain of the logarithmic function to the base a is the set of positive numbers.

Example 8 Evaluating Logarithms

a. $\log_2 8 = 3$ $\qquad\qquad 2^3 = 8$

b. $\log_{10} 100 = 2$ $\qquad\qquad 10^2 = 100$

c. $\log_{10} \frac{1}{10} = -1$ $\qquad\qquad 10^{-1} = \frac{1}{10}$

d. $\log_3 81 = 4$ $\qquad\qquad 3^4 = 81$

✓ Checkpoint 8

Evaluate each logarithm without using a calculator.

a. $\log_2 16$ $\qquad$ **b.** $\log_{10} \frac{1}{100}$ $\qquad$ **c.** $\log_2 \frac{1}{32}$ $\qquad$ **d.** $\log_5 125$ ■

Most calculators have only two logarithm keys—a natural logarithm key denoted by (LN) and a common logarithm key denoted by (LOG). Logarithms to other bases can be evaluated with the following change-of-base formula.

$$\log_a x = \frac{\ln x}{\ln a} \qquad\qquad \text{Change-of-base formula}$$

Example 9 Changing Bases to Evaluate Logarithms

Use the change-of-base formula and a calculator to evaluate each logarithm.

a. $\log_2 3$

b. $\log_3 6$

c. $\log_2(-1)$

SOLUTION In each case, use the change-of-base formula and a calculator.

a. $\log_2 3 = \dfrac{\ln 3}{\ln 2} \approx 1.585$ $\qquad \log_a x = \dfrac{\ln x}{\ln a}$

b. $\log_3 6 = \dfrac{\ln 6}{\ln 3} \approx 1.631$ $\qquad \log_a x = \dfrac{\ln x}{\ln a}$

c. $\log_2(-1)$ is not defined.

✓ Checkpoint 9

Use the change-of-base formula and a calculator to evaluate each logarithm.

a. $\log_2 5$ $\qquad$ **b.** $\log_3 18$ $\qquad$ **c.** $\log_4 80$ $\qquad$ **d.** $\log_{16} 0.25$ ■

To find derivatives of exponential or logarithmic functions to bases other than e, you can either convert to base e or use the differentiation rules shown below.

Other Bases and Differentiation

Let u be a differentiable function of x.

1. $\dfrac{d}{dx}[a^x] = (\ln a)a^x$ **2.** $\dfrac{d}{dx}[a^u] = (\ln a)a^u \dfrac{du}{dx}$

3. $\dfrac{d}{dx}[\log_a x] = \left(\dfrac{1}{\ln a}\right)\dfrac{1}{x}$ **4.** $\dfrac{d}{dx}[\log_a u] = \left(\dfrac{1}{\ln a}\right)\left(\dfrac{1}{u}\right)\dfrac{du}{dx}$

PROOF By definition, $a^x = e^{(\ln a)x}$. So, you can prove the first rule by letting $u = (\ln a)x$ and differentiating with base e to obtain

$$\frac{d}{dx}[a^x] = \frac{d}{dx}[e^{(\ln a)x}] = e^u \frac{du}{dx} = e^{(\ln a)x}(\ln a) = (\ln a)a^x.$$

 Example 10 **Finding a Rate of Change**

Radioactive carbon isotopes have a half-life of 5715 years. An object contains 1 gram of the isotopes. The amount A (in grams) that will be present after t years is

$$A = \left(\frac{1}{2}\right)^{t/5715}.$$

At what rate is the amount changing when $t = 10{,}000$ years?

SOLUTION The derivative of A with respect to t is

$$\frac{dA}{dt} = \left(\ln\frac{1}{2}\right)\left(\frac{1}{2}\right)^{t/5715}\left(\frac{1}{5715}\right).$$

When $t = 10{,}000$, the rate at which the amount is changing is

$$\left(\ln\frac{1}{2}\right)\left(\frac{1}{2}\right)^{10{,}000/5715}\left(\frac{1}{5715}\right) \approx -0.000036$$

which implies that the amount of isotopes in the object is decreasing at the rate of 0.000036 gram per year.

Checkpoint 10

Use a graphing utility to graph the model in Example 10. Describe the rate at which the amount is changing as time t increases.

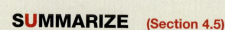
SUMMARIZE (Section 4.5)

1. State the derivative of the natural logarithmic function *(page 243)*. For examples of the derivative of the natural logarithmic function, see Examples 1, 2, 3, and 4.

2. State the derivative of the logarithmic function to base a *(page 248)*. For an example of a derivative of a logarithmic function to base a, see Example 10.

The following warm-up exercises involve skills that were covered in earlier sections. You will use these skills in the exercise set for this section. For additional help, review Sections 2.6, 2.7, and 4.4.

In Exercises 1–6, expand the logarithmic expression.

1. $\ln(x + 1)^2$

2. $\ln x(x + 1)$

3. $\ln \dfrac{x}{x + 1}$

4. $\ln\left(\dfrac{x}{x - 3}\right)^3$

5. $\ln \dfrac{4x(x - 7)}{x^2}$

6. $\ln x^3(x + 1)$

In Exercises 7 and 8, find dy/dx implicitly.

7. $y^2 + xy = 7$

8. $x^2y - xy^2 = 3x$

In Exercises 9 and 10, find the second derivative of f.

9. $f(x) = x^2(x + 1) - 3x^3$

10. $f(x) = -\dfrac{1}{x^2}$

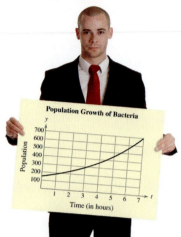

4.6 Exponential Growth and Decay

■ Use exponential growth and decay to model real-life situations.

Exponential Growth and Decay

In this section, you will learn to create models of *exponential growth and decay*. Real-life situations that involve exponential growth and decay deal with a substance or population whose *rate of change at any time t is proportional to the amount of the substance present at that time*. For example, the rate of decomposition of a radioactive substance is proportional to the amount of radioactive substance at a given instant. In its simplest form, this relationship is represented by the equation

Rate of change of *y* ⌐ is ⌐ proportional to *y*.

$$\frac{dy}{dt} = ky.$$

In this equation, k is a constant and y is a function of t. The solution of this equation is shown below.

In Exercise 23, you will use exponential growth to find the time it takes a population of bacteria to double.

Exponential Growth and Decay

If y is a positive quantity whose rate of change with respect to time is proportional to the quantity present at any time t, then y is of the form

$$y = Ce^{kt}$$

where C is the **initial value** and k is the **constant of proportionality. Exponential growth** is indicated by $k > 0$ and **exponential decay** by $k < 0$.

PROOF   Because the rate of change of y is proportional to y, you can write

$$\frac{dy}{dt} = ky.$$

You can see that $y = Ce^{kt}$ is a solution of this equation by differentiating to obtain $dy/dt = kCe^{kt}$ and substituting.

$y = Ce^{kt}$	Original equation
$\dfrac{dy}{dt} = kCe^{kt}$	Differentiate.
$= k(Ce^{kt})$	Rewrite.
$= ky$	Substitute y for Ce^{kt}.

STUDY TIP

In the model $y = Ce^{kt}$, C is called the "initial value" because, when $t = 0$,

$$y = Ce^{k(0)}$$
$$= C(1)$$
$$= C.$$

Radioactive decay is measured in terms of **half-life,** the number of years required for half of the atoms in a sample of radioactive material to decay. The half-lives of some common radioactive isotopes are as shown.

Uranium (^{238}U)	4,470,000,000 years
Plutonium (^{239}Pu)	24,100 years
Carbon (^{14}C)	5,715 years
Radium (^{226}Ra)	1,599 years
Einsteinium (^{254}Es)	276 days
Nobelium (^{257}No)	25 seconds

Example 1   Modeling Radioactive Decay

A sample contains 1 gram of radium. Will more than 0.5 gram of radium remain after 1000 years?

SOLUTION   Let y represent the mass (in grams) of the radium in the sample. Because the rate of decay is proportional to y, you can conclude that y is of the form $y = Ce^{kt}$, where t is the time in years. From the given information, you know that $y = 1$ when $t = 0$. Substituting these values into the model produces

$$1 = Ce^{k(0)} \qquad \text{Substitute 1 for } y \text{ and 0 for } t.$$

which implies that $C = 1$. Because radium has a half-life of 1599 years, you know that $y = \frac{1}{2}$ when $t = 1599$. Substituting these values into the model allows you to solve for k.

$$y = e^{kt} \qquad \text{Exponential decay model}$$
$$\tfrac{1}{2} = e^{k(1599)} \qquad \text{Substitute } \tfrac{1}{2} \text{ for } y \text{ and 1599 for } t.$$
$$\ln \tfrac{1}{2} = 1599k \qquad \text{Take natural log of each side.}$$
$$\tfrac{1}{1599} \ln \tfrac{1}{2} = k \qquad \text{Divide each side by 1599.}$$

So, $k \approx -0.0004335$, and the exponential decay model is

$$y = e^{-0.0004335t}.$$

To find the amount of radium remaining in the sample after 1000 years, substitute $t = 1000$ into the model.

$$y = e^{-0.0004335(1000)} \approx 0.648 \text{ gram}$$

Yes, more than 0.5 gram of radium will remain after 1000 years. The graph of the model is shown in Figure 4.19.

✓ Checkpoint 1

Use the model in Example 1 to determine the number of years required for a one-gram sample of radium to decay to 0.4 gram.   ■

Instead of approximating the value of k in Example 1, you could leave the value exact and obtain

$$y = e^{[(1/1599)\ln(1/2)]t}$$
$$= e^{\ln[(1/2)^{(t/1599)}]}$$
$$= \left(\frac{1}{2}\right)^{t/1599}.$$

This version of the model clearly shows the "half-life." When $t = 1599$, the value of y is $\frac{1}{2}$; when $t = 2(1599)$, the value of y is $\frac{1}{4}$; and so on.

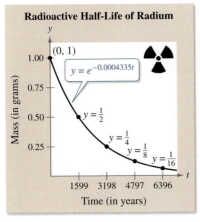

Radioactive Half-Life of Radium

$y = e^{-0.0004335t}$

$(0, 1)$

$y = \frac{1}{2}$

$y = \frac{1}{4}$

$y = \frac{1}{8}$

$y = \frac{1}{16}$

Mass (in grams)

Time (in years)

FIGURE 4.19

Guidelines for Modeling Exponential Growth and Decay

1. Use the given information to write *two* sets of conditions involving y and t.

2. Substitute the given conditions into the model $y = Ce^{kt}$ and use the results to solve for the constants C and k. (When one of the conditions involves $t = 0$, substitute that value first to solve for C.)

3. Use the model $y = Ce^{kt}$ to answer the question.

Example 2 **Modeling Population Growth**

ALGEBRA TUTOR

For help with the algebra in Example 2, see Example 1(c) in the *Chapter 4 Algebra Tutor* on page 256.

In a research experiment, a population of fruit flies is increasing in accordance with the exponential growth model. After 2 days, there are 100 flies, and after 4 days, there are 300 flies. How many flies will there be after 5 days?

SOLUTION   Let y be the number of flies at time t. From the given information, you know that $y = 100$ when $t = 2$ and $y = 300$ when $t = 4$. Substituting this information into the model $y = Ce^{kt}$ produces

$$100 = Ce^{2k} \quad \text{and} \quad 300 = Ce^{4k}.$$

To solve for k, solve for C in the first equation and substitute the result into the second equation.

$$300 = Ce^{4k} \qquad \text{Second equation}$$

$$300 = \left(\frac{100}{e^{2k}}\right)e^{4k} \qquad \text{Substitute } 100/e^{2k} \text{ for } C.$$

$$\frac{300}{100} = e^{2k} \qquad \text{Divide each side by 100.}$$

$$\ln 3 = 2k \qquad \text{Take natural log of each side.}$$

$$\frac{1}{2}\ln 3 = k \qquad \text{Solve for } k.$$

Using $k = \frac{1}{2}\ln 3 \approx 0.5493$, you can determine that

$$C \approx \frac{100}{e^{2(0.5493)}}$$

$$\approx 33.$$

So, the exponential growth model is

$$y = 33e^{0.5493t}$$

as shown in Figure 4.20. This implies that, after 5 days, the population is

$$y = 33e^{0.5493(5)}$$

$$\approx 514 \text{ flies.}$$

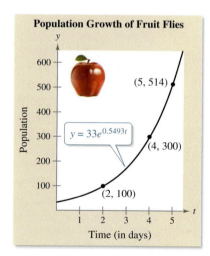

FIGURE 4.20

✓ Checkpoint 2

Find the exponential growth model for a population of fruit flies for which there are 100 flies after 2 days and 400 flies after 4 days.

Example 3 Modeling Compound Interest

Money is deposited in an account for which the interest is compounded continuously. The balance in the account doubles in 6 years. What is the annual interest rate?

SOLUTION The balance A in an account with continuously compounded interest is given by the exponential growth model

$A = Pe^{rt}$ Exponential growth model

where P is the original deposit, r is the annual interest rate (in decimal form), and t is the time (in years). From the given information, you know that

$A = 2P$

when $t = 6$, as shown in Figure 4.21. Use this information to solve for r.

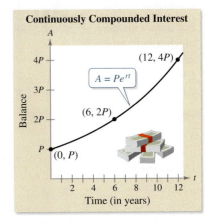

Continuously Compounded Interest

FIGURE 4.21

$$
\begin{aligned}
A &= Pe^{rt} &&\text{Exponential growth model} \\
2P &= Pe^{r(6)} &&\text{Substitute } 2P \text{ for } A \text{ and } 6 \text{ for } t. \\
2 &= e^{6r} &&\text{Divide each side by } P. \\
\ln 2 &= 6r &&\text{Take natural log of each side.} \\
\tfrac{1}{6} \ln 2 &= r &&\text{Divide each side by } 6.
\end{aligned}
$$

So, the annual interest rate is

$$
\begin{aligned}
r &= \tfrac{1}{6} \ln 2 \\
&\approx 0.1155
\end{aligned}
$$

or about 11.55%.

✓ Checkpoint 3

Find the annual interest rate for an account whose balance doubles in 8 years and for which the interest is compounded continuously. ■

Each of the examples in this section uses the exponential growth model $y = Ce^{kt}$, in which the base is e. Exponential growth, however, can be modeled with any base. That is, the model

$y = Ca^{bt}$

also represents exponential growth. (To see this, note that the model can be written in the form $y = Ce^{(\ln a)bt}$.) In some real-life settings, bases other than e are more convenient. For instance, in Example 1, knowing that the half-life of radium is 1599 years, you can immediately write the exponential decay model as

$$
y = \left(\frac{1}{2}\right)^{t/1599}.
$$

Using this model, the amount of radium left in the sample after 1000 years is

$$
y = \left(\frac{1}{2}\right)^{1000/1599}
$$

$$
\approx 0.648 \text{ gram}
$$

which is the same answer obtained in Example 1.

Example 4   **Modeling Sales**

Four months after discontinuing advertising on national television, a manufacturer notices that sales have dropped from 100,000 MP3 players per month to 80,000. Using an exponential pattern of decline, what will the sales be after another 4 months?

SOLUTION   Let y represent the number of MP3 players, let t represent the time (in months), and consider the exponential decay model

$$y = Ce^{kt}. \qquad \text{Exponential decay model}$$

From the given information, you know that $y = 100{,}000$ when $t = 0$. Using this information, you have

$$100{,}000 = Ce^0$$

which implies that $C = 100{,}000$. To solve for k, use the fact that $y = 80{,}000$ when $t = 4$.

$y = 100{,}000e^{kt}$	Exponential decay model
$80{,}000 = 100{,}000e^{k(4)}$	Substitute 80,000 for y and 4 for t.
$0.8 = e^{4k}$	Divide each side by 100,000.
$\ln 0.8 = 4k$	Take natural log of each side.
$\frac{1}{4}\ln 0.8 = k$	Divide each side by 4.

So, $k = \frac{1}{4}\ln 0.8 \approx -0.0558$, which means that the model is

$$y = 100{,}000e^{-0.0558t}.$$

After four more months ($t = 8$), you can expect sales to drop to

$$y = 100{,}000e^{-0.0558(8)}$$
$$\approx 64{,}000 \text{ MP3 players}$$

as shown in Figure 4.22.

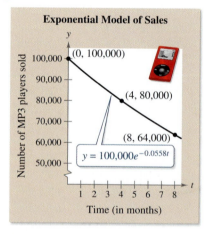

Exponential Model of Sales

$(0, 100{,}000)$
$(4, 80{,}000)$
$(8, 64{,}000)$
$y = 100{,}000e^{-0.0558t}$

Number of MP3 players sold

Time (in months)

FIGURE 4.22

ALGEBRA TUTOR

xy

For help with the algebra in Example 4, see Example 1(b) in the *Chapter 4 Algebra Tutor* on page 256.

✓**Checkpoint 4**

Use the model in Example 4 to determine when sales will drop to 50,000 MP3 players. ■

SUMMARIZE   (Section 4.6)

1. State the model used for exponential growth and decay *(page 250)*. For examples of the use of this model, see Examples 1, 2, 3, and 4.

2. State the guidelines for modeling exponential growth and decay *(page 252)*. For examples of the use of these guidelines, see Examples 2, 3, and 4.

3. Describe a real-life example of an exponential decay model *(pages 251 and 254, Examples 1 and 4)*.

4. Describe a real-life example of an exponential growth model *(pages 252 and 253, Examples 2 and 3)*.

SKILLS WARM UP 4.6

The following warm-up exercises involve skills that were covered in earlier sections. You will use these skills in the exercise set for this section. For additional help, review Sections 4.3 and 4.4.

In Exercises 1–4, solve the equation for k.

1. $12 = 24e^{4k}$

2. $10 = 3e^{5k}$

3. $25 = 16e^{-0.01k}$

4. $22 = 32e^{-0.02k}$

In Exercises 5–8, find the derivative of the function.

5. $y = 32e^{0.23t}$

6. $y = 18e^{0.072t}$

7. $y = 24e^{-1.4t}$

8. $y = 25e^{-0.001t}$

In Exercises 9–12, simplify the expression.

9. $e^{\ln 4}$

10. $4e^{\ln 3}$

11. $e^{\ln(2x+1)}$

12. $e^{\ln(x^2+1)}$

ENHANCED
WebAssign Access end-of-section exercises online at **www.webassign.net**

ALGEBRA TUTOR

xy

Solving Exponential and Logarithmic Equations

To find the extrema or points of inflection of an exponential or logarithmic function, you must know how to solve exponential and logarithmic equations. A few examples are given on page 240. Some additional examples are presented in this Algebra Tutor.

As with all equations, remember that your basic goal is to isolate the variable on one side of the equation. To do this, you use inverse operations. For instance, to isolate x in

$$e^x = 7$$

take the natural log of each side of the equation and use the property $\ln e^x = x$. Similarly, to isolate x in

$$\ln x = 5$$

exponentiate each side of the equation and use the property $e^{\ln x} = x$.

Example 1   **Solving Exponential Equations**

Solve each exponential equation.

a. $25 = 5e^{7t}$ **b.** $80{,}000 = 100{,}000e^{k(4)}$ **c.** $300 = \left(\dfrac{100}{e^{2k}}\right)e^{4k}$

SOLUTION

a.

$25 = 5e^{7t}$	Write original equation.
$5 = e^{7t}$	Divide each side by 5.
$\ln 5 = \ln e^{7t}$	Take natural log of each side.
$\ln 5 = 7t$	Apply the property $\ln e^a = a$.
$\dfrac{1}{7}\ln 5 = t$	Divide each side by 7.

b.

$80{,}000 = 100{,}000e^{k(4)}$	Example 4, page 254
$0.8 = e^{4k}$	Divide each side by 100,000.
$\ln 0.8 = \ln e^{4k}$	Take natural log of each side.
$\ln 0.8 = 4k$	Apply the property $\ln e^a = a$.
$\dfrac{1}{4}\ln 0.8 = k$	Divide each side by 4.

c.

$300 = \left(\dfrac{100}{e^{2k}}\right)e^{4k}$	Example 2, page 252
$300 = (100)\dfrac{e^{4k}}{e^{2k}}$	Rewrite product.
$300 = 100e^{4k-2k}$	To divide powers, subtract exponents.
$300 = 100e^{2k}$	Simplify.
$3 = e^{2k}$	Divide each side by 100.
$\ln 3 = \ln e^{2k}$	Take natural log of each side.
$\ln 3 = 2k$	Apply the property $\ln e^a = a$.
$\dfrac{1}{2}\ln 3 = k$	Divide each side by 2.

Example 2   Solving Logarithmic Equations

Solve each logarithmic equation.

a. $\ln x = 2$ **b.** $5 + 2 \ln x = 4$

c. $2 \ln 3x = 4$ **d.** $\ln x - \ln(x - 1) = 1$

SOLUTION

a. $\ln x = 2$ Write original equation.

    $e^{\ln x} = e^2$ Exponentiate each side.

       $x = e^2$ Apply the property $e^{\ln a} = a$.

b. $5 + 2 \ln x = 4$ Write original equation.

       $2 \ln x = -1$ Subtract 5 from each side.

        $\ln x = -\dfrac{1}{2}$ Divide each side by 2.

        $e^{\ln x} = e^{-1/2}$ Exponentiate each side.

           $x = e^{-1/2}$ Apply the property $e^{\ln a} = a$.

c. $2 \ln 3x = 4$ Write original equation.

       $\ln 3x = 2$ Divide each side by 2.

    $e^{\ln 3x} = e^2$ Exponentiate each side.

        $3x = e^2$ Apply the property $e^{\ln a} = a$.

         $x = \dfrac{1}{3}e^2$ Divide each side by 3.

d. $\ln x - \ln(x - 1) = 1$ Write original equation.

        $\ln \dfrac{x}{x - 1} = 1$ $\ln m - \ln n = \ln(m/n)$

     $e^{\ln[x/(x-1)]} = e^1$ Exponentiate each side.

          $\dfrac{x}{x - 1} = e^1$ Apply the property $e^{\ln a} = a$.

               $x = ex - e$ Multiply each side by $x - 1$.

         $x - ex = -e$ Subtract ex from each side.

      $x(1 - e) = -e$ Factor.

               $x = \dfrac{-e}{1 - e}$ Divide each side by $1 - e$.

               $x = \dfrac{e}{e - 1}$ Simplify.

STUDY TIP

Because the domain of a logarithmic function generally does not include all real numbers, be sure to check for extraneous solutions.

SUMMARY AND STUDY STRATEGIES

After studying this chapter, you should have acquired the following skills.
The exercise numbers are keyed to the Review Exercises that begin on page 260.
Answers to odd-numbered Review Exercises are given in the back of the text.*

Section 4.1

	Review Exercises
■ Use the properties of exponents to evaluate and simplify exponential expressions.	1, 2

$$a^0 = 1, \qquad a^x a^y = a^{x+y}, \qquad \frac{a^x}{a^y} = a^{x-y}, \qquad (a^x)^y = a^{xy}$$

$$(ab)^x = a^x b^x, \qquad \left(\frac{a}{b}\right)^x = \frac{a^x}{b^x}, \qquad a^{-x} = \frac{1}{a^x}$$

■ Sketch the graphs of exponential functions.	3–8
■ Use properties of exponents to answer questions about real-life situations.	9–12

Section 4.2

■ Use the properties of exponents to evaluate and simplify natural exponential expressions.	13, 14
■ Sketch the graphs of natural exponential functions.	15–18
■ Solve compound interest problems.	19–24

$$A = P(1 + r/n)^{nt}, \quad A = Pe^{rt}$$

■ Solve effective rate of interest problems.	25, 26

$$r_{eff} = (1 + r/n)^n - 1$$

■ Solve present value problems.	27, 28

$$P = \frac{A}{(1 + r/n)^{nt}}$$

■ Answer questions involving the natural exponential function as a real-life model.	29–34

Section 4.3

■ Find the derivatives of natural exponential functions.	35–40

$$\frac{d}{dx}[e^x] = e^x, \quad \frac{d}{dx}[e^u] = e^u \frac{du}{dx}$$

■ Find equations of the tangent lines to the graphs of natural exponential functions.	41–44
■ Use calculus to analyze the graphs of functions that involve the natural exponential function.	45–48

Section 4.4

■ Use the definition of the natural logarithmic function to write exponential equations in logarithmic form, and vice versa.	49–52

$$\ln x = b \quad \text{if and only if} \quad e^b = x.$$

---

* A wide range of valuable study aids are available to help you master the material in this chapter. The *Student Solutions Manual* includes step-by-step solutions to all odd-numbered exercises to help you review and prepare. The student website at *www.cengagebrain.com* offers algebra help and a *Graphing Technology Guide,* which contains step-by-step commands and instructions for a wide variety of graphing calculators.

Section 4.4 (continued)

■ Sketch the graphs of natural logarithmic functions.

■ Use properties of logarithms to simplify, expand, and condense logarithmic
expressions.

$$\ln xy = \ln x + \ln y, \quad \ln \frac{x}{y} = \ln x - \ln y, \quad \ln x^n = n \ln x$$

■ Use inverse properties of exponential and logarithmic functions to solve
exponential and logarithmic equations.

$$\ln e^x = x, \quad e^{\ln x} = x$$

■ Use properties of natural logarithms to answer questions about real-life situations.

Section 4.5

■ Find the derivatives of natural logarithmic functions.

$$\frac{d}{dx}[\ln x] = \frac{1}{x}, \quad \frac{d}{dx}[\ln u] = \frac{1}{u}\frac{du}{dx}$$

■ Use the definition of logarithms to evaluate logarithmic expressions involving
other bases.

$$\log_a x = b \quad \text{if and only if} \quad a^b = x$$

■ Use the change-of-base formula to evaluate logarithmic expressions involving
other bases.

$$\log_a x = \frac{\ln x}{\ln a}$$

■ Find the derivatives of exponential and logarithmic functions involving other bases.

$$\frac{d}{dx}[a^x] = (\ln a)a^x, \quad \frac{d}{dx}[a^u] = (\ln a)a^u\frac{du}{dx}$$

$$\frac{d}{dx}[\log_a x] = \left(\frac{1}{\ln a}\right)\frac{1}{x}, \quad \frac{d}{dx}[\log_a u] = \left(\frac{1}{\ln a}\right)\left(\frac{1}{u}\right)\frac{du}{dx}$$

■ Use calculus to analyze the graphs of functions that involve the natural
logarithmic function.

■ Use calculus to answer questions about real-life situations.

Section 4.6

■ Use exponential growth and decay to model real-life situations.

Study Strategies

■ **Classifying Differentiation Rules**   Differentiation rules fall into two basic classes: (1) general rules that apply to
all differentiable functions; and (2) specific rules that apply to special types of functions. At this point in the course,
you have studied six general rules: the Constant Rule, the Constant Multiple Rule, the Sum Rule, the Difference Rule,
the Product Rule, and the Quotient Rule. Although these rules were introduced in the context of algebraic functions,
remember that they also can be used with exponential and logarithmic functions. You have also studied three specific
rules: the Power Rule, the derivative of the natural exponential function, and the derivative of the natural logarithmic
function. Each of these rules comes in two forms: the "simple" version, such as $D_x[e^x] = e^x$, and the Chain Rule
version, such as $D_x[e^u] = e^u(du/dx)$.

■ **To Memorize or Not to Memorize?**   When studying mathematics, you need to memorize some formulas and rules.
Much of this will come from practice—the formulas that you use most often will be committed to memory. Some
formulas, however, are used only infrequently. With these, it is helpful to be able to *derive* the formula from a *known*
formula. For instance, knowing the Log Rule for differentiation and the change-of-base formula, $\log_a x = (\ln x)/(\ln a)$,
allows you to derive the formula for the derivative of a logarithmic function to base a.

Review Exercises

Applying Properties of Exponents   In Exercises 1 and 2, use the properties of exponents to simplify the expression.

1. (a) $(4^5)(4^2)$ (b) $(7^2)^3$

   (c) 2^{-4} (d) $\dfrac{3^8}{3^4}$

2. (a) $(5^4)(25^2)$ (b) $(9^{1/3})(3^{1/3})$

   (c) $\left(\dfrac{1}{3}\right)^{-3}$ (d) $(6^4)(6^{-5})$

Graphing Exponential Functions   In Exercises 3–8, sketch the graph of the function.

3. $f(x) = 9^{x/2}$
4. $g(x) = 16^{3x/2}$
5. $f(t) = \left(\frac{1}{6}\right)^t$
6. $g(t) = \left(\frac{1}{3}\right)^{-t}$
7. $f(x) = \left(\frac{1}{2}\right)^{2x} + 4$
8. $g(x) = \left(\frac{2}{3}\right)^{2x} + 1$

9. **Population Growth**   The resident populations P (in thousands) of Wisconsin from 2000 through 2009 can be modeled by the exponential function

   $$P(t) = 5382(1.0057)^t$$

   where t is the time in years, with $t = 0$ corresponding to 2000. Use the model to estimate the populations in the years (a) 2016 and (b) 2025.   *(Source: U.S. Census Bureau)*

10. **Revenue**   The revenues R (in millions of dollars) for Panera Bread Company from 2000 through 2009 can be modeled by the exponential function

    $$R(t) = 163.82(1.2924)^t$$

    where t is the time in years, with $t = 0$ corresponding to 2000. Use the model to estimate the sales in the years (a) 2014 and (b) 2017.   *(Source: Panera Bread Company)*

11. **Property Value**   Suppose that the value of a piece of property doubles every 12 years. If you buy the property for $55,000, its value t years after the date of purchase should be

    $$V(t) = 55,000(2)^{t/12}.$$

    Use the model to approximate the value of the property (a) 4 years and (b) 25 years after it is purchased.

12. **Inflation Rate**   Suppose the annual rate of inflation averages 2% over the next 10 years. With this rate of inflation, the approximate cost C of goods or services during any year in the decade will be given by

    $$C(t) = P(1.02)^t, \quad 0 \le t \le 10$$

    where t is time in years and P is the present cost. If the cost of a graphing calculator is presently $80, estimate the cost 10 years from now.

Applying Properties of Exponents   In Exercises 13 and 14, use the properties of exponents to simplify the expression.

13. (a) $(e^5)^2$ (b) $\dfrac{e^3}{e^5}$

    (c) $(e^4)(e^{3/2})$ (d) $(e^2)^{-4}$

14. (a) $(e^6)(e^{-3})$ (b) $(e^{-2})^{-5}$

    (c) $\left(\dfrac{e^6}{e^2}\right)^{-1}$ (d) $(e^3)^{4/3}$

Graphing Natural Exponential Functions   In Exercises 15–18, sketch the graph of the function.

15. $f(x) = e^{-x} + 1$
16. $g(x) = e^{2x} - 1$
17. $f(x) = 1 - e^x$
18. $g(x) = 2 + e^{x-1}$

Finding Account Balances   In Exercises 19–22, complete the table to determine the balance A for P dollars invested at rate r for t years, compounded n times per year.

n	1	2	4	12	365	Continuous compounding
A						

19. $P = \$1000, r = 4\%, t = 5$ years
20. $P = \$7000, r = 6\%, t = 20$ years
21. $P = \$3000, r = 3.5\%, t = 10$ years
22. $P = \$4500, r = 2\%, t = 25$ years

Comparing Account Balances   In Exercises 23 and 24, $2000 is deposited in an account. Decide which account, (a) or (b), will have the greater balance after 10 years.

23. (a) 5%, compounded continuously
    (b) 6%, compounded quarterly
24. (a) $6\frac{1}{2}\%$, compounded monthly
    (b) $6\frac{1}{4}\%$, compounded continuously

25. **Effective Rate**   Find the effective rate of interest corresponding to a nominal rate of 6% per year compounded (a) annually, (b) semiannually, (c) quarterly, and (d) monthly.

26. **Effective Rate**   Find the effective rate of interest corresponding to a nominal rate of 8.25% per year compounded (a) annually, (b) semiannually, (c) quarterly, and (d) monthly.

27. Present Value How much should be deposited in an account paying 5% interest compounded quarterly in order to have a balance of $12,000 three years from now?

28. Present Value How much should be deposited in an account paying 8% interest compounded monthly in order to have a balance of $20,000 five years from now?

29. Demand The demand function for a product is modeled by

$$p = 12,500 - \frac{10,000}{2 + e^{-0.001x}}.$$

Find the price p (in dollars) of the product when the quantity demanded is (a) $x = 1000$ units and (b) $x = 2500$ units. (c) What is the limit of the price as x increases without bound?

30. Demand The demand function for a product is modeled by

$$p = 8000\left(1 - \frac{5}{5 + e^{-0.002x}}\right).$$

Find the price p (in dollars) of the product when the quantity demanded is (a) $x = 1000$ units and (b) $x = 2500$ units. (c) What is the limit of the price as x increases without bound?

31. Profit The net profits P (in millions of dollars) of Medco Health Solutions from 2000 through 2009 are shown in the table.

Year	2000	2001	2002	2003	2004
Profit	216.8	256.6	361.6	425.8	481.6

Year	2005	2006	2007	2008	2009
Profit	602.0	729.8	912.0	1102.9	1280.3

A model for this data is given by $P = 223.89e^{0.1979t}$, where t represents the year, with $t = 0$ corresponding to 2000. *(Source: Medco Health Solutions, Inc.)*

(a) How well does the model fit the data?

(b) Find a linear model for the data. How well does the linear model fit the data? Which model, exponential or linear, is a better fit?

(c) Use both models to predict the net profit in 2015.

32. Population The populations P (in thousands) of Albuquerque, New Mexico from 2000 through 2009 can be modeled by $P = 450e^{0.019t}$, where t is the time in years, with $t = 0$ corresponding to 2000. *(Source: U.S. Census Bureau)*

(a) Find the populations in 2000, 2005, and 2009.

(b) Use the model to estimate the population in 2020.

33. Biology A lake is stocked with 500 fish, and the fish population P begins to increase according to the logistic growth model

$$P = \frac{10,000}{1 + 19e^{-t/5}}, \quad t \geq 0$$

where t is measured in months.

(a) Find the number of fish in the lake after 4 months.

(b) Use a graphing utility to graph the model. Find the number of months it takes for the population of fish to reach 4000.

(c) Does the population have a limit as t increases without bound? Explain your reasoning.

34. Medicine On a college campus of 5000 students, the spread of a flu virus through the student body is modeled by

$$P = \frac{5000}{1 + 4999e^{-0.8t}}, \quad t \geq 0$$

where P is the total number of infected people and t is the time, measured in days.

(a) Find the number of students infected after 5 days.

(b) Use a graphing utility to graph the model. Find the number of days it takes for 2000 students to become infected with the flu.

(c) According to this model, will all the students on campus become infected with the flu? Explain your reasoning.

Differentiating Exponential Functions In Exercises 35–40, find the derivative of the function.

35. $y = 4e^{x^2}$

36. $y = 4e^{\sqrt{x}}$

37. $y = \dfrac{x}{e^{2x}}$

38. $y = x^2e^x$

39. $y = \dfrac{5}{1 + e^{2x}}$

40. $y = \dfrac{10}{1 - 2e^x}$

Finding an Equation of a Tangent Line In Exercises 41–44, find an equation of the tangent line to the graph of the function at the given point.

41. $y = e^{2-x}$, $(2, 1)$

42. $y = e^{2x^2}$, $(1, e^2)$

43. $y = x^2e^{-x}$, $\left(1, \dfrac{1}{e}\right)$

44. $y = xe^x - e^x$, $(1, 0)$

Analyzing a Graph In Exercises 45–48, analyze and sketch the graph of the function. Label any relative extrema, points of inflection, and asymptotes.

45. $f(x) = x^3e^x$

46. $f(x) = \dfrac{e^x}{x^2}$

47. $f(x) = \dfrac{1}{xe^x}$

48. $f(x) = \dfrac{x^2}{e^x}$

Logarithmic and Exponential Forms of Equations

In Exercises 49–52, write the logarithmic equation as an exponential equation, or vice versa.

49. $\ln 12 = 2.4849\ldots$ **50.** $\ln 0.6 = -0.5108\ldots$

51. $e^{1.5} = 4.4816\ldots$ **52.** $e^{-4} = 0.0183\ldots$

Graphing Logarithmic Expressions

In Exercises 53–56, sketch the graph of the function.

53. $y = \ln(4 - x)$ **54.** $y = \ln x - 3$

55. $y = \ln \dfrac{x}{3}$ **56.** $y = -2 \ln x$

Expanding Logarithmic Expressions

In Exercises 57–62, use the properties of logarithms to rewrite the expression as a sum, difference, or multiple of logarithms.

57. $\ln \sqrt{x^2(x - 1)}$

58. $\ln \sqrt[3]{x^2 - 1}$

59. $\ln \dfrac{x^2}{(x + 1)^3}$

60. $\ln \dfrac{x^2}{x^2 + 1}$

61. $\ln\left(\dfrac{1 - x}{3x}\right)^3$

62. $\ln\left(\dfrac{x - 1}{x + 1}\right)^2$

Condensing Logarithmic Expressions

In Exercises 63–66, use the properties of logarithms to rewrite the expression as the logarithm of a single quantity.

63. $\ln(2x + 5) + \ln(x - 3)$

64. $\frac{1}{3} \ln(x^2 - 6) - 2 \ln(3x + 2)$

65. $4[\ln(x^3 - 1) + 2 \ln x - \ln(x - 5)]$

66. $\frac{1}{2}[\ln x + 3 \ln(x + 1) - \ln(x - 2)]$

Solving Exponential and Logarithmic Equations

In Exercises 67–80, solve for x.

67. $e^{\ln x} = 3$ **68.** $e^{\ln(x + 2)} = 5$

69. $\ln x = 3$ **70.** $\ln 5x = 2$

71. $\ln 2x - \ln(3x - 1) = 0$

72. $\ln x - \ln(x + 1) = 2$

73. $\ln x + \ln(x - 3) = 0$

74. $2 \ln x + \ln(x - 2) = 0$

75. $e^{-1.386x} = 0.25$

76. $e^{-0.01x} - 5.25 = 0$

77. $e^{2x - 1} - 6 = 0$

78. $4e^{2x - 3} - 5 = 0$

79. $100(1.21)^x = 110$

80. $500(1.075)^{120x} = 100,000$

81. Compound Interest A deposit of \$400 is made in an account that earns interest at an annual rate of 2.5%. How long will it take for the balance to double when the interest is compounded (a) annually, (b) monthly, (c) daily, and (d) continuously?

82. Hourly Earnings The average hourly wages w (in dollars) for private industry employees in the United States from 1990 through 2009 can be modeled by

$$w = 10.2e^{0.0315t}$$

where $t = 0$ corresponds to 1990. *(Source: U.S. Bureau of Labor Statistics)*

(a) What was the average hourly wage in 2000?

(b) In what year will the average hourly wage be \$23?

83. Learning Theory Students in a psychology experiment were given an exam and then retested monthly with equivalent exams. The average scores S (on a 100-point scale) for the students can be modeled by

$$S = 75 - 6 \ln(t + 1), \quad 0 \le t \le 12$$

where t is the time in months.

(a) What was the average score on the original exam?

(b) What was the average score after 4 months?

(c) After how many months was the average score 60?

84. Demand The demand function for a product is given by

$$p = 8000\left(1 - \dfrac{5}{5 + e^{-0.002x}}\right)$$

where p is the price per unit (in dollars) and x is the number of units sold. Find the numbers of units sold for prices of (a) $p = \$200$ and (b) $p = \$800$.

Differentiating a Logarithmic Function

In Exercises 85–98, find the derivative of the function.

85. $f(x) = \ln 3x^2$ **86.** $y = \ln \sqrt{x}$

87. $y = \ln \dfrac{x(x - 1)}{x - 2}$ **88.** $y = \ln \dfrac{x^2}{x + 1}$

89. $f(x) = \ln e^{2x + 1}$ **90.** $f(x) = \ln e^{x^2}$

91. $y = \dfrac{\ln x}{x^3}$ **92.** $y = \dfrac{x^2}{\ln x}$

93. $y = \ln(x^2 - 2)^{2/3}$

94. $y = \ln \sqrt[3]{x^3 + 1}$

95. $f(x) = \ln(x^2 \sqrt{x + 1})$

96. $f(x) = \ln \dfrac{x}{\sqrt{x + 1}}$

97. $y = \ln \dfrac{e^x}{1 + e^x}$

98. $y = \ln(e^{2x} \sqrt{e^{2x} - 1})$

Evaluating Logarithms In Exercises 99–102, evaluate the logarithm without using a calculator.

99. $\log_6 36$

100. $\log_2 32$

101. $\log_{10} 1$

102. $\log_4 \frac{1}{64}$

Changing Bases to Evaluate Logarithms In Exercises 103–106, use the change-of-base formula and a calculator to evaluate the logarithm.

103. $\log_5 13$

104. $\log_4 18$

105. $\log_{16} 64$

106. $\log_4 125$

Differentiating Functions of Other Bases In Exercises 107–112, find the derivative of the function.

107. $y = 5^{2x+1}$

108. $y = 8^{x^3}$

109. $y = \log_3(2x - 1)$

110. $y = \log_{16}(x^2 - 3x)$

111. $y = \log_{10} \dfrac{3}{x}$

112. $y = \log_2 \dfrac{1}{x^2}$

Analyzing a Graph In Exercises 113–116, analyze and sketch the graph of the function. Label any relative extrema, points of inflection, and asymptotes.

113. $y = \ln(x + 3)$

114. $y = \dfrac{8 \ln x}{x^2}$

115. $y = \ln \dfrac{10}{x + 2}$

116. $y = \ln \dfrac{x^2}{9 - x^2}$

117. Music The numbers of download music singles D (in millions) from 2004 through 2009 can be modeled by

$$D = -1671.88 + 1282 \ln t$$

where $t = 4$ corresponds to 2004. Find the rates of change of the number of download music singles in 2005 and 2008. *(Source: Recording Industry Association of America)*

118. Minimum Average Cost The cost of producing x units of a product is modeled by

$$C = 200 + 75x - 300 \ln x, \quad x \geq 1.$$

(a) Find the average cost function $\overline{C}$.

(b) Find the minimum average cost analytically. Use a graphing utility to confirm your result.

Modeling Exponential Growth and Decay In Exercises 119 and 120, find the exponential function

$$y = Ce^{kt}$$

that passes through the two given points.

119. $(0, 3)$, $(4, 1)$

120. $(1, 1)$, $(5, 5)$

Modeling Radioactive Decay In Exercises 121–126, complete the table for each radioactive isotope.

Isotope	Half-life (in years)	Initial quantity	Amount after 1000 years	Amount after 10,000 years
121. ^{226}Ra	1599	8 grams		
122. ^{226}Ra	1599		0.7 gram	
123. ^{14}C	5715			6 grams
124. ^{14}C	5715	5 grams		
125. ^{239}Pu	24,100		2.4 grams	
126. ^{239}Pu	24,100			7.1 grams

Modeling Compound Interest In Exercises 127–130, complete the table for an account in which interest is compounded continuously.

Initial investment	Annual rate	Time to double	Amount after 10 years	Amount after 25 years
127. $600	8%			
128. $2000		7 years		
129. $15,000			$18,321.04	
130.	4%		$11,934.60	

131. Medical Science Soon after an injection, the concentration D (in milligrams per milliliter) of a drug in a patient's bloodstream is 500 milligrams per milliliter. After 6 hours, 50 milligrams per milliliter of the drug remains in the bloodstream.

(a) Find an exponential model for the concentration D after t hours.

(b) What is the concentration of the drug after 4 hours?

132. Population Growth The number of a certain type of bacteria increases continuously at a rate proportional to the number present. After 2 hours, there are 200 bacteria, and after 4 hours, there are 300 bacteria.

(a) Find an exponential model given the population P after t hours.

(b) How many bacteria will there be after 7 hours?

(c) How long will it take for the population to double?

TEST YOURSELF

Take this test as you would take a test in class. When you are done, check your work against the answers given in the back of the book.

In Exercises 1–4, use the properties of exponents to simplify the expression.

1. $3^2(3^{-2})$

2. $\left(\dfrac{2^3}{2^{-5}}\right)^{-1}$

3. $(e^{1/2})(e^4)$

4. $(e^3)^4$

In Exercises 5–10, sketch the graph of the function.

5. $f(x) = 5^{x-2}$

6. $f(x) = 4^{-x}$

7. $f(x) = e^{x-3}$

8. $f(x) = 8 + \ln x^2$

9. $f(x) = \ln(x - 5)$

10. $f(x) = 0.5 \ln x$

In Exercises 11–13, use the properties of logarithms to rewrite the expression as a sum, difference, or multiple of logarithms.

11. $\ln \dfrac{3}{2}$

12. $\ln \sqrt{x + y}$

13. $\ln \dfrac{x + 1}{y}$

In Exercises 14–16, use the properties of logarithms to rewrite the expression as the logarithm of a single quantity.

14. $\ln y + \ln(x + 1)$

15. $3 \ln x - 2 \ln(x - 1)$

16. $\ln x + 4 \ln y - \frac{1}{2} \ln(z + 4)$

In Exercises 17–19, solve the equation.

17. $e^{x-1} = 9$

18. $10e^{2x+1} = 900$

19. $50(1.06)^x = 1500$

20. A deposit of $500 is made in an account that earns interest at an annual rate of 4%. How long will it take for the balance to double when the interest is compounded (a) annually, (b) monthly, (c) daily, and (d) continuously?

In Exercises 21–24, find the derivative of the function.

21. $y = e^{-3x} + 5$

22. $y = 7e^{x+2} + 2x$

23. $y = \ln(3 + x^2)$

24. $y = \ln \dfrac{5x}{x + 2}$

25. The revenues R (in millions of dollars) of skiing facilities in the United States from 2000 through 2008 can be modeled by

$$R = 1548e^{0.0617t}$$

where $t = 0$ corresponds to 2000.   *(Source: U.S. Census Bureau)*

(a) Use this model to estimate the revenues in 2006.

(b) At what rate were the revenues changing in 2006?

26. What percent of a present amount of radioactive radium (^{226}Ra) will remain after 1200 years? (The half-life of ^{226}Ra is 1599 years.)

27. A population is growing continuously at the rate of 1.75% per year. Find the time necessary for the population to double in size.

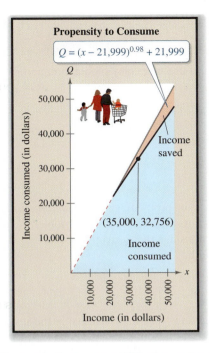

Propensity to Consume

$$Q = (x - 21{,}999)^{0.98} + 21{,}999$$

Income saved

(35,000, 32,756)

Income consumed

Income (in dollars)

Example 8 on page 280 shows how integration can be used to analyze the marginal propensity to consume.

5 Integration and Its Applications

5.1 Antiderivatives and Indefinite Integrals

■ Understand the definition of antiderivative and use indefinite integral notation for antiderivatives.

■ Use basic integration rules to find antiderivatives.

■ Use initial conditions to find particular solutions of indefinite integrals.

■ Use antiderivatives to solve real-life problems.

Horry County, South Carolina

In Exercise 69, you will use integration to find a model for the population of a county.

Antiderivatives

In Chapter 2, you were concerned primarily with the problem: *given a function, find its derivative.* Some important applications of calculus involve the inverse problem: *given a derivative, find the function.* For instance, consider the derivative $f'(x) = 3x^2$. To determine the function f, you might come up with

$$f(x) = x^3 \quad \text{because} \quad \frac{d}{dx}[x^3] = 3x^2.$$

This operation of determining the original function from its derivative is the inverse operation of differentiation. It is called **antidifferentiation.**

Definition of Antiderivative

A function F is an **antiderivative** of a function f when for every x in the domain of f, it follows that $F'(x) = f(x)$.

If $F(x)$ is an antiderivative of $f(x)$, then $F(x) + C$, where C is any constant, is also an antiderivative of $f(x)$. For example,

$$F(x) = x^3, \quad G(x) = x^3 - 5, \quad \text{and} \quad H(x) = x^3 + 0.3$$

are all antiderivatives of $3x^2$ because the derivative of each is $3x^2$. As it turns out, *all* antiderivatives of $3x^2$ are of the form $x^3 + C$. So, the process of antidifferentiation does not determine a single function, but rather a *family* of functions, each differing from the others by a constant.

The antidifferentiation process is also called **integration** and is denoted by

$$\int \qquad \text{Integral sign}$$

which is called an **integral sign.** The symbol

$$\int f(x)\, dx \qquad \text{Indefinite integral}$$

is the **indefinite integral** of $f(x)$, and it denotes the family of antiderivatives of $f(x)$. That is, if $F'(x) = f(x)$ for all x, then you can write

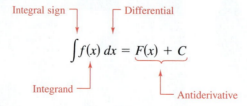

$$\int f(x)\, dx = \underline{F(x)} + C$$

where $f(x)$ is the **integrand** and C is the **constant of integration.** The differential dx in the indefinite integral identifies the variable of integration. That is, the symbol $\int f(x)\, dx$ denotes the "antiderivative of f with respect to x" just as the symbol dy/dx denotes the "derivative of y with respect to x."

STUDY TIP

In this text, the phrase "$F(x)$ is an antiderivative of $f(x)$" is used synonymously with "F is an antiderivative of f."

Finding Antiderivatives

The inverse relationship between the operations of integration and differentiation can be shown symbolically, as follows.

$$\frac{d}{dx}\left[\int f(x)\,dx\right] = f(x)$$

Differentiation is the inverse of integration.

$$\int f'(x)\,dx = f(x) + C$$

Integration is the inverse of differentiation.

This inverse relationship between integration and differentiation allows you to obtain integration formulas directly from differentiation formulas. The following summary lists the integration formulas that correspond to some of the differentiation formulas you have studied.

STUDY TIP

You will study the General Power Rule for integration in Section 5.2 and the Exponential and Log Rules in Section 5.3.

Basic Integration Rules

1. $\int k\,dx = kx + C,\quad k$ is a constant. **Constant Rule**

2. $\int kf(x)\,dx = k\int f(x)\,dx$ **Constant Multiple Rule**

3. $\int [f(x) + g(x)]\,dx = \int f(x)\,dx + \int g(x)\,dx$ **Sum Rule**

4. $\int [f(x) - g(x)]\,dx = \int f(x)\,dx - \int g(x)\,dx$ **Difference Rule**

5. $\int x^n\,dx = \frac{x^{n+1}}{n+1} + C,\quad n \neq -1$ **Simple Power Rule**

Be sure you see that the Simple Power Rule has the restriction that n cannot be -1. So, you *cannot* use the Simple Power Rule to evaluate the integral

$$\int \frac{1}{x}\,dx.$$

To evaluate this integral, you need the Log Rule, which is described in Section 5.3.

Example 1 Finding Indefinite Integrals

Find each indefinite integral.

a. $\int \frac{1}{2}\,dx$ **b.** $\int 1\,dx$ **c.** $\int -5\,dt$

SOLUTION

a. $\int \frac{1}{2}\,dx = \frac{1}{2}x + C$ **b.** $\int 1\,dx = x + C$ **c.** $\int -5\,dt = -5t + C$

STUDY TIP

Note in Example 1(b) that the integral $\int 1\,dx$ is usually shortened to the form $\int dx$.

✓ **Checkpoint 1**

Find each indefinite integral.

a. $\int 5\,dx$ **b.** $\int -1\,dr$ **c.** $\int 2\,dt$

Example 2   Finding an Indefinite Integral

$$\int 3x\, dx = 3 \int x\, dx \qquad \text{Constant Multiple Rule}$$

$$= 3 \int x^1\, dx \qquad \text{Rewrite } x \text{ as } x^1.$$

$$= 3\left(\frac{x^2}{2}\right) + C \qquad \text{Simple Power Rule with } n = 1$$

$$= \frac{3}{2}x^2 + C \qquad \text{Simplify.}$$

✓ **Checkpoint 2**

Find $\int 5x\, dx$.   ■

In finding indefinite integrals, a strict application of the basic integration rules tends to produce cumbersome constants of integration. For instance, in Example 2, you could have written

$$\int 3x\, dx = 3 \int x\, dx = 3\left(\frac{x^2}{2} + C\right) = \frac{3}{2}x^2 + 3C.$$

However, because C represents *any* constant, it is unnecessary to write $3C$ as the constant of integration. You can simply write $\frac{3}{2}x^2 + C$.

In Example 2, note that the general pattern of integration is similar to that of differentiation.

Original Integral:	Rewrite:	Integrate:	Simplify:
$\int 3x\, dx$	$3 \int x^1\, dx$	$3\left(\frac{x^2}{2}\right) + C$	$\frac{3}{2}x^2 + C$

Example 3   Rewriting Before Integrating

	Original Integral	*Rewrite*	*Integrate*	*Simplify*
a.	$\int \frac{1}{x^3}\, dx$	$\int x^{-3}\, dx$	$\frac{x^{-2}}{-2} + C$	$-\frac{1}{2x^2} + C$
b.	$\int \sqrt{x}\, dx$	$\int x^{1/2}\, dx$	$\frac{x^{3/2}}{3/2} + C$	$\frac{2}{3}x^{3/2} + C$

✓ **Checkpoint 3**

Find each indefinite integral.

a. $\int \frac{1}{x^2}\, dx$ **b.** $\int \sqrt[3]{x}\, dx$ ■

Remember that you can check your answer to an antidifferentiation problem by differentiating. For instance, in Example 3(b), you can confirm that $\frac{2}{3}x^{3/2} + C$ is the correct antiderivative by differentiating to obtain

$$\frac{d}{dx}\left[\frac{2}{3}x^{3/2} + C\right] = \left(\frac{2}{3}\right)\left(\frac{3}{2}\right)x^{1/2} = \sqrt{x}.$$

With the five basic integration rules, you can integrate *any* polynomial function, as demonstrated in the next example.

Example 4   Integrating Polynomial Functions

Find (a) $\int (x + 2)\, dx$ and (b) $\int (3x^4 - 5x^2 + x)\, dx$.

SOLUTION

a. $\int (x + 2)\, dx = \int x\, dx + \int 2\, dx$ Apply Sum Rule.

$$= \frac{x^2}{2} + C_1 + 2x + C_2 \qquad \text{Apply Simple Power and Constant Rules.}$$

$$= \frac{x^2}{2} + 2x + C \qquad C = C_1 + C_2$$

The second line in this solution is usually omitted.

b. $\int (3x^4 - 5x^2 + x)\, dx = 3\left(\frac{x^5}{5}\right) - 5\left(\frac{x^3}{3}\right) + \frac{x^2}{2} + C$

$$= \frac{3}{5}x^5 - \frac{5}{3}x^3 + \frac{1}{2}x^2 + C$$

✓ **Checkpoint 4**

Find (a) $\int (x + 4)\, dx$ and (b) $\int (4x^3 - 5x + 2)\, dx$.   ■

Example 5   Rewriting Before Integrating

Find $\int \dfrac{x - 1}{\sqrt{x}}\, dx$.

SOLUTION   Begin by rewriting the quotient in the integrand as a difference. Then rewrite each term using rational exponents.

$$\int \frac{x - 1}{\sqrt{x}}\, dx = \int \left(\frac{x}{\sqrt{x}} - \frac{1}{\sqrt{x}}\right) dx \qquad \text{Rewrite as a difference.}$$

$$= \int (x^{1/2} - x^{-1/2})\, dx \qquad \text{Rewrite using rational exponents.}$$

$$= \int x^{1/2}\, dx - \int x^{-1/2}\, dx \qquad \text{Apply Difference Rule.}$$

$$= \frac{x^{3/2}}{3/2} - \frac{x^{1/2}}{1/2} + C \qquad \text{Apply Simple Power Rule.}$$

$$= \frac{2}{3}x^{3/2} - 2x^{1/2} + C \qquad \text{Simplify.}$$

$$= \frac{2}{3}\sqrt{x}(x - 3) + C \qquad \text{Factor.}$$

✓ **Checkpoint 5**

Find $\int \dfrac{x + 2}{\sqrt{x}}\, dx$.   ■

STUDY TIP

When integrating quotients, remember *not* to integrate the numerator and denominator separately. For instance, in Example 5, be sure you understand that

$$\int \frac{x - 1}{\sqrt{x}}\, dx = \frac{2}{3}\sqrt{x}(x - 3) + C$$

is not the same as

$$\frac{\int (x - 1)\, dx}{\int \sqrt{x}\, dx} = \frac{\frac{1}{2}x^2 - x + C_1}{\frac{2}{3}x\sqrt{x} + C_2}.$$

ALGEBRA TUTOR *xy*

For help on the algebra in Example 5, see Example 1(a) in the *Chapter 5 Algebra Tutor*, on page 311.

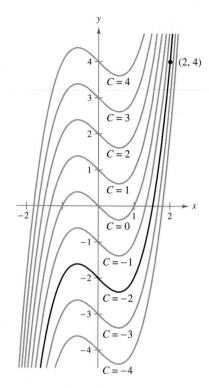

$F(x) = x^3 - x + C$

FIGURE 5.1

Particular Solutions

You have already seen that the equation $y = \int f(x)\,dx$ has many solutions, each differing from the others by a constant. This means that the graphs of any two antiderivatives of f are vertical translations of each other. For example, Figure 5.1 shows the graphs of several antiderivatives of the form

$$y = F(x) = \int (3x^2 - 1)\,dx = x^3 - x + C$$

for various integer values of C. Each of these antiderivatives is a solution of the *differential equation*

$$\frac{dy}{dx} = 3x^2 - 1.$$

A **differential equation** in x and y is an equation that involves x, y, and derivatives of y. The **general solution** of $dy/dx = 3x^2 - 1$ is $F(x) = x^3 - x + C$.

In many applications of integration, you are given enough information to determine a **particular solution.** To do this, you need to know the value of $F(x)$ for only one value of x. (This information is called an **initial condition.**) For example, in Figure 5.1, there is only one curve that passes through the point $(2, 4)$. To find this curve, use the information below.

$F(x) = x^3 - x + C$ General solution

$F(2) = 4$ Initial condition

By using the initial condition in the general solution, you can determine that $F(2) = 2^3 - 2 + C = 4$, which implies that $C = -2$. So, the particular solution is

$F(x) = x^3 - x - 2.$ Particular solution

Example 6   **Finding a Particular Solution**

Find the general solution of

$$F'(x) = 2x - 2$$

and find the particular solution that satisfies the initial condition $F(1) = 2$.

SOLUTION   Begin by integrating to find the general solution.

$F(x) = \int (2x - 2)\,dx$ Integrate $F'(x)$ to obtain $F(x)$.

$\qquad = x^2 - 2x + C$ General solution

Using the initial condition $F(1) = 2$, you can write

$$F(1) = 1^2 - 2(1) + C = 2$$

which implies that $C = 3$. So, the particular solution is

$F(x) = x^2 - 2x + 3.$ Particular solution

This solution is shown graphically in Figure 5.2. Note that each of the gray curves represents a solution of the equation $F'(x) = 2x - 2$. The black curve, however, is the only solution that passes through the point $(1, 2)$, which means that $F(x) = x^2 - 2x + 3$ is the only solution that satisfies the initial condition.

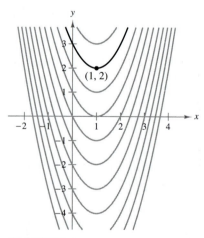

FIGURE 5.2

✓Checkpoint 6

Find the general solution of $F'(x) = 4x + 2$, and find the particular solution that satisfies the initial condition $F(1) = 8$.

Applications

In Chapter 2, you used the general position function (neglecting air resistance) for a falling object

$$s(t) = -16t^2 + v_0 t + s_0$$

where $s(t)$ is the height (in feet) and t is the time (in seconds). In the next example, integration is used to *derive* this function.

Example 7 Deriving a Position Function

A ball is thrown upward with an initial velocity of 64 feet per second from an initial height of 80 feet, as shown in Figure 5.3. Derive the position function giving the height s (in feet) as a function of the time t (in seconds). Will the ball be in the air for more than 5 seconds?

SOLUTION   Let $t = 0$ represent the initial time. Then the two given conditions can be written as

$$s(0) = 80 \qquad \text{Initial height is 80 feet.}$$
$$s'(0) = 64. \qquad \text{Initial velocity is 64 feet per second.}$$

Because the acceleration due to gravity is -32 feet per second per second, you can integrate the acceleration function to find the velocity function, as shown.

$$s''(t) = -32 \qquad \text{Acceleration due to gravity}$$
$$s'(t) = \int -32 \, dt \qquad \text{Integrate } s''(t) \text{ to obtain } s'(t).$$
$$= -32t + C_1 \qquad \text{Velocity function}$$

Using the initial velocity, you can conclude that $C_1 = 64$. Next, integrate the velocity function to find the position function.

$$s'(t) = -32t + 64 \qquad \text{Velocity function}$$
$$s(t) = \int (-32t + 64) \, dt \qquad \text{Integrate } s'(t) \text{ to obtain } s(t).$$
$$= -16t^2 + 64t + C_2 \qquad \text{Position function}$$

Using the initial height, it follows that $C_2 = 80$. So, the position function is given by

$$s(t) = -16t^2 + 64t + 80. \qquad \text{Position function}$$

To find the time when the ball hits the ground, set the position function equal to 0 and solve for t.

$$-16t^2 + 64t + 80 = 0 \qquad \text{Set } s(t) \text{ equal to zero.}$$
$$-16(t + 1)(t - 5) = 0 \qquad \text{Factor.}$$
$$t = -1, \quad t = 5 \qquad \text{Solve for } t.$$

Because the time must be positive, you can conclude that the ball hits the ground 5 seconds after it is thrown. So, the ball is not in the air for more than 5 seconds.

✓ Checkpoint 7

Derive the position function when a ball is thrown upward with an initial velocity of 32 feet per second from an initial height of 48 feet. When does the ball hit the ground? With what velocity does the ball hit the ground? ■

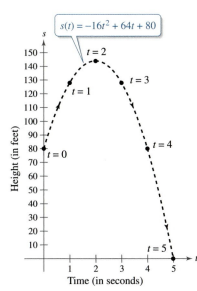

FIGURE 5.3

Example 8   **Finding a Cost Function**

The marginal cost of producing x units of a product is modeled by

$$\frac{dC}{dx} = 32 - 0.04x. \qquad \text{Marginal cost}$$

It costs \$50 to produce one unit. Find the total cost of producing 200 units.

SOLUTION   To find the cost function, integrate the marginal cost function.

$$C = \int (32 - 0.04x)\, dx \qquad \text{Integrate } \frac{dC}{dx} \text{ to obtain } C.$$

$$= 32x - 0.04\left(\frac{x^2}{2}\right) + K$$

$$= 32x - 0.02x^2 + K \qquad \text{Cost function}$$

To solve for K, use the initial condition $C = 50$ when $x = 1$.

$$50 = 32(1) - 0.02(1)^2 + K \qquad \text{Substitute 50 for } C \text{ and 1 for } x.$$

$$18.02 = K \qquad \text{Solve for } K.$$

So, the total cost function is given by

$$C = 32x - 0.02x^2 + 18.02 \qquad \text{Cost function}$$

which implies that the cost of producing 200 units is

$$C = 32(200) - 0.02(200)^2 + 18.02$$

$$= \$5618.02.$$

 Checkpoint 8

The marginal cost function for producing x units of a product is modeled by

$$\frac{dC}{dx} = 28 - 0.02x.$$

It costs \$40 to produce one unit. Find the total cost of producing 200 units. ■

STUDY TIP

In Example 8, note that K is used to represent the constant of integration rather than C. This is done to avoid confusion between the constant C and the cost function

$$C = 32x - 0.02x^2 + 18.02.$$

SUMMARIZE   (Section 5.1)

1. State the definition of antiderivative *(page 266)*. For examples of antiderivatives, see Examples 1, 2, 3, 4, and 5.

2. State the Constant Rule *(page 267)*. For an example of the Constant Rule, see Example 1.

3. State the Constant Multiple Rule *(page 267)*. For an example of the Constant Multiple Rule, see Example 2.

4. State the Sum Rule *(page 267)*. For an example of the Sum Rule, see Example 4.

5. State the Difference Rule *(page 267)*. For an example of the Difference Rule, see Example 5.

6. State the Simple Power Rule *(page 267)*. For examples of the Simple Power Rule, see Examples 2, 3, 4, and 5.

7. Describe a real-life example of how antidifferentiation can be used to find a cost function *(page 272, Example 8)*.

SKILLS WARM UP 5.1

The following warm-up exercises involve skills that were covered in a previous course or in earlier sections. You will use these skills in the exercise set for this section. For additional help, review Appendix Section A.3 and Section 1.2.

In Exercises 1–6, rewrite the expression using rational exponents.

1. $\dfrac{\sqrt{x}}{x}$

2. $\sqrt[3]{2x}\,(2x)$

3. $\sqrt{5x^3} + \sqrt{x^5}$

4. $\dfrac{1}{\sqrt{x}} + \dfrac{1}{\sqrt[3]{x^2}}$

5. $\dfrac{(x+1)^3}{\sqrt{x+1}}$

6. $\dfrac{\sqrt{x}}{\sqrt[3]{x}}$

In Exercises 7–10, let $(x, y) = (2, 2)$, and solve the equation for C.

7. $y = x^2 + 5x + C$

8. $y = 3x^3 - 6x + C$

9. $y = -16x^2 + 26x + C$

10. $y = -\tfrac{1}{4}x^4 - 2x^2 + C$

5.2 Integration by Substitution and the General Power Rule

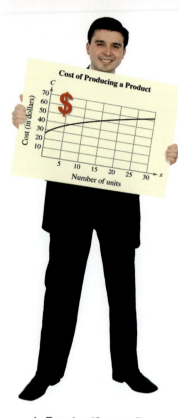

In Exercise 49, you will use integration to find a model for the cost of producing a product.

■ Use the General Power Rule to find indefinite integrals.
■ Use substitution to find indefinite integrals.
■ Use the General Power Rule to solve real-life problems.

The General Power Rule

In Section 5.1, you used the Simple Power Rule

$$\int x^n \, dx = \frac{x^{n+1}}{n+1} + C, \quad n \neq -1$$

to find antiderivatives of functions expressed as powers of x alone. In this section, you will study a technique for finding antiderivatives of more complicated functions.

To begin, consider how you might find the antiderivative of

$$2x(x^2 + 1)^3.$$

Because you are hunting for a function whose derivative is $2x(x^2 + 1)^3$, you might discover the antiderivative as shown.

$$\frac{d}{dx}[(x^2 + 1)^4] = 4(x^2 + 1)^3(2x) \qquad \text{Use Chain Rule.}$$

$$\frac{d}{dx}\left[\frac{(x^2 + 1)^4}{4}\right] = (x^2 + 1)^3(2x) \qquad \text{Divide both sides by 4.}$$

$$\frac{(x^2 + 1)^4}{4} + C = \int 2x(x^2 + 1)^3 \, dx \qquad \text{Write in integral form.}$$

The key to this solution is the presence of the factor $2x$ in the integrand. In other words, this solution works because $2x$ is precisely the derivative of $(x^2 + 1)$. Letting $u = x^2 + 1$, you can write

$$\int \overbrace{(x^2 + 1)^3}^{u^3} \underbrace{2x \, dx}_{du} = \int u^3 \, du$$
$$= \frac{u^4}{4} + C.$$

This is an example of the **General Power Rule** for integration.

General Power Rule for Integration

If u is a differentiable function of x, then

$$\int u^n \frac{du}{dx} \, dx = \int u^n \, du$$

$$= \frac{u^{n+1}}{n+1} + C, \quad n \neq -1.$$

When using the General Power Rule, you must first identify a factor u of the integrand that is raised to a power. Then, you must show that its derivative du/dx is also a factor of the integrand. This is demonstrated in Example 1.

Example 1 Applying the General Power Rule

Find each indefinite integral.

a. $\int 3(3x - 1)^4 \, dx$ **b.** $\int (2x + 1)(x^2 + x) \, dx$

c. $\int 3x^2 \sqrt{x^3 - 2} \, dx$ **d.** $\int \dfrac{-4x}{(1 - 2x^2)^2} \, dx$

SOLUTION

a. $\int 3(3x - 1)^4 \, dx = \int \overbrace{(3x - 1)^4}^{u^n} \overbrace{(3)}^{\frac{du}{dx}} \, dx$ Let $u = 3x - 1$.

$= \dfrac{(3x - 1)^5}{5} + C$ General Power Rule

b. $\int (2x + 1)(x^2 + x) \, dx = \int \overbrace{(x^2 + x)}^{u^n} \overbrace{(2x + 1)}^{\frac{du}{dx}} \, dx$ Let $u = x^2 + x$.

$= \dfrac{(x^2 + x)^2}{2} + C$ General Power Rule

STUDY TIP

Example 1(b) illustrates a case of the General Power Rule that is sometimes overlooked—when the power is $n = 1$. In this case, the rule takes the form

$\int u \dfrac{du}{dx} \, dx = \dfrac{u^2}{2} + C.$

c. $\int 3x^2 \sqrt{x^3 - 2} \, dx = \int \overbrace{(x^3 - 2)^{1/2}}^{u^n} \overbrace{(3x^2)}^{\frac{du}{dx}} \, dx$ Let $u = x^3 - 2$.

$= \dfrac{(x^3 - 2)^{3/2}}{3/2} + C$ General Power Rule

$= \dfrac{2}{3}(x^3 - 2)^{3/2} + C$ Simplify.

d. $\int \dfrac{-4x}{(1 - 2x^2)^2} \, dx = \int \overbrace{(1 - 2x^2)^{-2}}^{u^n} \overbrace{(-4x)}^{\frac{du}{dx}} \, dx$ Let $u = 1 - 2x^2$.

$= \dfrac{(1 - 2x^2)^{-1}}{-1} + C$ General Power Rule

$= -\dfrac{1}{1 - 2x^2} + C$ Simplify.

✓ **Checkpoint 1**

Find each indefinite integral.

a. $\int (3x^2 + 6)(x^3 + 6x)^2 \, dx$ **b.** $\int 2x \sqrt{x^2 - 2} \, dx$ ■

Remember that you can verify the result of an indefinite integral by differentiating the function. For instance, you can check the answer to Example 1(a) as follows.

$\dfrac{d}{dx}\left[\dfrac{(3x - 1)^5}{5} + C\right] = \left(\dfrac{1}{5}\right)(5)(3x - 1)^4(3)$ Apply Chain Rule.

$= 3(3x - 1)^4$ Simplify.

Many times, part of the derivative du/dx is missing from the integrand, and in *some* cases you can make the necessary adjustments to apply the General Power Rule.

ALGEBRA TUTOR xy

For help on the algebra in Example 2, see Example 1(b) in the *Chapter 5 Algebra Tutor*, on page 311.

Example 2   Multiplying and Dividing by a Constant

Find $\displaystyle \int x(3 - 4x^2)^2 \, dx$.

SOLUTION   Let $u = 3 - 4x^2$. To apply the General Power Rule, you need to create $du/dx = -8x$ as a factor of the integrand. You can accomplish this by multiplying and dividing by the constant -8.

$$\int x(3 - 4x^2)^2 \, dx = \int \left(-\frac{1}{8}\right) \overbrace{(3 - 4x^2)^2}^{u^n} \overbrace{(-8x)}^{\frac{du}{dx}} \, dx \qquad \text{Multiply and divide by } -8.$$

$$= -\frac{1}{8} \int (3 - 4x^2)^2 (-8x) \, dx \qquad \text{Factor } -\tfrac{1}{8} \text{ out of integrand.}$$

$$= \left(-\frac{1}{8}\right) \left[\frac{(3 - 4x^2)^3}{3}\right] + C \qquad \text{General Power Rule}$$

$$= -\frac{(3 - 4x^2)^3}{24} + C \qquad \text{Simplify.}$$

STUDY TIP

Try using the Chain Rule to check the result of Example 2. After differentiating

$$-\tfrac{1}{24}(3 - 4x^2)^3 + C$$

and simplifying, you should obtain the original integrand.

✓ **Checkpoint 2**

Find $\displaystyle \int x^3(3x^4 + 1)^2 \, dx$.   ■

Example 3   Multiplying and Dividing by a Constant

Find $\displaystyle \int (x^2 + 2x)^3(x + 1) \, dx$.

SOLUTION   Let $u = x^2 + 2x$. To apply the General Power Rule, you need to create $du/dx = 2x + 2$ as a factor of the integrand. You can accomplish this by multiplying and dividing by the constant 2.

$$\int (x^2 + 2x)^3(x + 1) \, dx = \int \left(\frac{1}{2}\right) \overbrace{(x^2 + 2x)^3}^{u^n} \overbrace{(2)(x + 1)}^{\frac{du}{dx}} \, dx \qquad \text{Multiply and divide by 2.}$$

$$= \frac{1}{2} \int (x^2 + 2x)^3(2x + 2) \, dx \qquad \text{Rewrite integrand.}$$

$$= \frac{1}{2} \left[\frac{(x^2 + 2x)^4}{4}\right] + C \qquad \text{General Power Rule}$$

$$= \frac{1}{8}(x^2 + 2x)^4 + C \qquad \text{Simplify.}$$

✓ **Checkpoint 3**

Find $\displaystyle \int (x^3 - 3x)^2(x^2 - 1) \, dx$.   ■

Example 4   A Failure of the General Power Rule

Find $\int -8(3 - 4x^2)^2 \, dx$.

SOLUTION   Let $u = 3 - 4x^2$. To apply the General Power Rule, you must create $du/dx = -8x$ as a factor of the integrand. In Examples 2 and 3, this was done by multiplying and dividing by a constant, and then factoring that constant out of the integrand. This strategy doesn't work with variables. That is,

$$\int -8(3 - 4x^2)^2 \, dx \neq \frac{1}{x} \int (3 - 4x^2)^2(-8x) \, dx.$$

To find this indefinite integral, you can expand the integrand and use the Simple Power Rule.

$$\int -8(3 - 4x^2)^2 \, dx = \int (-72 + 192x^2 - 128x^4) \, dx$$

$$= -72x + 64x^3 - \frac{128}{5}x^5 + C$$

✓ **Checkpoint 4**

Find $\int 2(3x^4 + 1)^2 \, dx$.

When an integrand contains an extra constant factor that is not needed as part of du/dx, you can simply move the factor outside the integral sign, as shown in the next example.

Example 5   Applying the General Power Rule

Find $\int 7x^2 \sqrt{x^3 + 1} \, dx$.

SOLUTION   Let $u = x^3 + 1$. Then you need to create $du/dx = 3x^2$ by multiplying and dividing by 3. The constant factor $\frac{7}{3}$ is not needed as part of du/dx, and can be moved outside the integral sign.

$$\int 7x^2 \sqrt{x^3 + 1} \, dx = \int 7x^2(x^3 + 1)^{1/2} \, dx \qquad \text{Rewrite with rational exponent.}$$

$$= \int \frac{7}{3}(x^3 + 1)^{1/2}(3x^2) \, dx \qquad \text{Multiply and divide by 3.}$$

$$= \frac{7}{3} \int (x^3 + 1)^{1/2}(3x^2) \, dx \qquad \text{Factor } \tfrac{7}{3} \text{ outside integral.}$$

$$= \frac{7}{3}\left[\frac{(x^3 + 1)^{3/2}}{3/2}\right] + C \qquad \text{General Power Rule}$$

$$= \frac{14}{9}(x^3 + 1)^{3/2} + C \qquad \text{Simplify.}$$

✓ **Checkpoint 5**

Find $\int 5x \sqrt{x^2 - 1} \, dx$.

Substitution

The integration technique used in Examples 1, 2, 3, and 5 depends on your ability to recognize or create an integrand of the form

$$u^n \frac{du}{dx}.$$

With more complicated integrands, it is difficult to recognize the steps needed to fit the integrand to a basic integration formula. When this occurs, an alternative procedure called **substitution** or **change of variables** can be helpful. With this procedure, you completely rewrite the integral in terms of u and du. That is, if $u = f(x)$, then $du = f'(x)\, dx$, and the General Power Rule takes the form

$$\int u^n \frac{du}{dx}\, dx = \int u^n\, du. \qquad \text{General Power Rule}$$

Example 6   Integration by Substitution

Find $\displaystyle\int \sqrt{1 - 3x}\, dx$.

SOLUTION   Begin by letting $u = 1 - 3x$. Then, $du/dx = -3$ and $du = -3\, dx$. This implies that

$$dx = -\frac{1}{3}\, du$$

and you can find the indefinite integral as shown.

$$
\begin{aligned}
\int \sqrt{1 - 3x}\, dx &= \int (1 - 3x)^{1/2}\, dx & &\text{Rewrite with rational exponent.}\\
&= \int u^{1/2}\left(-\frac{1}{3}\, du\right) & &\text{Substitute for } x \text{ and } dx.\\
&= -\frac{1}{3}\int u^{1/2}\, du & &\text{Factor } -\tfrac{1}{3} \text{ out of integrand.}\\
&= \left(-\frac{1}{3}\right)\!\left(\frac{u^{3/2}}{3/2}\right) + C & &\text{Apply Power Rule.}\\
&= -\frac{2}{9} u^{3/2} + C & &\text{Simplify.}\\
&= -\frac{2}{9}(1 - 3x)^{3/2} + C & &\text{Substitute } 1 - 3x \text{ for } u.
\end{aligned}
$$

You can check this result by differentiating.

$$
\begin{aligned}
\frac{d}{dx}\!\left[-\frac{2}{9}(1 - 3x)^{3/2} + C\right] &= \left(-\frac{2}{9}\right)\!\left(\frac{3}{2}\right)(1 - 3x)^{1/2}(-3)\\
&= \left(-\frac{1}{3}\right)(-3)(1 - 3x)^{1/2}\\
&= \sqrt{1 - 3x}
\end{aligned}
$$

✓ Checkpoint 6

Find $\displaystyle\int \sqrt{1 - 2x}\, dx$ by the method of substitution.   ■

The basic steps for integration by substitution are outlined in the guidelines below.

Guidelines for Integration by Substitution

1. Let u be a function of x (usually part of the integrand).

2. Solve for x and dx in terms of u and du.

3. Convert the entire integral to u-variable form.

4. After integrating, rewrite the antiderivative as a function of x.

5. Check your answer by differentiating.

Example 7 Integration by Substitution

Find $\displaystyle\int x\sqrt{x^2 - 1}\, dx$.

SOLUTION Consider the substitution $u = x^2 - 1$, which produces

$$du = 2x\, dx.$$

To create $2x\, dx$ as part of the integral, multiply and divide by 2.

$$
\begin{aligned}
\int x\sqrt{x^2 - 1}\, dx &= \frac{1}{2}\int \overbrace{(x^2 - 1)^{1/2}}^{u^{1/n}}\,\overbrace{2x\, dx}^{du} &&\text{Multiply and divide by 2.}\\[2mm]
&= \frac{1}{2}\int u^{1/2}\, du &&\text{Substitute for } x \text{ and } dx.\\[2mm]
&= \frac{1}{2}\left(\frac{u^{3/2}}{3/2}\right) + C &&\text{Apply Power Rule.}\\[2mm]
&= \frac{1}{3}u^{3/2} + C &&\text{Simplify.}\\[2mm]
&= \frac{1}{3}(x^2 - 1)^{3/2} + C &&\text{Substitute for } u.
\end{aligned}
$$

You can check this result by differentiating.

$$
\begin{aligned}
\frac{d}{dx}\left[\frac{1}{3}(x^2 - 1)^{3/2} + C\right] &= \frac{1}{3}\left(\frac{3}{2}\right)(x^2 - 1)^{1/2}(2x)\\[2mm]
&= \frac{1}{2}(2x)(x^2 - 1)^{1/2}\\[2mm]
&= x\sqrt{x^2 - 1}
\end{aligned}
$$

✓ Checkpoint 7

Find $\displaystyle\int x\sqrt{x^2 + 4}\, dx$ by the method of substitution. ■

To become efficient at integration, you should learn to use *both* techniques discussed in this section. For simpler integrals, you should use pattern recognition and create du/dx by multiplying and dividing by an appropriate constant. For more complicated integrals, you should use a formal change of variables, as shown in Examples 6 and 7. For the integrals in this section's exercise set, try working several of the problems twice—once with pattern recognition and once using formal substitution.

Extended Application: Propensity to Consume

In 2009, the U.S. poverty level for a family of four was about \$22,000. Families at or below the poverty level tend to consume 100% of their income—that is, they use all their income to purchase necessities such as food, clothing, and shelter. As income level increases, the average consumption tends to drop below 100%. For instance, a family earning \$25,000 may be able to save \$500 and so consume only \$24,500 (98%) of their income. As the income increases, the ratio of consumption to savings tends to decrease. The rate of change of consumption with respect to income is called the **marginal propensity to consume.**   *(Source: U.S. Census Bureau)*

 **Example 8**   **Analyzing Consumption**

For a family of four in 2009, the marginal propensity to consume income x (in dollars) can be modeled by

$$\frac{dQ}{dx} = \frac{0.98}{(x - 21{,}999)^{0.02}}, \quad x \geq 22{,}000$$

where Q represents the income consumed (in dollars). Use the model to estimate the amount consumed by a family of four whose 2009 income was \$35,000.

SOLUTION   Begin by integrating dQ/dx to find a model for the consumption Q.

$$Q = \int \frac{0.98}{(x - 21{,}999)^{0.02}}\, dx \qquad \text{Integrate } \frac{dQ}{dx} \text{ to obtain } Q.$$

$$= \int 0.98(x - 21{,}999)^{-0.02}\, dx \qquad \text{Rewrite.}$$

$$= (x - 21{,}999)^{0.98} + C \qquad \text{General Power Rule}$$

To solve for C, use the initial condition that $Q = 22{,}000$ when $x = 22{,}000$.

$$22{,}000 = (22{,}000 - 21{,}999)^{0.98} + C$$
$$22{,}000 = 1 + C$$
$$21{,}999 = C$$

So, you can use the model $Q = (x - 21{,}999)^{0.98} + 21{,}999$ to estimate that a family of four with an income of $x = 35{,}000$ consumed about

$$Q = (35{,}000 - 21{,}999)^{0.98} + 21{,}999 \approx \$32{,}756.$$

The graph of Q is shown in Figure 5.4.

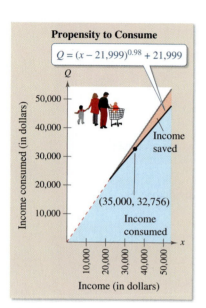

Propensity to Consume

$Q = (x - 21{,}999)^{0.98} + 21{,}999$

Income consumed (in dollars)

Income (in dollars)

(35,000, 32,756)

Income saved

Income consumed

FIGURE 5.4

✓ Checkpoint 8

According to the model in Example 8, at what income level would a family of four consume \$32,000?

SUMMARIZE   *(Section 5.2)*

1. State the General Power Rule for integration *(page 274)*. For examples of the General Power Rule, see Examples 1, 2, 3, and 5.

2. List the guidelines for integration by substitution *(page 279)*. For examples of integration by substitution, see Examples 6 and 7.

3. Describe a real-life example of how the General Power Rule can be used to analyze the marginal propensity to consume *(page 280, Example 8)*.

SKILLS WARM UP 5.2   The following warm-up exercises involve skills that were covered in earlier sections. You will use these skills in the exercise set for this section. For additional help, review Section 5.1.

In Exercises 1–9, find the indefinite integral.

1. $\displaystyle\int (2x^3 + 1)\, dx$

2. $\displaystyle\int (x^{1/2} + 3x - 4)\, dx$

3. $\displaystyle\int \frac{1}{x^2}\, dx$

4. $\displaystyle\int \frac{1}{3t^3}\, dt$

5. $\displaystyle\int (1 + 2t)t^{3/2}\, dt$

6. $\displaystyle\int \sqrt{x}\,(2x - 1)\, dx$

7. $\displaystyle\int \frac{5x^3 + 2}{x^2}\, dx$

8. $\displaystyle\int \frac{2x^2 - 5}{x^4}\, dx$

9. $\displaystyle\int \frac{8x^2 + 3}{\sqrt{x}}\, dx$

WebAssign   Access end-of-section exercises online at **www.webassign.net**

5.3 Exponential and Logarithmic Integrals

■ Use the Exponential Rule to find indefinite integrals.
■ Use the Log Rule to find indefinite integrals.

Using the Exponential Rule

Each of the differentiation rules for exponential functions has a corresponding integration rule.

Integrals of Exponential Functions

Let u be a differentiable function of x.

$$\int e^x \, dx = e^x + C \qquad \text{Simple Exponential Rule}$$

$$\int e^u \frac{du}{dx} \, dx = \int e^u \, du = e^u + C \qquad \text{General Exponential Rule}$$

Population of Bacteria

In Exercise 51, you will use integration to find a model for a population of bacteria.

Example 1   Integrating Exponential Functions

Find each indefinite integral.

a. $\displaystyle\int 2e^x \, dx$ **b.** $\displaystyle\int 2e^{2x} \, dx$ **c.** $\displaystyle\int (e^x + x) \, dx$

SOLUTION

a. $\displaystyle\int 2e^x \, dx = 2 \int e^x \, dx$ Constant Multiple Rule

$$= 2e^x + C \qquad \text{Simple Exponential Rule}$$

b. $\displaystyle\int 2e^{2x} \, dx = \int e^{2x}(2) \, dx$ Let $u = 2x$, then $\dfrac{du}{dx} = 2.$

$$= \int e^u \frac{du}{dx} \, dx \qquad \text{Substitute } u \text{ and } \dfrac{du}{dx}.$$

$$= e^u + C \qquad \text{General Exponential Rule}$$

$$= e^{2x} + C \qquad \text{Substitute for } u.$$

c. $\displaystyle\int (e^x + x) \, dx = \int e^x \, dx + \int x \, dx$ Sum Rule

$$= e^x + \frac{x^2}{2} + C \qquad \text{Simple Exponential and Power Rules}$$

You can check each of these results by differentiating. For instance, in part (a),

$$\frac{d}{dx}[2e^x + C] = 2e^x.$$

✓ Checkpoint 1

Find each indefinite integral.

a. $\displaystyle\int 3e^x \, dx$ **b.** $\displaystyle\int 5e^{5x} \, dx$ **c.** $\displaystyle\int (e^x - x) \, dx$

Example 2 **Integrating an Exponential Function**

Find $\displaystyle\int e^{3x+1}\,dx$.

SOLUTION Let $u = 3x + 1$; then $du/dx = 3$. You can introduce the missing factor of 3 in the integrand by multiplying and dividing by 3.

$$\int e^{3x+1}\,dx = \frac{1}{3}\int e^{3x+1}(3)\,dx \qquad \text{Multiply and divide by 3.}$$

$$= \frac{1}{3}\int e^{u}\frac{du}{dx}\,dx \qquad \text{Substitute } u \text{ and } \frac{du}{dx}.$$

$$= \frac{1}{3}e^{u} + C \qquad \text{General Exponential Rule}$$

$$= \frac{1}{3}e^{3x+1} + C \qquad \text{Substitute for } u.$$

✓ **Checkpoint 2**

Find $\displaystyle\int e^{2x+3}\,dx$. ■

Example 3 **Integrating an Exponential Function**

Find $\displaystyle\int 5xe^{-x^2}\,dx$.

SOLUTION Let $u = -x^2$; then $du/dx = -2x$. You can create the factor $-2x$ in the integrand by multiplying and dividing by -2.

$$\int 5xe^{-x^2}\,dx = \int\left(-\frac{5}{2}\right)e^{-x^2}(-2x)\,dx \qquad \text{Multiply and divide by } -2.$$

$$= -\frac{5}{2}\int e^{-x^2}(-2x)\,dx \qquad \text{Factor } -\frac{5}{2} \text{ out of the integrand.}$$

$$= -\frac{5}{2}\int e^{u}\frac{du}{dx}\,dx \qquad \text{Substitute } u \text{ and } \frac{du}{dx}.$$

$$= -\frac{5}{2}e^{u} + C \qquad \text{General Exponential Rule}$$

$$= -\frac{5}{2}e^{-x^2} + C \qquad \text{Substitute for } u.$$

✓ **Checkpoint 3**

Find $\displaystyle\int 4xe^{x^2}\,dx$. ■

Remember that you cannot introduce a missing *variable* in the integrand. For instance, you cannot find $\int e^{x^2}\,dx$ by multiplying and dividing by $2x$ and then factoring $1/(2x)$ out of the integrand. That is,

$$\int e^{x^2}\,dx \neq \frac{1}{2x}\int e^{x^2}(2x)\,dx.$$

Using the Log Rule

When the Power Rules for integration were introduced in Sections 5.1 and 5.2, you saw that they work for powers other than $n = -1$.

$$\int x^n \, dx = \frac{x^{n+1}}{n+1} + C, \quad n \neq -1 \qquad \text{Simple Power Rule}$$

$$\int u^n \frac{du}{dx} \, dx = \int u^n \, du = \frac{u^{n+1}}{n+1} + C, \quad n \neq -1 \qquad \text{General Power Rule}$$

The Log Rule for integration allows you to integrate functions of the form $\int x^{-1} \, dx$ and $\int u^{-1} \, du$.

STUDY TIP

Notice the absolute values in the Log Rule. For those special cases in which u or x cannot be negative, you can omit the absolute value. For instance, in Example 4(b), it is not necessary to write the antiderivative as $\ln|x^2| + C$ because x^2 cannot be negative.

Log Rule for Integration

Let u be a differentiable function of x.

$$\int \frac{1}{x} \, dx = \ln|x| + C \qquad \text{Simple Log Rule}$$

$$\int \frac{du/dx}{u} \, dx = \int \frac{1}{u} \, du = \ln|u| + C \qquad \text{General Log Rule}$$

You can verify each of these rules by differentiating. For instance, to verify that $d/dx[\ln|x|] = 1/x$, notice that

$$\frac{d}{dx}[\ln x] = \frac{1}{x} \quad \text{and} \quad \frac{d}{dx}[\ln(-x)] = \frac{-1}{-x} = \frac{1}{x}.$$

Example 4   Using the Log Rule for Integration

Find each indefinite integral.

a. $\displaystyle\int \frac{4}{x} \, dx$ **b.** $\displaystyle\int \frac{2x}{x^2} \, dx$ **c.** $\displaystyle\int \frac{3}{3x+1} \, dx$

SOLUTION

a. $\displaystyle\int \frac{4}{x} \, dx = 4 \int \frac{1}{x} \, dx$ Constant Multiple Rule

$\qquad\qquad = 4 \ln|x| + C$ Simple Log Rule

b. $\displaystyle\int \frac{2x}{x^2} \, dx = \int \frac{du/dx}{u} \, dx$ Let $u = x^2$; then $\dfrac{du}{dx} = 2x.$

$\qquad\qquad = \ln|u| + C$ General Log Rule

$\qquad\qquad = \ln x^2 + C$ Substitute for u.

c. $\displaystyle\int \frac{3}{3x+1} \, dx = \int \frac{du/dx}{u} \, dx$ Let $u = 3x + 1$; then $\dfrac{du}{dx} = 3.$

$\qquad\qquad = \ln|u| + C$ General Log Rule

$\qquad\qquad = \ln|3x+1| + C$ Substitute for u.

✓Checkpoint 4

Find each indefinite integral.

a. $\displaystyle\int \frac{2}{x} \, dx$ **b.** $\displaystyle\int \frac{3x^2}{x^3} \, dx$ **c.** $\displaystyle\int \frac{2}{2x+1} \, dx$

Example 5 Using the Log Rule for Integration

Find $\displaystyle\int \frac{1}{2x-1}\,dx$.

SOLUTION Let $u = 2x - 1$; then $du/dx = 2$. You can create the necessary factor of 2 in the integrand by multiplying and dividing by 2.

$$\int \frac{1}{2x-1}\,dx = \frac{1}{2}\int \frac{2}{2x-1}\,dx \qquad \text{Multiply and divide by 2.}$$

$$= \frac{1}{2}\int \frac{du/dx}{u}\,dx \qquad \text{Substitute } u \text{ and } \frac{du}{dx}.$$

$$= \frac{1}{2}\ln|u| + C \qquad \text{General Log Rule}$$

$$= \frac{1}{2}\ln|2x-1| + C \qquad \text{Substitute for } u.$$

✓ **Checkpoint 5**

Find $\displaystyle\int \frac{1}{4x+1}\,dx$.

Example 6 Using the Log Rule for Integration

Find $\displaystyle\int \frac{6x}{x^2+1}\,dx$.

SOLUTION Let $u = x^2 + 1$; then

$$\frac{du}{dx} = 2x.$$

You can create the necessary factor of $2x$ in the integrand by factoring a 3 out of the integrand.

$$\int \frac{6x}{x^2+1}\,dx = 3\int \frac{2x}{x^2+1}\,dx \qquad \text{Factor 3 out of integrand.}$$

$$= 3\int \frac{du/dx}{u}\,dx \qquad \text{Substitute } u \text{ and } \frac{du}{dx}.$$

$$= 3\ln|u| + C \qquad \text{General Log Rule}$$

$$= 3\ln(x^2+1) + C \qquad \text{Substitute for } u.$$

✓ **Checkpoint 6**

Find $\displaystyle\int \frac{3x}{x^2+4}\,dx$.

ALGEBRA TUTOR xy

For help on the algebra at the right, see Example 2(d) in the *Chapter 5 Algebra Tutor*, on page 312.

Integrals to which the Log Rule can be applied are often given in disguised form. For instance, when a rational function has a numerator of degree greater than or equal to that of the denominator, you should use long division to rewrite the integrand. Here is an example.

$$\int \frac{x^2+6x+1}{x^2+1}\,dx = \int \left(1 + \frac{6x}{x^2+1}\right)dx$$

$$= x + 3\ln(x^2+1) + C$$

The next example summarizes some additional situations in which it is helpful to rewrite the integrand in order to recognize the antiderivative.

ALGEBRA TUTOR *xy*

For help on the algebra in Example 7, see Example 2(a)–(c) in the *Chapter 5 Algebra Tutor*, on page 312.

Example 7   **Rewriting Before Integrating**

Find each indefinite integral.

a. $\displaystyle\int \frac{3x^2 + 2x - 1}{x^2}\,dx$ **b.** $\displaystyle\int \frac{1}{1 + e^{-x}}\,dx$ **c.** $\displaystyle\int \frac{x^2 + x + 1}{x - 1}\,dx$

SOLUTION

a. Begin by rewriting the integrand as the sum of three fractions.

$$\int \frac{3x^2 + 2x - 1}{x^2}\,dx = \int \left(\frac{3x^2}{x^2} + \frac{2x}{x^2} - \frac{1}{x^2}\right)dx$$

$$= \int \left(3 + \frac{2}{x} - \frac{1}{x^2}\right)dx$$

$$= 3x + 2\ln|x| + \frac{1}{x} + C$$

b. Begin by rewriting the integrand by multiplying and dividing by e^x.

$$\int \frac{1}{1 + e^{-x}}\,dx = \int \left(\frac{e^x}{e^x}\right)\frac{1}{1 + e^{-x}}\,dx$$

$$= \int \frac{e^x}{e^x + 1}\,dx$$

$$= \ln(e^x + 1) + C$$

c. Begin by dividing the numerator by the denominator.

$$\int \frac{x^2 + x + 1}{x - 1}\,dx = \int \left(x + 2 + \frac{3}{x - 1}\right)dx$$

$$= \frac{x^2}{2} + 2x + 3\ln|x - 1| + C$$

✓Checkpoint 7

Find each indefinite integral.

a. $\displaystyle\int \frac{4x^2 - 3x + 2}{x^2}\,dx$ **b.** $\displaystyle\int \frac{2}{e^{-x} + 1}\,dx$ **c.** $\displaystyle\int \frac{x^2 + 2x + 4}{x + 1}\,dx$

SUMMARIZE   (Section 5.3)

1. State the Simple Exponential Rule *(page 282)*. For an example of the Simple Exponential Rule, see Example 1.

2. State the General Exponential Rule *(page 282)*. For examples of the General Exponential Rule, see Examples 2 and 3.

3. State the Simple Log Rule *(page 284)*. For an example of the Simple Log Rule, see Example 4.

4. State the General Log Rule *(page 284)*. For examples of the General Log Rule, see Examples 5 and 6.

SKILLS WARM UP 5.3 The following warm-up exercises involve skills that were covered in earlier sections. You will use these skills in the exercise set for this section. For additional help, review Section 5.1.

In Exercises 1–4, use long division to rewrite the quotient.

1. $\dfrac{x^2 + 4x + 2}{x + 2}$

2. $\dfrac{x^2 - 6x + 9}{x - 4}$

3. $\dfrac{x^3 + 4x^2 - 30x - 4}{x^2 - 4x}$

4. $\dfrac{x^4 - x^3 + x^2 + 15x + 2}{x^2 + 5}$

In Exercises 5–8, find the indefinite integral.

5. $\displaystyle\int \left(x^3 + \frac{1}{x^2} \right) dx$

6. $\displaystyle\int \frac{x^2 + 2x}{x}\, dx$

7. $\displaystyle\int \frac{x^3 + 4}{x^2}\, dx$

8. $\displaystyle\int \frac{x + 3}{x^3}\, dx$

QUIZ YOURSELF

Take this quiz as you would take a quiz in class. When you are done, check your work against the answers given in the back of the book.

In Exercises 1–9, find the indefinite integral. Check your result by differentiation.

1. $\displaystyle\int 3\,dx$ **2.** $\displaystyle\int 10x\,dx$ **3.** $\displaystyle\int \frac{1}{x^5}\,dx$

4. $\displaystyle\int (x^2 - 2x + 15)\,dx$ **5.** $\displaystyle\int (6x + 1)^3(6)\,dx$ **6.** $\displaystyle\int x(5x^2 - 2)^4\,dx$

7. $\displaystyle\int (x^2 - 5x)(2x - 5)\,dx$ **8.** $\displaystyle\int \frac{3x^2}{(x^3 + 3)^3}\,dx$ **9.** $\displaystyle\int \sqrt{5x + 2}\,dx$

In Exercises 10 and 11, find the particular solution that satisfies the differential equation and initial condition.

10. $f'(x) = 16x; f(0) = 1$ **11.** $f'(x) = 9x^2 + 4; f(1) = 5$

12. The marginal cost function for producing x units of a product is modeled by

$$\frac{dC}{dx} = 16 - 0.06x.$$

It costs \$25 to produce one unit. Find (a) the cost function C (in dollars), (b) the fixed cost (when $x = 0$), and (c) the total cost of producing 500 units.

13. Find the equation of the function f whose graph passes through the point $(0, 1)$ and whose derivative is

$$f'(x) = 2x^2 + 1.$$

14. The number of bolts B produced by a foundry changes according to the model

$$\frac{dB}{dt} = \frac{250t}{\sqrt{t^2 + 36}}, \quad 0 \le t \le 40$$

where t is the time (in hours). Find the number of bolts produced in (a) 8 hours and (b) 40 hours.

In Exercises 15–17, use the Exponential Rule to find the indefinite integral.

15. $\displaystyle\int 5e^{5x+4}\,dx$ **16.** $\displaystyle\int 3x^2 e^{x^3}\,dx$ **17.** $\displaystyle\int (x - 3)e^{x^2 - 6x}\,dx$

In Exercises 18–20, use the Log Rule to find the indefinite integral.

18. $\displaystyle\int \frac{2}{2x - 1}\,dx$ **19.** $\displaystyle\int \frac{1}{3 - 8x}\,dx$ **20.** $\displaystyle\int \frac{x}{3x^2 + 4}\,dx$

21. The rate of change in sales for Advance Auto Parts from 2001 through 2009 can be modeled by

$$\frac{dS}{dt} = 26.32t + \frac{848.99}{t}$$

where S is the sales (in millions) and t is the time (in years), with $t = 1$ corresponding to 2001. In 2001, the sales for Advance Auto Parts were \$2517.6 million. *(Source: Advance Auto Parts, Inc.)*

(a) Find a model for the sales of Advance Auto Parts.

(b) Find the sales for Advance Auto Parts in 2008.

5.4 Area and the Fundamental Theorem of Calculus

■ Understand the relationship between area and definite integrals.
■ Evaluate definite integrals using the Fundamental Theorem of Calculus.
■ Use definite integrals to solve marginal analysis problems.
■ Find the average values of functions over closed intervals.
■ Use properties of even and odd functions to help evaluate definite integrals.
■ Find the amounts of annuities.

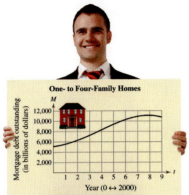

In Exercise 79 , you will use integration to find a model for the mortgage debt outstanding for one- to four-family homes.

Area and Definite Integrals

From your study of geometry, you know that area is a number that defines the size of a bounded region. For simple regions, such as rectangles, triangles, and circles, area can be found using geometric formulas.

In this section, you will learn how to use calculus to find the areas of nonstandard regions, such as the region R shown in Figure 5.5.

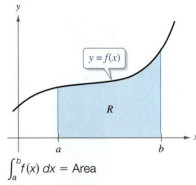

$$\int_a^b f(x)\, dx = \text{Area}$$

FIGURE 5.5

Definition of a Definite Integral

Let f be nonnegative and continuous on the closed interval $[a, b]$. The area of the region bounded by the graph of f, the x-axis, and the lines $x = a$ and $x = b$ is denoted by

$$\text{Area} = \int_a^b f(x)\, dx.$$

The expression $\int_a^b f(x)\, dx$ is called the **definite integral** from a to b, where a is the **lower limit of integration** and b is the **upper limit of integration.**

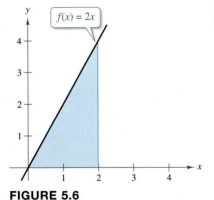

FIGURE 5.6

Example 1   Evaluating a Definite Integral Using a Geometric Formula

The definite integral

$$\int_0^2 2x\, dx$$

represents the area of the region bounded by the graph of $f(x) = 2x$, the x-axis, and the line $x = 2$, as shown in Figure 5.6. The region is triangular, with a height of 4 units and a base of 2 units. Using the formula for the area of a triangle, you have

$$\int_0^2 2x\, dx = \frac{1}{2}(\text{base})(\text{height}) = \frac{1}{2}(2)(4) = 4.$$

✓ **Checkpoint 1**

Evaluate the definite integral using a geometric formula. Illustrate your answer with an appropriate sketch.

$$\int_0^3 4x\, dx$$

The Fundamental Theorem of Calculus

Consider the function A, which denotes the area of the region shown in Figure 5.7.

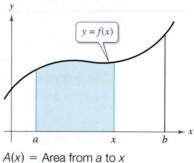

$A(x)$ = Area from a to x

FIGURE 5.7

To discover the relationship between A and f, let x increase by an amount Δx. This increases the area by ΔA. Let $f(m)$ and $f(M)$ denote the minimum and maximum values of f on the interval $[x, x + \Delta x]$.

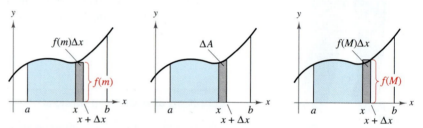

FIGURE 5.8

As indicated in Figure 5.8, you can write the inequality below.

$$f(m)\,\Delta x \leq \quad \Delta A \quad \leq f(M)\,\Delta x \qquad \text{See Figure 5.8.}$$

$$f(m) \leq \quad \frac{\Delta A}{\Delta x} \quad \leq f(M) \qquad \text{Divide each term by } \Delta x.$$

$$\lim_{\Delta x \to 0} f(m) \leq \lim_{\Delta x \to 0} \frac{\Delta A}{\Delta x} \leq \lim_{\Delta x \to 0} f(M) \qquad \text{Take limit of each term.}$$

$$f(x) \leq \quad A'(x) \quad \leq f(x) \qquad \text{Definition of derivative of } A(x).$$

So, $f(x) = A'(x)$, and $A(x) = F(x) + C$, where $F'(x) = f(x)$. Because $A(a) = 0$, it follows that $C = -F(a)$. So, $A(x) = F(x) - F(a)$, which implies that

$$A(b) = \int_a^b f(x)\,dx = F(b) - F(a).$$

This equation tells you that *if you can find an antiderivative for f,* then you can use the antiderivative to evaluate the definite integral $\int_a^b f(x)\,dx$. This result is called the **Fundamental Theorem of Calculus.**

The Fundamental Theorem of Calculus

If f is nonnegative and continuous on the closed interval $[a, b]$, then

$$\int_a^b f(x)\,dx = F(b) - F(a)$$

where F is any function such that $F'(x) = f(x)$ for all x in $[a, b]$.

Guidelines for Using the Fundamental Theorem of Calculus

1. The Fundamental Theorem of Calculus describes a way of *evaluating* a definite integral, not a procedure for finding antiderivatives.

2. In applying the Fundamental Theorem, it is helpful to use the notation

$$\int_a^b f(x)\, dx = F(x)\Big]_a^b = F(b) - F(a).$$

   For instance, to evaluate $\int_1^3 x^3\, dx$, you can write

$$\int_1^3 x^3\, dx = \frac{x^4}{4}\Big]_1^3$$

$$= \frac{3^4}{4} - \frac{1^4}{4}$$

$$= 20.$$

3. The constant of integration C can be dropped because

$$\int_a^b f(x)\, dx = \left[F(x) + C\right]_a^b$$

$$= [F(b) + C] - [F(a) + C]$$

$$= F(b) - F(a) + C - C$$

$$= F(b) - F(a).$$

In the development of the Fundamental Theorem of Calculus, f was assumed to be nonnegative on the closed interval $[a, b]$. As such, the definite integral was defined as an area. Now, with the Fundamental Theorem, the definition can be extended to include functions that are negative on all or part of the closed interval $[a, b]$. Specifically, if f is *any* function that is continuous on a closed interval $[a, b]$, then the **definite integral** of $f(x)$ from a to b is defined to be

$$\int_a^b f(x)\, dx = F(b) - F(a)$$

where F is an antiderivative of f. Remember that definite integrals do not necessarily represent areas and can be negative, zero, or positive.

STUDY TIP

Be sure you see the distinction between indefinite and definite integrals. The *indefinite integral*

$$\int f(x)\, dx$$

denotes a *family of functions*, each of which is an antiderivative of f, whereas the *definite integral*

$$\int_a^b f(x)\, dx$$

is a *number*.

Properties of Definite Integrals

Let f and g be continuous on the closed interval $[a, b]$.

1. $\displaystyle\int_a^b k f(x)\, dx = k \int_a^b f(x)\, dx,\quad k$ is a constant.

2. $\displaystyle\int_a^b [f(x) \pm g(x)]\, dx = \int_a^b f(x)\, dx \pm \int_a^b g(x)\, dx$

3. $\displaystyle\int_a^b f(x)\, dx = \int_a^c f(x)\, dx + \int_c^b f(x)\, dx,\quad a < c < b$

4. $\displaystyle\int_a^a f(x)\, dx = 0$

5. $\displaystyle\int_a^b f(x)\, dx = -\int_b^a f(x)\, dx$

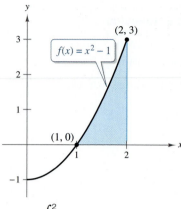

Area $= \int_1^2 (x^2 - 1)\, dx$

FIGURE 5.9

Example 2 **Finding Area by the Fundamental Theorem**

Find the area of the region bounded by the x-axis and the graph of

$$f(x) = x^2 - 1, \quad 1 \le x \le 2.$$

SOLUTION   Note that $f(x) \ge 0$ on the interval $1 \le x \le 2$, as shown in Figure 5.9. So, you can represent the area of the region by a definite integral. To find the area, use the Fundamental Theorem of Calculus.

$$\text{Area} = \int_1^2 (x^2 - 1)\, dx \qquad \text{Definition of definite integral}$$

$$= \left[\frac{x^3}{3} - x \right]_1^2 \qquad \text{Find antiderivative.}$$

$$= \left(\frac{2^3}{3} - 2 \right) - \left(\frac{1^3}{3} - 1 \right) \qquad \text{Apply Fundamental Theorem.}$$

$$= \frac{2}{3} - \left(-\frac{2}{3} \right)$$

$$= \frac{4}{3} \qquad \text{Simplify.}$$

So, the area of the region is $\frac{4}{3}$ square units.

✓ Checkpoint 2

Find the area of the region bounded by the x-axis and the graph of

$$f(x) = x^2 + 1, \quad 2 \le x \le 3. \qquad ■$$

Example 3 **Evaluating a Definite Integral**

Evaluate the definite integral

$$\int_0^1 (4t + 1)^2\, dt$$

and sketch the region whose area is represented by the integral.

SOLUTION

$$\int_0^1 (4t + 1)^2\, dt = \frac{1}{4} \int_0^1 (4t + 1)^2 (4)\, dt \qquad \text{Multiply and divide by 4.}$$

$$= \frac{1}{4} \left[\frac{(4t + 1)^3}{3} \right]_0^1 \qquad \text{Find antiderivative.}$$

$$= \frac{1}{4} \left[\left(\frac{5^3}{3} \right) - \left(\frac{1}{3} \right) \right] \qquad \text{Apply Fundamental Theorem.}$$

$$= \frac{1}{4} \left(\frac{124}{3} \right)$$

$$= \frac{31}{3} \qquad \text{Simplify.}$$

The region is shown in Figure 5.10.

✓ Checkpoint 3

Evaluate $\int_0^1 (2t + 3)^3\, dt.$ ■

STUDY TIP

It is easy to make errors in signs when evaluating definite integrals. To avoid such errors, enclose the values of the antiderivative at the upper and lower limits of integration in separate sets of parentheses, as shown in Example 2.

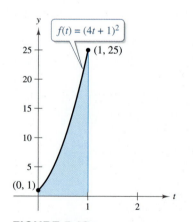

FIGURE 5.10

Example 4 **Evaluating Definite Integrals**

Evaluate each definite integral.

a. $\displaystyle\int_0^3 e^{2x}\, dx$ **b.** $\displaystyle\int_1^2 \frac{1}{x}\, dx$ **c.** $\displaystyle\int_1^4 -3\sqrt{x}\, dx$

SOLUTION

a. $\displaystyle\int_0^3 e^{2x}\, dx = \frac{1}{2}e^{2x}\Big]_0^3 = \frac{1}{2}(e^6 - e^0) \approx 201.21$

b. $\displaystyle\int_1^2 \frac{1}{x}\, dx = \ln x\Big]_1^2 = \ln 2 - \ln 1 = \ln 2 \approx 0.69$

c. $\displaystyle\int_1^4 -3\sqrt{x}\, dx = -3\int_1^4 x^{1/2}\, dx$ Rewrite with rational exponent.

$\displaystyle\qquad\qquad = -3\left[\frac{x^{3/2}}{3/2}\right]_1^4$ Find antiderivative.

$\displaystyle\qquad\qquad = -2x^{3/2}\Big]_1^4$

$\displaystyle\qquad\qquad = -2(4^{3/2} - 1^{3/2})$ Apply Fundamental Theorem.

$\displaystyle\qquad\qquad = -2(8 - 1)$

$\displaystyle\qquad\qquad = -14$ Simplify.

> **STUDY TIP**
>
> In Example 4(c), note that the value of a definite integral can be negative.

✓ **Checkpoint 4**

Evaluate each definite integral.

a. $\displaystyle\int_0^1 e^{4x}\, dx$ **b.** $\displaystyle\int_2^5 -\frac{1}{x}\, dx$ ■

Example 5 **Interpreting Absolute Value**

Evaluate $\displaystyle\int_0^2 |2x - 1|\, dx$.

SOLUTION The region represented by the definite integral is shown in Figure 5.11. From the definition of absolute value, you can write

$$|2x - 1| = \begin{cases} -(2x - 1), & x < \frac{1}{2} \\ 2x - 1, & x \ge \frac{1}{2} \end{cases}.$$

Using Property 3 of definite integrals, rewrite the integral as two definite integrals.

$$\int_0^2 |2x - 1|\, dx = \int_0^{1/2} -(2x - 1)\, dx + \int_{1/2}^2 (2x - 1)\, dx$$

$$= \left[-x^2 + x\right]_0^{1/2} + \left[x^2 - x\right]_{1/2}^2$$

$$= \left(-\frac{1}{4} + \frac{1}{2}\right) - (0 + 0) + (4 - 2) - \left(\frac{1}{4} - \frac{1}{2}\right)$$

$$= \frac{5}{2}$$

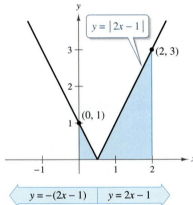

$y = |2x - 1|$

(2, 3)

(0, 1)

$y = -(2x - 1)$ $y = 2x - 1$

FIGURE 5.11

✓ **Checkpoint 5**

Evaluate $\displaystyle\int_0^5 |x - 2|\, dx$. ■

Marginal Analysis

You have already studied *marginal analysis* in the context of derivatives and differentials
(Sections 2.3 and 3.8). There, you were given a cost, revenue, or profit function, and
you used the derivative to approximate the additional cost, revenue, or profit obtained
by selling one additional unit. In this section, you will examine the reverse process.
That is, you will be given the marginal cost, marginal revenue, or marginal profit and
you will use a definite integral to find the exact increase or decrease in cost, revenue,
or profit obtained by selling one or several additional units.

For instance, you are asked to find the additional revenue obtained by increasing
sales from x_1 to x_2 units. When you know the revenue function R, you can find the
additional revenue by subtracting $R(x_1)$ from $R(x_2)$. When you don't know R, you can
use the marginal revenue function dR/dx to find the additional revenue by using a
definite integral.

$$\int_{x_1}^{x_2} \frac{dR}{dx} \, dx = R(x_2) - R(x_1)$$

 Example 6   Analyzing a Profit Function

The marginal profit for a product is modeled by

$$\frac{dP}{dx} = -0.0005x + 12.2.$$

a. Find the change in profit when sales increase from 100 to 101 units.

b. Find the change in profit when sales increase from 100 to 110 units.

SOLUTION

a. The change in profit obtained by increasing sales from 100 to 101 units is

$$\int_{100}^{101} \frac{dP}{dx} \, dx = \int_{100}^{101} (-0.0005x + 12.2) \, dx$$
$$= \left[-0.00025x^2 + 12.2x \right]_{100}^{101}$$
$$\approx \$12.15.$$

b. The change in profit obtained by increasing sales from 100 to 110 units is

$$\int_{100}^{110} \frac{dP}{dx} \, dx = \int_{100}^{110} (-0.0005x + 12.2) \, dx$$
$$= \left[-0.00025x^2 + 12.2x \right]_{100}^{110}$$
$$\approx \$121.48.$$

✓ **Checkpoint 6**

The marginal profit for a product is modeled by

$$\frac{dP}{dx} = -0.0002x + 14.2.$$

a. Find the change in profit when sales increase from 100 to 101 units.

b. Find the change in profit when sales increase from 100 to 110 units.   ■

Average Value

The *average value* of a function on a closed interval is defined below.

Definition of the Average Value of a Function

If f is continuous on $[a, b]$, then the **average value** of f on $[a, b]$ is

$$\text{Average value of } f \text{ on } [a, b] = \frac{1}{b - a}\int_a^b f(x)\, dx.$$

In Section 3.5, you studied the effects of production levels on cost using an average cost function. In the next example, you will study the effects of time on cost by using integration to find the average cost.

Example 7   Finding the Average Cost

The cost per unit c of producing MP3 players over a two-year period is modeled by

$$c = 0.005t^2 + 0.01t + 13.15, \quad 0 \le t \le 24$$

where t is the time (in months). Approximate the average cost per unit over the two-year period.

SOLUTION   The average cost can be found by integrating c over the interval $[0, 24]$.

$$\begin{aligned}
\text{Average cost per unit} &= \frac{1}{24}\int_0^{24}(0.005t^2 + 0.01t + 13.15)\, dt \\
&= \frac{1}{24}\left[\frac{0.005t^3}{3} + \frac{0.01t^2}{2} + 13.15t\right]_0^{24} \\
&= \frac{1}{24}(341.52) \\
&= \$14.23 \qquad \text{(See Figure 5.12.)}
\end{aligned}$$

✓ **Checkpoint 7**

Find the average cost per unit over a two-year period when the cost per unit c of inline skates is given by

$$c = 0.005t^2 + 0.02t + 12.5, \quad 0 \le t \le 24$$

where t is the time (in months). ■

You can use a spreadsheet, as shown at the left, to check the reasonableness of the average value found in Example 7. The spreadsheet assumes that one unit is produced each month, beginning with $t = 0$ and ending with $t = 24$. So, when $t = 0$, the cost is

$$c = 0.005(0)^2 + 0.01(0) + 13.15$$
$$= \$13.15$$

and when $t = 1$ the cost is

$$c = 0.005(1)^2 + 0.01(1) + 13.15$$
$$= \$13.165$$

and so on. Note in the spreadsheet that the cost increases each month, and the average of the 25 costs is \$14.25. So, you can conclude that the result of Example 7 is reasonable.

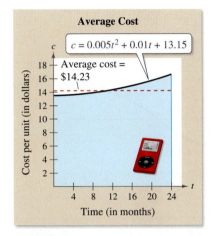

Average Cost

$c = 0.005t^2 + 0.01t + 13.15$

Average cost = \$14.23

Cost per unit (in dollars)

Time (in months)

FIGURE 5.12

	A	B
1	t	$c = 0.005t^2 + 0.01t + 13.15$
2	0	13.15
3	1	13.165
4	2	13.19
5	3	13.225
6	4	13.27
7	5	13.325
8	6	13.39
9	7	13.465
10	8	13.55
11	9	13.645
12	10	13.75
13	11	13.865
14	12	13.99
15	13	14.125
16	14	14.27
17	15	14.425
18	16	14.59
19	17	14.765
20	18	14.95
21	19	15.145
22	20	15.35
23	21	15.565
24	22	15.79
25	23	16.025
26	24	16.27
27		
28	**Sum**	356.25
29	**Average**	14.25

Even and Odd Functions

Several common functions have graphs that are symmetric with respect to the y-axis or the origin, as shown in Figure 5.13. If the graph of f is symmetric with respect to the y-axis, as in Figure 5.13(a), then

$$f(-x) = f(x) \qquad \text{Even function}$$

and f is called an **even** function. If the graph of f is symmetric with respect to the origin, as in Figure 5.13(b), then

$$f(-x) = -f(x) \qquad \text{Odd function}$$

and f is called an **odd** function.

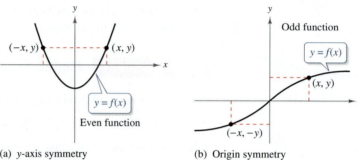

(a) y-axis symmetry   (b) Origin symmetry

FIGURE 5.13

Integration of Even and Odd Functions

1. If f is an *even* function, then $\displaystyle\int_{-a}^{a} f(x)\ dx = 2\int_{0}^{a} f(x)\ dx.$

2. If f is an *odd* function, then $\displaystyle\int_{-a}^{a} f(x)\ dx = 0.$

Example 8   **Integrating Even and Odd Functions**

Evaluate each definite integral.

a. $\displaystyle\int_{-2}^{2} x^2\ dx$ **b.** $\displaystyle\int_{-2}^{2} x^3\ dx$

SOLUTION

a. Because $f(x) = x^2$ is an even function,

$$\int_{-2}^{2} x^2\ dx = 2\int_{0}^{2} x^2\ dx = 2\left[\frac{x^3}{3}\right]_{0}^{2} = 2\left(\frac{8}{3} - 0\right) = \frac{16}{3}.$$

b. Because $f(x) = x^3$ is an odd function,

$$\int_{-2}^{2} x^3\ dx = 0.$$

✓ **Checkpoint 8**

Evaluate each definite integral.

a. $\displaystyle\int_{-1}^{1} x^4\ dx$ **b.** $\displaystyle\int_{-1}^{1} x^5\ dx$

Annuity

A sequence of equal payments made at regular time intervals over a period of time is called an **annuity.** Some examples of annuities are payroll savings plans, monthly home mortgage payments, and individual retirement accounts. The **amount of an annuity** is the sum of the payments plus the interest earned.

Amount of an Annuity

If c represents a continuous income function in dollars per year (where t is the time in years), r represents the interest rate compounded continuously, and T represents the term of the annuity in years, then the **amount of an annuity** is

$$\text{Amount of an annuity} = e^{rT} \int_0^T c(t)e^{-rt}\, dt.$$

 Example 9 **Finding the Amount of an Annuity**

You deposit $2000 each year for 15 years in an individual retirement account (IRA) paying 5% interest. How much will you have in your IRA after 15 years?

SOLUTION The income function for your deposit is

$$c(t) = 2000.$$

So, the amount of the annuity after 15 years will be

$$\begin{aligned}
\text{Amount of an annuity} &= e^{rT} \int_0^T c(t)e^{-rt}\, dt \\
&= e^{(0.05)(15)} \int_0^{15} 2000e^{-0.05t}\, dt \\
&= 2000e^{0.75}\left[-\frac{e^{-0.05t}}{0.05}\right]_0^{15} \\
&\approx \$44{,}680.00.
\end{aligned}$$

✓**Checkpoint 9**

You deposit $1000 each year in a savings account paying 4% interest. How much will be in the account after 10 years?

SUMMARIZE (Section 5.4)

1. State the definition of a definite integral *(page 289)*. For an example of a definite integral, see Example 1.

2. State the Fundamental Theorem of Calculus *(page 290)*. For examples of the Fundamental Theorem of Calculus, see Examples 2 and 3.

3. State the properties of definite integrals *(page 291)*. For examples of the properties, see Examples 4 and 5.

4. State the definition of the average value of a function *(page 295)*. For an example of finding the average value of a function, see Example 7.

5. State the rules for integrating even and odd functions *(page 296)*. For an example of integrating even and odd functions, see Example 8.

SKILLS WARM UP 5.4   The following warm-up exercises involve skills that were covered in earlier sections. You will use these skills in the exercise set for this section. For additional help, review Sections 5.1–5.3.

In Exercises 1–4, find the indefinite integral.

1. $\displaystyle\int (3x + 7)\, dx$

2. $\displaystyle\int \left(x^{3/2} + 2\sqrt{x}\right) dx$

3. $\displaystyle\int \frac{1}{5x}\, dx$

4. $\displaystyle\int e^{-6x}\, dx$

In Exercises 5–8, integrate the marginal function.

5. $\dfrac{dC}{dx} = 0.02x^{3/2} + 29{,}500$

6. $\dfrac{dR}{dx} = 9000 + 2x$

7. $\dfrac{dP}{dx} = 25{,}000 - 0.01x$

8. $\dfrac{dC}{dx} = 0.03x^2 + 4600$

ENHANCED
WebAssign   Access end-of-section exercises online at **www.webassign.net**

5.5 The Area of a Region Bounded by Two Graphs

■ Find the areas of regions bounded by two graphs.
■ Find consumer and producer surpluses.
■ Use the areas of regions bounded by two graphs to solve real-life problems.

Area of a Region Bounded by Two Graphs

With a few modifications, you can extend the use of definite integrals from finding the area of a region *under a graph* to finding the area of a region *bounded by two graphs*. To see how this is done, consider the region bounded by the graphs of

$$f, \quad g, \quad x = a, \quad \text{and} \quad x = b$$

as shown in Figure 5.14. If the graphs of both *f* and *g* lie above the *x*-axis, then you can interpret the area of the region between the graphs as the area of the region under the graph of *g* subtracted from the area of the region under the graph of *f*, as shown in Figure 5.14.

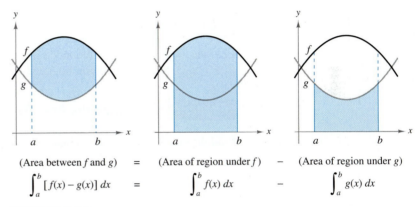

(Area between *f* and *g*)   =   (Area of region under *f*)   −   (Area of region under *g*)

$$\int_a^b [f(x) - g(x)]\, dx \quad = \quad \int_a^b f(x)\, dx \quad - \quad \int_a^b g(x)\, dx$$

FIGURE 5.14

Although Figure 5.14 depicts the graphs of *f* and *g* lying above the *x*-axis, this is not necessary, and the same integrand

$$[f(x) - g(x)]$$

can be used as long as both functions are continuous and $g(x) \le f(x)$ on the interval $[a, b]$.

Area of a Region Bounded by Two Graphs

If *f* and *g* are continuous on $[a, b]$ and $g(x) \le f(x)$ for all *x* in $[a, b]$, then the area of the region bounded by the graphs of

$$f, \quad g, \quad x = a, \quad \text{and} \quad x = b$$

(see Figure 5.15) is given by

$$A = \int_a^b [f(x) - g(x)]\, dx.$$

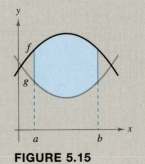

FIGURE 5.15

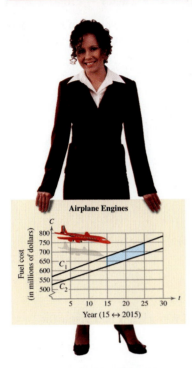

Airplane Engines

In Exercise 49, you will use integration to find the amount saved on fuel costs by switching to more efficient airplane engines.

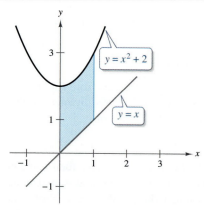

FIGURE 5.16

Example 1 **Finding the Area Bounded by Two Graphs**

Find the area of the region bounded by the graphs of $y = x^2 + 2$ and $y = x$ for $0 \leq x \leq 1$.

SOLUTION Begin by sketching the graphs of both functions, as shown in Figure 5.16. From the figure, you can see that $x \leq x^2 + 2$ for all x in $[0, 1]$. So, you can let $f(x) = x^2 + 2$ and $g(x) = x$. Then find the area as shown.

$$\text{Area} = \int_a^b [f(x) - g(x)]\, dx \qquad \text{Area between } f \text{ and } g$$

$$= \int_0^1 [(x^2 + 2) - (x)]\, dx \qquad \text{Substitute for } f \text{ and } g.$$

$$= \int_0^1 (x^2 - x + 2)\, dx$$

$$= \left[\frac{x^3}{3} - \frac{x^2}{2} + 2x \right]_0^1 \qquad \text{Find antiderivative.}$$

$$= \frac{11}{6} \text{ square units} \qquad \text{Apply Fundamental Theorem.}$$

✓**Checkpoint 1**

Find the area of the region bounded by the graphs of $y = x^2 + 1$ and $y = x$ for $0 \leq x \leq 2$. Sketch the region bounded by the graphs. ■

Example 2 **Finding the Area Between Intersecting Graphs**

Find the area of the region bounded by the graphs of $y = 2 - x^2$ and $y = x$.

SOLUTION Because the values of a and b are not given, you must determine them by finding the x-coordinates of the points of intersection of the two graphs. To do this, equate the two functions and solve for x.

$$2 - x^2 = x \qquad \text{Equate functions.}$$

$$-x^2 - x + 2 = 0 \qquad \text{Write in general form.}$$

$$-(x + 2)(x - 1) = 0 \qquad \text{Factor.}$$

$$x = -2, \, x = 1 \qquad \text{Solve for } x.$$

So, $a = -2$ and $b = 1$. In Figure 5.17, you can see that the graph of $f(x) = 2 - x^2$ lies above the graph of $g(x) = x$ for all x in the interval $[-2, 1]$.

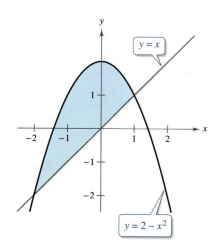

FIGURE 5.17

$$\text{Area} = \int_a^b [f(x) - g(x)]\, dx \qquad \text{Area between } f \text{ and } g$$

$$= \int_{-2}^1 [(2 - x^2) - (x)]\, dx \qquad \text{Substitute for } f \text{ and } g.$$

$$= \int_{-2}^1 (-x^2 - x + 2)\, dx$$

$$= \left[-\frac{x^3}{3} - \frac{x^2}{2} + 2x \right]_{-2}^1 \qquad \text{Find antiderivative.}$$

$$= \frac{9}{2} \text{ square units} \qquad \text{Apply Fundamental Theorem.}$$

✓**Checkpoint 2**

Find the area of the region bounded by the graphs of $y = 3 - x^2$ and $y = 2x$. ■

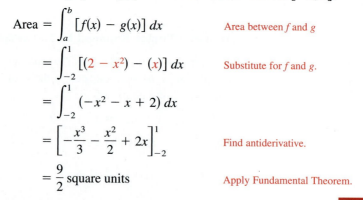

Example 3   **Finding an Area Below the x-Axis**

Find the area of the region bounded by the graph of

$$y = x^2 - 3x - 4$$

and the x-axis.

SOLUTION   Begin by finding the x-intercepts of the graph. To do this, set the function equal to zero and solve for x.

$x^2 - 3x - 4 = 0$	Set function equal to 0.
$(x - 4)(x + 1) = 0$	Factor.
$x = 4, x = -1$	Solve for x.

From Figure 5.18, you can see that $x^2 - 3x - 4 \leq 0$ for all x in the interval $[-1, 4]$.

TECH TUTOR

Most graphing utilities can display regions that are bounded by two graphs. For instance, to graph the region in Example 3, set the viewing window to $-1 \leq x \leq 4$ and $-7 \leq y \leq 1$. Consult your user's manual for specific keystrokes on how to shade the graph. You should obtain the graph below.

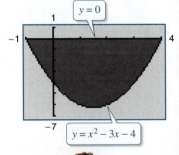

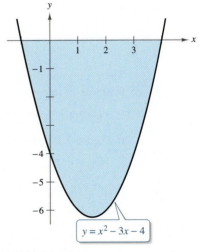

FIGURE 5.18

So, you can let

$$f(x) = 0 \quad \text{and} \quad g(x) = x^2 - 3x - 4$$

and find the area as shown.

$\text{Area} = \displaystyle\int_a^b [f(x) - g(x)]\, dx$	Area between f and g
$\quad = \displaystyle\int_{-1}^{4} [(0) - (x^2 - 3x - 4)]\, dx$	Substitute for f and g.
$\quad = \displaystyle\int_{-1}^{4} (-x^2 + 3x + 4)\, dx$	
$\quad = \left[-\dfrac{x^3}{3} + \dfrac{3x^2}{2} + 4x \right]_{-1}^{4}$	Find antiderivative.
$\quad = \dfrac{125}{6}$ square units	Apply Fundamental Theorem.

✓**Checkpoint 3**

Find the area of the region bounded by the graph of

$$y = x^2 - x - 2$$

and the x-axis.

Sometimes two graphs intersect at more than two points. To determine the area of the region bounded by two such graphs, you must find *all* points of intersection and check to see which graph is above the other in each interval determined by the points.

Example 4   Using Multiple Points of Intersection

Find the area of the region bounded by the graphs of

$$f(x) = 3x^3 - x^2 - 10x \quad \text{and} \quad g(x) = -x^2 + 2x.$$

SOLUTION   To find the points of intersection of the two graphs, set the functions equal to each other and solve for x.

$$f(x) = g(x) \qquad \text{Set } f(x) \text{ equal to } g(x).$$
$$3x^3 - x^2 - 10x = -x^2 + 2x \qquad \text{Substitute for } f(x) \text{ and } g(x).$$
$$3x^3 - 12x = 0 \qquad \text{Write in general form.}$$
$$3x(x^2 - 4) = 0$$
$$3x(x - 2)(x + 2) = 0 \qquad \text{Factor.}$$
$$x = 0, x = 2, x = -2 \qquad \text{Solve for } x.$$

These three points of intersection determine two intervals of integration:

$$[-2, 0] \quad \text{and} \quad [0, 2].$$

In Figure 5.19, you can see that

$$g(x) \le f(x)$$

for all x in the interval $[-2, 0]$, and that

$$f(x) \le g(x)$$

for all x in the interval $[0, 2]$. So, you must use two integrals to determine the area of the region bounded by the graphs of f and g: one for the interval $[-2, 0]$ and one for the interval $[0, 2]$.

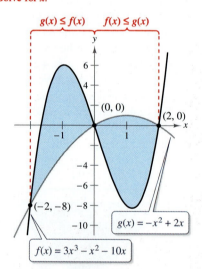

FIGURE 5.19

STUDY TIP

It is easy to make an error when calculating areas such as the one in Example 4. To check your solution, make a sketch of the region on graph paper and then use the grid on the graph paper to approximate the area. Try doing this with the graph shown in Figure 5.19. Is your approximation close to 24 square units?

$$\text{Area} = \int_{-2}^{0} [f(x) - g(x)] \, dx + \int_{0}^{2} [g(x) - f(x)] \, dx$$

$$= \int_{-2}^{0} (3x^3 - 12x) \, dx + \int_{0}^{2} (-3x^3 + 12x) \, dx$$

$$= \left[\frac{3x^4}{4} - 6x^2 \right]_{-2}^{0} + \left[-\frac{3x^4}{4} + 6x^2 \right]_{0}^{2}$$

$$= (0 - 0) - (12 - 24) + (-12 + 24) - (0 + 0)$$

$$= 24$$

So, the region has an area of 24 square units.

✓ Checkpoint 4

Find the area of the region bounded by the graphs of

$$f(x) = x^3 + 2x^2 - 3x \quad \text{and} \quad g(x) = x^2 + 3x.$$

Sketch a graph of the region.

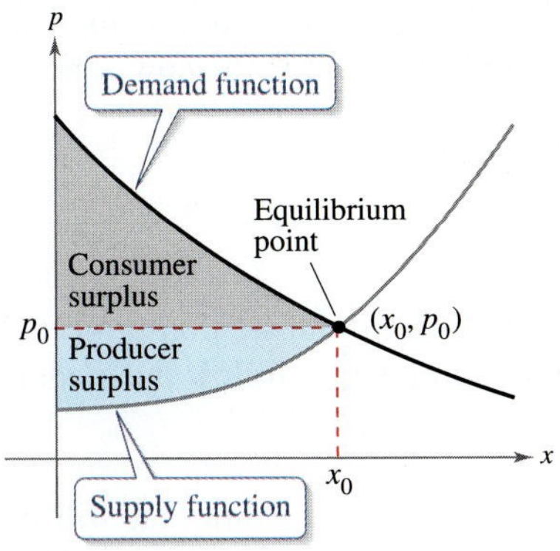

Consumer Surplus and Producer Surplus

In Section 1.2, you learned that a demand function relates the price of a product to the consumer demand. You also learned that a supply function relates the price of a product to producers' willingness to supply the product. The point (x_0, p_0) at which a demand function $p = D(x)$ and a supply function $p = S(x)$ intersect is the equilibrium point.

Economists call the area of the region bounded by the graph of the demand function, the horizontal line $p = p_0$, and the vertical line $x = 0$ the **consumer surplus,** as shown in Figure 5.20. Consumer surplus is the difference between the amount consumers would be willing to pay and the actual amount paid for a product. The area of the region bounded by the graph of the supply function, the horizontal line $p = p_0$, and the vertical line $x = 0$ is called the **producer surplus,** as shown in Figure 5.20. Producer surplus is the difference between the amount a producer receives for selling a product and the minimum price needed to get the producer to supply the product.

FIGURE 5.20

Example 5 Finding Surpluses

The demand and supply functions for a product are modeled by

Demand: $p = -0.36x + 9$ and *Supply:* $p = 0.14x + 2$

where p is the price (in dollars) and x is the number of units (in millions). Find the consumer and producer surpluses for this product.

SOLUTION By equating the demand and supply functions, you can determine that the point of equilibrium occurs when $x = 14$ (million) and the price is $3.96 per unit.

$$\text{Consumer surplus} = \int_0^{14} (\text{demand function} - \text{price})\, dx$$

$$= \int_0^{14} [(-0.36x + 9) - 3.96]\, dx$$

$$= \left[-0.18x^2 + 5.04x\right]_0^{14}$$

$$= 35.28$$

The consumer surplus is $35.28.

$$\text{Producer surplus} = \int_0^{14} (\text{price} - \text{supply function})\, dx$$

$$= \int_0^{14} [3.96 - (0.14x + 2)]\, dx$$

$$= \left[-0.07x^2 + 1.96x\right]_0^{14}$$

$$= 13.72$$

The producer surplus is $13.72. The consumer surplus and producer surplus are shown in Figure 5.21.

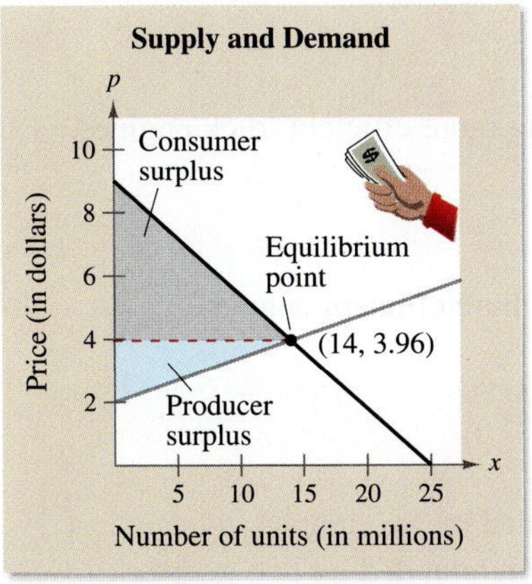

FIGURE 5.21

✓ Checkpoint 5

The demand and supply functions for a product are modeled by

Demand: $p = -0.2x + 8$ and *Supply:* $p = 0.1x + 2$

where p is the price (in dollars) and x is the number of units (in millions). Find the consumer and producer surpluses for this product. ∎

Application

In addition to consumer and producer surpluses, there are many other types of applications involving the area of a region bounded by two graphs. Example 6 shows one of these applications.

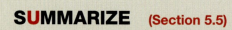 **Example 6** **Petroleum Consumption**

In the *Annual Energy Outlook*, the U.S. Energy Information Administration projected the consumption C (in quadrillions of Btu per year) of petroleum to follow the model

$$C_1 = 0.00078t^3 - 0.0445t^2 + 0.917t + 35.49, \quad 15 \le t \le 35$$

where $t = 15$ corresponds to 2015. Determine the amount of petroleum that will be saved when the actual consumption follows the model

$$C_2 = 0.0067t^2 - 0.211t + 40.95, \quad 15 \le t \le 35.$$

SOLUTION The petroleum saved can be represented as the area of the region between the graphs of C_1 and C_2, as shown in Figure 5.22.

$$
\begin{aligned}
\text{Petroleum saved} &= \int_{15}^{35} (C_1 - C_2)\, dt \\
&= \int_{15}^{35} (0.00078t^3 - 0.0512t^2 + 1.128t - 5.46)\, dt \\
&= \left[\frac{0.00078}{4}t^4 - \frac{0.0512}{3}t^3 + \frac{1.128}{2}t^2 - 5.46t \right]_{15}^{35} \\
&\approx 63.42
\end{aligned}
$$

So, about 63.42 quadrillion Btu of petroleum would be saved.

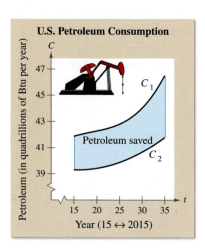

FIGURE 5.22

✓**Checkpoint 6**

The projected fuel cost C (in millions of dollars per year) for a trucking company from 2012 through 2024 is

$$C_1 = 2.21t + 5.6, \quad 12 \le t \le 24$$

where $t = 12$ corresponds to 2012. After purchasing more efficient truck engines, the company expects fuels costs to follow the model

$$C_2 = 2.04t + 4.7, \quad 12 \le t \le 24.$$

How much money will the company save with the more efficient engines?

SUMMARIZE   (Section 5.5)

1. State the definition of the area of a region bounded by two graphs *(page 299)*. For examples of finding the area of a region bounded by two graphs, see Examples 1, 2, 3, and 4.

2. Describe a real-life example of how finding the area of a region bounded by two graphs can be used to find the consumer and producer surpluses for a product *(page 303, Example 5)*.

3. Describe a real-life example of how finding the area of a region bounded by two graphs can be used to analyze petroleum consumption *(page 304, Example 6)*.

SKILLS WARM UP 5.5

The following warm-up exercises involve skills that were covered in earlier sections. You will use these skills in the exercise set for this section. For additional help, review Section 1.2.

In Exercises 1–4, simplify the expression.

1. $(-x^2 + 4x + 3) - (x + 1)$

2. $(-2x^2 + 3x + 9) - (-x + 5)$

3. $(-x^3 + 3x^2 - 1) - (x^2 - 4x + 4)$

4. $(3x + 1) - (-x^3 + 9x + 2)$

In Exercises 5–8, find the points of intersection of the graphs.

5. $f(x) = x^2 - 4x + 4, \ g(x) = 4$

6. $f(x) = -3x^2, \ g(x) = 6 - 9x$

7. $f(x) = x^2, \ g(x) = -x + 6$

8. $f(x) = \frac{1}{2}x^3, \ g(x) = 2x$

ENHANCED **WebAssign** Access end-of-section exercises online at **www.webassign.net**

5.6 The Definite Integral as the Limit of a Sum

■ Use the Midpoint Rule to approximate definite integrals.
■ Understand the definite integral as the limit of a sum.

The Midpoint Rule

In Section 5.4, you learned that you cannot use the Fundamental Theorem of Calculus to evaluate a definite integral unless you can find an antiderivative of the integrand. When you cannot find an antiderivative of an integrand, you can use an approximation technique. One such technique, the **Midpoint Rule,** is demonstrated in Example 1.

In Exercise 28, you will use the
Midpoint Rule to estimate
the surface area of a golf green.

Example 1   Approximating the Area of a Plane Region

Use the five rectangles in Figure 5.23 to approximate the area of the region bounded by the graph of $f(x) = -x^2 + 5$, the x-axis, and the lines $x = 0$ and $x = 2$.

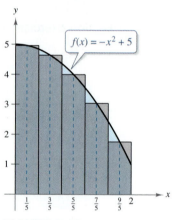

FIGURE 5.23

SOLUTION   You can find the heights of the five rectangles by evaluating f at the midpoint of each of the following intervals.

$$\left[0, \frac{2}{5}\right], \quad \left[\frac{2}{5}, \frac{4}{5}\right], \quad \left[\frac{4}{5}, \frac{6}{5}\right], \quad \left[\frac{6}{5}, \frac{8}{5}\right], \quad \left[\frac{8}{5}, \frac{10}{5}\right]$$

The width of each rectangle is $\frac{2}{5}$. So, the sum of the five areas is

$$\begin{aligned}
\text{Area} &\approx \frac{2}{5}f\left(\frac{1}{5}\right) + \frac{2}{5}f\left(\frac{3}{5}\right) + \frac{2}{5}f\left(\frac{5}{5}\right) + \frac{2}{5}f\left(\frac{7}{5}\right) + \frac{2}{5}f\left(\frac{9}{5}\right) \\
&= \frac{2}{5}\left[f\left(\frac{1}{5}\right) + f\left(\frac{3}{5}\right) + f\left(\frac{5}{5}\right) + f\left(\frac{7}{5}\right) + f\left(\frac{9}{5}\right)\right] \\
&= \frac{2}{5}\left(\frac{124}{25} + \frac{116}{25} + \frac{100}{25} + \frac{76}{25} + \frac{44}{25}\right) \\
&= \frac{920}{125} \\
&= 7.36.
\end{aligned}$$

✓ Checkpoint 1

Use four rectangles to approximate the area of the region bounded by the graph of $f(x) = x^2 + 1$, the x-axis, $x = 0$, and $x = 2$.

For the region in Example 1, you can find the exact area with a definite integral. That is,

$$\text{Area} = \int_0^2 (-x^2 + 5)\, dx = \frac{22}{3} \approx 7.33.$$

The approximation procedure used in Example 1 is the **Midpoint Rule.** You can use the Midpoint Rule to approximate *any* definite integral—not just those representing areas. The basic steps are summarized below.

Guidelines for Using the Midpoint Rule

To approximate the definite integral $\int_a^b f(x)\, dx$ with the Midpoint Rule, use the steps below.

1. Divide the interval $[a, b]$ into n subintervals, each of width

$$\Delta x = \frac{b - a}{n}.$$

2. Find the midpoint of each subinterval.

$$\text{Midpoints} = \{x_1, x_2, x_3, \ldots, x_n\}$$

3. Evaluate f at each midpoint and form the sum as shown.

$$\int_a^b f(x)\, dx \approx \frac{b - a}{n}[f(x_1) + f(x_2) + f(x_3) + \cdots + f(x_n)]$$

An important characteristic of the Midpoint Rule is that the approximation tends to improve as n increases. The table below shows the approximations for the area of the region described in Example 1 for various values of n. For example, when $n = 10$, the Midpoint Rule yields

$$\int_0^2 (-x^2 + 5)\, dx \approx \frac{2}{10}\left[f\left(\frac{1}{10}\right) + f\left(\frac{3}{10}\right) + \cdots + f\left(\frac{19}{10}\right)\right]$$

$$= 7.34.$$

n	5	10	15	20	25	30
Approximation	7.3600	7.3400	7.3363	7.3350	7.3344	7.3341

Note that as n increases, the approximation gets closer and closer to the exact value of the integral, which was found to be

$$\frac{22}{3} \approx 7.3333.$$

STUDY TIP

In Example 1, the Midpoint Rule is used to approximate an integral whose exact value can be found with the Fundamental Theorem of Calculus. This was done to illustrate the accuracy of the rule. In practice, of course, you would use the Midpoint Rule to approximate the values of definite integrals for which you cannot find an antiderivative. Examples 2 and 3 illustrate such integrals.

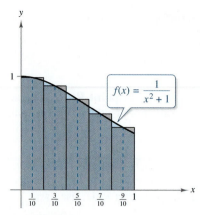

FIGURE 5.24

Example 2   **Using the Midpoint Rule**

Use the Midpoint Rule with $n = 5$ to approximate the area of the region bounded by the graph of

$$f(x) = \frac{1}{x^2 + 1}$$

the x-axis, and the lines $x = 0$ and $x = 1$.

SOLUTION   The region is shown in Figure 5.24. With $n = 5$, the interval $[0, 1]$ is divided into five subintervals.

$$\left[0, \frac{1}{5}\right], \quad \left[\frac{1}{5}, \frac{2}{5}\right], \quad \left[\frac{2}{5}, \frac{3}{5}\right], \quad \left[\frac{3}{5}, \frac{4}{5}\right], \quad \left[\frac{4}{5}, 1\right]$$

The midpoints of these intervals are $\frac{1}{10}, \frac{3}{10}, \frac{5}{10}, \frac{7}{10}$, and $\frac{9}{10}$. Because each subinterval has a width of $\Delta x = (1 - 0)/5 = \frac{1}{5}$, you can approximate the value of the definite integral as shown.

$$\int_0^1 \frac{1}{x^2 + 1}\, dx \approx \frac{1}{5}\left(\frac{1}{1.01} + \frac{1}{1.09} + \frac{1}{1.25} + \frac{1}{1.49} + \frac{1}{1.81}\right)$$

$$\approx 0.786$$

The actual area of this region is $\pi/4 \approx 0.785$. So, the approximation is off by about 0.001.

✓**Checkpoint 2**

Use the Midpoint Rule with $n = 4$ to approximate the area of the region bounded by the graph of $f(x) = 1/(x^2 + 2)$, the x-axis, and the lines $x = 0$ and $x = 1$.   ■

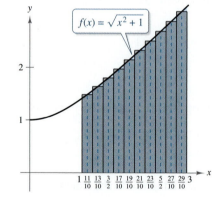

FIGURE 5.25

Example 3   **Using the Midpoint Rule**

Use the Midpoint Rule with $n = 10$ to approximate the area of the region bounded by the graph of $f(x) = \sqrt{x^2 + 1}$, the x-axis, and the lines $x = 1$ and $x = 3$.

SOLUTION   The region is shown in Figure 5.25. After dividing the interval $[1, 3]$ into 10 subintervals, you can determine that the midpoints of these intervals are

$$\frac{11}{10}, \quad \frac{13}{10}, \quad \frac{3}{2}, \quad \frac{17}{10}, \quad \frac{19}{10}, \quad \frac{21}{10}, \quad \frac{23}{10}, \quad \frac{5}{2}, \quad \frac{27}{10}, \quad \text{and} \quad \frac{29}{10}.$$

Because each subinterval has a width of $\Delta x = (3 - 1)/10 = \frac{1}{5}$, you can approximate the value of the definite integral as shown.

$$\int_1^3 \sqrt{x^2 + 1}\, dx \approx \frac{1}{5}\left[\sqrt{(1.1)^2 + 1} + \sqrt{(1.3)^2 + 1} + \cdots + \sqrt{(2.9)^2 + 1}\right]$$

$$\approx 4.504$$

It can be shown that the actual area is

$$\frac{1}{2}\left[3\sqrt{10} + \ln\left(3 + \sqrt{10}\right) - \sqrt{2} - \ln\left(1 + \sqrt{2}\right)\right] \approx 4.505.$$

So, the approximation is off by about 0.001.

✓**Checkpoint 3**

Use the Midpoint Rule with $n = 4$ to approximate the area of the region bounded by the graph of $f(x) = \sqrt{x^2 - 1}$, the x-axis, and the lines $x = 2$ and $x = 4$.   ■

STUDY TIP

The Midpoint Rule is necessary for solving certain real-life problems, such as measuring irregular areas like bodies of water (see Exercise 27).

The Definite Integral as the Limit of a Sum

Consider the closed interval $[a, b]$, divided into n subintervals whose midpoints are x_i and whose widths are $\Delta x = (b - a)/n$. In this section, you have seen that the midpoint approximation

$$\int_a^b f(x)\,dx \approx f(x_1)\,\Delta x + f(x_2)\,\Delta x + f(x_3)\,\Delta x + \cdots + f(x_n)\,\Delta x$$

$$= [f(x_1) + f(x_2) + f(x_3) + \cdots + f(x_n)]\,\Delta x$$

becomes better and better as n increases. In fact, the limit of this sum as n approaches infinity is exactly equal to the definite integral. That is,

$$\int_a^b f(x)\,dx = \lim_{n \to \infty} [f(x_1) + f(x_2) + f(x_3) + \cdots + f(x_n)]\,\Delta x.$$

It can be shown that this limit is valid as long as x_i is *any* point in the *i*th interval.

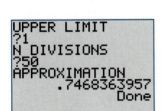

**UPPER LIMIT
?1
N DIVISIONS
?50
APPROXIMATION
          .7468363957
                    Done**

FIGURE 5.26

Example 4 — Approximating a Definite Integral

Use the Midpoint Rule program in Appendix E or a symbolic integration utility to approximate the definite integral

$$\int_0^1 e^{-x^2}\,dx.$$

SOLUTION   Using the Midpoint Rule program (see Figure 5.26), you can complete the following table.

n	10	20	30	40	50
Approximation	0.7471	0.7469	0.7469	0.7468	0.7468

From the table, it appears that

$$\int_0^1 e^{-x^2}\,dx \approx 0.7468.$$

Using a symbolic integration utility, the value of the integral is approximately 0.7468241328.

✓ Checkpoint 4

Use the Midpoint Rule program in Appendix E or a symbolic integration utility to approximate the definite integral

$$\int_0^1 e^{x^2}\,dx.$$

SUMMARIZE   (Section 5.6)

1. Describe how to approximate the area of a region using rectangles *(page 306, Example 1)*.

2. State the guidelines for using the Midpoint Rule *(page 307)*. For examples of using these guidelines, see Examples 2 and 3.

3. State the definite integral as the limit of a sum *(page 309)*.

SKILLS WARM UP 5.6   The following warm-up exercises involve skills that were covered in a previous course or in earlier sections. You will use these skills in the exercise set for this section. For additional help, review Appendix Section A.2 and Section 3.6.

In Exercises 1–6, find the midpoint of the interval.

1. $\left[0, \frac{1}{3}\right]$

2. $\left[\frac{1}{10}, \frac{2}{10}\right]$

3. $\left[\frac{3}{20}, \frac{4}{20}\right]$

4. $\left[1, \frac{7}{6}\right]$

5. $\left[2, \frac{31}{15}\right]$

6. $\left[\frac{26}{9}, 3\right]$

In Exercises 7–10, find the limit.

7. $\lim\limits_{x \to \infty} \dfrac{2x^2 + 4x - 1}{3x^2 - 2x}$

8. $\lim\limits_{x \to \infty} \dfrac{4x + 5}{7x - 5}$

9. $\lim\limits_{x \to \infty} \dfrac{x - 7}{x^2 + 1}$

10. $\lim\limits_{x \to \infty} \dfrac{5x^3 + 1}{x^3 + x^2 + 4}$

ENHANCED
WebAssign   Access end-of-section exercises online at **www.webassign.net**

ALGEBRA TUTOR

"Unsimplifying" an Algebraic Expression

In algebra it is often helpful to write an expression in its simplest form. In this chapter, you have seen that the reverse is often true in integration. That is, to fit an integrand to an integration formula, it often helps to "unsimplify" the expression. To do this, you use the same algebraic rules, but your goal is different. Here are some examples.

Example 1   Rewriting Algebraic Expressions

Rewrite each algebraic expression as indicated in the example.

a. $\dfrac{x-1}{\sqrt{x}}$

b. $x(3-4x^2)^2$

c. $7x^2\sqrt{x^3+1}$

d. $5xe^{-x^2}$

SOLUTION

a. $\dfrac{x-1}{\sqrt{x}} = \dfrac{x}{\sqrt{x}} - \dfrac{1}{\sqrt{x}}$
Example 5, page 269
Rewrite as two fractions.

$= \dfrac{x^1}{x^{1/2}} - \dfrac{1}{x^{1/2}}$
Rewrite with rational exponents.

$= x^{1-1/2} - x^{-1/2}$
Properties of exponents

$= x^{1/2} - x^{-1/2}$
Simplify exponent.

b. $x(3-4x^2)^2 = \dfrac{-8}{-8}x(3-4x^2)^2$
Example 2, page 276
Multiply and divide by -8.

$= \left(-\dfrac{1}{8}\right)(-8)x(3-4x^2)^2$
Regroup.

$= \left(-\dfrac{1}{8}\right)(3-4x^2)^2(-8x)$
Regroup.

c. $7x^2\sqrt{x^3+1} = 7x^2(x^3+1)^{1/2}$
Example 5, page 277
Rewrite with rational exponent.

$= \dfrac{3}{3}(7x^2)(x^3+1)^{1/2}$
Multiply and divide by 3.

$= \dfrac{7}{3}(3x^2)(x^3+1)^{1/2}$
Regroup.

$= \dfrac{7}{3}(x^3+1)^{1/2}(3x^2)$
Regroup.

d. $5xe^{-x^2} = \dfrac{-2}{-2}(5x)e^{-x^2}$
Example 3, page 283
Multiply and divide by -2.

$= \left(-\dfrac{5}{2}\right)(-2x)e^{-x^2}$
Regroup.

$= \left(-\dfrac{5}{2}\right)e^{-x^2}(-2x)$
Regroup.

Example 2   **Rewriting Algebraic Expressions**

Rewrite each algebraic expression.

a. $\dfrac{3x^2 + 2x - 1}{x^2}$

b. $\dfrac{1}{1 + e^{-x}}$

c. $\dfrac{x^2 + x + 1}{x - 1}$

d. $\dfrac{x^2 + 6x + 1}{x^2 + 1}$

SOLUTION

a. $\dfrac{3x^2 + 2x - 1}{x^2} = \dfrac{3x^2}{x^2} + \dfrac{2x}{x^2} - \dfrac{1}{x^2}$ Example 7(a), page 286
Rewrite as separate fractions.

$= 3 + \dfrac{2}{x} - x^{-2}$ Properties of exponents

$= 3 + 2\left(\dfrac{1}{x}\right) - x^{-2}$ Regroup.

b. $\dfrac{1}{1 + e^{-x}} = \left(\dfrac{e^x}{e^x}\right)\dfrac{1}{1 + e^{-x}}$ Example 7(b), page 286
Multiply and divide by e^x.

$= \dfrac{e^x}{e^x + e^x(e^{-x})}$ Multiply.

$= \dfrac{e^x}{e^x + e^{x-x}}$ Property of exponents

$= \dfrac{e^x}{e^x + e^0}$ Simplify exponent.

$= \dfrac{e^x}{e^x + 1}$ $e^0 = 1$

c. $\dfrac{x^2 + x + 1}{x - 1} = x + 2 + \dfrac{3}{x - 1}$ Example 7(c), page 286
Use long division as shown below.

$$\begin{array}{r} x + 2 \\ x - 1 \overline{)\, x^2 + x + 1} \\ \underline{x^2 - x} \\ 2x + 1 \\ \underline{2x - 2} \\ 3 \end{array}$$

d. $\dfrac{x^2 + 6x + 1}{x^2 + 1} = 1 + \dfrac{6x}{x^2 + 1}$ Bottom of page 285
Use long division as shown below.

$$\begin{array}{r} 1 \\ x^2 + 1 \overline{)\, x^2 + 6x + 1} \\ \underline{x^2 \qquad\; + 1} \\ 6x \end{array}$$

SUMMARY AND STUDY STRATEGIES

After studying this chapter, you should have acquired the following skills.
The exercise numbers are keyed to the Review Exercises that begin on page 315.
Answers to odd-numbered Review Exercises are given in the back of the text.*

Section 5.1   Review Exercises

■ Use basic integration rules to find indefinite integrals.   *1–14*

$$\int k \, dx = kx + C$$

$$\int kf(x) \, dx = k \int f(x) \, dx$$

$$\int [f(x) + g(x)] \, dx = \int f(x) \, dx + \int g(x) \, dx$$

$$\int [f(x) - g(x)] \, dx = \int f(x) \, dx - \int g(x) \, dx$$

$$\int x^n \, dx = \frac{x^{n+1}}{n+1} + C, \quad n \neq -1$$

■ Use initial conditions to find particular solutions of indefinite integrals.   *15–18*
■ Use antiderivatives to solve real-life problems.   *19, 20*

Section 5.2

■ Use the General Power Rule or integration by substitution to find indefinite integrals.   *21–32*

$$\int u^n \frac{du}{dx} \, dx = \int u^n \, du = \frac{u^{n+1}}{n+1} + C, \quad n \neq -1$$

■ Use the General Power Rule or integration by substitution to solve real-life problems.   *33, 34*

Section 5.3

■ Use the Exponential and Log Rules to find indefinite integrals.   *35–46*

$$\int e^x \, dx = e^x + C \qquad\qquad \int \frac{1}{x} \, dx = \ln|x| + C$$

$$\int e^u \frac{du}{dx} \, dx = \int e^u \, du = e^u + C \qquad \int \frac{du/dx}{u} \, dx = \int \frac{1}{u} \, du = \ln|u| + C$$

Section 5.4

■ Find the areas of regions using a geometric formula.   *47–50*
■ Use properties of definite integrals.   *51, 52*
■ Find the areas of regions bounded by the graph of a function and the *x*-axis.   *53–58*

---

* A wide range of valuable study aids are available to help you master the material in this chapter.
The *Student Solutions Manual* includes step-by-step solutions to all odd-numbered exercises to
help you review and prepare. The student website at *www.cengagebrain.com* offers algebra help
and a *Graphing Technology Guide*, which contains step-by-step commands and instructions for a
wide variety of graphing calculators.

Section 5.4   (continued)   Review Exercises

■ Use the Fundamental Theorem of Calculus to evaluate definite integrals.   59–70

$$\int_a^b f(x)\, dx = F(x)\Big]_a^b = F(b) - F(a), \quad \text{where} \quad F'(x) = f(x)$$

■ Find average values of functions over closed intervals.   71–76

$$\text{Average value} = \frac{1}{b - a} \int_a^b f(x)\, dx$$

■ Use properties of even and odd functions to help evaluate definite integrals.   77–80

Even function: $f(-x) = f(x)$

If f is an *even* function, then $\int_{-a}^a f(x)\, dx = 2\int_0^a f(x)\, dx$.

Odd function: $f(-x) = -f(x)$

If f is an *odd* function, then $\int_{-a}^a f(x)\, dx = 0$.

■ Find amounts of annuities.   81, 82
■ Use definite integrals to solve marginal analysis problems.   83, 84
■ Use average values to solve real-life problems.   85, 86

Section 5.5

■ Find areas of regions bounded by two graphs.   87–94

$$A = \int_a^b [f(x) - g(x)]\, dx$$

■ Find consumer and producer surpluses.   95–98
■ Use the areas of regions bounded by two graphs to solve real-life problems.   99–102

Section 5.6

■ Use the Midpoint Rule to approximate values of definite integrals.   103–112

$$\int_a^b f(x)\, dx \approx \frac{b - a}{n}[f(x_1) + f(x_2) + f(x_3) + \cdot\cdot\cdot + f(x_n)]$$

■ Use the Midpoint Rule to solve real-life problems.   113

Study Strategies

■ **Indefinite and Definite Integrals**   When evaluating integrals, remember that an indefinite integral is a *family of antiderivatives*, each differing by a constant C, whereas a definite integral is a *number*.

■ **Checking Antiderivatives by Differentiating**   When finding an antiderivative, remember that you can check your result by differentiating. For example, you can confirm that the antiderivative

$$\int (3x^3 - 4x)\, dx = \frac{3}{4}x^4 - 2x^2 + C \quad \text{is correct by differentiating to obtain} \quad \frac{d}{dx}\left[\frac{3}{4}x^4 - 2x^2 + C\right] = 3x^3 - 4x.$$

Because the derivative is equal to the original integrand, you know that the antiderivative is correct.

■ **Grouping Symbols and the Fundamental Theorem**   When using the Fundamental Theorem of Calculus to evaluate a definite integral, you can avoid sign errors by using grouping symbols. Here is an example.

$$\int_1^3 (x^3 - 9x)\, dx = \left[\frac{x^4}{4} - \frac{9x^2}{2}\right]_1^3 = \left[\frac{3^4}{4} - \frac{9(3^2)}{2}\right] - \left[\frac{1^4}{4} - \frac{9(1^2)}{2}\right] = \frac{81}{4} - \frac{81}{2} - \frac{1}{4} + \frac{9}{2} = -16$$

Review Exercises

Finding Indefinite Integrals In Exercises 1–14, find the indefinite integral. Check your result by differentiation.

1. $\int 16 \, dx$

2. $\int -9 \, dx$

3. $\int \frac{3}{5}x \, dx$

4. $\int 6x \, dx$

5. $\int 3x^2 \, dx$

6. $\int 8x^3 \, dx$

7. $\int (2x^2 + 5x) \, dx$

8. $\int (5 - 6x^2) \, dx$

9. $\int \frac{2}{3\sqrt[3]{x}} \, dx$

10. $\int 6x^{5/2} \, dx$

11. $\int \left(\sqrt[3]{x^4} + 3x\right) dx$

12. $\int \left(\frac{4}{\sqrt{x}} + \sqrt{x}\right) dx$

13. $\int \frac{2x^4 - 1}{\sqrt{x}} \, dx$

14. $\int \frac{1 - 3x}{x^2} \, dx$

Finding Particular Solutions In Exercises 15–18, find the particular solution that satisfies the differential equation and the initial condition.

15. $f'(x) = 12x; \ f(0) = -3$

16. $f'(x) = 3x + 1; \ f(2) = 6$

17. $f'(x) = 3x^2 - 8x; \ f(1) = 12$

18. $f'(x) = \sqrt{x}; \ f(9) = 4$

19. **Vertical Motion** An object is projected upward from the ground with an initial velocity of 80 feet per second. Express the height s (in feet) of the object as a function of the time t (in seconds). How long will the object be in the air? (Use $s''(t) = -32$ feet per second per second as the acceleration due to gravity.)

20. **Revenue** A company produces a new product for which the rate of change of the revenue can be modeled by

$$\frac{dR}{dt} = 0.675t^{3/2}, \quad 0 \le t \le 225$$

where t is the time (in weeks). When $t = 0$, $R = 0$.

(a) Find a model for the revenue function.

(b) What is the revenue after 20 weeks?

(c) When will the weekly revenue be $27,000?

Applying the General Power Rule In Exercises 21–32, find the indefinite integral. Check your result by differentiation.

21. $\int (x + 4)^3 \, dx$

22. $\int (x - 6)^{4/3} \, dx$

23. $\int (5x + 1)^4(5) \, dx$

24. $\int (x^3 + 1)^2(3x^2) \, dx$

25. $\int (1 + 5x)^2 \, dx$

26. $\int (6x - 2)^4 \, dx$

27. $\int x^2(3x^3 + 1)^2 \, dx$

28. $\int x(1 - 4x^2)^3 \, dx$

29. $\int \frac{x^2}{(2x^3 - 5)^3} \, dx$

30. $\int \frac{x^2}{(x^3 - 4)^2} \, dx$

31. $\int \frac{1}{\sqrt{5x - 1}} \, dx$

32. $\int \frac{4x}{\sqrt{1 - 3x^2}} \, dx$

33. **Production** The rate of change of the output of a small sawmill is modeled by

$$\frac{dP}{dt} = 2t(0.001t^2 + 0.5)^{1/4}, \quad 0 \le t \le 40$$

where t is the time (in hours) and P is the output (in board-feet). Find the numbers of board-feet produced in (a) 6 hours and (b) 12 hours.

34. Cost The marginal cost for a catering service to cater to x people can be modeled by

$$\frac{dC}{dx} = \frac{5x}{\sqrt{x^2 + 1000}}.$$

When $x = 225$, the cost C (in dollars) is $1136.06. Find the costs of catering to (a) 500 people and (b) 1000 people.

Using the Exponential and Log Rules In Exercises 35–46, use the Exponential Rule or the Log Rule to find the indefinite integral.

35. $\int 4e^{4x}\, dx$

36. $\int 3e^{-3x}\, dx$

37. $\int e^{-5x}\, dx$

38. $\int e^{6x}\, dx$

39. $\int 7xe^{3x^2}\, dx$

40. $\int (2t - 1)e^{t^2 - t}\, dt$

41. $\int \frac{1}{x - 6}\, dx$

42. $\int \frac{1}{1 - 4x}\, dx$

43. $\int \frac{4}{6x - 1}\, dx$

44. $\int \frac{5}{2x + 3}\, dx$

45. $\int \frac{x^2}{1 - x^3}\, dx$

46. $\int \frac{x - 4}{x^2 - 8x}\, dx$

Evaluating a Definite Integral Using a Geometric Formula In Exercises 47–50, sketch the region whose area is given by the definite integral. Then use a geometric formula to evaluate the integral.

47. $\int_0^3 2\, dx$

48. $\int_0^6 \frac{x}{2}\, dx$

49. $\int_0^4 (4 - x)\, dx$

50. $\int_{-4}^4 \sqrt{16 - x^2}\, dx$

51. Using Properties of Definite Integrals Given

$$\int_2^6 f(x)\, dx = 10 \quad \text{and} \quad \int_2^6 g(x)\, dx = 3$$

evaluate the definite integral.

(a) $\int_2^6 [f(x) + g(x)]\, dx$

(b) $\int_2^6 [f(x) - g(x)]\, dx$

(c) $\int_2^6 [2f(x) - 3g(x)]\, dx$

(d) $\int_2^6 5f(x)\, dx$

52. Using Properties of Definite Integrals Given

$$\int_0^3 f(x)\, dx = 4 \quad \text{and} \quad \int_3^6 f(x)\, dx = -1$$

evaluate the definite integral.

(a) $\int_0^6 f(x)\, dx$

(b) $\int_6^3 f(x)\, dx$

(c) $\int_4^4 f(x)\, dx$

(d) $\int_3^6 -10f(x)\, dx$

Finding Area by the Fundamental Theorem In Exercises 53–58, find the area of the region.

53. $f(x) = 4 - x^2$

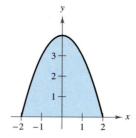

54. $f(x) = 9 - x^2$

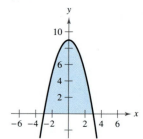

55. $f(x) = \frac{2}{x + 1}$

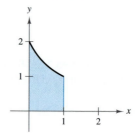

56. $f(x) = \frac{4}{\sqrt{x}}$

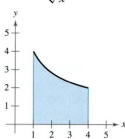

57. $f(x) = 2e^{x/2}$

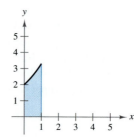

58. $f(x) = \frac{x - 1}{x}$

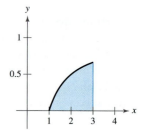

Evaluating a Definite Integral In Exercises 59–70, use the Fundamental Theorem of Calculus to evaluate the definite integral.

59. $\displaystyle\int_0^4 (2 + x)\, dx$

60. $\displaystyle\int_{-1}^1 (t^2 + 2)\, dt$

61. $\displaystyle\int_{-1}^1 (4t^3 - 2t)\, dt$

62. $\displaystyle\int_{-2}^2 (x^4 + 2x^2 - 5)\, dx$

63. $\displaystyle\int_{-2}^0 (x + 2)^3\, dx$

64. $\displaystyle\int_2^4 (2x - 3)^2\, dx$

65. $\displaystyle\int_0^3 \frac{1}{\sqrt{1 + x}}\, dx$

66. $\displaystyle\int_3^6 \frac{x}{3\sqrt{x^2 - 8}}\, dx$

67. $\displaystyle\int_3^9 \frac{5}{x}\, dx$

68. $\displaystyle\int_1^2 \left(\frac{1}{x^2} - \frac{1}{x^3}\right) dx$

69. $\displaystyle\int_0^{\ln 5} e^{x/5}\, dx$

70. $\displaystyle\int_{-1}^1 3x e^{x^2 - 1}\, dx$

Average Value of a Function In Exercises 71–76, find the average value of the function on the interval. Then find all x-values in the interval for which the function is equal to its average value.

71. $f(x) = 3x;\ [0, 2]$

72. $f(x) = x^2 + 2;\ [-3, 3]$

73. $f(x) = -2e^x;\ [0, 3]$

74. $f(x) = e^{5-x};\ [2, 5]$

75. $f(x) = \dfrac{1}{\sqrt{x}};\ [4, 9]$

76. $f(x) = \dfrac{1}{(x + 5)^2};\ [-1, 6]$

Integrating Even and Odd Functions In Exercises 77–80, evaluate the definite integral by using the properties of even and odd functions.

77. $\displaystyle\int_{-2}^2 6x^5\, dx$

78. $\displaystyle\int_{-4}^4 3x^4\, dx$

79. $\displaystyle\int_{-3}^3 (x^4 + x^2)\, dx$

80. $\displaystyle\int_{-1}^1 (x^3 - x)\, dx$

Finding the Amount of an Annuity In Exercises 81 and 82, find the amount of an annuity with income function $c(t)$, interest rate r, and term T.

81. $c(t) = \$3000,\ r = 6\%,\ T = 5$ years

82. $c(t) = \$1200,\ r = 7\%,\ T = 8$ years

83. Cost The marginal cost of serving an additional typical client at a law firm can be modeled by

$$\frac{dC}{dx} = 675 + 0.5x$$

where x is the number of clients. Find the change in cost C (in dollars) when x increases from 50 to 51 clients.

84. Profit The marginal profit obtained by selling x dollars of automobile insurance can be modeled by

$$\frac{dP}{dx} = 0.4\left(1 - \frac{5000}{x}\right), \quad x \geq 5000.$$

Find the change in the profit P (in dollars) when x increases from \$75,000 to \$100,000.

85. Compound Interest A deposit of \$500 is made in a savings account at an annual interest rate of 4%, compounded continuously. Find the average balance in the account during the first 2 years.

86. Revenue The rate of change in revenue for Texas Roadhouse from 2003 through 2009 can be modeled by

$$\frac{dR}{dt} = -11.5000t^2 + 142.140t - 294.91$$

where R is the revenue (in millions of dollars) and t is the time in years, with $t = 3$ corresponding to 2003. In 2006, the revenue for Texas Roadhouse was \$597.1 million. *(Source: Texas Roadhouse, Inc.)*

(a) Find the model for the revenue of Texas Roadhouse.

(b) What was the average revenue of Texas Roadhouse for 2003 through 2009?

Finding the Area Bounded by Two Graphs In Exercises 87–94, sketch the region bounded by the graphs of the functions and find the area of the region.

87. $y = \dfrac{1}{x^3},\ y = 0,\ x = 1,\ x = 3$

88. $y = x^2 + 4x - 5,\ y = 4x - 1$

89. $y = (x - 3)^2,\ y = 8 - (x - 3)^2$

90. $y = 4 - x,\ y = x^2 - 5x + 8,\ x = 0$

91. $y = \dfrac{4}{\sqrt{x + 1}},\ y = 0,\ x = 0,\ x = 8$

92. $y = \sqrt{x}(1 - x),\ y = 0$

93. $y = x,\ y = x^3$

94. $y = x^3 - 4x,\ y = -x^2 - 2x$

Consumer and Producer Surpluses In Exercises 95–98, find the consumer and producer surpluses by using the demand and supply functions, where p is the price (in dollars) and x is the number of units (in millions).

	Demand Function	Supply Function
95.	$p = 36 - 0.35x$	$p = 0.05x$
96.	$p = 200 - 0.2x$	$p = 50 + 1.3x$
97.	$p = 250 - x$	$p = 150 + x$
98.	$p = 500 - x$	$p = 1.25x + 162.5$

99. Revenue For the years 2015 through 2020, two models, R_1 and R_2, used to project the revenue (in millions of dollars) for a company are

$$R_1 = 24.3 + 8.24t$$

and

$$R_2 = 21.6 + 9.36t$$

where $t = 15$ corresponds to 2015. Which model projects the greater revenue? How much more total revenue does that model project over the six-year period?

100. Sales For the years 2000 through 2009, the sales (in millions of dollars) for Men's Wearhouse can be modeled by

$$R = \begin{cases} 23.596t^2 - 41.55t + 1310.5, & 0 \le t \le 6 \\ 38.7t^2 - 720.7t + 5261.2, & 6 < t \le 9 \end{cases}$$

where t is the year, with $t = 0$ corresponding to 2000. *(Source: Men's Wearhouse, Inc.)*

(a) Use a graphing utility to graph this model.

(b) Suppose the sales from 2007 through 2009 had continued to follow the model for 2000 through 2006. How much more or less would the sales have been for Men's Wearhouse?

101. Cost, Revenue, and Profit The revenue from a manufacturing process (in millions of dollars) is projected to follow the model

$$R = 70$$

for 10 years. Over the same period of time, the cost (in millions of dollars) is projected to follow the model

$$C = 30 + 0.3t^2$$

where t is the time (in years). Approximate the profit over the 10-year period.

102. Cost, Revenue, and Profit Repeat Exercise 101 for revenue and cost models given by

$$R = 70 + 0.1t$$

and

$$C = 30 + 0.3t^2.$$

Did the profit increase or decrease? Explain why.

Approximating the Area of a Plane Region In Exercises 103 and 104, use the rectangles to approximate the area of the region. Compare your result with the exact area obtained using a definite integral.

103. $f(x) = \dfrac{x}{3}$, $[0, 3]$

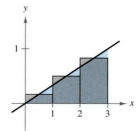

104. $f(x) = x^2 + 1$, $[0, 1]$

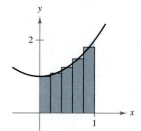

Using the Midpoint Rule In Exercises 105–108, use the Midpoint Rule with $n = 4$ to approximate the area of the region bounded by the graph of f and the x-axis over the interval. Sketch the region.

	Function	Interval
105.	$f(x) = x^2$	$[0, 2]$
106.	$f(x) = 2x - x^3$	$[0, 1]$
107.	$f(x) = (x^2 - 1)^2$	$[-1, 1]$
108.	$f(x) = \dfrac{3x}{x + 2}$	$[0, 4]$

Using the Midpoint Rule In Exercises 109–112, use the Midpoint Rule with $n = 6$ to approximate the area of the region bounded by the graph of f and the x-axis over the interval. Sketch the region.

	Function	Interval
109.	$f(x) = x + 3$	$[0, 3]$
110.	$f(x) = 9 - x^2$	$[-3, 3]$
111.	$f(x) = x\sqrt{x + 1}$	$[0, 2]$
112.	$f(x) = \dfrac{3}{x^2 + 1}$	$[-6, 6]$

113. Surface Area Use the Midpoint Rule to estimate the surface area of the swamp shown in the figure.

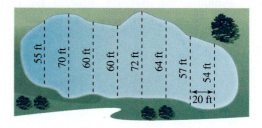

TEST YOURSELF

Take this test as you would take a test in class. When you are done, check your work against the answers given in the back of the book.

In Exercises 1–6, find the indefinite integral.

1. $\displaystyle\int (9x^2 - 4x + 13)\, dx$ **2.** $\displaystyle\int (x + 1)^2\, dx$

3. $\displaystyle\int 4x^3 \sqrt{x^4 - 7}\, dx$ **4.** $\displaystyle\int \frac{5x - 6}{\sqrt{x}}\, dx$

5. $\displaystyle\int 15e^{3x}\, dx$ **6.** $\displaystyle\int \frac{3}{4x - 1}\, dx$

In Exercises 7 and 8, find the particular solution that satisfies the differential equation and initial condition.

7. $f'(x) = 6x - 5;\ f(-1) = 6$ **8.** $f'(x) = e^x + 1;\ f(0) = 1$

In Exercises 9–14, evaluate the definite integral.

9. $\displaystyle\int_0^1 16x\, dx$ **10.** $\displaystyle\int_{-3}^3 (3 - 2x)\, dx$

11. $\displaystyle\int_{-1}^1 (x^3 + x^2)\, dx$ **12.** $\displaystyle\int_{-1}^2 \frac{2x}{\sqrt{x^2 + 1}}\, dx$

13. $\displaystyle\int_0^3 e^{4x}\, dx$ **14.** $\displaystyle\int_{-2}^3 \frac{1}{x + 3}\, dx$

15. The rate of change in sales of PetSmart from 2000 through 2009 can be modeled by

$$\frac{dS}{dt} = 226.912e^{0.1013t}$$

where S is the sales (in millions of dollars) and t is the time (in years), with $t = 0$ corresponding to 2000. In 2004, the sales of PetSmart were \$3363.5 million. *(Source: PetSmart, Inc.)*

(a) Find the model for the sales of PetSmart.

(b) What were the average sales for 2000 through 2009?

In Exercises 16 and 17, sketch the region bounded by the graphs of the functions and find the area of the region.

16. $f(x) = 6,\ g(x) = x^2 - x - 6$

17. $f(x) = \sqrt[3]{x},\ g(x) = x^2$

18. The demand and supply functions for a product are modeled by

Demand: $p = -0.625x + 10$ and   *Supply:* $p = 0.25x + 3$

where p is the price (in dollars) and x is the number of units (in millions). Find the consumer and producer surpluses for this product.

In Exercises 19 and 20, use the Midpoint Rule with $n = 4$ to approximate the area of the region bounded by the graph of f and the x-axis over the interval. Compare your result with the exact area. Sketch the region.

19. $f(x) = 3x^2,\quad [0, 1]$

20. $f(x) = x^2 + 1,\quad [-1, 1]$

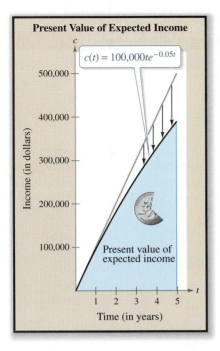

Present Value of Expected Income

$$c(t) = 100{,}000te^{-0.05t}$$

Present value of expected income

Time (in years)

6 Techniques of Integration

6.1   Integration by Parts and Present Value

6.2   Integration Tables

6.3   Numerical Integration

6.4   Improper Integrals

Example 7 on page 328 shows how integration by parts can be used to find the present value of a company's future income.

6.1 Integration by Parts and Present Value

■ Use integration by parts to find indefinite and definite integrals.
■ Find the present value of future income.

Integration by Parts

In this section, you will study an integration technique called **integration by parts.** This technique can be applied to a wide variety of functions and is particularly useful for integrands involving the products of algebraic and exponential or logarithmic functions. For instance, integration by parts works well with integrals such as

$$\int x^2 e^x \, dx \quad \text{and} \quad \int x \ln x \, dx.$$

Integration by parts is based on the Product Rule for differentiation.

$$\frac{d}{dx}[uv] = u\frac{dv}{dx} + v\frac{du}{dx} \qquad \text{Product Rule}$$

$$uv = \int u\frac{dv}{dx}\,dx + \int v\frac{du}{dx}\,dx \qquad \text{Integrate each side.}$$

$$uv = \int u\,dv + \int v\,du \qquad \text{Write in differential form.}$$

$$\int u\,dv = uv - \int v\,du \qquad \text{Rewrite.}$$

In Exercise 65, you will use integration by parts to find the average value of a memory model for children.

Integration by Parts

Let u and v be differentiable functions of x.

$$\int u\,dv = uv - \int v\,du$$

Note that the formula for integration by parts expresses the original integral in terms of another integral. Depending on the choices of u and dv, it may be easier to evaluate the second integral than the original one. Because the choices of u and dv are critical in the integration by parts process, the following guidelines are provided.

Guidelines for Integration by Parts

1. Try letting dv be the most complicated portion of the integrand that fits a basic integration rule. Then u will be the remaining factor(s) of the integrand.

2. Try letting u be the portion of the integrand whose derivative is a function simpler than u. Then dv will be the remaining factor(s) of the integrand.

Note that dv always includes the dx of the original integrand.

When using integration by parts, note that you can first choose dv or first choose u. After you choose, however, the choice of the other factor is determined—it must be the remaining portion of the integrand. Also note that dv must contain the differential dx of the original integral.

Example 1 **Integration by Parts**

Find $\int xe^x \, dx$.

SOLUTION To apply integration by parts, you must rewrite the original integral in the form $\int u \, dv$. That is, you must break $xe^x \, dx$ into two factors—one "part" representing u and the other "part" representing dv. There are several ways to do this.

$$\int \underbrace{(x)}_{u}\underbrace{(e^x \, dx)}_{dv} \qquad \int \underbrace{(e^x)}_{u}\underbrace{(x \, dx)}_{dv} \qquad \int \underbrace{(1)}_{u}\underbrace{(xe^x \, dx)}_{dv} \qquad \int \underbrace{(xe^x)}_{u}\underbrace{(dx)}_{dv}$$

The guidelines on the preceding page suggest the first option because $dv = e^x \, dx$ is the most complicated portion of the integrand that fits a basic integration formula *and* because the derivative of $u = x$ is simpler than x.

$$dv = e^x \, dx \qquad \Longrightarrow \qquad v = \int dv = \int e^x \, dx = e^x$$

$$u = x \qquad \Longrightarrow \qquad du = dx$$

Next, you can apply the integration by parts formula as shown.

$$\int u \, dv = uv - \int v \, du \qquad \text{\color{red}Integration by parts formula}$$

$$\int xe^x \, dx = xe^x - \int e^x \, dx \qquad \text{\color{red}Substitute.}$$

$$= xe^x - e^x + C \qquad \text{\color{red}Integrate } \int e^x \, dx.$$

You can check this result by differentiating.

$$\frac{d}{dx}[xe^x - e^x + C] = xe^x + e^x(1) - e^x = xe^x$$

✓ Checkpoint 1

Find $\int xe^{2x} \, dx$.

In Example 1, notice that you do not need to include a constant of integration when solving $v = \int e^x \, dx = e^x$. To see why this is true, try replacing e^x by $e^x + C_1$ in the solution.

$$\int xe^x \, dx = x(e^x + C_1) - \int (e^x + C_1) \, dx$$

$$= xe^x + C_1x - e^x - C_1x + C$$

$$= xe^x - e^x + C$$

After integrating, you can see that the terms involving C_1 subtract out.

TECH TUTOR

If you have access to a symbolic integration utility, try using it to solve several of the exercises in this section. Note that the form of the integral may be slightly different from what you obtain when solving the exercise by hand.

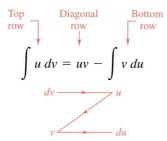

Example 2   Integration by Parts

Find $\displaystyle\int x^2 \ln x \, dx$.

SOLUTION In this case, x^2 is more easily integrated than $\ln x$. Furthermore, the derivative of $\ln x$ is simpler than $\ln x$. So, you should choose $dv = x^2 \, dx$.

$$dv = x^2 \, dx \qquad\Longrightarrow\qquad v = \int dv = \int x^2 \, dx = \frac{x^3}{3}$$

$$u = \ln x \qquad\Longrightarrow\qquad du = \frac{1}{x} \, dx$$

Next, apply the integration by parts formula.

$$\int u \, dv = uv - \int v \, du \qquad\qquad \text{Integration by parts formula}$$

$$\int x^2 \ln x \, dx = \frac{x^3}{3} \ln x - \int \left(\frac{x^3}{3}\right)\left(\frac{1}{x}\right) dx \qquad\qquad \text{Substitute.}$$

$$= \frac{x^3}{3} \ln x - \frac{1}{3}\int x^2 \, dx \qquad\qquad \text{Simplify.}$$

$$= \frac{x^3}{3} \ln x - \frac{x^3}{9} + C \qquad\qquad \text{Integrate.}$$

✓**Checkpoint 2**

Find $\displaystyle\int x \ln x \, dx$. ■

Example 3   Integrating by Parts with a Single Factor

Find $\displaystyle\int \ln x \, dx$.

SOLUTION This integrand is unusual because it has only one factor. In such cases, you should choose $dv = dx$ and choose u to be the single factor.

$$dv = dx \qquad\Longrightarrow\qquad v = \int dv = \int dx = x$$

$$u = \ln x \qquad\Longrightarrow\qquad du = \frac{1}{x} \, dx$$

Next, apply the integration by parts formula.

$$\int u \, dv = uv - \int v \, du \qquad\qquad \text{Integration by parts formula}$$

$$\int \ln x \, dx = x \ln x - \int (x)\left(\frac{1}{x}\right) dx \qquad\qquad \text{Substitute.}$$

$$= x \ln x - \int dx \qquad\qquad \text{Simplify.}$$

$$= x \ln x - x + C \qquad\qquad \text{Integrate.}$$

✓**Checkpoint 3**

Find $\displaystyle\int \ln 2x \, dx$. ■

Example 4 **Using Integration by Parts Repeatedly**

Find $\int x^2 e^x \, dx$.

SOLUTION The factors x^2 and e^x are both easy to integrate. Notice, however, that the derivative of x^2 becomes simpler, whereas the derivative of e^x does not. So, you should let $u = x^2$ and let $dv = e^x \, dx$.

$$dv = e^x \, dx \quad \Longrightarrow \quad v = \int dv = \int e^x \, dx = e^x$$

$$u = x^2 \quad \Longrightarrow \quad du = 2x \, dx$$

Next, apply the integration by parts formula.

$$\int x^2 e^x \, dx = x^2 e^x - \int 2x e^x \, dx \qquad \text{First application of integration by parts}$$

This first use of integration by parts has succeeded in simplifying the original integral, but the integral on the right still doesn't fit a basic integration rule. To evaluate that integral, you can apply integration by parts again. This time, let $u = 2x$ and $dv = e^x \, dx$.

$$dv = e^x \, dx \quad \Longrightarrow \quad v = \int dv = \int e^x \, dx = e^x$$

$$u = 2x \quad \Longrightarrow \quad du = 2 \, dx$$

Next, apply the integration by parts formula.

$$\int x^2 e^x \, dx = x^2 e^x - \int 2x e^x \, dx \qquad \text{First application of integration by parts}$$

$$= x^2 e^x - \left(2x e^x - \int 2 e^x \, dx \right) \qquad \text{Second application of integration by parts}$$

$$= x^2 e^x - 2x e^x + 2 e^x + C \qquad \text{Integrate.}$$

$$= e^x (x^2 - 2x + 2) + C \qquad \text{Simplify.}$$

You can confirm this result by differentiating.

$$\frac{d}{dx}[e^x(x^2 - 2x + 2) + C] = e^x(2x - 2) + (x^2 - 2x + 2)(e^x)$$

$$= 2x e^x - 2 e^x + x^2 e^x - 2x e^x + 2 e^x$$

$$= x^2 e^x$$

✓ **Checkpoint 4**

Find $\int x^3 e^x \, dx$. ■

When making repeated applications of integration by parts, you need to be careful not to interchange the substitutions in successive applications. For instance, in Example 4, the first substitution was $dv = e^x \, dx$ and $u = x^2$. If, in the second application, you had switched the substitution to $dv = 2x \, dx$ and $u = e^x$, you would have obtained

$$\int x^2 e^x \, dx = x^2 e^x - \int 2x e^x \, dx$$

$$= x^2 e^x - \left(x^2 e^x - \int x^2 e^x \, dx \right)$$

$$= \int x^2 e^x \, dx$$

thereby undoing the previous integration and returning to the *original* integral.

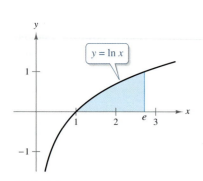

FIGURE 6.1

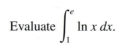

Evaluating a Definite Integral

Evaluate $\displaystyle\int_1^e \ln x \, dx$.

SOLUTION   Integration by parts was used to find the antiderivative of $\ln x$ in Example 3. Using this result, you can evaluate the definite integral as shown.

$$\int_1^e \ln x \, dx = \left[x \ln x - x \right]_1^e \qquad \text{Use result of Example 3.}$$

$$= (e \ln e - e) - (1 \ln 1 - 1) \qquad \text{Apply Fundamental Theorem.}$$

$$= (e - e) - (0 - 1)$$

$$= 1 \qquad \text{Simplify.}$$

The area represented by this definite integral is shown in Figure 6.1.

ALGEBRA TUTOR *xy*

For help on the algebra in Example 5, see Example 1 in the *Chapter 6 Algebra Tutor*, on page 352.

✓**Checkpoint 5**

Evaluate $\displaystyle\int_0^1 x^2 e^x \, dx$.   ■

Before starting the exercises in this section, remember that it is not enough to know *how* to use the various integration techniques. You also must know *when* to use them. Integration is first and foremost a problem of recognition—recognizing which formula or technique to apply to obtain an antiderivative. Often, a slight alteration of an integrand will necessitate the use of a different integration technique. Here are some examples.

Integral	*Technique*	*Antiderivative*
$\displaystyle\int x \ln x \, dx$	Integration by parts	$\dfrac{x^2}{2} \ln x - \dfrac{x^2}{4} + C$
$\displaystyle\int \dfrac{\ln x}{x} \, dx$	Power Rule: $\displaystyle\int u^n \dfrac{du}{dx} \, dx$	$\dfrac{(\ln x)^2}{2} + C$
$\displaystyle\int \dfrac{1}{x \ln x} \, dx$	Log Rule: $\displaystyle\int \dfrac{1}{u} \dfrac{du}{dx} \, dx$	$\ln |\ln x| + C$

As you gain experience in using integration by parts, your skill in determining u and dv will improve. The following summary lists several common integrals with suggestions for the choices of u and dv.

Summary of Common Integrals Using Integration by Parts

1. For integrals of the form

$$\int x^n e^{ax} \, dx$$

let $u = x^n$ and $dv = e^{ax} \, dx$. (See Examples 1 and 4.)

2. For integrals of the form

$$\int x^n \ln x \, dx$$

let $u = \ln x$ and $dv = x^n \, dx$. (See Examples 2 and 3.)

Present Value

Recall from Section 4.2 that the present value of a future payment is the amount that would have to be deposited today to produce the future payment. What is the present value of a future payment of $1000 one year from now? Because of inflation, $1000 today buys more than $1000 will buy a year from now. The definition below considers only the effect of inflation.

Present Value

If c represents a continuous income function in dollars per year and the annual rate of inflation is r, then the actual total income over t_1 years is

$$\text{Actual income over } t_1 \text{ years} = \int_0^{t_1} c(t)\, dt$$

and its **present value** is

$$\text{Present value} = \int_0^{t_1} c(t)e^{-rt}\, dt.$$

Ignoring inflation, the equation for present value also applies to an interest-bearing account, where the annual interest rate r is compounded continuously and c is an income function in dollars per year.

 Example 6 **Finding Present Value**

You have just won $1,000,000 in a state lottery. You will be paid an annuity of $50,000 a year for 20 years. When the annual rate of inflation is 6%, what is the present value of this income?

SOLUTION   The income function for your winnings is given by $c(t) = 50,000$. So,

$$\text{Actual income} = \int_0^{20} 50,000\, dt$$

$$= \left[\, 50,000t \,\right]_0^{20}$$

$$= \$1,000,000.$$

Because you do not receive this entire amount now, its present value is

$$\text{Present value} = \int_0^{20} 50,000e^{-0.06t}\, dt$$

$$= \left[\frac{50,000}{-0.06} e^{-0.06t} \right]_0^{20}$$

$$\approx \$582,338.$$

This present value represents the amount that the state must deposit now to cover your payments over the next 20 years. This shows why state lotteries are so profitable—for the states!

✓**Checkpoint 6**

Find the present value of the income from the lottery ticket in Example 6 when the annual rate of inflation is 7%.

Expected Income

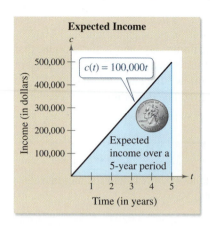

(a)

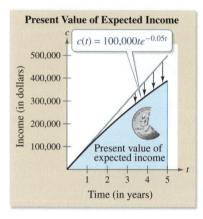

(b)

FIGURE 6.2

Example 7 **Finding Present Value**

A company expects its income during the next 5 years to be given by

$$c(t) = 100,000t, \quad 0 \le t \le 5. \qquad \text{See Figure 6.2(a).}$$

Assuming an annual inflation rate of 5%, can the company claim that the present value of this income is at least $1 million?

SOLUTION   The present value is

$$\text{Present value} = \int_0^5 100,000te^{-0.05t}\, dt = 100,000 \int_0^5 te^{-0.05t}\, dt.$$

Using integration by parts, let $dv = e^{-0.05t}\, dt$.

$$dv = e^{-0.05t}\, dt \quad \Longrightarrow \quad v = \int dv = \int e^{-0.05t}\, dt = -20e^{-0.05t}$$

$$u = t \quad \Longrightarrow \quad du = dt$$

This implies that

$$\int te^{-0.05t}\, dt = -20te^{-0.05t} + 20 \int e^{-0.05t}\, dt$$
$$= -20te^{-0.05t} - 400e^{-0.05t}$$
$$= -20e^{-0.05t}(t + 20).$$

So, the present value is

$$\text{Present value} = 100,000 \int_0^5 te^{-0.05t}\, dt \qquad \text{See Figure 6.2(b).}$$
$$= 100,000 \left[-20e^{-0.05t}(t + 20) \right]_0^5$$
$$\approx \$1,059,961.$$

Yes, the company can claim that the present value of its expected income during the next 5 years is at least $1 million.

─────────

✓**Checkpoint 7**

A company expects its income during the next 10 years to be given by $c(t) = 20,000t$, for $0 \le t \le 10$. Assuming an annual inflation rate of 5%, what is the present value of this income?   ■

SUMMARIZE   (Section 6.1)

1. State the integration by parts formula *(page 322)*. For examples of using this formula, see Examples 1, 2, 3, 4, and 7.

2. State the guidelines for integration by parts *(page 322)*. For an example of using these guidelines, see Example 1.

3. Give a summary of the common integrals using integration by parts *(page 326)*. For examples of these common integrals, see Examples 1, 2, 3, and 4.

4. Describe a real-life example of how integration by parts can be used to find the present value of an annuity *(page 327, Example 6)*.

SKILLS WARM UP 6.1

The following warm-up exercises involve skills that were covered in earlier sections. You will use these skills in the exercise set for this section. For additional help, review Sections 4.3, 4.5, and 5.5.

In Exercises 1–6, find $f'(x)$.

1. $f(x) = \ln(x + 1)$

2. $f(x) = \ln(x^2 - 1)$

3. $f(x) = e^{x^3}$

4. $f(x) = e^{-x^2}$

5. $f(x) = x^2 e^x$

6. $f(x) = xe^{-2x}$

In Exercises 7–10, find the area between the graphs of f and g.

7. $f(x) = -x^2 + 4, \ g(x) = x^2 - 4$

8. $f(x) = -x^2 + 2, \ g(x) = 1$

9. $f(x) = 4x, \ g(x) = x^2 - 5$

10. $f(x) = x^3 - 3x^2 + 2, \ g(x) = x - 1$

ENHANCED WebAssign Access end-of-section exercises online at **www.webassign.net**

6.2 Integration Tables

■ Use integration tables to find indefinite and definite integrals.
■ Use reduction formulas to find indefinite integrals.
■ Use integration tables to solve real-life problems.

In Exercise 59, you will use a formula from the integration table in Appendix C to find the total revenue of a new product during its first 2 years.

Integration Tables

You have studied several integration techniques that can be used with the basic integration formulas. Certainly these techniques and formulas do not cover every possible method for finding an antiderivative, but they do cover most of the important ones.

In this section, you will expand the list of integration formulas to form a table of integrals. As you add new integration formulas to the basic list, two effects occur. On one hand, it becomes increasingly difficult to memorize, or even become familiar with, the entire list of formulas. On the other hand, with a longer list you need fewer techniques for fitting an integral to one of the formulas on the list. The procedure of integrating by means of a long list of formulas is called **integration by tables.** (The table in Appendix C constitutes only a partial listing of integration formulas. Much longer lists exist, some of which contain several hundred formulas.)

Integration by tables should not be considered a trivial task. It requires considerable thought and insight, and it often requires substitution. Many people find a table of integrals to be a valuable supplement to the integration techniques discussed in this text. As you gain competence in the use of integration tables, you will improve in the use of the various integration techniques. In doing so, you should find that a combination of techniques and tables is the most versatile approach to integration.

Each integration formula in Appendix C can be developed using one or more of the techniques you have studied. You should try to verify several of the formulas. For instance, Formula 17

$$\int \frac{\sqrt{a + bu}}{u}\, du = 2\sqrt{a + bu} + a \int \frac{1}{u\sqrt{a + bu}}\, du \qquad \text{Formula 17}$$

can be verified using integration by parts, Formula 39

$$\int \frac{1}{1 + e^u}\, du = u - \ln(1 + e^u) + C \qquad \text{Formula 39}$$

can be verified using substitution, and Formula 44

$$\int (\ln u)^2\, du = u[2 - 2\ln u - (\ln u)^2] + C \qquad \text{Formula 44}$$

can be verified using integration by parts twice.

In the table of integrals in Appendix C, the formulas have been classified according to the form of the integrand. Several of the forms are listed below.

- Forms involving u^n
- Forms involving $a + bu$
- Forms involving $\sqrt{a + bu}$
- Forms involving $u^2 - a^2$
- Forms involving $\sqrt{u^2 \pm a^2}$
- Forms involving $\sqrt{a^2 - u^2}$
- Forms involving e^u
- Forms involving $\ln u$

TECH TUTOR

Throughout this section, remember that a symbolic integration utility can be used instead of integration tables. If you have access to such a utility, try using it to find the indefinite integrals in Examples 1 and 2.

Example 1 Using Integration Tables

Find $\displaystyle\int \frac{x}{\sqrt{x-1}}\,dx$.

SOLUTION Because the expression inside the radical is linear, you should consider forms involving $\sqrt{a+bu}$, as in Formula 19.

$$\int \frac{u}{\sqrt{a+bu}}\,du = -\frac{2(2a-bu)}{3b^2}\sqrt{a+bu} + C \qquad \text{Formula 19}$$

Using this formula, let $a = -1$, $b = 1$, and $u = x$. Then $du = dx$, and you obtain

$$\int \frac{x}{\sqrt{x-1}}\,dx = -\frac{2(-2-x)}{3}\sqrt{x-1} + C \qquad \text{Substitute values of } a, b, \text{ and } u.$$

$$= \frac{2}{3}(2+x)\sqrt{x-1} + C. \qquad \text{Simplify.}$$

✓ **Checkpoint 1**

Use the integration table in Appendix C to find

$$\int \frac{x}{\sqrt{2+x}}\,dx.$$

Example 2 Using Integration Tables

Find $\displaystyle\int x\sqrt{x^4-9}\,dx$.

SOLUTION Because it is not clear which formula to use, you can begin by letting $u = x^2$ and $du = 2x\,dx$. With these substitutions, you can write the integral as shown.

$$\int x\sqrt{x^4-9}\,dx = \frac{1}{2}\int \sqrt{(x^2)^2 - 9}\,(2x)\,dx \qquad \text{Multiply and divide by 2.}$$

$$= \frac{1}{2}\int \sqrt{u^2 - 9}\,du \qquad \text{Substitute } u \text{ and } du.$$

Now, it appears that you can use Formula 23.

$$\int \sqrt{u^2 - a^2}\,du = \frac{1}{2}\left(u\sqrt{u^2-a^2} - a^2 \ln\left|u + \sqrt{u^2-a^2}\right|\right) + C$$

Letting $a = 3$, you obtain

$$\int x\sqrt{x^4-9}\,dx = \frac{1}{2}\int \sqrt{u^2 - a^2}\,du$$

$$= \frac{1}{2}\left[\frac{1}{2}\left(u\sqrt{u^2-a^2} - a^2 \ln\left|u + \sqrt{u^2-a^2}\right|\right)\right] + C$$

$$= \frac{1}{4}\left(x^2\sqrt{x^4-9} - 9 \ln\left|x^2 + \sqrt{x^4-9}\right|\right) + C.$$

✓ **Checkpoint 2**

Use the integration table in Appendix C to find

$$\int \frac{\sqrt{x^2+16}}{x}\,dx.$$

Example 3 Using Integration Tables

Find $\displaystyle\int \frac{1}{x\sqrt{x+1}}\,dx$.

SOLUTION   Considering forms involving $\sqrt{a+bu}$, where $a = 1$, $b = 1$, and $u = x$, you can use Formula 15.

$$\int \frac{1}{u\sqrt{a+bu}}\,du = \frac{1}{\sqrt{a}}\ln\left|\frac{\sqrt{a+bu}-\sqrt{a}}{\sqrt{a+bu}+\sqrt{a}}\right| + C, \quad a > 0$$

So,

$$\int \frac{1}{x\sqrt{x+1}}\,dx = \int \frac{1}{u\sqrt{a+bu}}\,du$$

$$= \frac{1}{\sqrt{a}}\ln\left|\frac{\sqrt{a+bu}-\sqrt{a}}{\sqrt{a+bu}+\sqrt{a}}\right| + C$$

$$= \ln\left|\frac{\sqrt{x+1}-1}{\sqrt{x+1}+1}\right| + C.$$

✓ Checkpoint 3

Use the integration table in Appendix C to find $\displaystyle\int \frac{1}{x^2-4}\,dx$.

Example 4 Using Integration Tables

Evaluate $\displaystyle\int_0^2 \frac{x}{1+e^{-x^2}}\,dx$.

SOLUTION   Of the forms involving e^u, Formula 39

$$\int \frac{1}{1+e^u}\,du = u - \ln(1 + e^u) + C$$

seems most appropriate. To use this formula, let $u = -x^2$ and $du = -2x\,dx$.

$$\int \frac{x}{1+e^{-x^2}}\,dx = -\frac{1}{2}\int \frac{1}{1+e^{-x^2}}(-2x)\,dx$$

$$= -\frac{1}{2}\int \frac{1}{1+e^u}\,du$$

$$= -\frac{1}{2}\left[u - \ln(1 + e^u)\right] + C$$

$$= -\frac{1}{2}\left[-x^2 - \ln(1 + e^{-x^2})\right] + C$$

$$= \frac{1}{2}\left[x^2 + \ln(1 + e^{-x^2})\right] + C$$

So, the value of the definite integral is

$$\int_0^2 \frac{x}{1+e^{-x^2}}\,dx = \frac{1}{2}\left[x^2 + \ln(1 + e^{-x^2})\right]\Big|_0^2 \approx 1.66. \qquad \text{See Figure 6.3.}$$

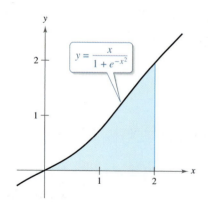

$$y = \frac{x}{1+e^{-x^2}}$$

FIGURE 6.3

✓ Checkpoint 4

Use the integration table in Appendix C to evaluate $\displaystyle\int_0^1 \frac{x^2}{1+e^{x^3}}\,dx$.

Reduction Formulas

Several of the formulas in the integration table have the form

$$\int f(x)\, dx = g(x) + \int h(x)\, dx$$

where the right side contains an integral. Such integration formulas are called **reduction formulas** because they reduce the original integral to the sum of a function and a simpler integral.

> **ALGEBRA TUTOR** xy
>
> For help on the algebra in Example 5, see Example 3 in the *Chapter 6 Algebra Tutor,* on page 353.

Example 5 Using a Reduction Formula

Find $\int x^2 e^x\, dx$.

SOLUTION Using Formula 38

$$\int u^n e^u\, du = u^n e^u - n\int u^{n-1} e^u\, du$$

you can let $u = x$ and $n = 2$. Then $du = dx$, and you can write

$$\int x^2 e^x\, dx = x^2 e^x - 2\int x e^x\, dx.$$

Then, using Formula 37

$$\int u e^u\, du = (u - 1)e^u + C$$

you can write

$$\int x^2 e^x\, dx = x^2 e^x - 2\int x e^x\, dx$$
$$= x^2 e^x - 2(x - 1)e^x + C$$
$$= x^2 e^x - 2x e^x + 2e^x + C$$
$$= e^x(x^2 - 2x + 2) + C.$$

You can check this result by differentiating.

$$\frac{d}{dx}[e^x(x^2 - 2x + 2) + C] = e^x(2x - 2) + (x^2 - 2x + 2)(e^x)$$
$$= 2x e^x - 2e^x + x^2 e^x - 2x e^x + 2e^x$$
$$= x^2 e^x$$

✓ Checkpoint 5

Use the integration table in Appendix C to find the indefinite integral $\int (\ln x)^2\, dx$. ■

> **TECH TUTOR**
>
> You have now studied two ways to find the indefinite integral in Example 5. Example 5 uses an integration table, and Example 4 in Section 6.1 uses integration by parts. A third way would be to use a symbolic integration utility.

Application

Integration can be used to find the probability that an event will occur. In such an application, the real-life situation is modeled by a *probability density function f*, and the probability that x will lie between a and b is represented by

$$P(a \leq x \leq b) = \int_a^b f(x)\, dx.$$

The probability $P(a \leq x \leq b)$ must be a number between 0 and 1.

 Example 6   **Finding a Probability**

A psychologist finds that the probability that a participant in a memory experiment will recall between a and b percent (in decimal form) of the material is

$$P(a \leq x \leq b) = \int_a^b \frac{1}{e-2} x^2 e^x\, dx, \quad 0 \leq a \leq b \leq 1.$$

Find the probability that a randomly chosen participant will recall between 0% and 87.5% of the material.

SOLUTION   You can use the Constant Multiple Rule to rewrite the integral as

$$\frac{1}{e-2} \int_a^b x^2 e^x\, dx.$$

Note that the integrand is the same as the one in Example 5. Use the result of Example 5 to find the probability with $a = 0$ and $b = 0.875$.

$$\frac{1}{e-2} \int_0^{0.875} x^2 e^x\, dx = \frac{1}{e-2} \left[e^x(x^2 - 2x + 2) \right]_0^{0.875}$$

$$\approx 0.608$$

So, the probability is about 60.8%, as indicated in Figure 6.4.

FIGURE 6.4

 Checkpoint 6

Use Example 6 to find the probability that a participant will recall between 0% and 62.5% of the material.

SUMMARIZE   (Section 6.2)

1. Describe what is meant by integration by tables *(page 330)*. For examples of integration by tables, see Examples 1, 2, 3, and 4.

2. Describe what is meant by a reduction formula *(page 333)*. For an example of a reduction formula, see Example 5.

3. Describe a real-life example of how integration by tables can be used to analyze the results of a memory experiment *(page 334, Example 6)*.

SKILLS WARM UP 6.2

The following warm-up exercises involve skills that were covered in a previous course or in earlier sections. You will use these skills in the exercise set for this section. For additional help, review Appendix Section A.4 and Section 6.1.

In Exercises 1–4, expand the expression.

1. $(x + 4)^2$

2. $(x - 1)^2$

3. $\left(x + \frac{1}{2}\right)^2$

4. $\left(x - \frac{1}{3}\right)^2$

In Exercises 5 and 6, use integration by parts to find the indefinite integral.

5. $\displaystyle\int 2xe^x \, dx$

6. $\displaystyle\int 3x^2 \ln x \, dx$

ENHANCED
WebAssign   Access end-of-section exercises online at **www.webassign.net**

QUIZ YOURSELF

Take this quiz as you would take a quiz in class. When you are done, check your work against the answers given in the back of the book.

In Exercises 1–6, use integration by parts to find the indefinite integral.

1. $\displaystyle\int xe^{5x}\,dx$ **2.** $\displaystyle\int \ln x^3\,dx$

3. $\displaystyle\int (x+1)\ln x\,dx$ **4.** $\displaystyle\int x\sqrt{x+3}\,dx$

5. $\displaystyle\int x\ln\sqrt{x}\,dx$ **6.** $\displaystyle\int x^2 e^{-2x}\,dx$

7. A manufacturing company forecasts that the demand x (in units) for its product over the next 5 years can be modeled by

$$x = 1000(45 + 20te^{-0.5t})$$

where t is the time in years.

(a) Find the total demand over the next 5 years.

(b) Find the average annual demand during the 5-year period.

8. A small business expects its income c during the next 7 years to be given by

$$c(t) = 32{,}000t, \quad 0 \le t \le 7.$$

(a) Find the actual income for the business over the 7 years.

(b) Assuming an annual inflation rate of 3.3%, what is the present value of this income?

In Exercises 9–14, use the integration table in Appendix C to find the indefinite integral.

9. $\displaystyle\int \frac{x}{1+2x}\,dx$ **10.** $\displaystyle\int \frac{1}{x(0.1+0.2x)}\,dx$

11. $\displaystyle\int \frac{\sqrt{x^2-16}}{x^2}\,dx$ **12.** $\displaystyle\int \frac{1}{x\sqrt{4+9x}}\,dx$

13. $\displaystyle\int \frac{2x}{1+e^{4x^2}}\,dx$ **14.** $\displaystyle\int 2x(x^2+1)e^{x^2+1}\,dx$

15. The revenue (in millions of dollars) for a new product is modeled by

$$R = \sqrt{144t^2 + 400}$$

where t is the time in years.

(a) Estimate the total revenue of the product over its first 3 years on the market.

(b) Estimate the total revenue of the product over its first 6 years on the market.

In Exercises 16–21, evaluate the definite integral.

16. $\displaystyle\int_{-2}^{0} xe^{x/2}\,dx$ **17.** $\displaystyle\int_{1}^{2} 5x\ln x\,dx$

18. $\displaystyle\int_{0}^{8} \frac{x}{\sqrt{x+8}}\,dx$ **19.** $\displaystyle\int_{1}^{e} (\ln x)^2\,dx$

20. $\displaystyle\int_{2}^{3} \frac{1}{x^2\sqrt{9-x^2}}\,dx$ **21.** $\displaystyle\int_{4}^{6} \frac{2x}{x^4-4}\,dx$

6.3 Numerical Integration

■ Use the Trapezoidal Rule to approximate definite integrals.

■ Use Simpson's Rule to approximate definite integrals.

■ Analyze the sizes of the errors when approximating definite integrals with the Trapezoidal Rule and Simpson's Rule.

In Exercise 43, you will use Simpson's Rule to find the average median age of the U.S. resident population from 2001 through 2009.

The Trapezoidal Rule

In Section 5.6, you studied one technique for approximating the value of a *definite* integral—the Midpoint Rule. In this section, you will study two other approximation techniques: the **Trapezoidal Rule** and **Simpson's Rule.**

To develop the Trapezoidal Rule, consider a function f that is nonnegative and continuous on the closed interval $[a, b]$. To approximate the area represented by

$$\int_a^b f(x)\,dx$$

partition the interval into n subintervals, each of width

$$\Delta x = \frac{b - a}{n}. \qquad \text{Width of each subinterval}$$

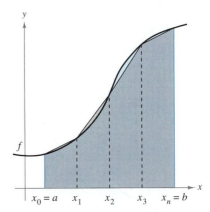

The area of the region can be approximated using four trapezoids.

FIGURE 6.5

Next, form n trapezoids, as shown in Figure 6.5. As you can see in Figure 6.6, the area of the first trapezoid is

$$\text{Area of first trapezoid} = \left(\frac{b - a}{n}\right)\left[\frac{f(x_0) + f(x_1)}{2}\right].$$

The areas of the other trapezoids follow a similar pattern, and the sum of the n areas is

$$\begin{aligned}
\text{Area} &= \left(\frac{b - a}{n}\right)\left[\frac{f(x_0) + f(x_1)}{2} + \frac{f(x_1) + f(x_2)}{2} + \cdots + \frac{f(x_{n-1}) + f(x_n)}{2}\right] \\
&= \left(\frac{b - a}{2n}\right)[f(x_0) + f(x_1) + f(x_1) + f(x_2) + \cdots + f(x_{n-1}) + f(x_n)] \\
&= \left(\frac{b - a}{2n}\right)[f(x_0) + 2f(x_1) + 2f(x_2) + \cdots + 2f(x_{n-1}) + f(x_n)].
\end{aligned}$$

Although this development assumes f to be continuous *and* nonnegative on $[a, b]$, the resulting formula is valid as long as f is continuous on $[a, b]$.

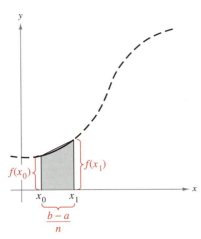

FIGURE 6.6

The Trapezoidal Rule

If f is continuous on $[a, b]$, then

$$\int_a^b f(x)\,dx \approx \left(\frac{b - a}{2n}\right)[f(x_0) + 2f(x_1) + \cdots + 2f(x_{n-1}) + f(x_n)].$$

Note that the coefficients in the Trapezoidal Rule have the following pattern.

$$1 \quad 2 \quad 2 \quad 2 \ldots 2 \quad 2 \quad 1$$

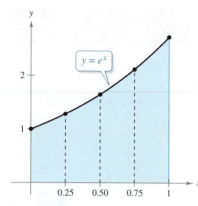

Four Subintervals
FIGURE 6.7

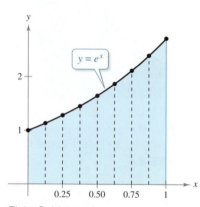

Eight Subintervals
FIGURE 6.8

Example 1   **Using the Trapezoidal Rule**

Use the Trapezoidal Rule to approximate $\int_0^1 e^x dx$. Compare the results for $n = 4$ and $n = 8$.

SOLUTION   When $n = 4$, the width of each subinterval is

$$\frac{1 - 0}{4} = \frac{1}{4}$$

and the endpoints of the subintervals are

$$x_0 = 0, \quad x_1 = \frac{1}{4}, \quad x_2 = \frac{1}{2}, \quad x_3 = \frac{3}{4}, \quad \text{and} \quad x_4 = 1$$

as indicated in Figure 6.7. So, by the Trapezoidal Rule,

$$\int_0^1 e^x dx = \frac{1}{8}(e^0 + 2e^{0.25} + 2e^{0.5} + 2e^{0.75} + e^1)$$

$$\approx 1.7272. \qquad \text{Approximation using } n = 4$$

When $n = 8$, the width of each subinterval is

$$\frac{1 - 0}{8} = \frac{1}{8}$$

and the endpoints of the subintervals are

$$x_0 = 0, \quad x_1 = \frac{1}{8}, \quad x_2 = \frac{1}{4}, \quad x_3 = \frac{3}{8}, \quad x_4 = \frac{1}{2},$$

$$x_5 = \frac{5}{8}, \quad x_6 = \frac{3}{4}, \quad x_7 = \frac{7}{8}, \quad \text{and} \quad x_8 = 1$$

as indicated in Figure 6.8. So, by the Trapezoidal Rule,

$$\int_0^1 e^x dx = \frac{1}{16}(e^0 + 2e^{0.125} + 2e^{0.25} + \cdots + 2e^{0.875} + e^1)$$

$$\approx 1.7205. \qquad \text{Approximation using } n = 8$$

Of course, for *this particular* integral, you could have found an antiderivative and used the Fundamental Theorem of Calculus to find the exact value of the definite integral. The exact value is

$$\int_0^1 e^x dx = e - 1 \qquad \text{Exact value}$$

which is approximately 1.718282.

✓**Checkpoint 1**

Use the Trapezoidal Rule with $n = 4$ to approximate

$$\int_0^1 e^{2x} dx.$$ ■

TECH TUTOR

A symbolic integration utility can be used to evaluate a definite integral. Use a symbolic integration utility to approximate the integral $\int_0^1 e^{x^2} dx$.

There are two important points that should be made concerning the Trapezoidal Rule. First, the approximation tends to become more accurate as n increases. For instance, in Example 1, when $n = 16$, the Trapezoidal Rule yields an approximation of 1.7188. Second, although you could have used the Fundamental Theorem of Calculus to evaluate the integral in Example 1, this theorem cannot be used to evaluate an integral as simple as $\int_0^1 e^{x^2} dx$. Yet the Trapezoidal Rule can be easily applied to estimate this integral.

Simpson's Rule

One way to view the Trapezoidal Rule is to say that f is approximated by a first-degree polynomial on each subinterval. In Simpson's Rule, f is approximated by a second-degree polynomial on each subinterval.

To develop Simpson's Rule, partition the interval $[a, b]$ into an *even number n* of subintervals, each of width

$$\Delta x = \frac{b - a}{n}.$$

On the subinterval $[x_0, x_2]$, approximate the function f by the second-degree polynomial $p(x)$ that passes through the points

$$(x_0, f(x_0)), \quad (x_1, f(x_1)), \quad \text{and} \quad (x_2, f(x_2))$$

as shown in Figure 6.9. The Fundamental Theorem of Calculus can be used to show that

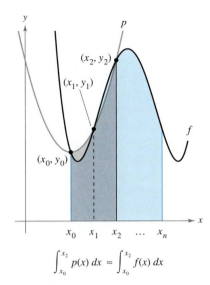

$$\int_{x_0}^{x_2} p(x)\, dx \approx \int_{x_0}^{x_2} f(x)\, dx$$

FIGURE 6.9

$$\int_{x_0}^{x_2} f(x)\, dx \approx \int_{x_0}^{x_2} p(x)\, dx$$

$$= \left(\frac{x_2 - x_0}{6}\right)\left[p(x_0) + 4p\left(\frac{x_0 + x_2}{2}\right) + p(x_2)\right]$$

$$= \frac{2[(b - a)/n]}{6}[p(x_0) + 4p(x_1) + p(x_2)]$$

$$= \left(\frac{b - a}{3n}\right)[f(x_0) + 4f(x_1) + f(x_2)].$$

Repeating this process on the subintervals $[x_{i-2}, x_i]$ produces

$$\int_a^b f(x)\, dx \approx \left(\frac{b - a}{3n}\right)[f(x_0) + 4f(x_1) + f(x_2) + f(x_2) + 4f(x_3) +$$

$$f(x_4) + \cdots + f(x_{n-2}) + 4f(x_{n-1}) + f(x_n)].$$

By grouping like terms, you can obtain the approximation shown below, which is known as Simpson's Rule. This rule is named after the English mathematician Thomas Simpson (1710–1761).

Simpson's Rule (*n* Is Even)

If f is continuous on $[a, b]$, then

$$\int_a^b f(x)\, dx \approx \left(\frac{b - a}{3n}\right)[f(x_0) + 4f(x_1) + 2f(x_2) + 4f(x_3) +$$

$$\cdots + 4f(x_{n-1}) + f(x_n)].$$

Note that the coefficients in Simpson's Rule have the following pattern.

$$1 \quad 4 \quad 2 \quad 4 \quad 2 \quad 4 \ldots 4 \quad 2 \quad 4 \quad 1$$

The Trapezoidal Rule and Simpson's Rule are necessary for solving certain real-life problems, such as approximating the present value of an income. You will see such problems in the exercise set for this section.

In Example 1, the Trapezoidal Rule was used to estimate the value of

$$\int_0^1 e^x \, dx.$$

The next example uses Simpson's Rule to approximate the same integral.

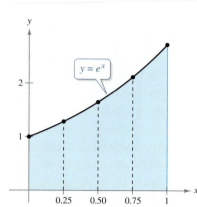

Four Subintervals

FIGURE 6.10

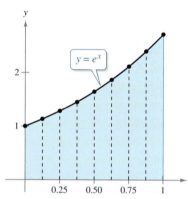

Eight Subintervals

FIGURE 6.11

Example 2   Using Simpson's Rule

Use Simpson's Rule to approximate

$$\int_0^1 e^x \, dx.$$

Compare the results for $n = 4$ and $n = 8$.

SOLUTION   When $n = 4$, the width of each subinterval is

$$\frac{1 - 0}{4} = \frac{1}{4}$$

and the endpoints of the subintervals are

$$x_0 = 0, \quad x_1 = \frac{1}{4}, \quad x_2 = \frac{1}{2}, \quad x_3 = \frac{3}{4}, \quad \text{and} \quad x_4 = 1$$

as indicated in Figure 6.10. So, by Simpson's Rule

$$\int_0^1 e^x \, dx = \frac{1}{12}(e^0 + 4e^{0.25} + 2e^{0.5} + 4e^{0.75} + e^1)$$

$$\approx 1.718319. \qquad \text{Approximation using } n = 4$$

When $n = 8$, the width of each subinterval is $(1 - 0)/8 = \frac{1}{8}$ and the endpoints of the subintervals are

$$x_0 = 0, \quad x_1 = \frac{1}{8}, \quad x_2 = \frac{1}{4}, \quad x_3 = \frac{3}{8}, \quad x_4 = \frac{1}{2},$$

$$x_5 = \frac{5}{8}, \quad x_6 = \frac{3}{4}, \quad x_7 = \frac{7}{8}, \quad \text{and} \quad x_8 = 1$$

as indicated in Figure 6.11. So, by Simpson's Rule

$$\int_0^1 e^x \, dx = \frac{1}{24}(e^0 + 4e^{0.125} + 2e^{0.25} + \cdots + 4e^{0.875} + e^1)$$

$$\approx 1.718284. \qquad \text{Approximation using } n = 8$$

Recall that the exact value of this integral is

$$\int_0^1 e^x \, dx = e - 1 \qquad \text{Exact value}$$

which is approximately

$$1.718282. \qquad \text{Approximate value}$$

So, with only eight subintervals, you obtained an approximation that is correct to the nearest 0.000002—an impressive result.

✓ Checkpoint 2

Use Simpson's Rule with $n = 4$ to approximate

$$\int_0^1 e^{2x} \, dx.$$

STUDY TIP

Comparing the results of Examples 1 and 2, you can see that for a given value of n, Simpson's Rule tends to be more accurate than the Trapezoidal Rule.

Error Analysis

In Examples 1 and 2, you were able to calculate the exact value of the integral and compare that value with the approximations to see how good they were. In practice, you need to have a different way of telling how good an approximation is: such a way is provided in the next result.

TECH TUTOR

A program for several models of graphing utilities that uses Simpson's Rule to approximate the definite integral $\int_a^b f(x)\, dx$ can be found in Appendix E.

Errors in the Trapezoidal Rule and Simpson's Rule

The errors E in approximating

$$\int_a^b f(x)\, dx$$

are as shown.

Trapezoidal Rule: $\quad |E| \le \dfrac{(b-a)^3}{12n^2}\Big[\max|f''(x)|\Big], \quad a \le x \le b$

Simpson's Rule: $\quad |E| \le \dfrac{(b-a)^5}{180n^4}\Big[\max|f^{(4)}(x)|\Big], \quad a \le x \le b$

This result indicates that the errors generated by the Trapezoidal Rule and Simpson's Rule have upper bounds dependent on the extreme values of

$$f''(x) \quad \text{and} \quad f^{(4)}(x)$$

in the interval $[a, b]$. Furthermore, the bounds for the errors can be made arbitrarily small by *increasing* n. To determine what value of n to choose, consider the steps below.

Trapezoidal Rule

1. Find $f''(x)$.

2. Find the maximum of $|f''(x)|$ on the interval $[a, b]$.

3. Set up the inequality

$$|E| \le \frac{(b-a)^3}{12n^2}\Big[\max|f''(x)|\Big].$$

4. For an error less than ϵ, solve for n in the inequality

$$\frac{(b-a)^3}{12n^2}\Big[\max|f''(x)|\Big] < \epsilon.$$

5. Partition $[a, b]$ into n subintervals and apply the Trapezoidal Rule.

Simpson's Rule

1. Find $f^{(4)}(x)$.

2. Find the maximum of $|f^{(4)}(x)|$ on the interval $[a, b]$.

3. Set up the inequality

$$|E| \le \frac{(b-a)^5}{180n^4}\Big[\max|f^{(4)}(x)|\Big].$$

4. For an error less than ϵ, solve for n in the inequality

$$\frac{(b-a)^5}{180n^4}\Big[\max|f^{(4)}(x)|\Big] < \epsilon.$$

5. Partition $[a, b]$ into n subintervals and apply Simpson's Rule.

ALGEBRA TUTOR xy

For help on the algebra in Example 3, see Example 4 in the Chapter 6 *Algebra Tutor*, on page 353.

Example 3   The Approximate Error in the Trapezoidal Rule

Use the Trapezoidal Rule to estimate the value of $\int_0^1 e^{-x^2}\,dx$ such that the error in the approximation of the integral is less than 0.01.

SOLUTION

1. Begin by finding the second derivative of $f(x) = e^{-x^2}$.

$$f(x) = e^{-x^2}$$
$$f'(x) = -2xe^{-x^2}$$
$$f''(x) = 4x^2e^{-x^2} - 2e^{-x^2}$$
$$= 2e^{-x^2}(2x^2 - 1)$$

2. f'' has only one critical number in the interval $[0, 1]$, and the maximum value of $|f''(x)|$ on this interval is $|f''(0)| = 2$.

3. The error E using the Trapezoidal Rule is bounded by

$$|E| \le \frac{(b-a)^3}{12n^2}(2) = \frac{1}{12n^2}(2) = \frac{1}{6n^2}.$$

4. To ensure that the approximation has an error of less than 0.01, you should choose n such that

$$\frac{1}{6n^2} < 0.01.$$

Solving for n, you can determine that n must be 5 or more.

5. Partition $[0, 1]$ into five subintervals, as shown in Figure 6.12. Then apply the Trapezoidal Rule to obtain

$$\int_0^1 e^{-x^2}\,dx = \frac{1}{10}\left(\frac{1}{e^0} + \frac{2}{e^{0.04}} + \frac{2}{e^{0.16}} + \frac{2}{e^{0.36}} + \frac{2}{e^{0.64}} + \frac{1}{e^1}\right) \approx 0.744.$$

So, with an error less than 0.01, you know that

$$0.734 \le \int_0^1 e^{-x^2}\,dx \le 0.754.$$

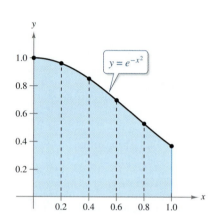

$y = e^{-x^2}$

FIGURE 6.12

✓ Checkpoint 3

Use the Trapezoidal Rule to estimate the value of

$$\int_0^1 \sqrt{1 + x^2}\,dx$$

such that the error in the approximation of the integral is less than 0.01. ■

SUMMARIZE   (Section 6.3)

1. State the Trapezoidal Rule *(page 337)*. For an example of the Trapezoidal Rule, see Example 1.

2. State Simpson's Rule *(page 339)*. For an example of Simpson's Rule, see Example 2.

3. State the approximate errors in the Trapezoidal Rule and Simpson's Rule *(page 341)*. For an example of using the approximate error in the Trapezoidal Rule, see Example 3.

SKILLS WARM UP 6.3   The following warm-up exercises involve skills that were covered in a previous course or in earlier sections. You will use these skills in the exercise set for this section. For additional help, review Appendix Section A.1, and Sections 2.2, 2.6, 3.2, 4.3, and 4.5.

In Exercises 1–6, find the indicated derivative.

1. $f(x) = \dfrac{1}{x}$, $f''(x)$

2. $f(x) = \ln(2x + 1)$, $f^{(4)}(x)$

3. $f(x) = 2 \ln x$, $f^{(4)}(x)$

4. $f(x) = x^3 - 2x^2 + 7x - 12$, $f''(x)$

5. $f(x) = e^{2x}$, $f^{(4)}(x)$

6. $f(x) = e^{x^2}$, $f''(x)$

In Exercises 7 and 8, find the absolute maximum of f on the interval.

7. $f(x) = -x^2 + 6x + 9$, $[0, 4]$

8. $f(x) = \dfrac{8}{x^3}$, $[1, 2]$

In Exercises 9 and 10, solve for n.

9. $\dfrac{1}{4n^2} < 0.001$

10. $\dfrac{1}{16n^4} < 0.0001$

6.4 Improper Integrals

■ Recognize improper integrals.
■ Evaluate improper integrals with infinite limits of integration.
■ Use improper integrals to solve real-life problems.
■ Find the present value of a perpetuity.

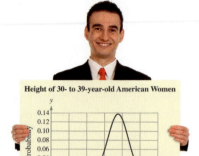

Height of 30- to 39-year-old American Women

In Exercise 27, you will evaluate an improper integral to determine the probability that a 30- to 39-year-old woman is 6 feet or taller.

Improper Integrals

The definition of a definite integral

$$\int_a^b f(x)\,dx$$

requires that the interval $[a, b]$ be finite. Furthermore, the Fundamental Theorem of Calculus, by which you have been evaluating definite integrals, requires that f be continuous on $[a, b]$. Some integrals do not satisfy these requirements because of one of the conditions below.

1. One or both of the limits of integration are infinite.

2. The function f has an infinite discontinuity in the interval $[a, b]$.

Integrals having either of these characteristics are called **improper integrals.** In this section, you will study integrals where one or both limits of integration are infinite. For instance, the integral

$$\int_0^\infty e^{-x}\,dx$$

is improper because one limit of integration is infinite, as indicated in Figure 6.13.

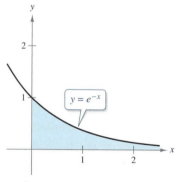

FIGURE 6.13

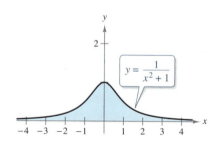

FIGURE 6.14

Similarly, the integral

$$\int_{-\infty}^\infty \frac{1}{x^2 + 1}\,dx$$

is improper because both limits of integration are infinite, as indicated in Figure 6.14.
The integrals

$$\int_1^5 \frac{1}{\sqrt{x-1}}\,dx \quad \text{and} \quad \int_{-2}^2 \frac{1}{(x+1)^2}\,dx$$

are improper because their integrands have an **infinite discontinuity**—that is, they approach infinity somewhere in the interval of integration. Evaluating an integral whose integrand has an infinite discontinuity is beyond the scope of this text.

Integrals with Infinite Limits of Integration

To see how to evaluate an improper integral, consider the integral shown in Figure 6.15.

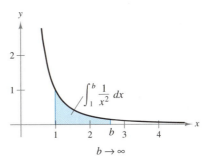

FIGURE 6.15

As long as b is a real number that is greater than 1 (no matter how large), this is a definite integral whose value is

$$\int_1^b \frac{1}{x^2}\,dx = \left[-\frac{1}{x}\right]_1^b = -\frac{1}{b} + 1 = 1 - \frac{1}{b}.$$

The table shows the values of this integral for several values of b.

b	2	5	10	100	1000	10,000
$\int_1^b \dfrac{1}{x^2}\,dx = 1 - \dfrac{1}{b}$	0.5000	0.8000	0.9000	0.9900	0.9990	0.9999

From this table, it appears that the value of the integral is approaching a limit as b increases without bound. This limit is denoted by the *improper integral* shown below.

$$\int_1^\infty \frac{1}{x^2}\,dx = \lim_{b\to\infty} \int_1^b \frac{1}{x^2}\,dx = \lim_{b\to\infty}\left(1 - \frac{1}{b}\right) = 1$$

This improper integral can be interpreted as the area of the *unbounded* region between the graph of $f(x) = 1/x^2$ and the x-axis (to the right of $x = 1$).

Improper Integrals (Infinite Limits of Integration)

1. If f is continuous on the interval $[a, \infty)$, then

$$\int_a^\infty f(x)\,dx = \lim_{b\to\infty} \int_a^b f(x)\,dx.$$

2. If f is continuous on the interval $(-\infty, b]$, then

$$\int_{-\infty}^b f(x)\,dx = \lim_{a\to-\infty} \int_a^b f(x)\,dx.$$

3. If f is continuous on the interval $(-\infty, \infty)$, then

$$\int_{-\infty}^\infty f(x)\,dx = \int_{-\infty}^c f(x)\,dx + \int_c^\infty f(x)\,dx$$

where c is any real number.

In the first two cases, if the limit exists, then the improper integral **converges;** otherwise, the improper integral **diverges.** In the third case, the integral on the left diverges when either one of the integrals on the right diverges.

Example 1   **Evaluating an Improper Integral**

Determine the convergence or divergence of $\int_1^\infty \frac{1}{x}\, dx$.

SOLUTION   Begin by applying the definition of an improper integral.

$$\int_1^\infty \frac{1}{x}\, dx = \lim_{b \to \infty} \int_1^b \frac{1}{x}\, dx \qquad \text{Definition of improper integral}$$

$$= \lim_{b \to \infty} \left[\ln x \right]_1^b \qquad \text{Find antiderivative.}$$

$$= \lim_{b \to \infty} (\ln b - 0) \qquad \text{Apply Fundamental Theorem.}$$

$$= \infty \qquad \text{Evaluate limit.}$$

Because the limit is infinite, the improper integral diverges.

✓**Checkpoint 1**

Determine the convergence or divergence of each improper integral.

a. $\int_1^\infty \frac{1}{x^3}\, dx$ **b.** $\int_1^\infty \frac{1}{\sqrt{x}}\, dx$ ■

As you begin to work with improper integrals, you will find that integrals that appear to be similar can have very different values. For instance, consider the two improper integrals

$$\int_1^\infty \frac{1}{x}\, dx = \infty \qquad \text{Divergent integral}$$

and

$$\int_1^\infty \frac{1}{x^2}\, dx = 1. \qquad \text{Convergent integral}$$

The first integral diverges and the second converges to 1. Graphically, this means that the areas shown in Figure 6.16 are very different. The region lying between the graph of

$$y = \frac{1}{x}$$

and the *x*-axis (for $x \geq 1$) has an *infinite* area, and the region lying between the graph of

$$y = \frac{1}{x^2}$$

and the *x*-axis (for $x \geq 1$) has a *finite* area.

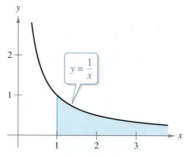

Diverges (infinite area)

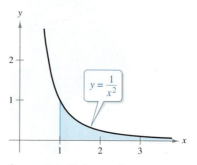

Converges (finite area)

FIGURE 6.16

ALGEBRA TUTOR *xy*

For help on the algebra in Example 2, see Example 2(a) in the *Chapter 6 Algebra Tutor*, on page 352.

Example 2 **Evaluating an Improper Integral**

Evaluate the improper integral.

$$\int_{-\infty}^{0} \frac{1}{(1 - 2x)^{3/2}} \, dx$$

SOLUTION Begin by applying the definition of an improper integral.

$$\int_{-\infty}^{0} \frac{1}{(1 - 2x)^{3/2}} \, dx = \lim_{a \to -\infty} \int_{a}^{0} \frac{1}{(1 - 2x)^{3/2}} \, dx \qquad \text{Definition of improper integral}$$

$$= \lim_{a \to -\infty} \left[\frac{1}{\sqrt{1 - 2x}} \right]_{a}^{0} \qquad \text{Find antiderivative.}$$

$$= \lim_{a \to -\infty} \left(1 - \frac{1}{\sqrt{1 - 2a}} \right) \qquad \text{Apply Fundamental Theorem.}$$

$$= 1 - 0 \qquad \text{Evaluate limit.}$$

$$= 1 \qquad \text{Simplify.}$$

So, the improper integral converges to 1. As shown in Figure 6.17, this implies that the region lying between the graph of $y = 1/(1 - 2x)^{3/2}$ and the *x*-axis (for $x \le 0$) has an area of 1 square unit.

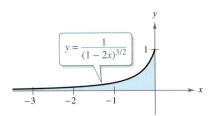

$$y = \frac{1}{(1 - 2x)^{3/2}}$$

FIGURE 6.17

✓ **Checkpoint 2**

Evaluate the improper integral, if possible.

$$\int_{-\infty}^{0} \frac{1}{(x - 1)^2} \, dx$$

ALGEBRA TUTOR *xy*

For help on the algebra in Example 3, see Example 2(b) in the *Chapter 6 Algebra Tutor*, on page 352.

Example 3 **Evaluating an Improper Integral**

Evaluate the improper integral.

$$\int_{0}^{\infty} 2xe^{-x^2} \, dx$$

SOLUTION Begin by applying the definition of an improper integral.

$$\int_{0}^{\infty} 2xe^{-x^2} \, dx = \lim_{b \to \infty} \int_{0}^{b} 2xe^{-x^2} \, dx \qquad \text{Definition of improper integral}$$

$$= \lim_{b \to \infty} \left[-e^{-x^2} \right]_{0}^{b} \qquad \text{Find antiderivative.}$$

$$= \lim_{b \to \infty} \left(-e^{-b^2} + 1 \right) \qquad \text{Apply Fundamental Theorem.}$$

$$= 0 + 1 \qquad \text{Evaluate limit.}$$

$$= 1 \qquad \text{Simplify.}$$

So, the improper integral converges to 1. As shown in Figure 6.18, this implies that the region lying between the graph of $y = 2xe^{-x^2}$ and the *x*-axis (for $x \ge 0$) has an area of 1 square unit.

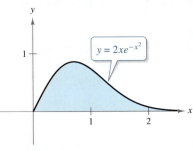

$$y = 2xe^{-x^2}$$

FIGURE 6.18

✓ **Checkpoint 3**

Evaluate the improper integral, if possible.

$$\int_{-\infty}^{0} e^{2x} \, dx$$

Application

In Section 4.3, you studied the graph of the *normal probability density function*

$$f(x) = \frac{1}{\sigma\sqrt{2\pi}}e^{-(x-\mu)^2/(2\sigma^2)}.$$

This function is used in statistics to represent a population that is normally distributed with a mean of μ and a standard deviation of σ. Specifically, when an outcome x is chosen at random from the population, the probability that x will have a value between a and b is

$$P(a \le x \le b) = \int_a^b \frac{1}{\sigma\sqrt{2\pi}}e^{-(x-\mu)^2/(2\sigma^2)}dx.$$

As shown in Figure 6.19, the probability $P(-\infty < x < \infty)$ is

$$P(-\infty < x < \infty) = \int_{-\infty}^{\infty} \frac{1}{\sigma\sqrt{2\pi}}e^{-(x-\mu)^2/(2\sigma^2)}dx = 1.$$

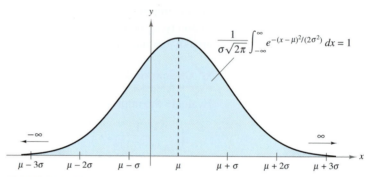

FIGURE 6.19

 Example 4   Finding a Probability

The mean height of American men (from 20 to 29 years old) is 69 inches, and the standard deviation is 3 inches. A 20- to 29-year-old man is chosen at random from the population. What is the probability that he is 6 feet tall or taller? *(Source: U.S. National Center for Health Statistics)*

SOLUTION   Note that the mean and standard deviation are given in inches and the height of the man chosen at random is given in feet. To calculate the probability, you need to use the same units for these quantities. Because it is easier to convert feet to inches, use 72 inches (1 foot = 12 inches) for the man's height. So, the probability can be written as $P(72 \le x < \infty)$. Using a mean of $\mu = 69$ and a standard deviation of $\sigma = 3$, the probability $P(72 \le x < \infty)$ is given by the improper integral

$$P(72 \le x < \infty) = \int_{72}^{\infty} \frac{1}{3\sqrt{2\pi}}e^{-(x-69)^2/18}dx.$$

Using a symbolic integration utility, you can approximate the value of this integral to be 0.158. So, the probability that the man is 6 feet tall or taller is about 15.8%.

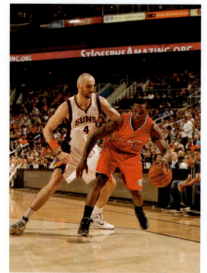

Many professional basketball players are over $6\frac{1}{2}$ feet tall. When a man is chosen at random from the population, the probability that he is $6\frac{1}{2}$ feet tall or taller is less than half of one percent.

✓ **Checkpoint 4**

Use Example 4 to find the probability that a 20- to 29-year-old man chosen at random from the population is 6 feet 6 inches tall or taller. ■

Present Value of a Perpetuity

Recall from Section 6.1 that for an interest-bearing account, the present value over t_1 years is

$$\text{Present value} = \int_0^{t_1} c(t)e^{-rt}\, dt$$

where c represents a continuous income function (in dollars per year) and the annual interest rate r is compounded continuously. If the size of an annuity's payment is a constant number of dollars P, then $c(t)$ is equal to P and the present value is

$$\text{Present value} = \int_0^{t_1} Pe^{-rt}\, dt = P\int_0^{t_1} e^{-rt}\, dt. \qquad \begin{array}{l}\text{\color{red}Present value of an annuity}\\ \text{\color{red}with payment } P\end{array}$$

Consider an annuity, such as a scholarship fund, that pays the same amount each year *forever*. Because the annuity continues indefinitely, the number of years t_1 approaches infinity. Such an annuity is called a **perpetual annuity** or a **perpetuity.** This situation can be represented by the following improper integral.

$$\text{Present value} = P\int_0^{\infty} e^{-rt}\, dt \qquad \begin{array}{l}\text{\color{red}Present value of a perpetuity}\\ \text{\color{red}with payment } P\end{array}$$

This integral is simplified as follows.

$$P\int_0^{\infty} e^{-rt}\, dt = P\lim_{b\to\infty}\int_0^{b} e^{-rt}\, dt \qquad \text{\color{red}Definition of improper integral.}$$

$$= P\lim_{b\to\infty}\left[-\frac{e^{-rt}}{r}\right]_0^{b} \qquad \text{\color{red}Find antiderivative.}$$

$$= P\lim_{b\to\infty}\left(-\frac{e^{-rb}}{r}+\frac{1}{r}\right) \qquad \text{\color{red}Apply Fundamental Theorem.}$$

$$= P\left(0+\frac{1}{r}\right) \qquad \text{\color{red}Evaluate limit.}$$

$$= \frac{P}{r} \qquad \text{\color{red}Simplify.}$$

So, the improper integral converges to P/r. As shown in Figure 6.20, this implies that the region lying between the graph of

$$y = Pe^{-rt}$$

and the t-axis for $t \geq 0$ has an area equal to the annual payment P divided by the annual interest rate r.

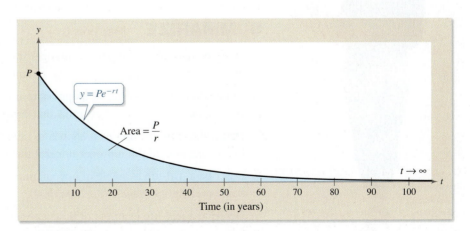

FIGURE 6.20

The present value of a perpetuity is defined as follows.

Present Value of a Perpetuity

If P represents the size of each annual payment in dollars and the annual interest rate is r (compounded continuously), then the present value of a perpetuity is

$$\text{Present value} = P \int_0^\infty e^{-rt}\, dt = \frac{P}{r}.$$

This definition is useful in determining the amount of money needed to start an endowment, such as a scholarship fund, as shown in Example 5.

 Example 5 **Finding Present Value**

You want to start a scholarship fund at your alma mater. You plan to give one $9000 scholarship annually beginning one year from now, and you have at most $120,000 to start the fund. You also want the scholarship to be given out indefinitely. Assuming an annual interest rate of 8% (compounded continuously), do you have enough money for the scholarship fund?

SOLUTION   To answer this question, you must find the present value of the scholarship fund. Because the scholarship is to be given out each year indefinitely, the time period is infinite. The fund is a perpetuity with $P = 9000$ and $r = 0.08$. The present value is

$$\text{Present value} = \frac{P}{r}$$

$$= \frac{9000}{0.08}$$

$$= 112{,}500.$$

The amount you need to start the scholarship fund is $112,500. Yes, you have enough money to start the scholarship fund.

✓ Checkpoint 5

In Example 5, do you have enough money to start a scholarship fund that pays $10,000 annually? Explain why or why not.   ■

SUMMARIZE   (Section 6.4)

1. Describe the different types of improper integrals *(page 344)*. For examples of evaluating improper integrals, see Examples 1, 2, and 3.

2. Define the term *converges* as it applies to improper integrals *(page 345)*. For examples of improper integrals that converge, see Examples 2 and 3.

3. Define the term *diverges* as it applies to improper integrals *(page 345)*. For an example of an improper integral that diverges, see Example 1.

4. Describe a real-life example of how an improper integral can be used to find a probability *(page 348, Example 4)*.

5. Describe a real-life example of how an improper integral can be used to find the present value of a perpetuity *(page 350, Example 5)*.

SKILLS WARM UP 6.4

The following warm-up exercises involve skills that were covered in earlier sections. You will use these skills in the exercise set for this section. For additional help, review Sections 1.5, 4.1, and 4.4.

In Exercises 1–6, find the limit.

1. $\lim\limits_{x \to 2}(2x + 5)$

2. $\lim\limits_{x \to 1}\left(\dfrac{1}{x} + 2x^2\right)$

3. $\lim\limits_{x \to -4}\dfrac{x + 4}{x^2 - 16}$

4. $\lim\limits_{x \to 0}\dfrac{x^2 - 2x}{x^3 + 3x^2}$

5. $\lim\limits_{x \to 1}\dfrac{1}{\sqrt{x - 1}}$

6. $\lim\limits_{x \to -3}\dfrac{x^2 + 2x - 3}{x + 3}$

In Exercises 7–10, evaluate the expression (a) when $x = b$ and (b) when $x = 0$.

7. $\dfrac{4}{3}(2x - 1)^3$

8. $\dfrac{1}{x - 5} + \dfrac{3}{(x - 2)^2}$

9. $\ln(5 - 3x^2) - \ln(x + 1)$

10. $e^{3x^2} + e^{-3x^2}$

ALGEBRA TUTOR

xy

Algebra and Integration Techniques

Integration techniques involve many different algebraic skills. For a definite integral, you need a variety of algebraic skills to apply the Fundamental Theorem of Calculus and evaluate the resulting expression. Study the examples in this Algebra Tutor. Be sure that you understand the algebra used in each step.

Example 1   Evaluating an Expression

Evaluate the expression

$$(e \ln e - e) - (1 \ln 1 - 1).$$

SOLUTION   Recall that

$$\ln e = 1 \quad \text{because} \quad e^1 = e$$

and

$$\ln 1 = 0 \quad \text{because} \quad e^0 = 1.$$

$$(e \ln e - e) - (1 \ln 1 - 1)$$ Example 5, page 326
$$= [e(1) - e] - [1(0) - 1]$$ Logarithmic properties
$$= (e - e) - (0 - 1)$$ Multiply.
$$= 0 - (-1)$$ Simplify.
$$= 1$$ Simplify.

Example 2   Evaluating Expressions

Find the limit.

a. $\displaystyle\lim_{a \to -\infty} \left(1 - \frac{1}{\sqrt{1 - 2a}} \right)$

b. $\displaystyle\lim_{b \to \infty} \left(-e^{-b^2} + 1 \right)$

SOLUTION

a. $\displaystyle\lim_{a \to -\infty} \left(1 - \frac{1}{\sqrt{1 - 2a}} \right)$ Example 2, page 347

$$= \lim_{a \to -\infty} 1 - \lim_{a \to -\infty} \left(\frac{1}{\sqrt{1 - 2a}} \right)$$ $\displaystyle\lim_{x \to -\infty} [f(x) - g(x)] = \lim_{x \to -\infty} f(x) - \lim_{x \to -\infty} g(x)$

$$= 1 - 0$$ Evaluate limits.

$$= 1$$ Simplify.

b. $\displaystyle\lim_{b \to \infty} \left(-e^{-b^2} + 1 \right)$ Example 3, page 347

$$= \lim_{b \to \infty} \left(-e^{-b^2} \right) + \lim_{b \to \infty} 1$$ $\displaystyle\lim_{x \to \infty} [f(x) + g(x)] = \lim_{x \to \infty} f(x) + \lim_{x \to \infty} g(x)$

$$= \lim_{b \to \infty} \left(\frac{1}{-e^{b^2}} \right) + \lim_{b \to \infty} 1$$ Rewrite with positive exponent.

$$= 0 + 1$$ Evaluate limits.

$$= 1$$ Simplify.

Example 3   Algebra and Integration Techniques

Simplify the expression

$$x^2 e^x - 2(x - 1)e^x.$$

SOLUTION

$$x^2 e^x - 2(x - 1)e^x \qquad \text{Example 5, page 333}$$
$$= x^2 e^x - 2(xe^x - e^x) \qquad \text{Multiply factors.}$$
$$= x^2 e^x - 2xe^x + 2e^x \qquad \text{Multiply factors.}$$
$$= e^x(x^2 - 2x + 2) \qquad \text{Factor.}$$

Example 4   Solving a Rational Inequality

Solve the rational inequality

$$\frac{1}{6n^2} < 0.01$$

for n, where n is a positive integer.

SOLUTION

$$\frac{1}{6n^2} < 0.01 \qquad \text{Example 3, page 342}$$

$$\frac{1}{6n^2} < \frac{1}{100} \qquad \text{Rewrite decimal as a fraction.}$$

$$\frac{100}{6n^2} < 1 \qquad \text{Multiply each side by 100.}$$

$$100 < 6n^2 \qquad \text{Multiply each side by } 6n^2 \; (n > 0).$$

$$\frac{100}{6} < n^2 \qquad \text{Divide each side by 6.}$$

$$\frac{50}{3} < n^2 \qquad \text{Simplify.}$$

$$\sqrt{\frac{50}{3}} < n \qquad \text{Take positive square root of each side } (n > 0).$$

Because n is a positive integer and

$$\sqrt{\frac{50}{3}} \approx 4.08$$

n must be 5 or more. You can check this result using a graphing utility. Let $y_1(x) = 1/6x^2$ and $y_2(x) = 0.01$. Then use the *intersect* feature (see figure) to determine that $x \approx 4.08$. So, the solution found algebraically is correct.

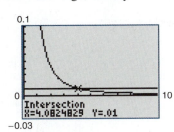

SUMMARY AND STUDY STRATEGIES

After studying this chapter, you should have acquired the following skills.
The exercise numbers are keyed to the Review Exercises that begin on page 356.
Answers to odd-numbered Review Exercises are given in the back of the text.*

Section 6.1	Review Exercises

■ Use integration by parts to find indefinite and definite integrals.   *1–12*

$$\int u\,dv = uv - \int v\,du$$

For integrals of the form

$$\int x^n e^{ax}\,dx$$

let $u = x^n$ and $dv = e^{ax}\,dx$.

For integrals of the form

$$\int x^n \ln x\,dx$$

let $u = \ln x$ and $dv = x^n\,dx$.

■ Find the present value of future income.   *13–18*

$$\text{Actual income over } t_1 \text{ years} = \int_0^{t_1} c(t)\,dt$$

$$\text{Present value} = \int_0^{t_1} c(t)e^{-rt}\,dt$$

Section 6.2

■ Use integration tables to find indefinite integrals.   *19–32*
■ Use integration tables to solve real-life problems.   *33, 34*

Section 6.3

■ Use the Trapezoidal Rule and Simpson's Rule to approximate definite integrals.   *35–46*

Trapezoidal Rule:

$$\int_a^b f(x)\,dx \approx \left(\frac{b-a}{2n}\right)[f(x_0) + 2f(x_1) + \cdots + 2f(x_{n-1}) + f(x_n)]$$

Simpson's Rule:

$$\int_a^b f(x)\,dx \approx \left(\frac{b-a}{3n}\right)[f(x_0) + 4f(x_1) + 2f(x_2) + 4f(x_3) + \cdots + 4f(x_{n-1}) + f(x_n)]$$

---

* A wide range of valuable study aids are available to help you master the material in this chapter.
The *Student Solutions Manual* includes step-by-step solutions to all odd-numbered exercises to
help you review and prepare. The student website at *www.cengagebrain.com* offers algebra help
and a *Graphing Technology Guide*, which contains step-by-step commands and instructions for a
wide variety of graphing calculators.

Section 6.3   (continued)   Review Exercises

■ Analyze the sizes of the errors when approximating definite integrals with the   *47–50*
Trapezoidal Rule and Simpson's Rule.

Errors in the Trapezoidal Rule:

$$|E| \leq \frac{(b-a)^3}{12n^2}[\max|f''(x)|], \quad a \leq x \leq b$$

Errors in Simpson's Rule:

$$|E| \leq \frac{(b-a)^5}{180n^4}[\max|f^{(4)}(x)|], \quad a \leq x \leq b$$

Section 6.4

■ Evaluate improper integrals with infinite limits of integration.   *51–56*

$$\int_a^\infty f(x)\,dx = \lim_{b\to\infty} \int_a^b f(x)\,dx$$

$$\int_{-\infty}^b f(x)\,dx = \lim_{a\to-\infty} \int_a^b f(x)\,dx$$

$$\int_{-\infty}^\infty f(x)\,dx = \int_{-\infty}^c f(x)\,dx + \int_c^\infty f(x)\,dx$$

■ Find the area of an unbounded region.   *57–60*

■ Find the present value of a perpetuity.   *61–64*

$$\text{Present value} = P\int_0^\infty e^{-rt}\,dt = \frac{P}{r}$$

Study Strategies

■ **Use a Variety of Approaches**   To be efficient at finding antiderivatives, you need to use a variety of approaches.

1. Check to see whether the integral fits one of the basic integration formulas—you should have these formulas memorized.

2. Try an integration technique such as substitution or integration by parts to rewrite the integral in a form that fits one of the basic integration formulas.

3. Use a table of integrals.

4. Use a symbolic integration utility.

■ **Use Numerical Integration**   When solving a definite integral, remember that you cannot apply the Fundamental Theorem of Calculus unless you can find an antiderivative of the integrand. This is not always possible—even with a symbolic integration utility. In such cases, you can use a numerical technique such as the Midpoint Rule, the Trapezoidal Rule, or Simpson's Rule to approximate the value of the integral.

Review Exercises

Integration by Parts In Exercises 1–8, use integration by parts to find the indefinite integral.

1. $\displaystyle\int \frac{\ln x}{\sqrt{x}}\, dx$

2. $\displaystyle\int x \ln 4x\, dx$

3. $\displaystyle\int (x + 1)e^x\, dx$

4. $\displaystyle\int xe^{-3x}\, dx$

5. $\displaystyle\int x\sqrt{x - 5}\, dx$

6. $\displaystyle\int \frac{x}{\sqrt{x + 8}}\, dx$

7. $\displaystyle\int 2x^2 e^{2x}\, dx$

8. $\displaystyle\int (\ln x)^3\, dx$

Evaluating Definite Integrals In Exercises 9–12, use integration by parts to evaluate the definite integral.

9. $\displaystyle\int_1^e 6x \ln x\, dx$

10. $\displaystyle\int_0^4 \ln(1 + 3x)\, dx$

11. $\displaystyle\int_0^1 \frac{x}{e^{x/4}}\, dx$

12. $\displaystyle\int_0^2 x^2 e^{3x}\, dx$

Finding Present Value In Exercises 13–16, find the present value of the income given by c (in dollars) over t_1 years at the given annual inflation rate r.

13. $c = 20{,}000$, $r = 4\%$, $t_1 = 5$ years

14. $c = 10{,}000 + 1500t$, $r = 6\%$, $t_1 = 10$ years

15. $c = 24{,}000t$, $r = 5\%$, $t_1 = 10$ years

16. $c = 20{,}000 + 100e^{t/2}$, $r = 5\%$, $t_1 = 5$ years

17. Present Value A company expects its income c during the next 4 years to be modeled by

$$c = 200{,}000 + 50{,}000t, \quad 0 \le t \le 4.$$

(a) Find the actual income for the business over the 4 years.

(b) Assuming an annual inflation rate of 6%, what is the present value of this income?

18. Present Value A company expects its income c during the next 7 years to be modeled by

$$c = 400{,}000 + 175{,}000t, \quad 0 \le t \le 7.$$

(a) Find the actual income for the business over the 7 years.

(b) Assuming an annual inflation rate of 4%, what is the present value of this income?

Using Integration Tables In Exercises 19–22, use the indicated formula from the integration table in Appendix C to find the indefinite integral.

19. $\displaystyle\int \frac{x^2}{2 + 3x}\, dx$, Formula 6

20. $\displaystyle\int \frac{1}{1 + e^{6x}}\, dx$, Formula 40

21. $\displaystyle\int \sqrt{x^2 - 16}\, dx$, Formula 23

22. $\displaystyle\int x^5 \ln x\, dx$, Formula 43

Using Integration Tables In Exercises 23–32, use the integration table in Appendix C to find the indefinite integral.

23. $\displaystyle\int \frac{x}{(2 + 3x)^2}\, dx$

24. $\displaystyle\int \frac{x}{\sqrt{2 + 3x}}\, dx$

25. $\displaystyle\int \frac{\sqrt{x^2 + 25}}{x}\, dx$

26. $\displaystyle\int \frac{1}{x(4 + 3x)}\, dx$

27. $\displaystyle\int \frac{1}{x^2 - 4}\, dx$

28. $\displaystyle\int (\ln 3x)^2\, dx$

29. $\displaystyle\int \frac{x}{\sqrt{1 + x}}\, dx$

30. $\displaystyle\int \frac{1}{x^2\sqrt{16 - x^2}}\, dx$

31. $\displaystyle\int \frac{\sqrt{1 + x}}{x}\, dx$

32. $\displaystyle\int \frac{1}{(x^2 - 9)^2}\, dx$

33. Probability The probability of recalling between a and b percent (in decimal form) of the material learned in a memory experiment is modeled by

$$P(a \le x \le b) = \int_a^b \frac{96}{11}\left(\frac{x}{\sqrt{9 + 16x}}\right) dx,$$

$$0 \le a \le b \le 1.$$

What are the probabilities of recalling (a) between 0% and 80% and (b) between 0% and 50% of the material?

34. Probability The probability of locating between a and b percent of the oil and gas deposits (in decimal form) in a region is modeled by

$$P(a \le x \le b) = \int_a^b 1.5x^2 e^{x^{1.5}}\, dx, \quad 0 \le a \le b \le 1.$$

What are the probabilities of locating (a) between 40% and 60% and (b) between 0% and 50% of the deposits?

Using the Trapezoidal Rule and Simpson's Rule In Exercises 35–40, use the Trapezoidal Rule and Simpson's Rule to approximate the value of the definite integral for the indicated value of n. Compare these results with the exact value of the definite integral. Round your answers to four decimal places.

35. $\int_1^3 \frac{1}{x^2} \, dx, \ n = 4$

36. $\int_0^2 (x^2 + 1) \, dx, \ n = 8$

37. $\int_1^2 \frac{1}{x^3} \, dx, \ n = 8$

38. $\int_1^2 x^3 \, dx, \ n = 4$

39. $\int_0^4 e^{-x/2} \, dx, \ n = 4$

40. $\int_0^8 \sqrt{x + 3} \, dx, \ n = 8$

Using the Trapezoidal Rule and Simpson's Rule In Exercises 41–46, approximate the value of the definite integral using (a) the Trapezoidal Rule and (b) Simpson's Rule for the indicated value of n. Round your answers to three decimal places.

41. $\int_1^2 \frac{1}{1 + \ln x} \, dx, \ n = 4$

42. $\int_0^2 \frac{1}{\sqrt{1 + x^3}} \, dx, \ n = 8$

43. $\int_0^1 \frac{x^{3/2}}{2 - x^2} \, dx, \ n = 4$

44. $\int_0^1 e^{x^2} \, dx, \ n = 6$

45. $\int_0^8 \frac{3}{x^2 + 2} \, dx, \ n = 8$

46. $\int_0^1 \sqrt{1 - x} \, dx, \ n = 4$

Error Analysis In Exercises 47 and 48, use the error formulas to find bounds for the error in approximating the definite integral using (a) the Trapezoidal Rule and (b) Simpson's Rule.

47. $\int_0^2 e^{2x} \, dx, \ n = 4$ **48.** $\int_2^4 \frac{1}{x - 1} \, dx, \ n = 8$

Error Analysis In Exercises 49 and 50, use the error formulas to find n such that the error in the approximation of the definite integral is less than 0.0001 using (a) the Trapezoidal Rule and (b) Simpson's Rule.

49. $\int_0^3 x^2 \, dx$ **50.** $\int_0^5 e^{x/5} \, dx$

Evaluating an Improper Integral In Exercises 51–56, determine whether the improper integral diverges or converges. Evaluate the integral if it converges.

51. $\int_{-\infty}^{-1} \frac{1}{x^5} \, dx$ **52.** $\int_1^\infty \frac{1}{\sqrt{x}} \, dx$

53. $\int_{-\infty}^0 \frac{1}{\sqrt[3]{8 - x}} \, dx$ **54.** $\int_0^\infty e^{-2x} \, dx$

55. $\int_1^\infty \frac{\ln x}{x} \, dx$ **56.** $\int_0^\infty \frac{e^x}{1 + e^x} \, dx$

Area of a Region In Exercises 57–60, find the area of the unbounded shaded region.

57. $y = e^{-x/4}$

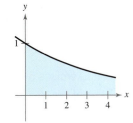

58. $y = \frac{2x}{x^2 + 2}$

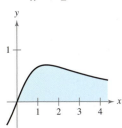

59. $y = 4xe^{-2x^2}$

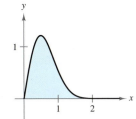

60. $y = \frac{3}{(1 - 3x)^{2/3}}$

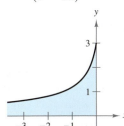

Endowment In Exercises 61 and 62, determine the amount of money required to set up a charitable endowment that pays the amount P each year indefinitely for the annual interest rate r compounded continuously.

61. $P = \$8000, r = 3\%$

62. $P = \$15,000, r = 5\%$

63. Scholarship Fund You want to start a scholarship fund at your alma mater. You plan to give one $21,000 scholarship annually beginning one year from now and you have at most $325,000 to start the fund. You also want the scholarship to be given out indefinitely. Assuming an annual interest rate of 7% compounded continuously, do you have enough money for the scholarship fund?

64. Present Value You are considering buying a franchise that yields a continuous income stream of $100,000 per year. Assuming an annual interest rate of 6% compounded continuously, what is the present value of the franchise (a) for 15 years and (b) forever?

TEST YOURSELF

Take this test as you would take a test in class. When you are done, check your work against the answers given in the back of the book.

In Exercises 1–3, use integration by parts to find the indefinite integral.

1. $\displaystyle\int xe^{x+1}\,dx$ **2.** $\displaystyle\int 9x^2 \ln x\,dx$ **3.** $\displaystyle\int x^2\,e^{-x/3}\,dx$

4. The revenue R (in millions of dollars) for P. F. Chang's China Bistro from 2001 through 2009 can be modeled by

$$R = 295.1 + 147.66\sqrt{t}\,\ln t, \quad 1 \le t \le 9$$

where t is the year, with $t = 1$ corresponding to 2001. *(Source: P. F. Chang's China Bistro)*

(a) Find the total revenue for the years 2001 through 2009.

(b) Find the average revenue for the years 2001 through 2009.

In Exercises 5–7, use the integration table in Appendix C to find the indefinite integral.

5. $\displaystyle\int \frac{x}{(7+2x)^2}\,dx$ **6.** $\displaystyle\int \frac{3x^2}{1+e^{x^3}}\,dx$ **7.** $\displaystyle\int \frac{2x^3}{\sqrt{1+5x^2}}\,dx$

In Exercises 8–10, use integration by parts or the integration table in Appendix C to evaluate the definite integral.

8. $\displaystyle\int_0^1 \ln(3-2x)\,dx$ **9.** $\displaystyle\int_3^6 \frac{x}{\sqrt{x-2}}\,dx$ **10.** $\displaystyle\int_{-3}^{-1} \frac{\sqrt{x^2+16}}{x}\,dx$

11. Use the Trapezoidal Rule with $n = 4$ to approximate

$$\int_2^5 (x^2 - 2x)\,dx.$$

Compare your result with the exact value of the definite integral.

12. Use Simpson's Rule with $n = 4$ to approximate

$$\int_0^1 9xe^{3x}\,dx.$$

Compare your result with the exact value of the definite integral.

In Exercises 13–15, determine whether the improper integral diverges or converges. Evaluate the integral if it converges.

13. $\displaystyle\int_0^\infty e^{-3x}\,dx$ **14.** $\displaystyle\int_0^9 \frac{2}{\sqrt{x}}\,dx$ **15.** $\displaystyle\int_{-\infty}^0 \frac{1}{(4x-1)^{2/3}}\,dx$

16. A magazine publisher offers two subscription plans. Plan A is a one-year subscription for $19.95. Plan B is a lifetime subscription (lasting indefinitely) for $149.

(a) A subscriber considers using plan A indefinitely. Assuming an annual inflation rate of 4%, find the present value of the money the subscriber will spend using plan A.

(b) Based on your answer to part (a), which plan should the subscriber use? Explain.

7 Functions of Several Variables

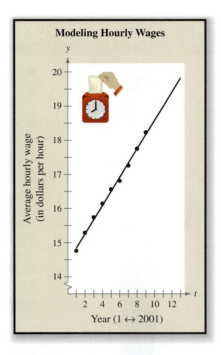

Modeling Hourly Wages

Average hourly wage (in dollars per hour)

Year (1 ↔ 2001)

Example 3 on page 407 shows how least squares regression analysis can be used to find the best-fitting line that models hourly wages for production workers in manufacturing industries.

7.1 The Three-Dimensional Coordinate System

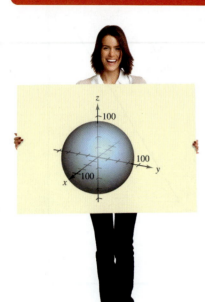

In Exercise 57, you will model the shape of a spherical building using the standard equation of a sphere.

- ■ Plot points in space.
- ■ Find distances between points in space and find midpoints of line segments in space.
- ■ Write the standard forms of the equations of spheres and find the centers and radii of spheres.
- ■ Sketch the coordinate plane traces of surfaces.

The Three-Dimensional Coordinate System

Recall from Section 1.1 that the Cartesian plane is determined by two perpendicular real number lines called the x-axis and the y-axis. These axes together with their point of intersection (the origin) allow you to develop a two-dimensional coordinate system for identifying points in a plane. To identify a point in space, you must introduce a third dimension to the model. The geometry of this three-dimensional model is called **solid analytic geometry.**

You can construct a **three-dimensional coordinate system** by passing a z-axis perpendicular to both the x- and y-axes at the origin. Figure 7.1 shows the positive portion of each coordinate axis. Taken as pairs, the axes determine three **coordinate planes:** the **xy-plane,** the **xz-plane,** and the **yz-plane.** These three coordinate planes separate the three-dimensional coordinate system into eight **octants.** The first octant is the one for which all three coordinates are positive. In this three-dimensional system, a point P in space is determined by an ordered triple (x, y, z), where x, y, and z are as follows.

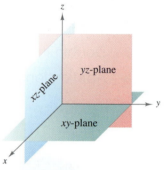

FIGURE 7.1

$x =$ directed distance from yz-plane to P

$y =$ directed distance from xz-plane to P

$z =$ directed distance from xy-plane to P

Example 1   Plotting Points in Space

Plot the points in the same three-dimensional coordinate system.

a. $(2, -3, 3)$ **b.** $(-2, 6, 2)$

c. $(1, 4, 0)$ **d.** $(2, 2, -3)$

SOLUTION To plot the point $(2, -3, 3)$, notice that

$$x = 2, \quad y = -3, \quad \text{and} \quad z = 3.$$

To help visualize the point, locate the point $(2, -3)$ in the xy-plane (denoted by a cross in Figure 7.2). The point lies three units above the cross. You can plot the other points in a similar manner, as shown in Figure 7.2.

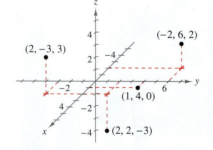

FIGURE 7.2

✓ Checkpoint 1

Plot the points in the same three-dimensional coordinate system.

a. $(2, 5, 1)$ **b.** $(-2, -4, 3)$ **c.** $(4, 0, -5)$

The Distance and Midpoint Formulas

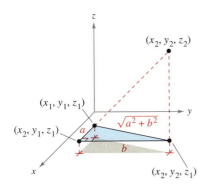

Many of the formulas established for the two-dimensional coordinate system can be extended to three dimensions. For example, to find the distance between two points in space, you can use the Pythagorean Theorem twice, as shown in Figure 7.3. By doing this, you will obtain the formula for the distance between two points in space.

Distance Formula in Space

The distance between the points (x_1, y_1, z_1) and (x_2, y_2, z_2) is

$$d = \sqrt{(x_2 - x_1)^2 + (y_2 - y_1)^2 + (z_2 - z_1)^2}.$$

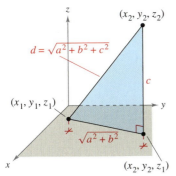

FIGURE 7.3

Example 2 Finding the Distance Between Two Points

Find the distance between $(1, 0, 2)$ and $(2, 4, -3)$.

SOLUTION

$$
\begin{aligned}
d &= \sqrt{(x_2 - x_1)^2 + (y_2 - y_1)^2 + (z_2 - z_1)^2} &&\text{Write Distance Formula.}\\
&= \sqrt{(2 - 1)^2 + (4 - 0)^2 + (-3 - 2)^2} &&\text{Substitute.}\\
&= \sqrt{1 + 16 + 25} &&\text{Simplify.}\\
&= \sqrt{42} &&\text{Simplify.}
\end{aligned}
$$

✓**Checkpoint 2**

Find the distance between $(2, 3, -1)$ and $(0, 5, 3)$. ■

Notice the similarity between the Distance Formula in the plane and the Distance Formula in space. The Midpoint Formulas in the plane and in space are also similar.

Midpoint Formula in Space

The midpoint of the line segment joining the points (x_1, y_1, z_1) and (x_2, y_2, z_2) is

$$\text{Midpoint} = \left(\frac{x_1 + x_2}{2}, \frac{y_1 + y_2}{2}, \frac{z_1 + z_2}{2}\right).$$

Example 3 Using the Midpoint Formula

Find the midpoint of the line segment joining

$$(5, -2, 3) \quad \text{and} \quad (0, 4, 4).$$

SOLUTION Using the Midpoint Formula, the midpoint is

$$\left(\frac{5 + 0}{2}, \frac{-2 + 4}{2}, \frac{3 + 4}{2}\right) = \left(\frac{5}{2}, 1, \frac{7}{2}\right)$$

as shown in Figure 7.4.

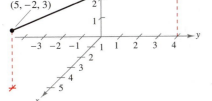

FIGURE 7.4

✓**Checkpoint 3**

Find the midpoint of the line segment joining

$$(3, -2, 0) \quad \text{and} \quad (-8, 6, -4).$$ ■

The Equation of a Sphere

A **sphere** with center at (h, k, l) and radius r is defined to be the set of all points (x, y, z) such that the distance between (x, y, z) and (h, k, l) is r, as shown in Figure 7.5. Using the Distance Formula, this condition can be written as

$$\sqrt{(x - h)^2 + (y - k)^2 + (z - l)^2} = r.$$

By squaring both sides of this equation, you obtain the standard equation of a sphere.

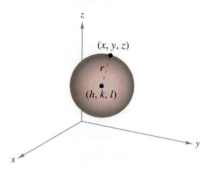

Sphere: Radius r, Center (h, k, l)

FIGURE 7.5

Standard Equation of a Sphere

The **standard equation of a sphere** with center at (h, k, l) and radius r is

$$(x - h)^2 + (y - k)^2 + (z - l)^2 = r^2.$$

Example 4   **Finding the Equation of a Sphere**

Find the standard equation of the sphere with center at $(2, 4, 3)$ and radius 3. Does this sphere intersect the xy-plane?

SOLUTION

$$(x - h)^2 + (y - k)^2 + (z - l)^2 = r^2 \qquad \text{Write standard equation.}$$
$$(x - 2)^2 + (y - 4)^2 + (z - 3)^2 = 3^2 \qquad \text{Substitute.}$$
$$(x - 2)^2 + (y - 4)^2 + (z - 3)^2 = 9 \qquad \text{Simplify.}$$

In Figure 7.6, note that the center of the sphere lies three units above the xy-plane. The sphere has a radius of 3, so it must intersect the xy-plane—at the point $(2, 4, 0)$.

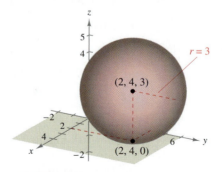

FIGURE 7.6

✓ **Checkpoint 4**

Find the standard equation of the sphere with center at $(4, 3, 2)$ and radius 5.

Example 5   **Finding the Equation of a Sphere**

Find the equation of the sphere that has the points $(3, -2, 6)$ and $(-1, 4, 2)$ as endpoints of a diameter.

SOLUTION   By the Midpoint Formula, the center of the sphere is

$$(h, k, l) = \left(\frac{3 + (-1)}{2}, \frac{-2 + 4}{2}, \frac{6 + 2}{2} \right)$$ Apply Midpoint Formula.

$$= (1, 1, 4).$$ Simplify.

By the Distance Formula, the radius is

$$r = \sqrt{(3 - 1)^2 + (-2 - 1)^2 + (6 - 4)^2}$$ Apply Distance Formula.

$$= \sqrt{4 + 9 + 4}$$ Simplify.

$$= \sqrt{17}.$$ Simplify.

So, the standard equation of the sphere is

$$(x - h)^2 + (y - k)^2 + (z - l)^2 = r^2$$ Write formula for a sphere.

$$(x - 1)^2 + (y - 1)^2 + (z - 4)^2 = 17.$$ Substitute.

✓**Checkpoint 5**

Find the equation of the sphere that has the points $(-2, 5, 7)$ and $(4, 1, -3)$ as endpoints of a diameter. ▪

Example 6   **Finding the Center and Radius of a Sphere**

Find the center and radius of the sphere whose equation is

$$x^2 + y^2 + z^2 - 2x + 4y - 6z + 8 = 0.$$

SOLUTION   You can obtain the standard equation of the sphere by completing the square. To do this, begin by grouping terms with the same variable. Then add "the square of half the coefficient of each linear term" to each side of the equation. So, to complete the square of $(x^2 - 2x)$, add $\left[\frac{1}{2}(-2)\right]^2 = 1$ to each side. To complete the square of $(y^2 + 4y)$, add $\left[\frac{1}{2}(4)\right]^2 = 4$ to each side. To complete the square of $(z^2 - 6z)$, add $\left[\frac{1}{2}(-6)\right]^2 = 9$ to each side.

$$x^2 + y^2 + z^2 - 2x + 4y - 6z + 8 = 0$$

$$(x^2 - 2x + \quad) + (y^2 + 4y + \quad) + (z^2 - 6z + \quad) = -8$$

$$(x^2 - 2x + 1) + (y^2 + 4y + 4) + (z^2 - 6z + 9) = -8 + 1 + 4 + 9$$

$$(x - 1)^2 + (y + 2)^2 + (z - 3)^2 = 6$$

So, the center of the sphere is $(1, -2, 3)$, and its radius is $\sqrt{6}$, as shown in Figure 7.7.

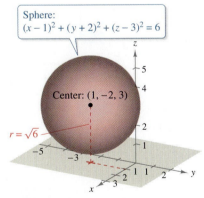

Sphere:
$(x - 1)^2 + (y + 2)^2 + (z - 3)^2 = 6$

Center: $(1, -2, 3)$

$r = \sqrt{6}$

FIGURE 7.7

✓**Checkpoint 6**

Find the center and radius of the sphere whose equation is

$$x^2 + y^2 + z^2 + 6x - 8y + 2z - 10 = 0.$$ ▪

Note in Example 6 that the points satisfying the equation of the sphere are "surface points," not "interior points." In general, the collection of points satisfying an equation involving x, y, and z is called a **surface in space.**

Traces of Surfaces

Finding the intersection of a surface with one of the three coordinate planes (or with a plane parallel to one of the three coordinate planes) helps visualize the surface. Such an intersection is called a **trace** of the surface. For example, the *xy*-trace of a surface consists of all points that are common to both the surface *and* the *xy*-plane. Similarly, the *xz*-trace of a surface consists of all points that are common to both the surface and the *xz*-plane.

Example 7 Finding a Trace of a Surface

Sketch the *xy*-trace of the sphere given by $(x - 3)^2 + (y - 2)^2 + (z + 4)^2 = 5^2$.

SOLUTION To find the *xy*-trace of this surface, use the fact that every point in the *xy*-plane has a *z*-coordinate of zero. By substituting $z = 0$ into the original equation, the resulting equation will represent the intersection of the surface with the *xy*-plane.

$$(x - 3)^2 + (y - 2)^2 + (z + 4)^2 = 5^2 \qquad \text{Write original equation.}$$
$$(x - 3)^2 + (y - 2)^2 + (0 + 4)^2 = 25 \qquad \text{Let } z = 0 \text{ to find } xy\text{-trace.}$$
$$(x - 3)^2 + (y - 2)^2 + 16 = 25 \qquad \text{Simplify.}$$
$$(x - 3)^2 + (y - 2)^2 = 9 \qquad \text{Subtract 16 from each side.}$$
$$(x - 3)^2 + (y - 2)^2 = 3^2 \qquad \text{Equation of circle}$$

From this equation, you can see that the *xy*-trace is a circle of radius 3, as shown in Figure 7.8.

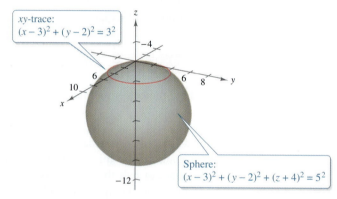

xy-trace:
$(x - 3)^2 + (y - 2)^2 = 3^2$

Sphere:
$(x - 3)^2 + (y - 2)^2 + (z + 4)^2 = 5^2$

FIGURE 7.8

✓ Checkpoint 7

Find the equation of the *xy*-trace of the sphere given by

$$(x + 1)^2 + (y - 2)^2 + (z + 3)^2 = 5^2.$$

SUMMARIZE (Section 7.1)

1. State the Distance Formula in space *(page 361)*. For an example of the Distance Formula in space, see Example 2.

2. State the Midpoint Formula in space *(page 361)*. For an example of the Midpoint Formula in space, see Example 3.

3. State the standard equation of a sphere *(page 362)*. For examples of finding equations of spheres, see Examples 4 and 5.

4. Explain what is meant by the trace of a surface *(page 364)*. For an example of finding the trace of a surface, see Example 7.

SKILLS WARM UP 7.1

The following warm-up exercises involve skills that were covered in earlier sections. You will use these skills in the exercise set for this section. For additional help, review Sections 1.1 and 1.2.

In Exercises 1–4, find the distance between the points.

1. $(5, 1), (3, 5)$

2. $(2, 3), (-1, -1)$

3. $(-5, 4), (-5, -4)$

4. $(-3, 6), (-3, -2)$

In Exercises 5–8, find the midpoint of the line segment connecting the points.

5. $(2, 5), (6, 9)$

6. $(-1, -2), (3, 2)$

7. $(-6, 0), (6, 6)$

8. $(-4, 3), (2, -1)$

In Exercises 9 and 10, write the standard form of the equation of the circle.

9. Center: $(2, 3)$; radius: 2

10. Endpoints of a diameter: $(4, 0), (-2, 8)$

ENHANCED
WebAssign Access end-of-section exercises online at **www.webassign.net**

7.2 Surfaces in Space

■ Sketch planes in space.
■ Draw planes in space with different numbers of intercepts.
■ Classify quadric surfaces in space.

Equations of Planes in Space

In Section 7.1, you studied one type of surface in space—a sphere. In this section, you will study a second type—a plane in space. The **general equation of a plane** in space is

$$ax + by + cz = d.$$ General equation of a plane

Note the similarity of this equation to the general equation of a line in the plane. In fact, when you intersect the plane represented by this equation with each of the three coordinate planes, you will obtain traces that are lines, as shown in Figure 7.9.

In Figure 7.9, the points where the plane intersects the three coordinate axes are the x-, y-, and z-intercepts of the plane. By connecting these three points, you can form a triangular region, which helps you visualize the plane in space.

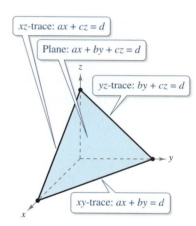

xz-trace: $ax + cz = d$

Plane: $ax + by + cz = d$

yz-trace: $by + cz = d$

xy-trace: $ax + by = d$

FIGURE 7.9

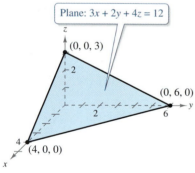

In Exercise 49, you will use the dimensions of Earth to write an equation of an ellipsoid that models its shape.

Plane: $3x + 2y + 4z = 12$

(0, 0, 3)

(0, 6, 0)

(4, 0, 0)

Sketch Made by Connecting Intercepts: (4, 0, 0), (0, 6, 0), (0, 0, 3)

FIGURE 7.10

Example 1   Sketching a Plane in Space

Find the x-, y-, and z-intercepts of the plane given by

$$3x + 2y + 4z = 12.$$

Then sketch the plane.

SOLUTION   To find the x-intercept, let both y and z be zero.

$$3x + 2(0) + 4(0) = 12$$ Substitute 0 for y and z.
$$3x = 12$$ Simplify.
$$x = 4$$ Solve for x.

So, the x-intercept is $(4, 0, 0)$. To find the y-intercept, let x and z be zero and conclude that $y = 6$. So, the y-intercept is $(0, 6, 0)$. Similarly, by letting x and y be zero, you can determine that $z = 3$ and that the z-intercept is $(0, 0, 3)$. Figure 7.10 shows the triangular portion of the plane formed by connecting the three intercepts

$$(4, 0, 0), \quad (0, 6, 0), \quad \text{and} \quad (0, 0, 3).$$

✓Checkpoint 1

Find the x-, y-, and z-intercepts of the plane given by

$$2x + 4y + z = 8.$$

Then sketch the plane.

Drawing Planes in Space

The planes shown in Figures 7.9 and 7.10 have three intercepts. When this occurs, you can draw the plane by sketching the triangular region formed by connecting the three intercepts.

It is possible for a plane in space to have fewer than three intercepts. This occurs when one or more of the coefficients in the equation $ax + by + cz = d$ is zero. Figure 7.11 shows some planes in space that have only one intercept, and Figure 7.12 shows some that have only two intercepts. In each figure, note the use of dashed lines and shading to give the illusion of three dimensions.

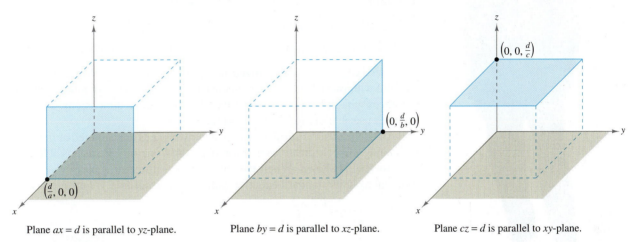

Plane $ax = d$ is parallel to yz-plane.   Plane $by = d$ is parallel to xz-plane.   Plane $cz = d$ is parallel to xy-plane.

Planes Parallel to Coordinate Planes
FIGURE 7.11

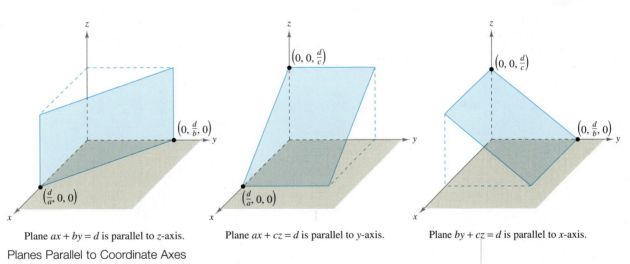

Plane $ax + by = d$ is parallel to z-axis.   Plane $ax + cz = d$ is parallel to y-axis.   Plane $by + cz = d$ is parallel to x-axis.

Planes Parallel to Coordinate Axes
FIGURE 7.12

When an equation of a plane has a missing variable, such as

$$2x + z = 1 \qquad \text{See Figure 7.13.}$$

the plane must be *parallel to the axis* represented by the missing variable, as shown in Figure 7.12. When two variables are missing from an equation of a plane, the plane is *parallel to the coordinate plane* represented by the missing variables, as shown in Figure 7.11.

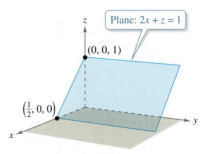

Plane $2x + z = 1$ is parallel to the y-axis.
FIGURE 7.13

Quadric Surfaces

A third common type of surface in space is a **quadric surface.** Quadric surfaces are the three-dimensional analogs of conic sections. The equation of a quadric surface in space is a second-degree equation in three variables, such as

$$Ax^2 + By^2 + Cz^2 + Dx + Ey + Fz + G = 0.$$ Second-degree equation

There are six basic types of quadric surfaces.

1. Elliptic cone

2. Elliptic paraboloid

3. Hyperbolic paraboloid

4. Ellipsoid

5. Hyperboloid of one sheet

6. Hyperboloid of two sheets

The six types are summarized on the next two pages. Notice that each surface is pictured with two types of three-dimensional sketches. The computer-generated sketches use traces with hidden lines to give the illusion of three dimensions. The artist-rendered sketches use shading to create the same illusion.

All of the quadric surfaces on the next two pages are centered at the origin and have axes along the coordinate axes. Moreover, only one of several possible orientations of each surface is shown. When the surface has a different center or is oriented along a different axis, its standard equation will change accordingly. For instance, the ellipsoid

$$\frac{x^2}{1^2} + \frac{y^2}{3^2} + \frac{z^2}{2^2} = 1$$

has $(0, 0, 0)$ as its center, but the ellipsoid

$$\frac{(x - 2)^2}{1^2} + \frac{(y + 1)^2}{3^2} + \frac{(z - 4)^2}{2^2} = 1$$

has $(2, -1, 4)$ as its center. A computer-generated graph of the first ellipsoid is shown in Figure 7.14.

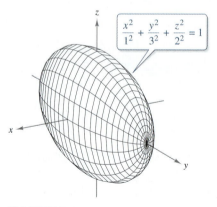

FIGURE 7.14

TECH TUTOR

If you have access to a three-dimensional graphing utility, try using it to graph the surface in Figure 7.14. When you do this, you will discover that sketching surfaces in space is not a simple task—even with a graphing utility.

Elliptic Cone

$$\frac{x^2}{a^2} + \frac{y^2}{b^2} - \frac{z^2}{c^2} = 0$$

Trace	Plane
Ellipse	Parallel to xy-plane
Hyperbola	Parallel to xz-plane
Hyperbola	Parallel to yz-plane

The axis of the cone corresponds to the variable whose coefficient is negative. The traces in the coordinate planes parallel to this axis are intersecting lines.

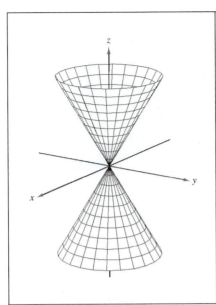

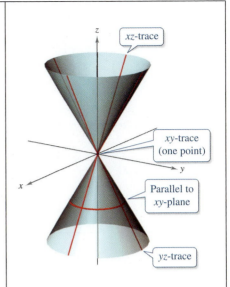

Elliptic Paraboloid

$$z = \frac{x^2}{a^2} + \frac{y^2}{b^2}$$

Trace	Plane
Ellipse	Parallel to xy-plane
Parabola	Parallel to xz-plane
Parabola	Parallel to yz-plane

The axis of the paraboloid corresponds to the variable raised to the first power.

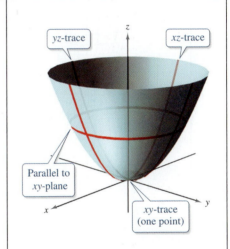

Hyperbolic Paraboloid

$$z = \frac{y^2}{b^2} - \frac{x^2}{a^2}$$

Trace	Plane
Hyperbola	Parallel to xy-plane
Parabola	Parallel to xz-plane
Parabola	Parallel to yz-plane

The axis of the paraboloid corresponds to the variable raised to the first power.

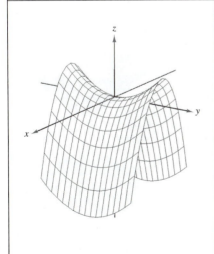

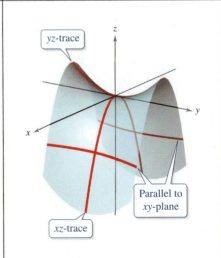

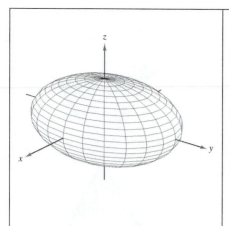

Ellipsoid

$$\frac{x^2}{a^2} + \frac{y^2}{b^2} + \frac{z^2}{c^2} = 1$$

Trace	Plane
Ellipse	Parallel to xy-plane
Ellipse	Parallel to xz-plane
Ellipse	Parallel to yz-plane

The surface is a sphere when the coefficients a, b, and c are equal and nonzero.

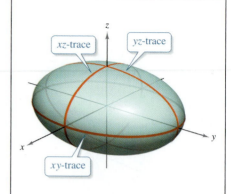

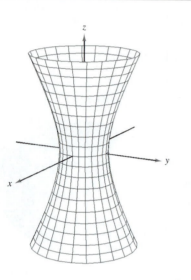

Hyperboloid of One Sheet

$$\frac{x^2}{a^2} + \frac{y^2}{b^2} - \frac{z^2}{c^2} = 1$$

Trace	Plane
Ellipse	Parallel to xy-plane
Hyperbola	Parallel to xz-plane
Hyperbola	Parallel to yz-plane

The axis of the hyperboloid corresponds to the variable whose coefficient is negative.

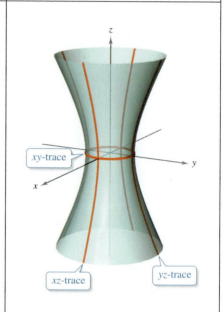

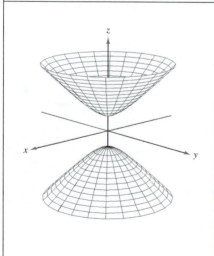

Hyperboloid of Two Sheets

$$\frac{z^2}{c^2} - \frac{x^2}{a^2} - \frac{y^2}{b^2} = 1$$

Trace	Plane
Ellipse	Parallel to xy-plane
Hyperbola	Parallel to xz-plane
Hyperbola	Parallel to yz-plane

The axis of the hyperboloid corresponds to the variable whose coefficient is positive. There is no trace in the coordinate plane perpendicular to this axis.

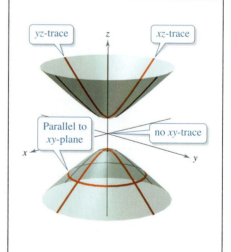

When classifying quadric surfaces, note that the two types of paraboloids have one variable raised to the first power. The other four types of quadric surfaces have equations that are of second degree in *all* three variables.

Example 2   Classifying a Quadric Surface

Describe the traces of the surface given by $x - y^2 - z^2 = 0$ in the xy-plane, the xz-plane, and the plane given by $x = 1$. Then classify the surface.

SOLUTION   Because x is raised only to the first power, the surface is a paraboloid whose axis is the x-axis. In standard form, the equation is $x = y^2 + z^2$. The traces in the xy-plane, the xz-plane, and the plane given by $x = 1$ are as shown.

Trace in xy-plane $(z = 0)$:	$x = y^2$	Parabola
Trace in xz-plane $(y = 0)$:	$x = z^2$	Parabola
Trace in plane $x = 1$:	$y^2 + z^2 = 1$	Circle

These three traces are shown in Figure 7.15.

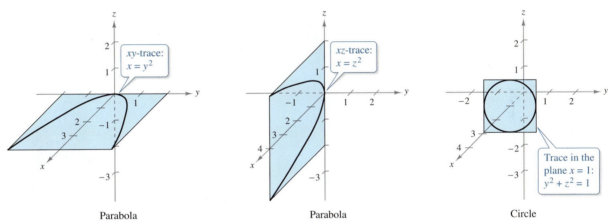

FIGURE 7.15

From the traces, you can see that the surface is an elliptic (or circular) paraboloid, as shown in Figure 7.16.

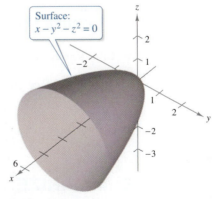

Elliptic Paraboloid
FIGURE 7.16

✓ Checkpoint 2

Describe the traces of the surface given by $x^2 + y^2 - z^2 = 1$ in the xy-plane, the yz-plane, the xz-plane, and the plane given by $z = 3$. Then classify the surface. ■

Example 3 **Classifying Quadric Surfaces**

Classify the surface given by each equation.

a. $x^2 - 4y^2 - 4z^2 - 4 = 0$

b. $x^2 + 4y^2 + z^2 - 4 = 0$

SOLUTION

a. The equation $x^2 - 4y^2 - 4z^2 - 4 = 0$ can be written in standard form as

$$\frac{x^2}{4} - y^2 - z^2 = 1. \qquad \text{Standard form}$$

From the standard form, you can see that the graph is a hyperboloid of two sheets, with the x-axis as its axis, as shown in Figure 7.17(a).

b. The equation $x^2 + 4y^2 + z^2 - 4 = 0$ can be written in standard form as

$$\frac{x^2}{4} + y^2 + \frac{z^2}{4} = 1. \qquad \text{Standard form}$$

From the standard form, you can see that the graph is an ellipsoid, as shown in Figure 7.17(b).

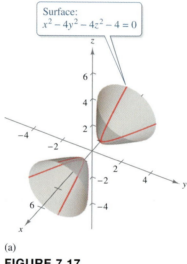

(a)   (b)

FIGURE 7.17

✓ **Checkpoint 3**

Classify the surface given by each equation.

a. $4x^2 + 9y^2 - 36z = 0$

b. $36x^2 + 16y^2 - 144z^2 = 0$

SUMMARIZE (Section 7.2)

1. State the general equation of a plane in space *(page 366)*. For an example of sketching a plane in space, see Example 1.

2. List the six basic types of quadric surfaces *(page 368)*. For examples of classifying quadric surfaces, see Examples 2 and 3.

SKILLS WARM UP 7.2   The following warm-up exercises involve skills that were covered in earlier sections. You will use these skills in the exercise set for this section. For additional help, review Sections 1.2 and 7.1.

In Exercises 1–4, find the x- and y-intercepts of the function.

1. $3x + 4y = 12$

2. $6x + y = -8$

3. $-2x + y = -2$

4. $-x - y = 5$

In Exercises 5 and 6, write the equation of the sphere in standard form.

5. $16x^2 + 16y^2 + 16z^2 = 4$

6. $9x^2 + 9y^2 + 9z^2 = 36$

ENHANCED
WebAssign   Access end-of-section exercises online at **www.webassign.net**

7.3 Functions of Several Variables

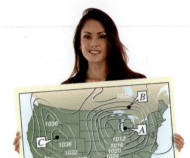

- Evaluate functions of several variables.
- Find the domains and ranges of functions of two variables.
- Read contour maps and sketch level curves of functions of two variables.
- Use functions of several variables to answer questions about real-life situations.

Functions of Several Variables

So far in this text, you have studied functions of a single independent variable. Many quantities in science, business, and technology, however, are functions not of one, but of two or more variables. For instance, the demand function for a product is often dependent on the price *and* the advertising, rather than on the price alone. The notation for a function of two or more variables is similar to that for a function of a single variable. Here are two examples.

$$z = f\underbrace{(x, y)}_{\text{2 variables}} = x^2 + xy \qquad \text{Function of two variables}$$

and

$$w = f\underbrace{(x, y, z)}_{\text{3 variables}} = x + 2y - 3z \qquad \text{Function of three variables}$$

In Exercise 49, you will read a weather map and identify areas of high and low pressure.

Definition of a Function of Two Variables

Let D be a set of ordered pairs of real numbers. If to each ordered pair (x, y) in D there corresponds a unique real number $f(x, y)$, then f is called a **function of x and y.** The set D is the **domain** of f, and the corresponding set of values for $f(x, y)$ is the **range** of f. Functions of three, four, or more variables are defined similarly.

For the function given by

$$z = f(x, y)$$

x and y are called the **independent variables** and z is called the **dependent variable.**

Example 1   Evaluating Functions of Several Variables

a. For $f(x, y) = 2x^2 - y^2$, you can evaluate $f(2, 3)$ as shown.

$$f(2, 3) = 2(2)^2 - (3)^2 = 8 - 9 = -1$$

b. For $f(x, y, z) = e^x(y + z)$, you can evaluate $f(0, -1, 4)$ as shown.

$$f(0, -1, 4) = e^0(-1 + 4) = (1)(3) = 3$$

✓ Checkpoint 1

Find the indicated function values.

a. For $f(x, y) = x^2 + 2xy$, find $f(2, -1)$.

b. For $f(x, y, z) = \dfrac{2x^2 z}{y^3}$, find $f(-3, 2, 1)$.

The Domain and Range of a Function of Two Variables

A function of two variables can be represented graphically as a surface in space by letting

$$z = f(x, y).$$ Function of two variables

When sketching the graph of a function of x and y, remember that even though the graph is three-dimensional, the domain of the function is two-dimensional—it consists of the points in the xy-plane for which the function is defined. As with functions of a single variable, unless specifically restricted, the domain of a function of two variables is assumed to be the set of all points (x, y) for which the defining equation has meaning. In other words, to each point (x, y) in the domain of f there corresponds a point (x, y, z) on the surface, and conversely, to each point (x, y, z) on the surface there corresponds a point (x, y) in the domain of f.

Example 2   Finding the Domain and Range of a Function

Find the domain and range of the function

$$f(x, y) = \sqrt{64 - x^2 - y^2}.$$

SOLUTION   Because no restrictions are given, the domain is assumed to be the set of all points for which the defining equation makes sense.

$64 - x^2 - y^2 \geq 0$ Quantity inside radical must be nonnegative.

$-x^2 - y^2 \geq -64$ Subtract 64 from each side.

$x^2 + y^2 \leq 64$ Multiply each side by -1 and reverse the inequality symbol.

So, the domain of f is the set of all points that lie on or inside the circle given by

$x^2 + y^2 \leq 8^2$ Domain of the function

as shown in Figure 7.18. The range of f is the set

$0 \leq z \leq 8.$ Range of the function

As shown in Figure 7.19, the graph of the function is a hemisphere.

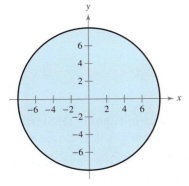

Domain of $f(x, y) = \sqrt{64 - x^2 - y^2}$

FIGURE 7.18

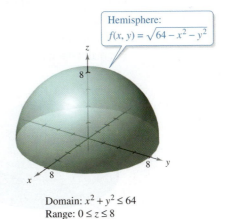

Hemisphere:
$f(x, y) = \sqrt{64 - x^2 - y^2}$

Domain: $x^2 + y^2 \leq 64$
Range: $0 \leq z \leq 8$

FIGURE 7.19

✓ Checkpoint 2

Consider the function

$$f(x, y) = \sqrt{9 - x^2 - y^2}.$$

a. Find the domain of f.

b. Find the range of f.

Contour Maps and Level Curves

A **contour map** of a surface is created by *projecting* traces, taken in evenly spaced planes that are parallel to the xy-plane, onto the xy-plane. Each projection is a **level curve** of the surface.

Contour maps are used to create weather, topographical, and population density maps. For instance, Figure 7.20(a) shows a graph of a "mountain and valley" surface given by $z = f(x, y)$. Each of the level curves in Figure 7.20(b) represents the intersection of the surface $z = f(x, y)$ with a plane $z = c$, where $c = 828, 830, \ldots, 854$.

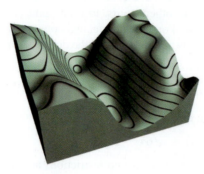

(a) Surface

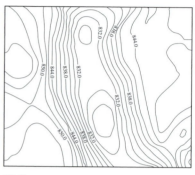

(b) Contour map

FIGURE 7.20

Example 3   **Sketching a Contour Map**

The hemisphere given by $f(x, y) = \sqrt{64 - x^2 - y^2}$ is shown in Figure 7.21. Sketch a contour map of this surface using level curves corresponding to $c = 0, 1, 2, \ldots, 8$.

SOLUTION   For each value of c, the equation given by $f(x, y) = c$ is a circle (or point) in the xy-plane. For instance, when $c_1 = 0$, the level curve is

$$x^2 + y^2 = 8^2 \qquad \text{Circle of radius 8}$$

which is a circle of radius 8. Figure 7.22 shows the nine level curves for the hemisphere.

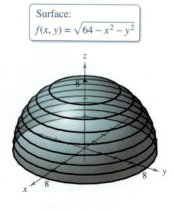

Hemisphere

FIGURE 7.21

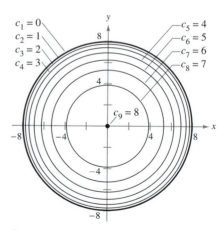

Contour map

FIGURE 7.22

✓ **Checkpoint 3**

Describe the level curves of $f(x, y) = \sqrt{9 - x^2 - y^2}$. Sketch the level curves for $c = 0, 1, 2,$ and 3.

Applications

The **Cobb-Douglas production function** is used in economics to represent the numbers of units produced by varying amounts of labor and capital. Let x represent the number of units of labor and let y represent the number of units of capital. Then, the number of units produced is modeled by

$$f(x, y) = Cx^a y^{1-a}$$

where C and a are constants, with $0 < a < 1$.

 Example 4 **Using a Production Function**

A manufacturer estimates that its production (measured in units of a product) can be modeled by $f(x, y) = 100x^{0.6}y^{0.4}$, where the labor x is measured in person-hours and the capital y is measured in thousands of dollars.

a. What is the production level when $x = 1000$ and $y = 500$?

b. What is the production level when $x = 2000$ and $y = 1000$?

c. How does doubling the amounts of labor and capital from part (a) to part (b) affect the production?

SOLUTION

a. When $x = 1000$ and $y = 500$, the production level is

$$f(1000, 500) = 100(1000)^{0.6}(500)^{0.4} \approx 75{,}786 \text{ units.}$$

b. When $x = 2000$ and $y = 1000$, the production level is

$$f(2000, 1000) = 100(2000)^{0.6}(1000)^{0.4} \approx 151{,}572 \text{ units.}$$

c. When the amounts of labor and capital are doubled, the production level also doubles. In Exercise 44, you are asked to show that this is characteristic of the Cobb-Douglas production function.

A contour map of this function is shown in Figure 7.23. Note that the level curves occur at increments of 10,000.

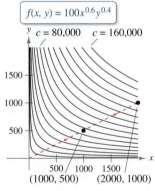

Level Curves (at Increments of 10,000)
FIGURE 7.23

 **Checkpoint 4**

Use the Cobb-Douglas production function in Example 4 to find the production levels when $x = 1500$ and $y = 1000$ and when $x = 1000$ and $y = 1500$. Use your results to determine which variable has a greater influence on production. ■

Example 5   **Finding Monthly Payments**

The monthly payment M for an installment loan of P dollars taken out over t years at an annual interest rate of r is given by

$$M = f(P, r, t) = \frac{\dfrac{Pr}{12}}{1 - \left[\dfrac{1}{1 + (r/12)}\right]^{12t}}.$$

a. Find the monthly payment for a home mortgage of $100,000 taken out for 30 years at an annual interest rate of 7%.

b. Find the monthly payment for a car loan of $22,000 taken out for 5 years at an annual interest rate of 8%.

SOLUTION

a. When $P = \$100{,}000$, $r = 0.07$, and $t = 30$, the monthly payment is

$$M = f(100{,}000, 0.07, 30)$$

$$= \frac{\dfrac{(100{,}000)(0.07)}{12}}{1 - \left[\dfrac{1}{1 + (0.07/12)}\right]^{12(30)}}$$

$$\approx \$665.30.$$

b. When $P = \$22{,}000$, $r = 0.08$, and $t = 5$, the monthly payment is

$$M = f(22{,}000, 0.08, 5)$$

$$= \frac{\dfrac{(22{,}000)(0.08)}{12}}{1 - \left[\dfrac{1}{1 + (0.08/12)}\right]^{12(5)}}$$

$$\approx \$446.08.$$

✓**Checkpoint 5**

a. Find the monthly payment M for a home mortgage of $100,000 taken out for 30 years at an annual interest rate of 3%.

b. Find the total amount of money you will pay for the mortgage.

SUMMARIZE   (Section 7.3)

1. State the definition of a function of two variables *(page 374)*. For an example of evaluating a function of two variables, see Example 1.

2. Describe how a contour map of a surface is created *(page 376)*. For an example of sketching a contour map, see Example 3.

3. State the Cobb-Douglas production function *(page 377)*. For an example of using the Cobb-Douglas production function, see Example 4.

4. Describe a real-life example of how a function of several variables can be used to find the monthly payment for a loan *(page 378, Example 5)*.

SKILLS WARM UP 7.3

The following warm-up exercises involve skills that were covered in a previous course or in earlier sections. You will use these skills in the exercise set for this section. For additional help, review Appendix Section A.3 and Section 1.4.

In Exercises 1–4, evaluate the function when $x = -3$.

1. $f(x) = 5 - 2x$

2. $f(x) = -x^2 + 4x + 5$

3. $y = \sqrt{4x^2 - 3x + 4}$

4. $y = \sqrt[3]{34 - 4x + 2x^2}$

In Exercises 5–8, find the domain of the function.

5. $f(x) = 5x^2 + 3x - 2$

6. $g(x) = \dfrac{1}{2x} - \dfrac{2}{x + 3}$

7. $h(y) = \sqrt{y - 5}$

8. $f(y) = \sqrt{y^2 - 5}$

In Exercises 9 and 10, evaluate the expression.

9. $(476)^{0.65}$

10. $(251)^{0.35}$

 Access end-of-section exercises online at **www.webassign.net**

7.4 Partial Derivatives

- Find the first partial derivatives of functions of two variables.
- Find the slopes of surfaces in the *x*- and *y*-directions and use partial derivatives to answer questions about real-life situations.
- Find the partial derivatives of functions of several variables.
- Find higher-order partial derivatives.

Functions of Two Variables

Real-life applications of functions of several variables are often concerned with how changes in one of the variables will affect the values of the functions. For instance, an economist who wants to determine the effect of a tax increase on the economy might make calculations using different tax rates while holding all other variables, such as unemployment, constant.

You can follow a similar procedure to find the rate of change of a function *f* with respect to one of its independent variables. That is, you find the derivative of *f* with respect to one independent variable while holding the other variable(s) constant. This process is called **partial differentiation,** and each derivative is called a **partial derivative.** A function of several variables has as many partial derivatives as it has independent variables.

In Exercise 60, you will use partial derivatives to find the marginal revenues of a pharmaceutical corporation at two locations that produce the same medicine.

Partial Derivatives of a Function of Two Variables

If $z = f(x, y)$, then the **first partial derivatives of f with respect to x and y** are the functions $\partial z/\partial x$ and $\partial z/\partial y$, defined as shown.

$$\frac{\partial z}{\partial x} = \lim_{\Delta x \to 0} \frac{f(x + \Delta x, y) - f(x, y)}{\Delta x} \qquad \text{\textcolor{red}{y is held constant.}}$$

$$\frac{\partial z}{\partial y} = \lim_{\Delta y \to 0} \frac{f(x, y + \Delta y) - f(x, y)}{\Delta y} \qquad \text{\textcolor{red}{x is held constant.}}$$

STUDY TIP

The notation $\partial z/\partial x$ is read as "the partial derivative of z with respect to x," and $\partial z/\partial y$ is read as "the partial derivative of z with respect to y."

This definition indicates that if $z = f(x, y)$, then to find $\partial z/\partial x$, you *consider y to be constant* and differentiate with respect to *x*. Similarly, to find $\partial z/\partial y$, you *consider x to be constant* and differentiate with respect to *y*.

Example 1   Finding Partial Derivatives

Find $\partial z/\partial x$ and $\partial z/\partial y$ for the function $z = 3x - x^2y^2 + 2x^3y$.

SOLUTION

$$\frac{\partial z}{\partial x} = 3 - 2xy^2 + 6x^2y \qquad \text{\textcolor{red}{Hold y constant and differentiate with respect to x.}}$$

$$\frac{\partial z}{\partial y} = -2x^2y + 2x^3 \qquad \text{\textcolor{red}{Hold x constant and differentiate with respect to y.}}$$

✓**Checkpoint 1**

Find $\dfrac{\partial z}{\partial x}$ and $\dfrac{\partial z}{\partial y}$ for $z = 2x^2 - 4x^2y^3 + y^4$.

Notation for First Partial Derivatives

The first partial derivatives of $z = f(x, y)$ are denoted by

$$\frac{\partial z}{\partial x} = f_x(x, y) = z_x = \frac{\partial}{\partial x}[f(x, y)]$$

and

$$\frac{\partial z}{\partial y} = f_y(x, y) = z_y = \frac{\partial}{\partial y}[f(x, y)].$$

The values of the first partial derivatives at the point (a, b) are denoted by

$$\frac{\partial z}{\partial x}\bigg|_{(a, b)} = f_x(a, b)$$

and

$$\frac{\partial z}{\partial y}\bigg|_{(a, b)} = f_y(a, b).$$

TECH TUTOR

Symbolic differentiation utilities can be used to find partial derivatives of a function of two variables. Try using a symbolic differentiation utility to find the first partial derivatives of the function in Example 2.

Example 2   **Finding and Evaluating Partial Derivatives**

Find the first partial derivatives of

$$f(x, y) = xe^{x^2 y}$$

and evaluate each at the point $(1, \ln 2)$.

SOLUTION   To find the first partial derivative with respect to x, hold y constant and differentiate using the Product Rule.

$$f_x(x, y) = x\frac{\partial}{\partial x}[e^{x^2 y}] + e^{x^2 y}\frac{\partial}{\partial x}[x] \qquad \text{Apply Product Rule.}$$

$$= xe^{x^2 y}(2xy) + e^{x^2 y} \qquad \text{y is held constant.}$$

$$= e^{x^2 y}(2x^2 y + 1) \qquad \text{Simplify.}$$

At the point $(1, \ln 2)$, the value of this derivative is

$$f_x(1, \ln 2) = e^{(1)^2(\ln 2)}[2(1)^2(\ln 2) + 1] \qquad \text{Substitute for x and y.}$$

$$= 2(2 \ln 2 + 1) \qquad \text{Simplify.}$$

$$\approx 4.773. \qquad \text{Use a calculator.}$$

To find the first partial derivative with respect to y, hold x constant and differentiate to obtain

$$f_y(x, y) = xe^{x^2 y}(x^2) \qquad \text{Apply Constant Multiple Rule.}$$

$$= x^3 e^{x^2 y}. \qquad \text{Simplify.}$$

At the point $(1, \ln 2)$, the value of this derivative is

$$f_y(1, \ln 2) = (1)^3 e^{(1)^2(\ln 2)} \qquad \text{Substitute for x and y.}$$

$$= 2. \qquad \text{Simplify.}$$

✓**Checkpoint 2**

Find the first partial derivatives of

$$f(x, y) = x^2 y^3$$

and evaluate each at the point $(1, 2)$.

Graphical Interpretation of Partial Derivatives

Earlier in the text, you studied graphical interpretations of the derivative of a function of a single variable. There, you found that $f'(x_0)$ represents the slope of the tangent line to the graph of $y = f(x)$ at the point (x_0, y_0). The partial derivatives of a function of two variables also have useful graphical interpretations. Consider the function

$$z = f(x, y). \qquad \text{Function of two variables}$$

As shown in Figure 7.24(a), the graph of this function is a surface in space. If $y = y_0$, then

$$z = f(x, y_0) \qquad \text{Function of one variable}$$

is a function of one variable. The graph of this function is the curve that is the intersection of the plane $y = y_0$ and the surface $z = f(x, y)$. On this curve, the partial derivative

$$f_x(x, y_0) \qquad \text{Slope in } x\text{-direction}$$

represents the slope in the plane $y = y_0$, as shown in Figure 7.24(a). Similarly, if $x = x_0$, then

$$z = f(x_0, y) \qquad \text{Function of one variable}$$

is a function of one variable. Its graph is the intersection of the plane $x = x_0$ and the surface $z = f(x, y)$. On this curve, the partial derivative

$$f_y(x_0, y) \qquad \text{Slope in } y\text{-direction}$$

represents the slope in the plane $x = x_0$, as shown in Figure 7.24(b).

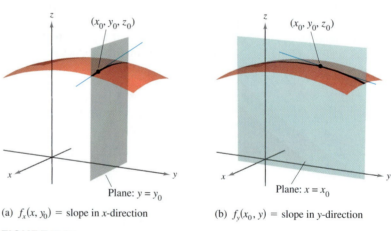

(a) $f_x(x, y_0) = $ slope in x-direction

(b) $f_y(x_0, y) = $ slope in y-direction

FIGURE 7.24

Informally, $f_x(x_0, y_0)$ and $f_y(x_0, y_0)$ at the point (x_0, y_0, z_0) denote the **slopes of the surface in the x- and y-directions,** respectively.

Guidelines for Finding the Slopes of a Surface at a Point

Let (x_0, y_0, z_0) be a point on the surface of

$$z = f(x, y).$$

1. Find the partial derivatives of f with respect to x and y.

2. The slope of the x-direction at (x_0, y_0, z_0) is $f_x(x_0, y_0)$.

3. The slope of the y-direction at (x_0, y_0, z_0) is $f_y(x_0, y_0)$.

Example 3 **Finding Slopes in the x- and y-Directions**

Find the slopes of the surface given by

$$f(x, y) = -\frac{x^2}{2} - y^2 + \frac{25}{8}$$

at the point $\left(\frac{1}{2}, 1, 2\right)$ in

a. the x-direction.

b. the y-direction.

SOLUTION

a. To find the slope in the x-direction, hold y constant and differentiate with respect to x to obtain

$$f_x(x, y) = -x. \qquad \text{Partial derivative with respect to } x$$

At the point $\left(\frac{1}{2}, 1, 2\right)$, the slope in the x-direction is

$$f_x\left(\frac{1}{2}, 1\right) = -\frac{1}{2} \qquad \text{Slope in } x\text{-direction}$$

as shown in Figure 7.25(a).

b. To find the slope in the y-direction, hold x constant and differentiate with respect to y to obtain

$$f_y(x, y) = -2y. \qquad \text{Partial derivative with respect to } y$$

At the point $\left(\frac{1}{2}, 1, 2\right)$, the slope in the y-direction is

$$f_y\left(\frac{1}{2}, 1\right) = -2 \qquad \text{Slope in } y\text{-direction}$$

as shown in Figure 7.25(b).

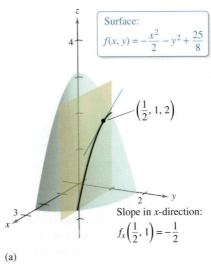

Surface:
$$f(x, y) = -\frac{x^2}{2} - y^2 + \frac{25}{8}$$

$\left(\frac{1}{2}, 1, 2\right)$

Slope in x-direction:
$$f_x\left(\frac{1}{2}, 1\right) = -\frac{1}{2}$$

(a)

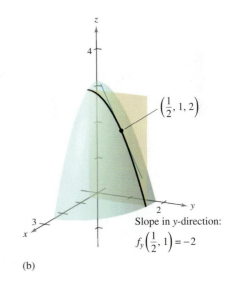

$\left(\frac{1}{2}, 1, 2\right)$

Slope in y-direction:
$$f_y\left(\frac{1}{2}, 1\right) = -2$$

(b)

FIGURE 7.25

✓ **Checkpoint 3**

Find the slopes of the surface given by

$$f(x, y) = 4x^2 + 9y^2 + 36$$

at the point $(1, -1, 49)$ in

a. the x-direction.

b. the y-direction.

In 2010, Subway was chosen as the number one franchise by *Entrepreneur* magazine. In early 2011, Subway had more than 34,000 franchises worldwide. What type of product would be complementary to a Subway sandwich? What type of product would be a substitute?

Consumer products in the same market or in related markets can be classified as **complementary** or **substitute products.** When two products have a complementary relationship, an increase in the sale of one product will be accompanied by an increase in the sale of the other product. For instance, Blu-ray™ players and Blu-ray™ discs have a complementary relationship.

When two products have a substitute relationship, an increase in the sale of one product will be accompanied by a decrease in the sale of the other product. For instance, Blu-ray Disc™ players and DVD players both compete in the same home entertainment market, and you would expect a drop in the price of one to be a deterrent to the sale of the other.

Example 4   Examining Demand Functions

The demand functions for two products are represented by

$$x_1 = f(p_1, p_2) \quad \text{and} \quad x_2 = g(p_1, p_2)$$

where p_1 and p_2 are the prices per unit for the two products, and x_1 and x_2 are the numbers of units sold. The graphs of two different demand functions for x_1 are shown in Figure 7.26. Use them to classify the products as complementary or substitute products.

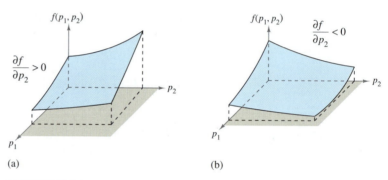

FIGURE 7.26

SOLUTION

a. Notice that Figure 7.26(a) represents the demand for the *first product*. From the graph of this function, you can see that for a fixed price p_1, an increase in p_2 results in an increase in the demand for the first product. Remember that an increase in p_2 will also result in a decrease in the demand for the second product. So, when $\partial f/\partial p_2 > 0$, the two products have a *substitute* relationship.

b. Notice that Figure 7.26(b) represents a different demand for the *first product*. From the graph of this function, you can see that for a fixed price p_1, an increase in p_2 results in a decrease in the demand for the first product. Remember that an increase in p_2 will also result in a decrease in the demand for the second product. So, when $\partial f/\partial p_2 < 0$, the two products have a *complementary* relationship.

✓ Checkpoint 4

Determine if the demand functions below describe a complementary or a substitute product relationship.

$$x_1 = 100 - 2p_1 + 1.5p_2$$

$$x_2 = 145 + \tfrac{1}{2}p_1 - \tfrac{3}{4}p_2$$

Functions of Three Variables

The concept of a partial derivative can be extended naturally to functions of three or more variables. For instance, the function

$$w = f(x, y, z)$$ Function of three variables

has three partial derivatives, each of which is formed by considering two of the variables to be constant. That is, to define the partial derivative of w with respect to x, consider y *and* z to be constant and write

$$\frac{\partial w}{\partial x} = f_x(x, y, z) = \lim_{\Delta x \to 0} \frac{f(x + \Delta x, y, z) - f(x, y, z)}{\Delta x}.$$

To define the partial derivative of w with respect to y, consider x *and* z to be constant and write

$$\frac{\partial w}{\partial y} = f_y(x, y, z) = \lim_{\Delta y \to 0} \frac{f(x, y + \Delta y, z) - f(x, y, z)}{\Delta y}.$$

To define the partial derivative of w with respect to z, consider x *and* y to be constant and write

$$\frac{\partial w}{\partial z} = f_z(x, y, z) = \lim_{\Delta z \to 0} \frac{f(x, y, z + \Delta z) - f(x, y, z)}{\Delta z}.$$

Example 5 Finding Partial Derivatives of a Function

Find the three partial derivatives of the function

$$w = xe^{xy + 2z}.$$

SOLUTION   Holding y and z constant, you obtain

$$\frac{\partial w}{\partial x} = x\frac{\partial}{\partial x}\left[e^{xy + 2z}\right] + e^{xy + 2z}\frac{\partial}{\partial x}[x]$$ Apply Product Rule.

$$= x(e^{xy + 2z})(y) + e^{xy + 2z}(1)$$ Hold y and z constant.

$$= (xy + 1)e^{xy + 2z}.$$ Simplify.

Holding x and z constant, you obtain

$$\frac{\partial w}{\partial y} = x\frac{\partial}{\partial y}\left[e^{xy + 2z}\right]$$ Apply Constant Multiple Rule.

$$= x(e^{xy + 2z})(x)$$ Hold x and z constant.

$$= x^2 e^{xy + 2z}.$$ Simplify.

Holding x and y constant, you obtain

$$\frac{\partial w}{\partial z} = x\frac{\partial}{\partial z}\left[e^{xy + 2z}\right]$$ Apply Constant Multiple Rule.

$$= x(e^{xy + 2z})(2)$$ Hold x and y constant.

$$= 2xe^{xy + 2z}.$$ Simplify.

✓Checkpoint 5

Find the three partial derivatives of the function

$$w = x^2 y \ln(xz).$$ ■

In Example 5, the Product Rule is used only when finding the partial derivative with respect to x. For $\partial w/\partial y$ and $\partial w/\partial z$, x is considered to be constant, so the Constant Multiple Rule is used.

TECH TUTOR

A symbolic differentiation utility can be used to find the partial derivatives of a function of three or more variables. Try using a symbolic differentiation utility to find the partial derivative $f_y(x, y, z)$ for the function in Example 5.

Higher-Order Partial Derivatives

As with ordinary derivatives, it is possible to take second-, third-, and higher-order partial derivatives of a function of several variables, provided such derivatives exist. Higher-order derivatives are denoted by the order in which the differentiation occurs. For instance, there are four different ways to find a second partial derivative of $z = f(x, y)$.

1. $\dfrac{\partial}{\partial x}\left(\dfrac{\partial f}{\partial x}\right) = \dfrac{\partial^2 f}{\partial x^2} = f_{xx}$ Differentiate twice with respect to x.

2. $\dfrac{\partial}{\partial y}\left(\dfrac{\partial f}{\partial y}\right) = \dfrac{\partial^2 f}{\partial y^2} = f_{yy}$ Differentiate twice with respect to y.

3. $\dfrac{\partial}{\partial y}\left(\dfrac{\partial f}{\partial x}\right) = \dfrac{\partial^2 f}{\partial y \partial x} = f_{xy}$ Differentiate first with respect to x and then with respect to y.

4. $\dfrac{\partial}{\partial x}\left(\dfrac{\partial f}{\partial y}\right) = \dfrac{\partial^2 f}{\partial x \partial y} = f_{yx}$ Differentiate first with respect to y and then with respect to x.

The third and fourth cases are **mixed partial derivatives.** Notice that with the two types of notation for mixed partials, different conventions are used for indicating the order of differentiation. For instance, the partial derivative

$$\frac{\partial}{\partial y}\left(\frac{\partial f}{\partial x}\right) = \frac{\partial^2 f}{\partial y \partial x}$$ Right-to-left order

indicates differentiation with respect to x first, but the partial derivative

$$(f_y)_x = f_{yx}$$ Left-to-right order

indicates differentiation with respect to y first. To remember this, note that in each case you differentiate first with respect to the variable "nearest" f.

Example 6   Finding Second Partial Derivatives

Find the second partial derivatives of

$$f(x, y) = 3xy^2 - 2y + 5x^2 y^2$$

and determine the value of $f_{xy}(-1, 2)$.

SOLUTION   Begin by finding the first partial derivatives.

$$f_x(x, y) = 3y^2 + 10xy^2 \qquad\qquad f_y(x, y) = 6xy - 2 + 10x^2 y$$

Then, differentiating with respect to x and y produces

$$f_{xx}(x, y) = 10y^2, \qquad\qquad f_{yy}(x, y) = 6x + 10x^2,$$

$$f_{xy}(x, y) = 6y + 20xy, \qquad\qquad f_{yx}(x, y) = 6y + 20xy.$$

Finally, the value of $f_{xy}(x, y)$ at the point $(-1, 2)$ is

$$f_{xy}(-1, 2) = 6(2) + 20(-1)(2) = 12 - 40 = -28.$$

✓ Checkpoint 6

Find the second partial derivatives of

$$f(x, y) = 4x^2 y^2 + 2x + 4y^2.$$ ■

Notice in Example 6 that the two mixed partials are equal. It can be shown that when a function has continuous second partial derivatives, then the order in which the partial derivatives are taken is irrelevant.

A function of two variables has two first partial derivatives and four second partial derivatives. For a function of three variables, there are three first partials

$$f_x, \quad f_y, \quad \text{and} \quad f_z$$

and nine second partials

$$f_{xx}, \quad f_{xy}, \quad f_{xz}, \quad f_{yx}, \quad f_{yy}, \quad f_{yz}, \quad f_{zx}, \quad f_{zy}, \quad \text{and} \quad f_{zz}$$

of which six are mixed partials. To find partial derivatives of order three and higher, follow the same pattern used to find second partial derivatives. For instance, if $z = f(x, y)$, then

$$z_{xxx} = \frac{\partial}{\partial x}\left(\frac{\partial^2 f}{\partial x^2}\right) = \frac{\partial^3 f}{\partial x^3} \quad \text{and} \quad z_{xxy} = \frac{\partial}{\partial y}\left(\frac{\partial^2 f}{\partial x^2}\right) = \frac{\partial^3 f}{\partial y \partial x^2}.$$

Example 7   Finding Second Partial Derivatives

Find the second partial derivatives of

$$f(x, y, z) = ye^x + x \ln z.$$

SOLUTION   Begin by finding the first partial derivatives.

$$f_x(x, y, z) = ye^x + \ln z, \quad f_y(x, y, z) = e^x, \quad f_z(x, y, z) = \frac{x}{z}$$

Then, differentiate with respect to x, y, and z to find the nine second partial derivatives.

$$f_{xx}(x, y, z) = ye^x, \quad f_{xy}(x, y, z) = e^x, \quad f_{xz}(x, y, z) = \frac{1}{z}$$

$$f_{yx}(x, y, z) = e^x, \quad f_{yy}(x, y, z) = 0, \quad f_{yz}(x, y, z) = 0$$

$$f_{zx}(x, y, z) = \frac{1}{z}, \quad f_{zy}(x, y, z) = 0, \quad f_{zz}(x, y, z) = -\frac{x}{z^2}$$

✓ Checkpoint 7

Find the second partial derivatives of $f(x, y, z) = xe^y + 2xz + y^2$.   ■

SUMMARIZE   (Section 7.4)

1. State the definition of partial derivatives of a function of two variables *(page 380)*. For an example of finding the partial derivatives of a function of two variables, see Example 1.

2. Describe the notation used for first partial derivatives *(page 381)*. For examples of this notation, see Examples 1 and 2.

3. State the guidelines for finding the slopes of a surface at a point *(page 382)*. For an example of finding slopes, see Example 3.

4. Describe a real-life example of how partial derivatives can be used to examine the demand functions of two products *(page 384, Example 4)*.

5. Explain how to find the partial derivatives of a function of three variables *(page 385)*. For an example of finding the partial derivatives of a function of three variables, see Example 5.

6. List the different ways to find the second partial derivatives of a function of two variables *(page 386)*. For an example of finding the second partial derivatives of a function of two variables, see Example 6.

In Exercises 1–8, find the derivative of the function.

1. $f(x) = \sqrt{x^2 + 3}$

2. $g(x) = (3 - x^2)^3$

3. $g(t) = te^{2t+1}$

4. $f(x) = e^{2x}\sqrt{1 - e^{2x}}$

5. $f(x) = \ln(3 - 2x)$

6. $u(t) = \ln\sqrt{t^3 - 6t}$

7. $g(x) = \dfrac{5x^2}{(4x - 1)^2}$

8. $f(x) = \dfrac{(x + 2)^3}{(x^2 - 9)^2}$

In Exercises 9 and 10, evaluate the derivative at the point (2, 4).

9. $f(x) = x^2 e^{x-2}$

10. $g(x) = x\sqrt{x^2 - x + 2}$

7.5 Extrema of Functions of Two Variables

■ Understand the relative extrema of functions of two variables.
■ Use the First-Partials Test to find the relative extrema of functions of two variables.
■ Use the Second-Partials Test to find the relative extrema of functions of two variables.
■ Use relative extrema to answer questions about real-life situations.

Relative Extrema

Earlier in the text, you learned how to use derivatives to find the relative minimum and relative maximum values of a function of a single variable. In this section, you will learn how to use partial derivatives to find the relative minimum and relative maximum values of a function of two variables.

In Exercise 43, you will find the dimensions of a rectangular package of maximum volume that can be sent by a shipping company.

Relative Extrema of a Function of Two Variables

Let f be a function defined on a region containing (x_0, y_0). The function f has a **relative maximum** at (x_0, y_0) when there is a circular region R centered at (x_0, y_0) such that

$$f(x, y) \le f(x_0, y_0) \qquad \textcolor{red}{f \text{ has a relative maximum at } (x_0, y_0).}$$

for all (x, y) in R. The function f has a **relative minimum** at (x_0, y_0) when there is a circular region R centered at (x_0, y_0) such that

$$f(x, y) \ge f(x_0, y_0) \qquad \textcolor{red}{f \text{ has a relative minimum at } (x_0, y_0).}$$

for all (x, y) in R.

To say that f has a relative maximum at (x_0, y_0) means that the point (x_0, y_0, z_0) is at least as high as all nearby points on the graph of $z = f(x, y)$. Similarly, f has a relative minimum at (x_0, y_0) when (x_0, y_0, z_0) is at least as low as all nearby points on the graph. (See Figure 7.27.)

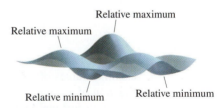

Relative Extrema
FIGURE 7.27

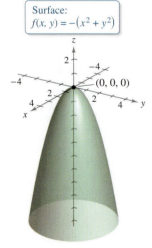

Surface:
$f(x, y) = -(x^2 + y^2)$

$(0, 0, 0)$

f has an absolute maximum at $(0, 0, 0)$.
FIGURE 7.28

As in single-variable calculus, you need to distinguish between relative extrema and absolute extrema of a function of two variables. The number $f(x_0, y_0)$ is an absolute maximum of f in the region R when it is greater than or equal to all other function values in the region. (An absolute minimum of f in a region is defined similarly.) For instance, the function

$$f(x, y) = -(x^2 + y^2)$$

is a paraboloid, opening downward, with vertex at $(0, 0, 0)$. (See Figure 7.28.) The number

$$f(0, 0) = 0$$

is an absolute maximum of the function over the entire xy-plane.

The First-Partials Test for Relative Extrema

To locate the relative extrema of a function of two variables, you can use a procedure that is similar to the First-Derivative Test used for functions of a single variable.

First-Partials Test for Relative Extrema

If f has a relative extremum at (x_0, y_0) on an open region R in the xy-plane, and the first partial derivatives of f exist in R, then

$$f_x(x_0, y_0) = 0$$

and

$$f_y(x_0, y_0) = 0$$

as shown in Figure 7.29.

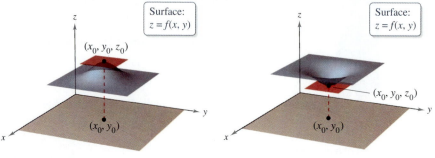

Relative maximum

Relative minimum

FIGURE 7.29

An *open* region in the xy-plane is similar to an open interval on the real number line. For instance, the region R consisting of the interior of the circle $x^2 + y^2 = 1$ is an open region. If the region R consists of the interior of the circle *and* the points on the circle, then it is a *closed* region.

A point (x_0, y_0) is a **critical point** of f when $f_x(x_0, y_0)$ or $f_y(x_0, y_0)$ is undefined or when

$$f_x(x_0, y_0) = 0 \quad \text{and} \quad f_y(x_0, y_0) = 0. \qquad \textcolor{red}{\text{Critical point}}$$

The First-Partials Test states that if the first partial derivatives exist, then you need only examine values of $f(x, y)$ at critical points to find the relative extrema. As is true for a function of a single variable, however, the critical points of a function of two variables do not always yield relative extrema. For instance, the point $(0, 0)$ is a critical point of the surface shown in Figure 7.30, but $f(0, 0)$ is not a relative extremum of the function. Such points are called **saddle points** of the function.

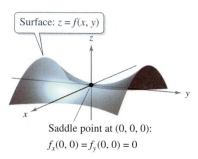

Saddle point at $(0, 0, 0)$:
$$f_x(0, 0) = f_y(0, 0) = 0$$

FIGURE 7.30

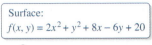

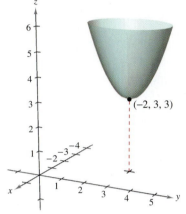

FIGURE 7.31

Example 1 Finding Relative Extrema

Find the relative extrema of

$$f(x, y) = 2x^2 + y^2 + 8x - 6y + 20.$$

SOLUTION Begin by finding the first partial derivatives of f.

$$f_x(x, y) = 4x + 8 \quad \text{and} \quad f_y(x, y) = 2y - 6$$

Because these partial derivatives are defined for all points in the xy-plane, the only critical points are those for which both first partial derivatives are zero. To locate these points, set $f_x(x, y)$ and $f_y(x, y)$ equal to 0, and solve the resulting system of equations.

$$4x + 8 = 0 \qquad \text{Set } f_x(x, y) \text{ equal to 0.}$$
$$2y - 6 = 0 \qquad \text{Set } f_y(x, y) \text{ equal to 0.}$$

The solution of this system is $x = -2$ and $y = 3$. So, the point $(-2, 3)$ is the only critical number of f. From the graph of the function, shown in Figure 7.31, you can see that this critical point yields a relative minimum of the function. So, the function has only one relative extremum, which is

$$f(-2, 3) = 3. \qquad \text{Relative minimum}$$

✓ Checkpoint 1

Find the relative extrema of

$$f(x, y) = x^2 + 2y^2 + 16x - 8y + 8.$$ ■

Example 1 shows a relative minimum occurring at one type of critical point—the type for which both $f_x(x, y)$ and $f_y(x, y)$ are zero. The next example shows a relative maximum that occurs at the other type of critical point—the type for which either $f_x(x, y)$ or $f_y(x, y)$ is undefined.

Example 2 Finding Relative Extrema

Find the relative extrema of

$$f(x, y) = 1 - (x^2 + y^2)^{1/3}.$$

SOLUTION Begin by finding the first partial derivatives of f.

$$f_x(x, y) = -\frac{2x}{3(x^2 + y^2)^{2/3}} \quad \text{and} \quad f_y(x, y) = -\frac{2y}{3(x^2 + y^2)^{2/3}}$$

These partial derivatives are defined for all points in the xy-plane *except* the point $(0, 0)$. So, $(0, 0)$ is a critical point of f. Moreover, this is the only critical point, because there are no other values of x and y for which either partial derivative is undefined or for which both partial derivatives are zero. From the graph of the function, shown in Figure 7.32, you can see that this critical point yields a relative maximum of the function. So, the function has only one relative extremum, which is

$$f(0, 0) = 1. \qquad \text{Relative maximum}$$

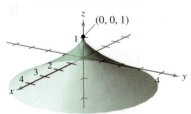

$f_x(x, y)$ and $f_y(x, y)$ are undefined at $(0, 0)$.

FIGURE 7.32

✓ Checkpoint 2

Find the relative extrema of

$$f(x, y) = \sqrt{1 - \frac{x^2}{16} - \frac{y^2}{4}}.$$ ■

The Second-Partials Test for Relative Extrema

For functions such as those in Examples 1 and 2, you can determine the *types* of extrema at the critical points by sketching the graph of the function. For more complicated functions, a graphical approach is not so easy to use. The **Second-Partials Test** is an analytical test that can be used to determine whether a critical number yields a relative minimum, a relative maximum, or neither.

Second-Partials Test for Relative Extrema

Let f have continuous second partial derivatives on an open region containing (a, b) for which

$$f_x(a, b) = 0 \quad \text{and} \quad f_y(a, b) = 0.$$

To test for relative extrema of f, consider the quantity

$$d = f_{xx}(a, b)f_{yy}(a, b) - [f_{xy}(a, b)]^2.$$

1. If $d > 0$ and $f_{xx}(a, b) > 0$, then f has a **relative minimum** at (a, b).

2. If $d > 0$ and $f_{xx}(a, b) < 0$, then f has a **relative maximum** at (a, b).

3. If $d < 0$, then $(a, b, f(a, b))$ is a **saddle point**.

4. The test gives no information when $d = 0$.

Note in the Second-Partials Test that if $d > 0$, then $f_{xx}(a, b)$ and $f_{yy}(a, b)$ must have the same sign. So, you can replace $f_{xx}(a, b)$ with $f_{yy}(a, b)$ in the first two parts of the test.

ALGEBRA TUTOR xy

For help in solving the system of equations

$$y - x^3 = 0$$
$$x - y^3 = 0$$

in Example 3, see Example 1(a) in the *Chapter 7 Algebra Tutor*, on page 423.

Example 3   Applying the Second-Partials Test

Find the relative extrema and saddle points of $f(x, y) = xy - \frac{1}{4}x^4 - \frac{1}{4}y^4$.

SOLUTION   Begin by finding the critical points of f. Because

$$f_x(x, y) = y - x^3 \quad \text{and} \quad f_y(x, y) = x - y^3$$

are defined for all points in the xy-plane, the only critical points are those for which both first partial derivatives are zero. By solving the equations

$$y - x^3 = 0 \quad \text{and} \quad x - y^3 = 0$$

simultaneously, you can determine that the critical points are $(1, 1)$, $(-1, -1)$, and $(0, 0)$. Furthermore, because

$$f_{xx}(x, y) = -3x^2, \quad f_{yy}(x, y) = -3y^2, \quad \text{and} \quad f_{xy}(x, y) = 1$$

you can use the quantity $d = f_{xx}(a, b)f_{yy}(a, b) - [f_{xy}(a, b)]^2$ to classify the critical points, as shown.

Critical Point	d	$f_{xx}(x, y)$	Conclusion
$(1, 1)$	$(-3)(-3) - 1 = 8$	-3	Relative maximum
$(-1, -1)$	$(-3)(-3) - 1 = 8$	-3	Relative maximum
$(0, 0)$	$(0)(0) - 1 = -1$	0	Saddle point

The graph of f is shown in Figure 7.33.

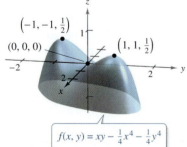

$\left(-1, -1, \frac{1}{2}\right)$
$(0, 0, 0)$
$\left(1, 1, \frac{1}{2}\right)$

$f(x, y) = xy - \frac{1}{4}x^4 - \frac{1}{4}y^4$

FIGURE 7.33

✓ **Checkpoint 3**

Find the relative extrema and saddle points of $f(x, y) = \dfrac{y^2}{16} - \dfrac{x^2}{4}$.

Applications

 Example 4 Finding a Maximum Profit

A company makes two substitute products whose demand functions are given by

$$x_1 = 200(p_2 - p_1)$$ Demand for product 1
$$x_2 = 500 + 100p_1 - 180p_2$$ Demand for product 2

where p_1 and p_2 are the prices per unit (in dollars) and x_1 and x_2 are the numbers of units sold. The costs of producing the two products are \$0.50 and \$0.75 per unit, respectively. Find the prices that will yield a maximum profit.

SOLUTION The cost function is

$$C = 0.5x_1 + 0.75x_2$$ Write cost function.
$$= 0.5(200)(p_2 - p_1) + 0.75(500 + 100p_1 - 180p_2)$$ Substitute.
$$= 375 - 25p_1 - 35p_2.$$ Simplify.

The revenue function is

$$R = p_1x_1 + p_2x_2$$ Write revenue function.
$$= p_1(200)(p_2 - p_1) + p_2(500 + 100p_1 - 180p_2)$$ Substitute.
$$= -200p_1^2 - 180p_2^2 + 300p_1p_2 + 500p_2.$$ Simplify.

This implies that the profit function is

$$P = R - C$$ Write profit function.
$$= -200p_1^2 - 180p_2^2 + 300p_1p_2 + 500p_2 - (375 - 25p_1 - 35p_2)$$
$$= -200p_1^2 - 180p_2^2 + 300p_1p_2 + 25p_1 + 535p_2 - 375.$$

Next, find the first partial derivatives of P.

$$\frac{\partial P}{\partial p_1} = -400p_1 + 300p_2 + 25 \qquad \frac{\partial P}{\partial p_2} = 300p_1 - 360p_2 + 535$$

By setting the first partial derivatives equal to zero and solving the equations

$$-400p_1 + 300p_2 + 25 = 0$$
$$300p_1 - 360p_2 + 535 = 0$$

simultaneously, you can conclude that the solution is $p_1 \approx \$3.14$ and $p_2 \approx \$4.10$. From the graph of P shown in Figure 7.34, you can see that this critical number yields a maximum. So, the maximum profit is

$$P(p_1, p_2) \approx P(3.14, 4.10) = \$761.48.$$

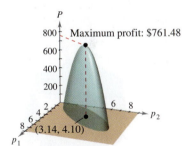

FIGURE 7.34

STUDY TIP

In Example 4, you can check that the two products are substitutes by observing that x_1 increases as p_2 increases and x_2 increases as p_1 increases.

ALGEBRA TUTOR xy

For help in solving the system of equations in Example 4, see Example 1(b) in the *Chapter 7 Algebra Tutor*, on page 423.

✓ **Checkpoint 4**

Find the prices that will yield a maximum profit for the products in Example 4 when the costs of producing the two products are \$0.75 and \$0.50 per unit, respectively. ■

In Example 4, to convince yourself that the maximum profit is \$761.48, try substituting other prices, such as $p_1 = \$2$ and $p_2 = \$3$, into the profit function. For each pair of prices, you will obtain a profit that is less than \$761.48.

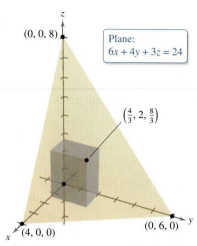

FIGURE 7.35

Plane:
$6x + 4y + 3z = 24$

$(0, 0, 8)$

$\left(\frac{4}{3}, 2, \frac{8}{3}\right)$

$(4, 0, 0)$

$(0, 6, 0)$

ALGEBRA TUTOR xy

For help in solving the system of equations

$$y(24 - 12x - 4y) = 0$$
$$x(24 - 6x - 8y) = 0$$

in Example 5, see Example 2(a) in the *Chapter 7 Algebra Tutor*, on page 424.

Example 5   **Finding a Maximum Volume**

A rectangular box is resting on the xy-plane with one vertex at the origin. The opposite vertex lies in the plane

$$6x + 4y + 3z = 24$$

as shown in Figure 7.35. Find the maximum volume of such a box.

SOLUTION   Let x, y, and z represent the length, width, and height of the box. Because one vertex of the box lies in the plane given by $6x + 4y + 3z = 24$ or

$$z = \tfrac{1}{3}(24 - 6x - 4y) \qquad \text{Solve for } z.$$

you can write the volume of the box as a function of two variables.

$$
\begin{aligned}
V &= xyz & \text{Volume} = (\text{width})(\text{length})(\text{height}) \\
&= xy\left(\tfrac{1}{3}\right)(24 - 6x - 4y) & \text{Substitute for } z. \\
&= \tfrac{1}{3}(24xy - 6x^2y - 4xy^2) & \text{Simplify.}
\end{aligned}
$$

Next, find the first partial derivatives of V.

$$
\begin{aligned}
V_x &= \tfrac{1}{3}(24y - 12xy - 4y^2) & \text{Partial with respect to } x \\
&= \tfrac{1}{3}y(24 - 12x - 4y) & \text{Factor.} \\
V_y &= \tfrac{1}{3}(24x - 6x^2 - 8xy) & \text{Partial with respect to } y \\
&= \tfrac{1}{3}x(24 - 6x - 8y) & \text{Factor.}
\end{aligned}
$$

By solving the equations

$$
\begin{aligned}
\tfrac{1}{3}y(24 - 12x - 4y) &= 0 & \text{Set } V_x \text{ equal to 0.} \\
\tfrac{1}{3}x(24 - 6x - 8y) &= 0 & \text{Set } V_y \text{ equal to 0.}
\end{aligned}
$$

simultaneously, you can conclude that the solutions are $(0, 0)$, $(0, 6)$, $(4, 0)$, and $\left(\frac{4}{3}, 2\right)$. Using the Second-Partials Test, you can determine that the maximum volume occurs when the width is $x = \frac{4}{3}$ and the length is $y = 2$. For these values, the height of the box is

$$z = \tfrac{1}{3}\left[24 - 6\left(\tfrac{4}{3}\right) - 4(2)\right] = \tfrac{8}{3}.$$

So, the maximum volume is

$$V = xyz = \left(\tfrac{4}{3}\right)(2)\left(\tfrac{8}{3}\right) = \tfrac{64}{9} \text{ cubic units.}$$

✓**Checkpoint 5**

Find the maximum volume of a box that is resting on the xy-plane with one vertex at the origin and the opposite vertex in the plane $2x + 4y + z = 8$. ■

SUMMARIZE   (Section 7.5)

1. State the definition of relative extrema of a function of two variables *(page 389)*. For examples of relative extrema, see Examples 1 and 2.

2. State the First-Partials Test for relative extrema *(page 390)*. For examples of using the First-Partials Test, see Examples 1 and 2.

3. State the Second-Partials Test for relative extrema *(page 392)*. For examples of using the Second-Partials Test, see Examples 3 and 5.

4. Describe a real-life example of how relative extrema can be used to find a company's maximum profit *(page 393, Example 4)*.

SKILLS WARM UP 7.5

The following warm-up exercises involve skills that were covered in earlier sections. You will use these skills in the exercise set for this section. For additional help, review Section 7.4.

In Exercises 1–8, solve the system of equations.

1. $\begin{cases} 5x = 15 \\ 3x - 2y = 5 \end{cases}$

2. $\begin{cases} \frac{1}{2}y = 3 \\ -x + 5y = 19 \end{cases}$

3. $\begin{cases} x + y = 5 \\ x - y = -3 \end{cases}$

4. $\begin{cases} x + y = 8 \\ 2x - y = 4 \end{cases}$

5. $\begin{cases} 2x - y = 8 \\ 3x - 4y = 7 \end{cases}$

6. $\begin{cases} 2x - 4y = 14 \\ 3x + y = 7 \end{cases}$

7. $\begin{cases} x^2 + x = 0 \\ 2yx + y = 0 \end{cases}$

8. $\begin{cases} 3y^2 + 6y = 0 \\ xy + x + 2 = 0 \end{cases}$

In Exercises 9–14, find all first and second partial derivatives of the function.

9. $z = 4x^3 - 3y^2$

10. $z = 2x^5 - y^3$

11. $z = x^4 - \sqrt{xy} + 2y$

12. $z = 2x^2 - 3xy + y^2$

13. $z = ye^{xy^2}$

14. $z = xe^{xy}$

QUIZ YOURSELF

Take this quiz as you would take a quiz in class. When you are done, check your work against the answers given in the back of the book.

In Exercises 1–3, (a) plot the points in a three-dimensional coordinate system, (b) find the distance between the points, and (c) find the coordinates of the midpoint of the line segment joining the points.

1. $(1, 3, 2), (-1, 2, 0)$ **2.** $(-1, 3, 4), (5, 1, -6)$ **3.** $(0, -3, 3), (3, 0, -3)$

In Exercises 4 and 5, find the standard equation of the sphere.

4. Center: $(2, -1, 3)$; radius: 4

5. Endpoints of a diameter: $(0, 3, 1), (2, 5, -5)$

6. Find the center and radius of the sphere whose equation is

$$x^2 + y^2 + z^2 - 8x - 2y - 6z - 23 = 0.$$

In Exercises 7–9, find the intercepts and sketch the graph of the plane.

7. $2x + 3y + z = 6$ **8.** $x - 2z = 4$ **9.** $y = 3$

In Exercises 10–12, classify the quadric surface.

10. $\dfrac{x^2}{4} + \dfrac{y^2}{9} + \dfrac{z^2}{16} = 1$ **11.** $z^2 - x^2 - y^2 = 25$ **12.** $81z - 9x^2 - y^2 = 0$

In Exercises 13–15, find $f(1, 0)$ and $f(4, -1)$.

13. $f(x, y) = x - 9y^2$ **14.** $f(x, y) = \sqrt{4x^2 + y}$ **15.** $f(x, y) = \ln(x - 2y)$

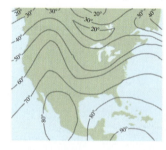

Figure for 16

16. The contour map shows level curves of equal temperature (isotherms), measured in degrees Fahrenheit, across North America on a spring day. Use the map to find the approximate range of temperatures in

(a) the Great Lakes region.

(b) the United States.

(c) Mexico.

In Exercises 17–20, find the first partial derivatives and evaluate each at the point $(-2, 3)$.

17. $f(x, y) = x^2 + 2y^2 - 3x - y + 1$ **18.** $f(x, y) = \dfrac{3x - y^2}{x + y}$

19. $f(x, y) = x^3 e^{2y}$ **20.** $f(x, y) = \ln(2x + 7y)$

In Exercises 21 and 22, find the critical points, relative extrema, and saddle points of the function.

21. $f(x, y) = 3x^2 + y^2 - 2xy - 6x + 2y$

22. $f(x, y) = -x^3 + 4xy - 2y^2 + 1$

23. A company manufactures two types of wood burning stoves: a freestanding model and a fireplace-insert model. The total cost (in thousands of dollars) for producing x freestanding stoves and y fireplace-insert stoves can be modeled by

$$C(x, y) = \tfrac{1}{16}x^2 + y^2 - 10x - 40y + 820.$$

Find the values of x and y that minimize the total cost. What is this cost?

7.6 Lagrange Multipliers

■ Understand the Method of Lagrange Multipliers.
■ Use Lagrange multipliers to solve constrained optimization problems.

Lagrange Multipliers with One Constraint

In Example 5 in Section 7.5, you were asked to find the dimensions of the rectangular box of maximum volume that would fit in the first octant beneath the plane

$$6x + 4y + 3z = 24$$

as shown again in Figure 7.36. Another way of stating this problem is to say that you are asked to find the maximum of

$$V = xyz \qquad \text{Objective function}$$

subject to the constraint

$$6x + 4y + 3z - 24 = 0. \qquad \text{Constraint}$$

This type of problem is called a **constrained optimization** problem. In Section 7.5, you answered this question by solving for z in the constraint equation and then rewriting V as a function of two variables.

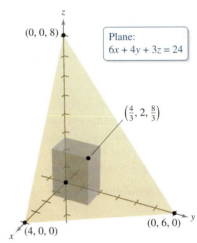

Plane:
$6x + 4y + 3z = 24$

$(0, 0, 8)$
$\left(\frac{4}{3}, 2, \frac{8}{3}\right)$
$(0, 6, 0)$
$(4, 0, 0)$

FIGURE 7.36

In this section, you will study a different (and often better) way to solve constrained optimization problems. This method involves the use of variables called **Lagrange multipliers,** named after the French mathematician Joseph Louis Lagrange (1736–1813).

In Exercise 37, you will use Lagrange multipliers to find the dimensions that will minimize the cost of fencing in two corrals.

Method of Lagrange Multipliers

If $f(x, y)$ has a maximum or minimum subject to the constraint $g(x, y) = 0$, then it will occur at one of the critical numbers of the function F defined by

$$F(x, y, \lambda) = f(x, y) - \lambda g(x, y).$$

The variable λ (the lowercase Greek letter lambda) is called a **Lagrange multiplier.** To find the minimum or maximum of f, use the following steps.

1. Solve the following system of equations.

$$F_x(x, y, \lambda) = 0 \qquad F_y(x, y, \lambda) = 0 \qquad F_\lambda(x, y, \lambda) = 0$$

2. Evaluate f at each solution point obtained in the first step. The greatest value yields the maximum of f subject to the constraint $g(x, y) = 0$, and the least value yields the minimum of f subject to the constraint $g(x, y) = 0$.

When using the Method of Lagrange Multipliers for functions of three variables, F has the form

$$F(x, y, z, \lambda) = f(x, y, z) - \lambda g(x, y, z).$$

The system of equations used in Step 1 is

$$F_x(x, y, z, \lambda) = 0 \quad F_y(x, y, z, \lambda) = 0 \quad F_z(x, y, z, \lambda) = 0 \quad F_\lambda(x, y, z, \lambda) = 0.$$

The Method of Lagrange Multipliers gives you a way of finding critical points but does not tell you whether these points yield minima, maxima, or neither. To make this distinction, you must rely on the context of the problem.

Constrained Optimization Problems

> ### Example 1   Using Lagrange Multipliers

Find the maximum of

$$V = xyz \qquad \text{Objective function}$$

subject to the constraint

$$6x + 4y + 3z - 24 = 0. \qquad \text{Constraint}$$

SOLUTION   First, let $f(x, y, z) = xyz$ and $g(x, y, z) = 6x + 4y + 3z - 24$. Then, define a new function F as

$$F(x, y, z, \lambda) = f(x, y, z) - \lambda g(x, y, z)$$
$$= xyz - \lambda(6x + 4y + 3z - 24).$$

To find the critical numbers of F, begin by finding the partial derivatives of F with respect to x, y, z, and λ. Then, set the partial derivatives equal to zero.

$$F_x(x, y, z, \lambda) = yz - 6\lambda \qquad\Longrightarrow\qquad yz - 6\lambda = 0$$
$$F_y(x, y, z, \lambda) = xz - 4\lambda \qquad\Longrightarrow\qquad xz - 4\lambda = 0$$
$$F_z(x, y, z, \lambda) = xy - 3\lambda \qquad\Longrightarrow\qquad xy - 3\lambda = 0$$
$$F_\lambda(x, y, z, \lambda) = -6x - 4y - 3z + 24 \qquad\Longrightarrow\qquad -6x - 4y - 3z + 24 = 0$$

Solving for λ in the first equation produces

$$yz - 6\lambda = 0 \qquad\Longrightarrow\qquad \lambda = \frac{yz}{6}.$$

Substituting for λ in the second and third equations produces the following.

$$xz - 4\left(\frac{yz}{6}\right) = 0 \qquad\Longrightarrow\qquad y = \frac{3}{2}x$$

$$xy - 3\left(\frac{yz}{6}\right) = 0 \qquad\Longrightarrow\qquad z = 2x$$

Next, substitute for y and z in the equation $F_\lambda(x, y, z, \lambda) = 0$ and solve for x.

$$F_\lambda(x, y, z, \lambda) = 0$$
$$-6x - 4y - 3z + 24 = 0$$
$$-6x - 4\left(\tfrac{3}{2}x\right) - 3(2x) + 24 = 0$$
$$-18x = -24$$
$$x = \tfrac{4}{3}$$

Using this x-value, you can conclude that the critical values are $x = \frac{4}{3}$, $y = 2$, and $z = \frac{8}{3}$, which implies that the maximum is

$$V = xyz \qquad \text{Write objective function.}$$
$$= \left(\frac{4}{3}\right)(2)\left(\frac{8}{3}\right) \qquad \text{Substitute values of } x, y, \text{ and } z.$$
$$= \frac{64}{9} \text{ cubic units.} \qquad \text{Maximum volume}$$

STUDY TIP

Example 1 shows how Lagrange multipliers can be used to solve the same problem that was solved in Example 5 in Section 7.5.

ALGEBRA TUTOR *xy*

The most difficult aspect of many Lagrange multiplier problems is the complicated algebra needed to solve the system of equations arising from

$$F(x, y, \lambda) = f(x, y) - \lambda g(x, y).$$

There is no general way to proceed in every case, so you should study the examples carefully, and refer to the *Chapter 7 Algebra Tutor* on pages 423 and 424.

✓ Checkpoint 1

Find the maximum volume of $V = xyz$ subject to the constraint

$$2x + 4y + z - 8 = 0.$$

Example 2 **Finding a Maximum Production Level**

A manufacturer's production is modeled by the Cobb-Douglas function

$$f(x, y) = 100x^{3/4}y^{1/4} \qquad \text{Objective function}$$

where x represents the units of labor (at \$150 per unit) and y represents the units of capital (at \$250 per unit). The total costs for labor and capital cannot exceed \$50,000. Will the manufacturer's maximum production level exceed 16,000 units?

SOLUTION Because total labor and capital expenses cannot exceed \$50,000, the constraint is

$$150x + 250y = 50,000 \qquad \text{Constraint}$$
$$150x + 250y - 50,000 = 0. \qquad \text{Write in general form.}$$

To find the maximum production level, begin by writing the function

$$F(x, y, \lambda) = 100x^{3/4}y^{1/4} - \lambda(150x + 250y - 50,000).$$

Then find the partial derivatives of F with respect to x, y, and λ.

$$F_x(x, y, \lambda) = 75x^{-1/4}y^{1/4} - 150\lambda$$
$$F_y(x, y, \lambda) = 25x^{3/4}y^{-3/4} - 250\lambda$$
$$F_\lambda(x, y, \lambda) = -150x - 250y + 50,000$$

For some industrial applications, a simple robot can cost more than a year's wages and benefits for one employee. So, manufacturers must carefully balance the amount of money spent on labor and capital.

Next, set the partial derivatives equal to zero to obtain the following system of equations.

$$75x^{-1/4}y^{1/4} - 150\lambda = 0 \qquad \text{Equation 1}$$
$$25x^{3/4}y^{-3/4} - 250\lambda = 0 \qquad \text{Equation 2}$$
$$-150x - 250y + 50,000 = 0 \qquad \text{Equation 3}$$

By solving for λ in the first equation

$$75x^{-1/4}y^{1/4} - 150\lambda = 0 \qquad \text{Equation 1}$$
$$\lambda = \tfrac{1}{2}x^{-1/4}y^{1/4} \qquad \text{Solve for } \lambda.$$

and substituting for λ in Equation 2, you obtain

$$25x^{3/4}y^{-3/4} - 250\left(\tfrac{1}{2}\right)x^{-1/4}y^{1/4} = 0 \qquad \text{Substitute in Equation 2.}$$
$$25x - 125y = 0 \qquad \text{Multiply by } x^{1/4}y^{3/4}.$$
$$x = 5y. \qquad \text{Solve for } x.$$

So, $x = 5y$. By substituting for x in Equation 3, you obtain

$$-150(5y) - 250y + 50,000 = 0 \qquad \text{Substitute in Equation 3.}$$
$$-1000y = -50,000 \qquad \text{Simplify.}$$
$$y = 50 \qquad \text{Solve for } y.$$

When $y = 50$ units of capital, it follows that $x = 5(50) = 250$ units of labor. So, the maximum production level is

$$f(250, 50) = 100(250)^{3/4}(50)^{1/4} \qquad \text{Substitute for } x \text{ and } y.$$
$$\approx 16,719 \text{ units.} \qquad \text{Maximum production level}$$

You can conclude that the maximum production level will exceed 16,000 units.

TECH TUTOR

You can use a spreadsheet to solve constrained optimization problems. Try using a spreadsheet to solve the problem in Example 2. (Consult the user's manual of a spreadsheet software program for specific instructions on how to solve a constrained optimization problem.)

✓ Checkpoint 2

In Example 2, suppose that each labor unit costs \$200 and each capital unit costs \$250. Find the maximum production level when labor and capital cannot exceed \$50,000.

Economists call the Lagrange multiplier obtained in a production function the **marginal productivity of money.** For instance, in Example 2, the marginal productivity of money when $x = 250$ and $y = 50$ is

$$\lambda = \tfrac{1}{2}x^{-1/4}y^{1/4} = \tfrac{1}{2}(250)^{-1/4}(50)^{1/4} \approx 0.334.$$

This means that for each additional dollar spent on production, approximately 0.334 additional unit of the product can be produced.

 Example 3 **Finding a Maximum Production Level**

The manufacturer in Example 2 now has \$70,000 available for labor and capital. What is the maximum number of units that can be produced?

SOLUTION You could rework the entire problem, as demonstrated in Example 2. However, because the only change in the problem is the availability of additional money to spend on labor and capital, you can use the fact that the marginal productivity of money is

$$\lambda \approx 0.334.$$

Because an additional \$20,000 is available and the maximum production level in Example 2 was 16,719 units, you can conclude that the maximum production level is now

$$16,719 + (0.334)(20,000) \approx 23,400 \text{ units.}$$

Try using the procedure demonstrated in Example 2 to confirm this result.

 **Checkpoint 3**

The manufacturer in Example 2 now has \$80,000 available for labor and capital. What is the maximum number of units that can be produced? ■

TECH TUTOR

You can use a three-dimensional graphing utility to confirm graphically the results of Examples 2 and 3. Begin by graphing the surface $f(x, y) = 100x^{3/4}y^{1/4}$. Then graph the vertical plane given by $150x + 250y = 50,000$. As shown below, the maximum production level corresponds to the highest point on the intersection of the surface and the plane.

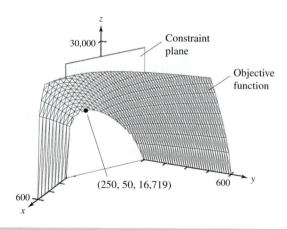

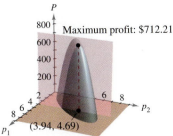

FIGURE 7.37

In Example 4 in Section 7.5, you found the maximum profit for two substitute products whose demand functions are given by

$$x_1 = 200(p_2 - p_1) \qquad \text{Demand for product 1}$$
$$x_2 = 500 + 100p_1 - 180p_2. \qquad \text{Demand for product 2}$$

With this model, the total demand, $x_1 + x_2$, is completely determined by the prices p_1 and p_2. In many real-life situations, this assumption is too simplistic; regardless of the prices of the substitute brands, the annual total demands for some products, such as toothpaste, are relatively constant. In such situations, the total demand is **limited,** and variations in price do not affect the total demand as much as they affect the market share of the substitute brands.

Example 4   Finding a Maximum Profit

A company makes two substitute products whose demand functions are given by

$$x_1 = 200(p_2 - p_1) \qquad \text{Demand for product 1}$$
$$x_2 = 500 + 100p_1 - 180p_2 \qquad \text{Demand for product 2}$$

where p_1 and p_2 are the prices per unit (in dollars) and x_1 and x_2 are the numbers of units sold. The costs of producing the two products are \$0.50 and \$0.75 per unit, respectively. The total demand is limited to 200 units per year. Find the prices that will yield a maximum profit.

SOLUTION   From Example 4 in Section 7.5, the profit function is modeled by

$$P = -200p_1^2 - 180p_2^2 + 300p_1p_2 + 25p_1 + 535p_2 - 375.$$

The total demand for the two products is

$$x_1 + x_2 = 200(p_2 - p_1) + 500 + 100p_1 - 180p_2$$
$$= 200p_2 - 200p_1 + 500 + 100p_1 - 180p_2$$
$$= -100p_1 + 20p_2 + 500.$$

Because the total demand is limited to 200 units, the constraint is

$$-100p_1 + 20p_2 + 500 = 200. \qquad \text{Constraint}$$

Using Lagrange multipliers, you can determine that the maximum profit occurs when $p_1 \approx \$3.94$ and $p_2 \approx \$4.69$. This corresponds to an annual profit of about \$712.21.

✓Checkpoint 4

In Example 4, find the prices that will yield a maximum profit when the total demand is limited to 250 units per year.   ■

SUMMARIZE   (Section 7.6)

1. Explain the Method of Lagrange Multipliers *(page 397)*. For an example of how to use Lagrange multipliers, see Example 1.

2. Describe a real-life example of using Lagrange multipliers to find a manufacturer's maximum production level *(page 399, Example 2)*.

3. Describe a real-life example of using Lagrange multipliers to find a company's maximum profit *(page 401, Example 4)*.

SKILLS WARM UP 7.6   The following warm-up exercises involve skills that were covered in earlier sections. You will use these skills in the exercise set for this section. For additional help, review Section 7.4.

In Exercises 1–6, solve the system of linear equations.

1. $\begin{cases} 4x - 6y = 3 \\ 2x + 3y = 2 \end{cases}$

2. $\begin{cases} 6x - 6y = 5 \\ -3x - y = 1 \end{cases}$

3. $\begin{cases} 5x - y = 25 \\ x - 5y = 15 \end{cases}$

4. $\begin{cases} 4x - 9y = 5 \\ -x + 8y = -2 \end{cases}$

5. $\begin{cases} 2x - y + z = 3 \\ 2x + 2y + z = 4 \\ -x + 2y + 3z = -1 \end{cases}$

6. $\begin{cases} -x - 4y + 6z = -2 \\ x - 3y - 3z = 4 \\ 3x + y + 3z = 0 \end{cases}$

In Exercises 7–10, find all first partial derivatives.

7. $f(x, y) = x^2y + xy^2$

8. $f(x, y) = 25(xy + y^2)^2$

9. $f(x, y, z) = x(x^2 - 2xy + yz)$

10. $f(x, y, z) = z(xy + xz + yz)$

ENHANCED
WebAssign   Access end-of-section exercises online at **www.webassign.net**

7.7 Least Squares Regression Analysis

■ Find the sum of the squared errors for mathematical models.

■ Find the least squares regression lines for data.

Measuring the Accuracy of a Mathematical Model

When seeking a mathematical model to fit data, the goals are simplicity and accuracy. For instance, a simple linear model for the points shown in Figure 7.38(a) is

$$f(x) = 1.9x - 5. \qquad \text{Linear model}$$

Figure 7.38(b), however, shows that by choosing the slightly more complicated quadratic model

$$g(x) = 0.20x^2 - 0.7x + 1 \qquad \text{Quadratic model}$$

you can obtain significantly greater accuracy.*

In Exercise 14, you will find the least squares regression line that models the demand of a tool at a hardware store in terms of the price.

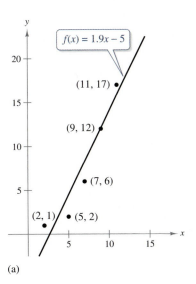

(a)

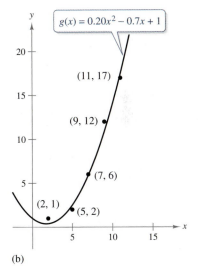

(b)

FIGURE 7.38

To measure how well the model

$$y = f(x)$$

fits a collection of points, sum the squares of the differences between the actual y-values and the model's y-values. This sum is called the **sum of the squared errors** and is denoted by S. Graphically, S can be interpreted as the sum of the squares of the vertical distances between the graph of f and the given points in the plane, as shown in Figure 7.39. If the model is a perfect fit, then

$$S = 0.$$

However, when a perfect fit is not feasible, you should use a model that minimizes S.

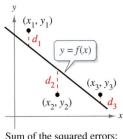

Sum of the squared errors:
$$S = d_1^2 + d_2^2 + d_3^2$$

FIGURE 7.39

---

* An analytic method for finding a quadratic model for a collection of data is not given in this text. You can perform this task using a graphing utility or a spreadsheet software program that has a built-in program for finding the least squares regression quadratic.

Definition of the Sum of the Squared Errors

The **sum of the squared errors** for the model $y = f(x)$ with respect to the points

$$(x_1, y_1), (x_2, y_2), \ldots, (x_n, y_n)$$

is given by

$$S = [f(x_1) - y_1]^2 + [f(x_2) - y_2]^2 + \cdots + [f(x_n) - y_n]^2.$$

Example 1 Finding the Sum of the Squared Errors

Find the sum of the squared errors for the linear model

$$f(x) = 1.9x - 5 \qquad \text{Linear model}$$

and the quadratic model

$$g(x) = 0.20x^2 - 0.7x + 1 \qquad \text{Quadratic model}$$

(see Figure 7.38) with respect to the points

$$(2, 1), (5, 2), (7, 6), (9, 12), (11, 17).$$

SOLUTION   Begin by evaluating each model at the given x-values, as shown in the table.

x	2	5	7	9	11
Actual y-values	1	2	6	12	17
Linear model, $f(x)$	-1.2	4.5	8.3	12.1	15.9
Quadratic model, $g(x)$	0.4	2.5	5.9	10.9	17.5

For the linear model f, the sum of the squared errors is

$$S = (-1.2 - 1)^2 + (4.5 - 2)^2 + (8.3 - 6)^2 + (12.1 - 12)^2 + (15.9 - 17)^2$$
$$= 17.6.$$

Similarly, the sum of the squared errors for the quadratic model g is

$$S = (0.4 - 1)^2 + (2.5 - 2)^2 + (5.9 - 6)^2 + (10.9 - 12)^2 + (17.5 - 17)^2$$
$$= 2.08.$$

✓**Checkpoint 1**

Find the sum of the squared errors for the linear model

$$f(x) = 2.9x - 6 \qquad \text{Linear model}$$

and the quadratic model

$$g(x) = 0.20x^2 + 0.5x - 1 \qquad \text{Quadratic model}$$

with respect to the points

$$(2, 1), (4, 5), (6, 9), (8, 16), (10, 24).$$

Then decide which model is a better fit.   ■

In Example 1, note that the sum of the squared errors for the quadratic model is less than the sum of the squared errors for the linear model, which confirms that the quadratic model is a better fit.

Least Squares Regression Line

The sum of the squared errors can be used to determine which of several models is the best fit for a collection of data. In general, if the sum of the squared errors of f is less than the sum of the squared errors of g, then f is said to be a better fit for the data than g. In regression analysis, you consider all possible models of a certain type. The one that is defined to be the best-fitting model is the one with the least sum of the squared errors. Example 2 shows how to use the optimization techniques described in Section 7.5 to find the best-fitting linear model for a collection of data.

Example 2   Finding the Best Linear Model

Find the values of a and b such that the linear model

$$f(x) = ax + b$$

has a minimum sum of the squared errors for the points

$$(-3, 0), (-1, 1), (0, 2), (2, 3).$$

ALGEBRA TUTOR
xy

For help in solving the system of equations in Example 2, see Example 2(b) in the *Chapter 7 Algebra Tutor*, on page 424.

SOLUTION   The sum of the squared errors is

$$S = [f(x_1) - y_1]^2 + [f(x_2) - y_2]^2 + [f(x_3) - y_3]^2 + [f(x_4) - y_4]^2$$
$$= (-3a + b - 0)^2 + (-a + b - 1)^2 + (b - 2)^2 + (2a + b - 3)^2$$
$$= 14a^2 - 4ab + 4b^2 - 10a - 12b + 14.$$

To find the values of a and b for which S is a minimum, you can use the techniques described in Section 7.5. That is, find the partial derivatives of S.

$$\frac{\partial S}{\partial a} = 28a - 4b - 10 \qquad \text{Differentiate with respect to } a.$$

$$\frac{\partial S}{\partial b} = -4a + 8b - 12 \qquad \text{Differentiate with respect to } b.$$

Next, set each partial derivative equal to zero.

$$28a - 4b - 10 = 0 \qquad \text{Set } \partial S/\partial a \text{ equal to 0.}$$
$$-4a + 8b - 12 = 0 \qquad \text{Set } \partial S/\partial b \text{ equal to 0.}$$

The solution of this system of linear equations is

$$a = \frac{8}{13} \quad \text{and} \quad b = \frac{47}{26}.$$

So, the best-fitting linear model for the given points is

$$f(x) = \frac{8}{13}x + \frac{47}{26}.$$

The graph of this model is shown in Figure 7.40.

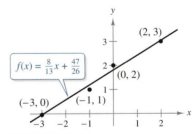

$$f(x) = \frac{8}{13}x + \frac{47}{26}$$

FIGURE 7.40

✓ Checkpoint 2

Find the values of a and b such that the linear model

$$f(x) = ax + b$$

has a minimum sum of the squared errors for the points

$$(-2, 0), (0, 2), (2, 5), (4, 7).$$

The line in Example 2 is called the **least squares regression line** for the given data. The solution shown in Example 2 can be generalized to find a formula for the least squares regression line. Consider the linear model

$$f(x) = ax + b$$

and the points

$$(x_1, y_1), (x_2, y_2), \ldots, (x_n, y_n).$$

The sum of the squared errors is

$$
\begin{aligned}
S &= [f(x_1) - y_1]^2 + [f(x_2) - y_2]^2 + \cdots + [f(x_n) - y_n]^2 \\
&= (ax_1 + b - y_1)^2 + (ax_2 + b - y_2)^2 + \cdots + (ax_n + b - y_n)^2.
\end{aligned}
$$

To minimize S, set the partial derivatives $\partial S/\partial a$ and $\partial S/\partial b$ equal to zero and solve for a and b. The results are summarized below.

The Least Squares Regression Line

The **least squares regression line** for the points

$$(x_1, y_1), (x_2, y_2), \ldots, (x_n, y_n)$$

is $f(x) = ax + b$, where

$$
a = \frac{n\sum_{i=1}^{n} x_i y_i - \sum_{i=1}^{n} x_i \sum_{i=1}^{n} y_i}{n\sum_{i=1}^{n} x_i^2 - \left(\sum_{i=1}^{n} x_i\right)^2}
\quad \text{and} \quad
b = \frac{1}{n}\left(\sum_{i=1}^{n} y_i - a\sum_{i=1}^{n} x_i\right).
$$

The summation notation

$$\sum_{i=1}^{n} x_i$$

where Σ is the Greek letter sigma, is used to indicate the sum of the numbers

$$x_1 + x_2 + \cdots + x_n.$$

Similarly,

$$\sum_{i=1}^{n} x_i y_i = x_1 y_1 + x_2 y_2 + \ldots + x_n y_n, \quad \sum_{i=1}^{n} x_i^2 = x_1^2 + x_2^2 + \ldots + x_n^2,$$

and so on.

In the formula for the least squares regression line, note that if the *x*-values are symmetrically spaced about zero, then

$$\sum_{i=1}^{n} x_i = 0$$

and the formulas for a and b simplify to

$$a = \frac{n\sum_{i=1}^{n} x_i y_i}{n\sum_{i=1}^{n} x_i^2} \quad \text{and} \quad b = \frac{1}{n}\sum_{i=1}^{n} y_i.$$

Note also that only the *development* of the least squares regression line involves partial derivatives. The *application* of this formula is a matter of computing the values of a and b. This task is performed much more simply on a calculator or a computer than by hand.

Example 3   **Modeling Hourly Wages**

The average hourly wages y (in dollars per hour) for production workers in manufacturing industries from 2001 through 2009 are shown in the table. Find the least squares regression line for the data and use the result to estimate the average hourly wage in 2013.   *(Source: U.S. Bureau of Labor Statistics)*

Year	2001	2002	2003	2004	2005	2006	2007	2008	2009
y	14.76	15.29	15.74	16.14	16.56	16.81	17.26	17.75	18.23

Modeling Hourly Wages

FIGURE 7.41

SOLUTION   Let t represent the year, with $t = 1$ corresponding to 2001. Then, you need to find the linear model that best fits the points

$(1, 14.76)$, $(2, 15.29)$, $(3, 15.74)$, $(4, 16.14)$, $(5, 16.56)$, $(6, 16.81)$, $(7, 17.26)$, $(8, 17.75)$, $(9, 18.23)$.

Using a calculator with a built-in least squares regression program, you can determine that the best-fitting line is

$$y = 0.416t + 14.42.$$ Best-fitting line

With this model, you can estimate the 2013 average hourly wage, using $t = 13$, to be

$$y = 0.416(13) + 14.42$$
$$= 19.828$$
$$\approx \$19.83 \text{ per hour.}$$

This result is shown graphically in Figure 7.41.

✓ **Checkpoint 3**

The numbers of cellular phone subscribers y (in thousands) for the years 2000 through 2009 are shown in the table. Find the least squares regression line for the data and use the result to estimate the number of subscribers in 2013. Let t represent the year, with $t = 0$ corresponding to 2000.   *(Source: CTIA-The Wireless Association)*

Year	2000	2001	2002	2003	2004
y	190,478	128,375	140,767	158,722	182,140

Year	2005	2006	2007	2008	2009
y	207,896	233,041	255,396	270,334	285,646

SUMMARIZE   (Section 7.7)

1. State the definition of the sum of the squared errors *(page 404)*. For an example of finding the sum of the squared errors, See Example 1.

2. State the definition of the least squares regression line *(page 406)*. For an example of finding the least squares regression line, see Example 2.

3. Describe a real-life example of modeling hourly wages using the least squares regression line *(page 407, Example 3)*.

SKILLS WARM UP 7.7

The following warm-up exercises involve skills that were covered in a previous course or in earlier sections. You will use these skills in the exercise set for this section. For additional help, review Appendix Section A.3 and Section 7.4.

In Exercises 1 and 2, evaluate the expression.

1. $(2.5 - 1)^2 + (3.25 - 2)^2 + (4.1 - 3)^2$

2. $(1.1 - 1)^2 + (2.08 - 2)^2 + (2.95 - 3)^2$

In Exercises 3 and 4, find the partial derivatives of S.

3. $S = a^2 + 6b^2 - 4a - 8b - 4ab + 6$

4. $S = 4a^2 + 9b^2 - 6a - 4b - 2ab + 8$

In Exercises 5–10, evaluate the sum.

5. $\displaystyle\sum_{i=1}^{5} i$

6. $\displaystyle\sum_{i=1}^{6} 2i$

7. $\displaystyle\sum_{i=1}^{4} \frac{1}{i}$

8. $\displaystyle\sum_{i=1}^{3} i^2$

9. $\displaystyle\sum_{i=1}^{6} (2 - i)^2$

10. $\displaystyle\sum_{i=1}^{5} (30 - i^2)$

WebAssign   Access end-of-section exercises online at **www.webassign.net**

7.8 Double Integrals and Area in the Plane

■ Evaluate double integrals.
■ Use double integrals to find the areas of regions.

Double Integrals

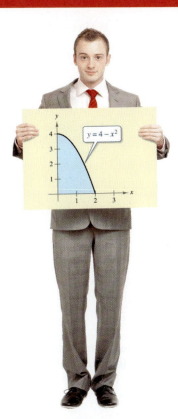

In Exercise 29, you will use a double integral to find the area of a region.

In Section 7.4, you learned that it is meaningful to differentiate functions of several variables with respect to one variable while holding the other variable(s) constant. You can *integrate* functions of several variables by a similar procedure. For instance, if you are given the partial derivative

$$f_x(x, y) = 2xy \qquad \text{Partial with respect to } x$$

then, by holding y constant, you can integrate with respect to x to obtain

$$f(x, y) = \int f_x(x, y)\, dx \qquad \text{Integrate with respect to } x.$$

$$= \int 2xy\, dx \qquad \text{Hold } y \text{ constant.}$$

$$= y \int 2x\, dx \qquad \text{Factor out constant } y.$$

$$= y(x^2) + C(y) \qquad \text{Antiderivative of } 2x \text{ is } x^2.$$

$$= x^2y + C(y). \qquad C(y) \text{ is a function of } y.$$

This procedure is called **partial integration with respect to x.** Note that the "constant of integration" $C(y)$ is a function of y, because y is fixed during integration with respect to x. Similarly, if you are given the partial derivative

$$f_y(x, y) = x^2 + 2 \qquad \text{Partial with respect to } y$$

then, by holding x constant, you can integrate with respect to y to obtain

$$f(x, y) = \int f_y(x, y)\, dy \qquad \text{Integrate with respect to } y.$$

$$= \int (x^2 + 2)\, dy \qquad \text{Hold } x \text{ constant.}$$

$$= (x^2 + 2) \int dy \qquad \text{Factor out constant } x^2 + 2.$$

$$= (x^2 + 2)(y) + C(x) \qquad \text{Antiderivative of 1 is } y.$$

$$= x^2y + 2y + C(x). \qquad C(x) \text{ is a function of } x.$$

In this case, the "constant of integration" $C(x)$ is a function of x, because x is fixed during integration with respect to y.

To evaluate a definite integral of a function of several variables, you can apply the Fundamental Theorem of Calculus to one variable while holding the other variable(s) constant, as shown.

$$\int_1^{2y} 2xy\, dx = x^2y \Big]_1^{2y} = (2y)^2y - (1)^2y = 4y^3 - y$$

x is the variable of integration and y is fixed.

Replace x by the limits of integration.

The result is a function of y.

Note that you omit the constant of integration, just as you do for a definite integral of a function of one variable.

Example 1 Finding Partial Integrals

a. $\int_1^x (2x^2y^{-2} + 2y)\, dy = \left[\dfrac{-2x^2}{y} + y^2 \right]_1^x$ Hold x constant.

$$= \left(\dfrac{-2x^2}{x} + x^2 \right) - \left(\dfrac{-2x^2}{1} + 1 \right)$$

$$= 3x^2 - 2x - 1$$

b. $\int_y^{5y} \sqrt{x - y}\, dx = \left[\dfrac{2}{3}(x - y)^{3/2} \right]_y^{5y}$ Hold y constant.

$$= \dfrac{2}{3}[(5y - y)^{3/2} - (y - y)^{3/2}]$$

$$= \dfrac{16}{3}y^{3/2}$$

✓ Checkpoint 1

Find each partial integral.

a. $\displaystyle\int_1^x (4xy + y^3)\, dy$ **b.** $\displaystyle\int_y^{y^2} \dfrac{1}{x + y}\, dx$ ■

In Example 1(a), note that the definite integral defines a function of x and can *itself* be integrated. An "integral of an integral" is called a **double integral.** With a function of two variables, there are two types of double integrals.

$$\int_a^b \int_{g_1(x)}^{g_2(x)} f(x, y)\, dy\, dx = \int_a^b \left[\int_{g_1(x)}^{g_2(x)} f(x, y)\, dy \right] dx$$

$$\int_a^b \int_{g_1(y)}^{g_2(y)} f(x, y)\, dx\, dy = \int_a^b \left[\int_{g_1(y)}^{g_2(y)} f(x, y)\, dx \right] dy$$

STUDY TIP

Notice that the difference between the two types of double integrals is the order in which the integration is performed, $dy\, dx$ or $dx\, dy$.

Example 2 Evaluating a Double Integral

$$\int_1^2 \int_0^x (2xy + 3)\, dy\, dx = \int_1^2 \left[\int_0^x (2xy + 3)\, dy \right] dx$$

$$= \int_1^2 \left[xy^2 + 3y \right]_0^x dx$$

$$= \int_1^2 (x^3 + 3x)\, dx$$

$$= \left[\dfrac{x^4}{4} + \dfrac{3x^2}{2} \right]_1^2$$

$$= \left(\dfrac{2^4}{4} + \dfrac{3(2^2)}{2} \right) - \left(\dfrac{1^4}{4} + \dfrac{3(1^2)}{2} \right)$$

$$= \dfrac{33}{4}$$

TECH TUTOR

A symbolic integration utility can be used to evaluate double integrals. To do this, you need to enter the integrand, then integrate twice—once with respect to one of the variables and then with respect to the other variable. Use a symbolic integration utility to evaluate the double integral in Example 2.

✓ Checkpoint 2

Evaluate $\displaystyle\int_1^2 \int_0^x (5x^2y - 2)\, dy\, dx$. ■

Finding Area with a Double Integral

One of the simplest applications of a double integral is finding the area of a plane region. For instance, consider the region R that is bounded by

$$a \leq x \leq b \quad \text{and} \quad g_1(x) \leq y \leq g_2(x)$$

as shown in Figure 7.42. Using the techniques described in Section 5.5, you know that the area of R is

$$\int_a^b [g_2(x) - g_1(x)]\, dx. \qquad \text{Area of } R$$

This same area is also given by the double integral

$$\int_a^b \int_{g_1(x)}^{g_2(x)} dy\, dx \qquad \text{Area of } R$$

because

$$\int_a^b \int_{g_1(x)}^{g_2(x)} dy\, dx = \int_a^b \left[y \right]_{g_1(x)}^{g_2(x)} dx = \int_a^b [g_2(x) - g_1(x)]\, dx.$$

Figure 7.43 shows the two basic types of plane regions whose areas can be determined by a double integral.

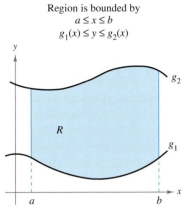

Region is bounded by
$a \leq x \leq b$
$g_1(x) \leq y \leq g_2(x)$

FIGURE 7.42

Determining Area in the Plane by Double Integrals

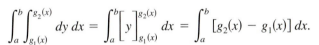

Region is bounded by
$a \leq x \leq b$
$g_1(x) \leq y \leq g_2(x)$

Region is bounded by
$c \leq y \leq d$
$h_1(y) \leq x \leq h_2(y)$

$$\text{Area} = \int_a^b \int_{g_1(x)}^{g_2(x)} dy\, dx$$

$$\text{Area} = \int_c^d \int_{h_1(y)}^{h_2(y)} dx\, dy$$

FIGURE 7.43

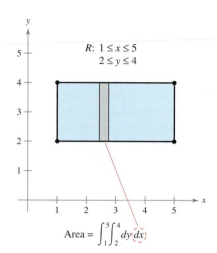

FIGURE 7.44

Example 3 Finding Area with a Double Integral

Use a double integral to find the area of the rectangular region shown in Figure 7.44.

SOLUTION The bounds for x are $1 \leq x \leq 5$ and the bounds for y are $2 \leq y \leq 4$. So, the area of the region is

$$\int_1^5 \int_2^4 dy\, dx = \int_1^5 \left[y \right]_2^4 dx \qquad \text{Integrate with respect to } y.$$

$$= \int_1^5 (4 - 2)\, dx \qquad \text{Apply Fundamental Theorem of Calculus.}$$

$$= \int_1^5 2\, dx \qquad \text{Simplify.}$$

$$= \left[2x \right]_1^5 \qquad \text{Integrate with respect to } x.$$

$$= 10 - 2 \qquad \text{Apply Fundamental Theorem of Calculus.}$$

$$= 8 \text{ square units.} \qquad \text{Simplify.}$$

You can confirm this by noting that the rectangle measures two units by four units.

✔**Checkpoint 3**

Use a double integral to find the area of the rectangular region shown in Example 3 by integrating with respect to x and then with respect to y. ■

Example 4 Finding Area with a Double Integral

Use a double integral to find the area of the region bounded by the graphs of

$$y = x^2 \quad \text{and} \quad y = x^3.$$

SOLUTION As shown in Figure 7.45, the two graphs intersect when $x = 0$ and $x = 1$. Choosing x to be the outer variable, the bounds for x are $0 \leq x \leq 1$. On the interval $0 \leq x \leq 1$, the region is bounded above by $y = x^2$ and below by $y = x^3$. So, the bounds for y are

$$x^3 \leq y \leq x^2.$$

This implies that the area of the region is

$$\int_0^1 \int_{x^3}^{x^2} dy\, dx = \int_0^1 \left[y \right]_{x^3}^{x^2} dx \qquad \text{Integrate with respect to } y.$$

$$= \int_0^1 (x^2 - x^3)\, dx \qquad \text{Apply Fundamental Theorem of Calculus.}$$

$$= \left[\frac{x^3}{3} - \frac{x^4}{4} \right]_0^1 \qquad \text{Integrate with respect to } x.$$

$$= \frac{1}{3} - \frac{1}{4} \qquad \text{Apply Fundamental Theorem of Calculus.}$$

$$= \frac{1}{12} \text{ square unit.} \qquad \text{Simplify.}$$

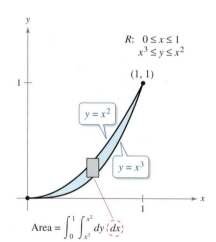

FIGURE 7.45

✔**Checkpoint 4**

Use a double integral to find the area of the region bounded by the graphs of

$$y = 2x \quad \text{and} \quad y = x^2.$$ ■

In setting up double integrals, the most difficult task is likely to be determining the correct limits of integration. This can be simplified by making a sketch of the region R and identifying the appropriate bounds for x and y.

Example 5 Changing the Order of Integration

For the double integral

$$\int_0^2 \int_{y^2}^4 dx\, dy$$

a. sketch the region R whose area is represented by the integral,

b. rewrite the integral so that x is the outer variable, and

c. show that both orders of integration yield the same value.

SOLUTION

a. From the limits of integration, you know that

$$y^2 \le x \le 4 \qquad \text{Inner limits of integration}$$

which means that the region R is bounded on the left by the parabola $x = y^2$ and on the right by the line $x = 4$. Furthermore, because

$$0 \le y \le 2 \qquad \text{Outer limits of integration}$$

you know that the region lies above the x-axis, as shown in Figure 7.46.

b. If you interchange the order of integration so that x is the outer variable, then x will have constant bounds of integration given by

$$0 \le x \le 4. \qquad \text{Outer limits of integration}$$

By solving for y in the equation $x = y^2$, you can conclude that the bounds for y are

$$0 \le y \le \sqrt{x} \qquad \text{Inner limits of integration}$$

as shown in Figure 7.47. So, with x as the outer variable, the integral can be written as

$$\int_0^4 \int_0^{\sqrt{x}} dy\, dx.$$

c. Integrating with respect to x, you have

$$\int_0^2 \int_{y^2}^4 dx\, dy = \int_0^2 \left[x \right]_{y^2}^4 dy = \int_0^2 (4 - y^2)\, dy = \left[4y - \frac{y^3}{3} \right]_0^2 = \frac{16}{3}.$$

Integrating with respect to y, you have

$$\int_0^4 \int_0^{\sqrt{x}} dy\, dx = \int_0^4 \left[y \right]_0^{\sqrt{x}} dx = \int_0^4 \sqrt{x}\, dx = \left[\frac{2}{3} x^{3/2} \right]_0^4 = \frac{16}{3}.$$

So, both orders of integration yield the same value.

✓ Checkpoint 5

For the double integral $\displaystyle\int_0^2 \int_{2y}^4 dx\, dy$,

a. sketch the region R whose area is represented by the integral,

b. rewrite the integral so that x is the outer variable, and

c. show that both orders of integration yield the same result.

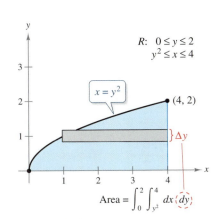

$R: \ 0 \le y \le 2$
$\quad\ y^2 \le x \le 4$

$x = y^2$

$(4, 2)$

$\} \Delta y$

$\text{Area} = \displaystyle\int_0^2 \int_{y^2}^4 dx\, (dy)$

FIGURE 7.46

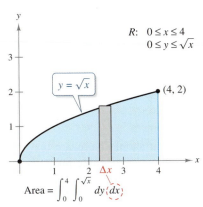

$R: \ 0 \le x \le 4$
$\quad\ 0 \le y \le \sqrt{x}$

$y = \sqrt{x}$

$(4, 2)$

Δx

$\text{Area} = \displaystyle\int_0^4 \int_0^{\sqrt{x}} dy\, (dx)$

FIGURE 7.47

Example 6 Finding Area with a Double Integral

Use a double integral to calculate the area denoted by

$$\int_R\int dA$$

where R is the region bounded by $y = x$ and $y = x^2 - x$.

SOLUTION Begin by sketching the region R, as shown in Figure 7.48. From the sketch, you can see that vertical rectangles of width dx are more convenient than horizontal ones. So, x is the outer variable of integration and its constant bounds are $0 \le x \le 2$. This implies that the bounds for y are $x^2 - x \le y \le x$, and the area is given by

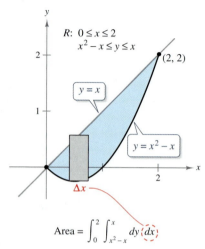

$R:\ 0 \le x \le 2$
$x^2 - x \le y \le x$

$y = x$

$(2, 2)$

$y = x^2 - x$

Δx

Area $= \int_0^2 \int_{x^2-x}^x dy\,dx$

FIGURE 7.48

$$\int_R\int dA = \int_0^2 \int_{x^2-x}^x dy\,dx \qquad \text{Substitute bounds for region.}$$

$$= \int_0^2 \Big[y\Big]_{x^2-x}^x dx \qquad \text{Integrate with respect to } y.$$

$$= \int_0^2 \big[x - (x^2 - x)\big] dx \qquad \text{Apply Fundamental Theorem of Calculus.}$$

$$= \int_0^2 (2x - x^2)\,dx \qquad \text{Simplify.}$$

$$= \Big[x^2 - \frac{x^3}{3}\Big]_0^2 \qquad \text{Integrate with respect to } x.$$

$$= 4 - \frac{8}{3} \qquad \text{Apply Fundamental Theorem of Calculus.}$$

$$= \frac{4}{3} \text{ square units.} \qquad \text{Simplify.}$$

✓ Checkpoint 6

Use a double integral to calculate the area denoted by

$$\int_R\int dA$$

where R is the region bounded by $y = 2x + 3$ and $y = x^2$.

As you are working the exercises for this section, you should be aware that the primary uses of double integrals will be discussed in Section 7.9. Double integrals by way of areas in the plane have been introduced so that you can gain practice in finding the limits of integration. When setting up a double integral, remember that your first step should be to sketch the region R. After doing this, you have two choices of integration orders: $dx\,dy$ or $dy\,dx$.

SUMMARIZE (Section 7.8)

1. Describe a procedure for finding a partial integral with respect to one variable *(page 409)*. For an example of finding a partial integral with respect to x or to y, see Example 1.

2. Explain how to determine the area of a region in the plane using a double integral *(page 411)*. For examples of finding area using a double integral, see Examples 3, 4, and 5.

SKILLS WARM UP 7.8

The following warm-up exercises involve skills that were covered in earlier sections. You will use these skills in the exercise set for this section. For additional help, review Sections 5.2–5.5.

In Exercises 1–12, evaluate the definite integral.

1. $\displaystyle\int_0^1 dx$

2. $\displaystyle\int_0^2 3\,dy$

3. $\displaystyle\int_1^4 2x^2\,dx$

4. $\displaystyle\int_0^1 2x^3\,dx$

5. $\displaystyle\int_1^2 (x^3 - 2x + 4)\,dx$

6. $\displaystyle\int_0^2 (4 - y^2)\,dy$

7. $\displaystyle\int_1^2 \frac{2}{7x^2}\,dx$

8. $\displaystyle\int_1^4 \frac{2}{\sqrt{x}}\,dx$

9. $\displaystyle\int_0^2 \frac{2x}{x^2 + 1}\,dx$

10. $\displaystyle\int_2^e \frac{1}{y - 1}\,dy$

11. $\displaystyle\int_0^2 xe^{x^2 + 1}\,dx$

12. $\displaystyle\int_0^1 e^{-2y}\,dy$

In Exercises 13–16, sketch the region bounded by the graphs of the equations.

13. $y = x,\ y = 0,\ x = 3$

14. $y = x,\ y = 3,\ x = 0$

15. $y = 4 - x^2,\ y = 0,\ x = 0$

16. $y = x^2,\ y = 4x$

ENHANCED WebAssign   Access end-of-section exercises online at **www.webassign.net**

7.9 Applications of Double Integrals

■ Use double integrals to find the volumes of solids.
■ Use double integrals to find the average values of real-life models.

Volume of a Solid Region

In Section 7.8, you used double integrals as an alternative way to find the area of a plane region. In this section, you will study the primary uses of double integrals: to find the volume of a solid region and to find the average value of a function.

Consider a function $z = f(x, y)$ that is continuous and nonnegative over a region R. Let S be the solid region that lies between the xy-plane and the surface

$$z = f(x, y)$$

directly above the region R, as shown in Figure 7.49. You can find the volume of S by integrating

$$f(x, y)$$

over the region R.

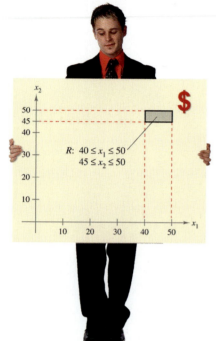

In Exercise 31, you will use a double integral to find the average weekly profit of a company.

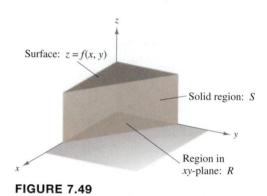

FIGURE 7.49

Determining Volume with Double Integrals

If R is a bounded region in the xy-plane and f is continuous and nonnegative over R, then the **volume of the solid** region between the surface

$$z = f(x, y)$$

and R is given by the double integral

$$\int_R \int f(x, y)\, dA$$

where $dA = dx\, dy$ or $dA = dy\, dx$.

You can use the following guidelines when finding the volume of a solid.

Guidelines for Finding the Volume of a Solid

1. Write the equation of the surface in the form

   $$z = f(x, y)$$

   and sketch the solid region.

2. Sketch the region R in the xy-plane and determine the order and limits of integration.

3. Evaluate the double integral

   $$\int_R \int f(x, y)\, dA$$

   using the order and limits determined in the second step.

Finding the Volume of a Solid

Find the volume of the solid region bounded in the first octant by the plane

$$z = 2 - x - 2y.$$

SOLUTION

1. The equation of the surface is already in the form $z = f(x, y)$. A graph of the solid region is shown in Figure 7.50.

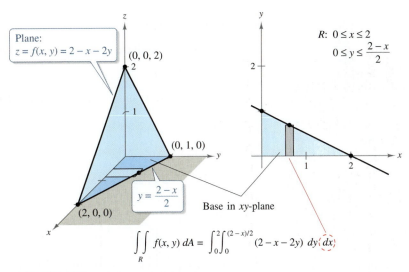

$$\iint_R f(x, y)\, dA = \int_0^2 \int_0^{(2-x)/2} (2 - x - 2y)\, dy\, dx$$

FIGURE 7.50

2. Sketch the region R in the xy-plane. In Figure 7.50, you can see that the region R is bounded by the lines $x = 0$, $y = 0$, and $y = \frac{1}{2}(2 - x)$. One way to set up the double integral is to choose x as the outer variable. With that choice, the bounds for x are $0 \le x \le 2$ and the bounds for y are $0 \le y \le \frac{1}{2}(2 - x)$.

3. The volume of the solid region is

$$
\begin{aligned}
V &= \int_0^2 \int_0^{(2-x)/2} (2 - x - 2y)\, dy\, dx \\
&= \int_0^2 \left[(2 - x)y - y^2 \right]_0^{(2-x)/2} dx \\
&= \int_0^2 \left\{ (2 - x)\left(\frac{1}{2}\right)(2 - x) - \left[\frac{1}{2}(2 - x)\right]^2 \right\} dx \\
&= \frac{1}{4} \int_0^2 (2 - x)^2\, dx \\
&= \frac{1}{4} \left[-\frac{1}{3}(2 - x)^3 \right]_0^2 \\
&= \frac{2}{3} \text{ cubic unit.}
\end{aligned}
$$

✓ **Checkpoint 1**

Find the volume of the solid region bounded in the first octant by the plane $z = 4 - 2x - y$.

Example 1 uses $dy\, dx$ as the order of integration. The other order, $dx\, dy$, as indicated in Figure 7.51, produces the same result. Try verifying this.

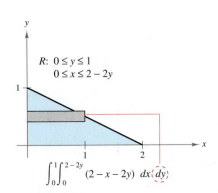

$$\int_0^1 \int_0^{2-2y} (2 - x - 2y)\, dx\, dy$$

FIGURE 7.51

In Example 1, the problem could be solved with either order of integration. Moreover, had you used the order $dx\,dy$, you would have obtained a double integral of comparable difficulty. There are, however, some occasions in which one order of integration is much more convenient than the other. Example 2 shows such a case.

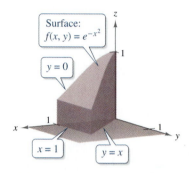

Surface:
$f(x, y) = e^{-x^2}$

$y = 0$

$x = 1$ $y = x$

FIGURE 7.52

Example 2   Comparing Different Orders of Integration

Find the volume of the solid region bounded by the surface

$$f(x, y) = e^{-x^2} \qquad \text{Surface}$$

and the planes $z = 0$, $y = 0$, $y = x$, and $x = 1$, as shown in Figure 7.52.

SOLUTION   In the xy-plane, the bounds of region R are the lines

$$y = 0, \quad x = 1, \quad \text{and} \quad y = x.$$

The two possible orders of integration are indicated in Figure 7.53.

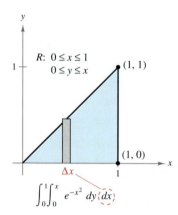

$R: 0 \le x \le 1$
$\quad 0 \le y \le x$
$(1, 1)$

$(1, 0)$

Δx

$\displaystyle\int_0^1 \int_0^x e^{-x^2}\, dy\, dx$

$R: 0 \le y \le 1$
$\quad y \le x \le 1$
$(1, 1)$

Δy

$(1, 0)$

$\displaystyle\int_0^1 \int_y^1 e^{-x^2}\, dx\, dy$

FIGURE 7.53

By setting up the corresponding integrals, you can see that the order $dy\,dx$ produces an integral that is easier to evaluate than the order $dx\,dy$.

$$V = \int_0^1 \int_0^x e^{-x^2}\, dy\, dx$$

$$= \int_0^1 \left[e^{-x^2} y \right]_0^x dx$$

$$= \int_0^1 x e^{-x^2}\, dx$$

$$= \left[-\frac{1}{2} e^{-x^2} \right]_0^1$$

$$= -\frac{1}{2}\left(\frac{1}{e} - 1\right)$$

$$\approx 0.316 \text{ cubic unit}$$

TECH TUTOR

Use a symbolic integration utility to evaluate the double integral in Example 2.

✓ **Checkpoint 2**

Find the volume under the surface

$$f(x, y) = e^{x^2}$$

bounded by the xz-plane and the planes $y = 2x$ and $x = 1$.

In the guidelines for finding the volume of a solid given at the beginning of this section, the first step suggests that you sketch the three-dimensional solid region. This is a good suggestion, but it is not always feasible and is not as important as making a sketch of the two-dimensional region R.

Example 3 Finding the Volume of a Solid

Find the volume of the solid bounded above by the surface

$$f(x, y) = 6x^2 - 2xy$$

and below by the plane region R shown in Figure 7.54.

SOLUTION   Because the region R is bounded by the parabola

$$y = 3x - x^2$$

and the line

$$y = x$$

the limits for y are $x \le y \le 3x - x^2$. The limits for x are $0 \le x \le 2$, and the volume of the solid is

$$V = \int_0^2 \int_x^{3x-x^2} (6x^2 - 2xy)\, dy\, dx$$

$$= \int_0^2 \left[6x^2 y - xy^2 \right]_x^{3x-x^2} dx$$

$$= \int_0^2 \left[(18x^3 - 6x^4 - 9x^3 + 6x^4 - x^5) - (6x^3 - x^3) \right] dx$$

$$= \int_0^2 (4x^3 - x^5)\, dx$$

$$= \left[x^4 - \frac{x^6}{6} \right]_0^2$$

$$= \frac{16}{3} \text{ cubic units.}$$

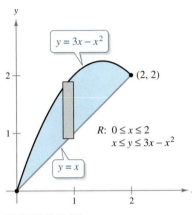

FIGURE 7.54

✓ **Checkpoint 3**

Find the volume of the solid bounded above by the surface

$$f(x, y) = 4x^2 + 2xy$$

and below by the plane region bounded by $y = x^2$ and $y = 2x$.

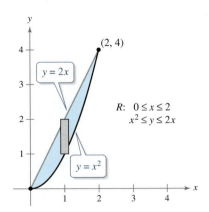

A *population density function*

$$p = f(x, y)$$

is a model that describes the density (in people per square unit) of a region. To find the population of a region R, evaluate the double integral

$$\int_R \int f(x, y) \, dA.$$

Example 4 Finding the Population of a City

The population density (in people per square mile) of the city shown in Figure 7.55 can be modeled by

$$f(x, y) = \frac{50,000}{x + |y| + 1}$$

where x and y are measured in miles. Approximate the city's population. Is the city's average population density less than 10,000 people per square mile?

SOLUTION Because the model involves the absolute value of y, it follows that the population density is symmetrical about the x-axis. So, the population in the first quadrant is equal to the population in the fourth quadrant. This means that you can find the total population by doubling the population in the first quadrant.

$$\text{Population} = 2 \int_0^4 \int_0^5 \frac{50,000}{x + y + 1} \, dy \, dx$$

$$= 100,000 \int_0^4 \int_0^5 \frac{1}{x + y + 1} \, dy \, dx$$

$$= 100,000 \int_0^4 \left[\ln(x + y + 1) \right]_0^5 dx$$

$$= 100,000 \int_0^4 \left[\ln(x + 6) - \ln(x + 1) \right] dx$$

$$= 100,000 \left[(x + 6) \ln(x + 6) - (x + 6) - (x + 1) \ln(x + 1) + (x + 1) \right]_0^4$$

$$= 100,000 \left[(x + 6) \ln(x + 6) - (x + 1) \ln(x + 1) - 5 \right]_0^4$$

$$= 100,000 [10 \ln(10) - 5 \ln(5) - 5 - 6 \ln(6) + 5]$$

$$\approx 422,810 \text{ people}$$

So, the city's population is about 422,810. Because the city covers a region 4 miles wide and 10 miles long, its area is 40 square miles. So, the average population density is

$$\text{Average population density} = \frac{422,810}{40}$$

$$\approx 10,570 \text{ people per square mile.}$$

So, you can conclude that the city's average population density is not less than 10,000 people per square mile.

✓ Checkpoint 4

In Example 4, what integration technique was used to integrate

$$\int \left[\ln(x + 6) - \ln(x + 1) \right] dx?$$

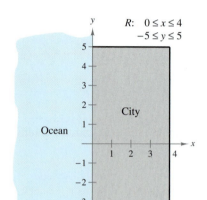

$R: \quad 0 \le x \le 4$
$\quad -5 \le y \le 5$

Ocean

City

FIGURE 7.55

Average Value of a Function over a Region

Average Value of a Function over a Region

If f is integrable over the plane region R with area A, then its **average value** over R is

$$\text{Average value} = \frac{1}{A} \int_R \int f(x, y) \, dA.$$

 **Example 5**   **Finding Average Profit**

A manufacturer determines that the profit for selling x units of one product and y units of a second product is modeled by

$$P = -(x - 200)^2 - (y - 100)^2 + 5000.$$

The weekly sales for product 1 vary between 150 and 200 units, and the weekly sales for product 2 vary between 80 and 100 units. Estimate the average weekly profit for the two products.

SOLUTION   Because $150 \le x \le 200$ and $80 \le y \le 100$, you can estimate the weekly profit to be the average of the profit function over the rectangular region shown in Figure 7.56. Because the area of this rectangular region is $(50)(20) = 1000$, it follows that the average profit V is

$$V = \frac{1}{1000} \int_{150}^{200} \int_{80}^{100} [-(x - 200)^2 - (y - 100)^2 + 5000] \, dy \, dx$$

$$= \frac{1}{1000} \int_{150}^{200} \left[-(x - 200)^2 y - \frac{(y - 100)^3}{3} + 5000y \right]_{80}^{100} dx$$

$$= \frac{1}{1000} \int_{150}^{200} \left[-20(x - 200)^2 - \frac{292{,}000}{3} \right] dx$$

$$= \frac{1}{3000} \left[-20(x - 200)^3 + 292{,}000x \right]_{150}^{200}$$

$$\approx \$4033.$$

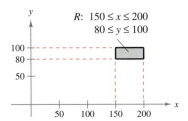

R: $150 \le x \le 200$
      $80 \le y \le 100$

FIGURE 7.56

✓ Checkpoint 5

Find the average value of $f(x, y) = 4 - \frac{1}{2}x - \frac{1}{2}y$ over the region $0 \le x \le 2$ and $0 \le y \le 2$.

SUMMARIZE   (Section 7.9)

1. State the volume of a solid region using double integrals *(page 416)*. For examples of finding the volume of a solid, see Examples 1, 2, and 3.

2. Give the guidelines for finding the volume of a solid *(page 416)*. For examples of using these guidelines, see Examples 1, 2, and 3.

3. Describe a real-life example of how a double integral can be used to find a city's population *(page 420, Example 4)*.

4. State the average value of a function over a region *(page 421)*. For an example of finding the average value of a function, see Example 5.

SKILLS WARM UP 7.9   The following warm-up exercises involve skills that were covered in earlier sections. You will use these skills in the exercise set for this section. For additional help, review Sections 5.4 and 7.8.

In Exercises 1–4, sketch the region that is described.

1. $0 \le x \le 2,\ 0 \le y \le 1$

2. $1 \le x \le 3,\ 2 \le y \le 3$

3. $0 \le x \le 4,\ 0 \le y \le 2x - 1$

4. $0 \le x \le 2,\ 0 \le y \le x^2$

In Exercises 5–10, evaluate the double integral.

5. $\displaystyle \int_0^1 \int_1^2 dy\, dx$

6. $\displaystyle \int_0^3 \int_1^3 dx\, dy$

7. $\displaystyle \int_0^1 \int_0^x x\, dy\, dx$

8. $\displaystyle \int_0^4 \int_1^y y\, dx\, dy$

9. $\displaystyle \int_1^3 \int_x^{x^2} 2\, dy\, dx$

10. $\displaystyle \int_0^1 \int_x^{-x^2+2} dy\, dx$

ALGEBRA TUTOR

Solving Systems of Equations

Three of the sections in this chapter (7.5, 7.6, and 7.7) involve solutions of systems of equations. These systems can be linear or nonlinear, as shown below.

Nonlinear System in Two Variables

$$\begin{cases} 4x + 3y = 6 \\ x^2 - y = 4 \end{cases}$$

Linear System in Three Variables

$$\begin{cases} -x + 2y + 4z = 2 \\ 2x - y + z = 0 \\ 6x + 2z = 3 \end{cases}$$

There are many techniques for solving a system of linear equations. Two of the more common ones are listed here.

1. *Substitution:* Solve for one of the variables in one of the equations and substitute the value into another equation.

2. *Elimination:* Add multiples of one equation to a second equation to eliminate a variable in the second equation.

Example 1   Solving Systems of Equations

Solve each system of equations.

a. $\begin{cases} y - x^3 = 0 \\ x - y^3 = 0 \end{cases}$ b. $\begin{cases} -400p_1 + 300p_2 = -25 \\ 300p_1 - 360p_2 = -535 \end{cases}$

SOLUTION

a. Example 3, page 392

$$\begin{cases} y - x^3 = 0 \\ x - y^3 = 0 \end{cases}$$ Equation 1
    Equation 2

$$y = x^3$$ Solve for y in Equation 1.

$$x - (x^3)^3 = 0$$ Substitute x^3 for y in Equation 2.

$$x - x^9 = 0$$ $(x^m)^n = x^{mn}$

$$x(x - 1)(x + 1)(x^2 + 1)(x^4 + 1) = 0$$ Factor.

$$x = 0$$ Set factors equal to zero.

$$x = 1$$ Set factors equal to zero.

$$x = -1$$ Set factors equal to zero.

b. Example 4, page 393

$$\begin{cases} -400p_1 + 300p_2 = -25 \\ 300p_1 - 360p_2 = -535 \end{cases}$$ Equation 1
    Equation 2

$$p_2 = \tfrac{1}{12}(16p_1 - 1)$$ Solve for p_2 in Equation 1.

$$300p_1 - 360\left(\tfrac{1}{12}\right)(16p_1 - 1) = -535$$ Substitute for p_2 in Equation 2.

$$300p_1 - 30(16p_1 - 1) = -535$$ Multiply factors.

$$-180p_1 = -565$$ Combine like terms.

$$p_1 = \tfrac{113}{36} \approx 3.14$$ Divide each side by -180.

$$p_2 = \tfrac{1}{12}\left[16\left(\tfrac{113}{36}\right) - 1\right]$$ Find p_2 by substituting p_1.

$$p_2 \approx 4.10$$ Solve for p_2.

Example 2   Solving Systems of Equations

Solve each system of equations.

a. $\begin{cases} y(24 - 12x - 4y) = 0 \\ x(24 - 6x - 8y) = 0 \end{cases}$

b. $\begin{cases} 28a - 4b = 10 \\ -4a + 8b = 12 \end{cases}$

SOLUTION

a. Example 5, page 394

Before solving this system of equations, factor 4 out of the first equation and factor 2 out of the second equation.

$$\begin{cases} y(24 - 12x - 4y) = 0 & \text{Original Equation 1} \\ x(24 - 6x - 8y) = 0 & \text{Original Equation 2} \end{cases}$$

$$\begin{cases} y(4)(6 - 3x - y) = 0 & \text{Factor 4 out of Equation 1.} \\ x(2)(12 - 3x - 4y) = 0 & \text{Factor 2 out of Equation 2.} \end{cases}$$

$$\begin{cases} y(6 - 3x - y) = 0 & \text{Equation 1} \\ x(12 - 3x - 4y) = 0 & \text{Equation 2} \end{cases}$$

In each equation, either factor can be 0, so you obtain four different linear systems. For the first system, substitute $y = 0$ into the second equation to obtain $x = 4$.

$$\begin{cases} y = 0 \\ 12 - 3x - 4y = 0 \end{cases} \qquad (4, 0) \text{ is a solution.}$$

You can solve the second system by the method of elimination.

$$\begin{cases} 6 - 3x - y = 0 \\ 12 - 3x - 4y = 0 \end{cases} \qquad \left(\tfrac{4}{3}, 2\right) \text{ is a solution.}$$

The third system is already solved.

$$\begin{cases} y = 0 \\ x = 0 \end{cases} \qquad (0, 0) \text{ is a solution.}$$

You can solve the last system by substituting $x = 0$ into the first equation to obtain $y = 6$.

$$\begin{cases} 6 - 3x - y = 0 \\ x = 0 \end{cases} \qquad (0, 6) \text{ is a solution.}$$

b. Example 2, page 405

$$\begin{cases} 28a - 4b = 10 & \text{Equation 1} \\ -4a + 8b = 12 & \text{Equation 2} \end{cases}$$

$$\begin{aligned} -2a + 4b &= 6 & \text{Divide Equation 2 by 2.} \\ 26a &= 16 & \text{Add new equation to Equation 1.} \\ a &= \tfrac{8}{13} & \text{Divide each side by 26.} \\ 28\left(\tfrac{8}{13}\right) - 4b &= 10 & \text{Substitute for } a \text{ in Equation 1.} \\ b &= \tfrac{47}{26} & \text{Solve for } b. \end{aligned}$$

SUMMARY AND STUDY STRATEGIES

After studying this chapter, you should have acquired the following skills.
The exercise numbers are keyed to the Review Exercises that begin on page 427.
Answers to odd-numbered Review Exercises are given in the back of the text.*

Section 7.1   Review Exercises

- Plot points in space.   *1, 2*
- Find the distance between two points in space.   *3, 4*

$$d = \sqrt{(x_2 - x_1)^2 + (y_2 - y_1)^2 + (z_2 - z_1)^2}$$

- Find the midpoint of a line segment in space.   *5, 6*

$$\text{Midpoint} = \left(\frac{x_1 + x_2}{2}, \frac{y_1 + y_2}{2}, \frac{z_1 + z_2}{2} \right)$$

- Write the standard forms of the equations of spheres.   *7–10*

$$(x - h)^2 + (y - k)^2 + (z - l)^2 = r^2$$

- Find the centers and radii of spheres.   *11, 12*
- Sketch the coordinate plane traces of spheres.   *13, 14*

Section 7.2

- Sketch planes in space.   *15–18*
- Classify quadric surfaces in space.   *19–26*

Section 7.3

- Evaluate functions of several variables.   *27, 28*
- Find the domains and ranges of functions of two variables.   *29–32*
- Sketch level curves of functions of two variables.   *33–36*
- Use functions of several variables to answer questions about real-life situations.   *37–40*

Section 7.4

- Find the first partial derivatives of functions of several variables.   *41–50*
- Find the slopes of surfaces in the *x*- and *y*-directions.   *51–54*
- Find the second partial derivatives of functions of several variables.   *55–60*
- Use partial derivatives to answer questions about real-life situations.   *61, 62*

Section 7.5

- Find the relative extrema of functions of two variables.   *63–70*
- Use relative extrema to answer questions about real-life situations.   *71, 72*

---

* A wide range of valuable study aids are available to help you master the material in this chapter.
 The *Student Solutions Manual* includes step-by-step solutions to all odd-numbered exercises to
 help you review and prepare. The student website at *www.cengagebrain.com* offers algebra help
 and a *Graphing Technology Guide*, which contains step-by-step commands and instructions for
 a wide variety of graphing calculators.

Section 7.6

■ Use Lagrange multipliers to find extrema of functions of several variables.   73–78

■ Use Lagrange multipliers to answer questions about real-life situations.   79, 80

Section 7.7

■ Find the least squares regression line, $y = ax + b$, for data.   81, 82

$$a = \left[n\sum_{i=1}^{n} x_i y_i - \sum_{i=1}^{n} x_i \sum_{i=1}^{n} y_i \right] \Big/ \left[n\sum_{i=1}^{n} x_i^2 - \left(\sum_{i=1}^{n} x_i \right)^2 \right], \quad b = \frac{1}{n}\left(\sum_{i=1}^{n} y_i - a\sum_{i=1}^{n} x_i \right)$$

■ Use least squares regression lines to model real-life data.   83, 84

Section 7.8

■ Evaluate double integrals.   85–88

■ Use double integrals to find the areas of regions.   89–92

Section 7.9

■ Use double integrals to find the volumes of solids.   93–98

$$\text{Volume} = \int_R \int f(x, y) \, dA$$

■ Use double integrals to find the average values of functions.   99–103

$$\text{Average value} = \frac{1}{A} \int_R \int f(x, y) \, dA$$

Study Strategies

■ **Comparing Two Dimensions with Three Dimensions**   Many of the formulas and techniques in this chapter are generalizations of formulas and techniques used in earlier chapters of the text. Here are several examples.

Two-Dimensional Coordinate System	Three-Dimensional Coordinate System
Distance Formula $d = \sqrt{(x_2 - x_1)^2 + (y_2 - y_1)^2}$	*Distance Formula* $d = \sqrt{(x_2 - x_1)^2 + (y_2 - y_1)^2 + (z_2 - z_1)^2}$
Midpoint Formula $\text{Midpoint} = \left(\dfrac{x_1 + x_2}{2}, \dfrac{y_1 + y_2}{2} \right)$	*Midpoint Formula* $\text{Midpoint} = \left(\dfrac{x_1 + x_2}{2}, \dfrac{y_1 + y_2}{2}, \dfrac{z_1 + z_2}{2} \right)$
Equation of Circle $(x - h)^2 + (y - k)^2 = r^2$	*Equation of Sphere* $(x - h)^2 + (y - k)^2 + (z - l)^2 = r^2$
Equation of Line $ax + by = c$	*Equation of Plane* $ax + by + cz = d$
Derivative of $y = f(x)$ $\dfrac{dy}{dx} = \lim_{\Delta x \to 0} \dfrac{f(x + \Delta x) - f(x)}{\Delta x}$	*Partial Derivative of* $z = f(x, y)$ $\dfrac{\partial z}{\partial x} = \lim_{\Delta x \to 0} \dfrac{f(x + \Delta x, y) - f(x, y)}{\Delta x}$
Area of Region $A = \displaystyle\int_a^b f(x) \, dx$	*Volume of Region* $V = \displaystyle\int_R \int f(x, y) \, dA$

Review Exercises

Plotting Points in Space In Exercises 1 and 2, plot the points in the same three-dimensional coordinate system.

1. $(2, -1, 4), (-1, 3, -3), (-2, -2, 1), (3, 1, 2)$
2. $(1, -2, -3), (-4, -3, 5), \left(4, \frac{5}{2}, 1\right), (-2, 2, 2)$

Finding the Distance Between Two Points In Exercises 3 and 4, find the distance between the two points.

3. $(1, 0, 2), (3, 5, 8)$
4. $(-4, 1, 5), (1, 3, 7)$

Using the Midpoint Formula In Exercises 5 and 6, find the midpoint of the line segment joining the two points.

5. $(2, 6, 4), (-4, 2, 8)$
6. $(5, 0, 7), (-1, -2, 9)$

Finding the Equation of a Sphere In Exercises 7–10, find the standard equation of the sphere.

7. Center: $(0, 1, 0)$; radius: 5
8. Center: $(4, -5, 3)$; radius: 10
9. Diameter endpoints: $(3, -4, -1), (1, 0, -5)$
10. Diameter endpoints: $(3, 4, 0), (5, 8, 2)$

Finding the Center and Radius of a Sphere In Exercises 11 and 12, find the center and radius of the sphere.

11. $x^2 + y^2 + z^2 - 8x + 4y - 6z - 20 = 0$
12. $x^2 + y^2 + z^2 + 4y - 10z - 7 = 0$

Finding the Trace of a Surface In Exercises 13 and 14, sketch the *xy*-trace of the sphere.

13. $(x + 2)^2 + (y - 1)^2 + (z - 3)^2 = 25$
14. $(x - 1)^2 + (y + 3)^2 + (z - 6)^2 = 72$

Sketching a Plane in Space In Exercises 15–18, find the intercepts and sketch the graph of the plane.

15. $x + 2y + 3z = 6$ 16. $2y + z = 4$
17. $3x - 6z = 12$ 18. $4x - y + 2z = 8$

Classifying a Quadric Surface In Exercises 19–26, classify the quadric surface.

19. $x^2 + y^2 + z^2 - 2x + 4y - 6z + 5 = 0$
20. $16x^2 + 16y^2 - 9z^2 = 0$
21. $x^2 + \dfrac{y^2}{16} + \dfrac{z^2}{9} = 1$
22. $x^2 - \dfrac{y^2}{16} - \dfrac{z^2}{9} = 1$
23. $z = \dfrac{x^2}{9} + y^2$
24. $-4x^2 + y^2 + z^2 = 4$
25. $z = \sqrt{x^2 + y^2}$
26. $z = x^2 - \dfrac{y^2}{4}$

Evaluating Functions of Several Variables In Exercises 27 and 28, find the function values.

27. $f(x, y) = xy^2$
   (a) $f(2, 3)$ (b) $f(0, 1)$
   (c) $f(-5, 7)$ (d) $f(-2, -4)$

28. $f(x, y) = \dfrac{x^2}{y}$
   (a) $f(6, 9)$ (b) $f(8, 4)$ (c) $f(t, 2)$ (d) $f(r, r)$

Finding the Domain and Range of a Function In Exercises 29–32, find the domain and range of the function.

29. $f(x, y) = \sqrt{1 - x^2 - y^2}$
30. $f(x, y) = x^2 + y^2 - 3$
31. $f(x, y) = e^{xy}$
32. $f(x, y) = \dfrac{1}{x + y}$

Sketching a Contour Map In Exercises 33–36, describe the level curves of the function. Sketch a contour map of the surface using level curves for the given *c*-values.

Function	c-Values
33. $z = 10 - 2x - 5y$	$c = 0, 2, 4, 5, 10$
34. $z = \sqrt{9 - x^2 - y^2}$	$c = 0, 1, 2, 3$
35. $z = (xy)^2$	$c = 1, 4, 9, 12, 16$
36. $z = y - x^2$	$c = 0, \pm 1, \pm 2$

37. **Meteorology** The contour map shown below represents the average yearly precipitation for Oklahoma. (*Source: National Climatic Data Center*)

 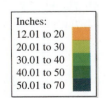

Inches:	
12.01 to 20	
20.01 to 30	
30.01 to 40	
40.01 to 50	
50.01 to 70	

(a) Do the level curves correspond to equally spaced levels of precipitation? Explain.

(b) Describe how to obtain a more detailed contour map.

38. Chemistry The acidity of rainwater is measured in units called pH, and smaller pH values are increasingly acidic. The map shows the curves of equal pH and gives evidence that downwind of heavily industrialized areas, the acidity has been increasing. Using the level curves on the map, determine the direction of the prevailing winds in the northeastern United States.

39. Earnings per Share The earnings per share z (in dollars) for Hewlett-Packard from 2003 through 2010 can be modeled by

$$z = -4.51 + 0.046x + 0.060y$$

where x is the sales (in billions of dollars) and y is the shareholder's equity (in billions of dollars). *(Source: Hewlett-Packard Company)*

(a) Find the earnings per share when $x = 100$ and $y = 40$.

(b) Which of the two variables in this model has the greater influence on the earnings per share? Explain.

40. Shareholder's Equity The shareholder's equity z (in billions of dollars) for Wal-Mart from 2000 through 2010 can be modeled by

$$z = 1.54 + 0.116x + 0.122y$$

where x is the net sales (in billions of dollars) and y is the total assets (in billions of dollars). *(Source: Wal-Mart Stores, Inc.)*

(a) Find the shareholder's equity when $x = 300$ and $y = 130$.

(b) Which of the two variables in this model has the greater influence on shareholder's equity? Explain.

Finding Partial Derivatives In Exercises 41–50, find the first partial derivatives.

41. $f(x, y) = x^2y + 3xy + 2x - 5y$

42. $f(x, y) = 4xy + xy^2 - 3x^2y$

43. $z = \dfrac{x^2}{y^2}$

44. $z = (xy + 2x + 4y)^2$

45. $f(x, y) = \ln(5x + 4y)$

46. $f(x, y) = \ln\sqrt{2x + 3y}$

47. $f(x, y) = xe^y + ye^x$

48. $f(x, y) = x^2e^{-2y}$

49. $w = xyz^2$

50. $w = 3xy - 5xz + 2yz$

Finding Slopes in the x- and y-Directions In Exercises 51–54, find the slopes of the surface at the given point in (a) the x-direction and (b) the y-direction.

51. $z = 3xy$
    $(-2, -3, 18)$

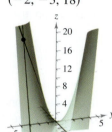

52. $z = y^2 - x^2$
    $(1, 2, 3)$

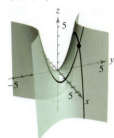

53. $z = 8 - x^2 - y^2$
    $(1, 1, 6)$

54. $z = \sqrt{100 - x^2 - y^2}$
    $(0, 6, 8)$

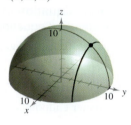

Finding Second Partial Derivatives In Exercises 55–60, find all second partial derivatives.

55. $f(x, y) = 3x^2 - xy + 2y^3$

56. $f(x, y) = \dfrac{y}{x + y}$

57. $f(x, y) = \sqrt{1 + x + y}$

58. $f(x, y) = x^2e^{-y^2}$

59. $f(x, y, z) = xy + 5x^2yz^3 - 3y^3z$

60. $f(x, y, z) = \dfrac{3yz}{x + z}$

61. Marginal Cost A company manufactures two models of skis: cross-country skis and downhill skis. The cost function for producing x pairs of cross-country skis and y pairs of downhill skis is given by

$$C = 15(xy)^{1/3} + 99x + 139y + 2293.$$

(a) Find the marginal costs ($\partial C/\partial x$ and $\partial C/\partial y$) when $x = 500$ and $y = 250$.

(b) When additional production is required, which model of skis results in the cost increasing at a higher rate? How can this be determined from the cost model?

62. Marginal Revenue At a baseball stadium, souvenir hats are sold at two locations. If x_1 and x_2 are the numbers of baseball hats sold at location 1 and location 2, respectively, then the total revenue for the hats is modeled by

$$R = 15x_1 + 16x_2 - \frac{1}{10}x_1^2 - \frac{1}{10}x_2^2 - \frac{1}{100}x_1x_2.$$

When $x_1 = 50$ and $x_2 = 40$, find

(a) the marginal revenue for location 1, $\partial R/\partial x_1$.

(b) the marginal revenue for location 2, $\partial R/\partial x_2$.

Applying the Second-Partials Test In Exercises 63–70, find the critical points, relative extrema, and saddle points of the function.

63. $f(x, y) = x^2 + 2y^2$

64. $f(x, y) = x^3 - 3xy + y^2$

65. $f(x, y) = 1 - (x + 2)^2 + (y - 3)^2$

66. $f(x, y) = e^x - x + y^2$

67. $f(x, y) = x^3 + y^2 - xy$

68. $f(x, y) = y^2 + xy + 3y - 2x + 5$

69. $f(x, y) = x^3 + y^3 - 3x - 3y + 2$

70. $f(x, y) = -x^2 - y^2$

71. Revenue A company manufactures and sells two products. The demand functions for the products are given by

$$p_1 = 100 - x_1 \quad \text{and} \quad p_2 = 200 - 0.5x_2$$

where p_1 and p_2 are the prices per unit (in dollars) and x_1 and x_2 are the numbers of units sold. The total revenue function is given by

$$R = x_1p_1 + x_2p_2.$$

Find x_1 and x_2 so as to maximize revenue.

72. Profit A company manufactures a product at two locations. The cost of producing x_1 units at location 1 is

$$C_1 = 0.03x_1^2 + 4x_1 + 300$$

and the cost of producing x_2 units at location 2 is

$$C_2 = 0.05x_2^2 + 7x_2 + 175.$$

The product sells for $10 per unit. Find the quantity that should be produced at each location to maximize the profit

$$P = 10(x_1 + x_2) - C_1 - C_2.$$

Using Lagrange Multipliers In Exercises 73–78, use Lagrange multipliers to find the given extremum. In each case, assume that the variables are positive.

73. Maximize $f(x, y) = 2xy$.

Constraint: $2x + y = 12$

74. Maximize $f(x, y) = 2x + 3xy + y$.

Constraint: $x + 2y = 29$

75. Minimize $f(x, y) = x^2 + y^2$.

Constraint: $x + y = 4$

76. Minimize $f(x, y) = 3x^2 - y^2$.

Constraint: $2x - 2y + 5 = 0$

77. Maximize $f(x, y, z) = xyz$.

Constraint: $x + 2y + z - 4 = 0$

78. Maximize $f(x, y, z) = x^2z + yz$.

Constraint: $2x + y + z = 5$

79. Cost A manufacturer has an order for 1000 units of wooden benches that can be produced at two locations. Let x_1 and x_2 be the numbers of units produced at the two locations. The cost function is modeled by

$$C = 0.25x_1^2 + 10x_1 + 0.15x_2^2 + 12x_2.$$

Use Lagrange multipliers to find the number of units that should be produced at each location to minimize the cost.

80. Production The production function for a manufacturer is given by

$$f(x, y) = 4x + xy + 2y$$

where x is the number of units of labor (at $20 per unit) and y is the number of units of capital (at $4 per unit). The total cost for labor and capital cannot exceed $2000. Use Lagrange multipliers to find the maximum production level for this manufacturer.

Finding the Least Squares Regression Line In Exercises 81 and 82, find the least squares regression line for the given points. Then plot the points and sketch the regression line.

81. $(-2, -3), (-1, -1), (1, 2), (3, 2)$

82. $(-3, -1), (-2, -1), (0, 0), (1, 1), (2, 1)$

83. Demand A store manager wants to know the demand y for a digital camera as a function of price x. The monthly sales for four different prices of the digital camera are listed in the table.

Price, x	$80	$90	$100	$110
Demand, y	140	117	91	63

(a) Use the regression capabilities of a graphing utility or a spreadsheet to find the least squares regression line for the data.

(b) Estimate the demand when the price is $85.

(c) What price will create a demand of 200 cameras?

84. Work Force The number of men x (in millions) and the number of women y (in millions) in the labor force from 2001 through 2010 are shown in the table. *(Source: U.S. Bureau of Labor Statistics)*

Year	2001	2002	2003	2004	2005
Men, x	76.9	77.5	78.2	79.0	80.0
Women, y	66.8	67.4	68.3	68.4	69.3

Year	2006	2007	2008	2009	2010
Men, x	81.3	82.1	82.5	82.1	82.0
Women, y	70.2	71.0	71.8	72.0	71.9

(a) Use the regression capabilities of a graphing utility or a spreadsheet to find the least squares regression line for the data.

(b) Estimate the number of women in the labor force when there are 80 million men in the labor force.

Evaluating a Double Integral In Exercises 85–88, evaluate the double integral.

85. $\displaystyle\int_0^1 \int_0^{1+x} (4x - 2y)\, dy\, dx$

86. $\displaystyle\int_{-3}^3 \int_0^4 (x - y^2)\, dx\, dy$

87. $\displaystyle\int_1^2 \int_1^{2y} \frac{x}{y^2}\, dx\, dy$

88. $\displaystyle\int_0^4 \int_0^{\sqrt{16-x^2}} 2x\, dy\, dx$

Finding Area with a Double Integral In Exercises 89–92, use a double integral to find the area of the region bounded by the graphs of the equations.

89. $y = 9 - x^2,\ y = 5$

90. $y = \dfrac{4}{x},\ y = 0,\ x = 1,\ x = 4$

91. $y = \sqrt{x + 3},\ y = \dfrac{1}{3}x + 1$

92. $y = x^2 - 2x - 2,\ y = -x$

Finding the Volume of a Solid In Exercises 93–96, use a double integral to find the volume of the specified solid.

93.

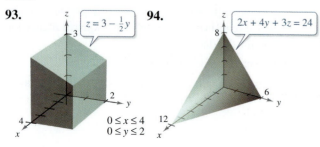

94.

95.

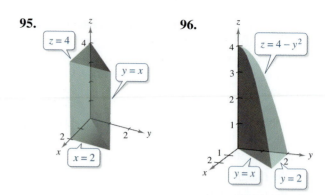

96.

Finding the Volume of a Solid In Exercises 97 and 98, use a double integral to find the volume of the solid bounded by the graphs of the equations.

97. $z = (xy)^2,\ z = 0,\ y = 0,\ y = 4,\ x = 0,\ x = 4$

98. $z = x + y,\ z = 0,\ x = 0,\ x = 3,\ y = x,\ y = 0$

Average Value of a Function over a Region In Exercises 99 and 100, find the average value of $f(x, y)$ over the region R.

99. $f(x, y) = xy$

R: rectangle with vertices $(0, 0),\ (4, 0),\ (4, 3),\ (0, 3)$

100. $f(x, y) = x^2 + 2xy + y^2$

R: rectangle with vertices $(0, 0),\ (2, 0),\ (2, 5),\ (0, 5)$

101. Average Weekly Profit A firm's weekly profit (in dollars) in marketing two products is given by

$$P = 150x_1 + 400x_2 - x_1^2 - 5x_2^2 - 2x_1x_2 - 3000$$

where x_1 and x_2 represent the numbers of units of each product sold weekly. Estimate the average weekly profit when x_1 varies between 30 and 40 units and x_2 varies between 40 and 50 units.

102. Average Revenue A company sells two products whose demand functions are given by

$$x_1 = 500 - 2.5p_1 \quad \text{and} \quad x_2 = 750 - 3p_2.$$

So, the total revenue is given by

$$R = x_1 p_1 + x_2 p_2.$$

Estimate the average revenue when price p_1 varies between \$25 and \$50 and price p_2 varies between \$75 and \$125.

103. Real Estate The value of real estate (in dollars per square foot) for a city is given by

$$f(x, y) = 0.003x^{2/3}y^{3/4}$$

where x and y are measured in feet. What is the average value of real estate inside the rectangular area defined by the vertices $(0, 0),\ (5280, 0),\ (5280, 3960),$ and $(0, 3960)$?

TEST YOURSELF

Take this test as you would take a test in class. When you are done, check your work against the answers given in the back of the book.

In Exercises 1–3, (a) plot the points in a three-dimensional coordinate system, (b) find the distance between the points, and (c) find the midpoint of the line segment joining the points.

1. $(1, -3, 0), (3, -1, 0)$ **2.** $(-2, 2, 3), (-4, 0, 2)$ **3.** $(3, -7, 2), (5, 11, -6)$

4. Find the center and radius of the sphere whose equation is

$$x^2 + y^2 + z^2 - 20x + 10y - 10z + 125 = 0.$$

In Exercise 5–7, classify the quadric surface.

5. $4x^2 + 2y^2 - z^2 = 16$ **6.** $36x^2 + 9y^2 - 4z^2 = 0$

7. $4x^2 - y^2 - 16z = 0$

In Exercises 8–10, find $f(3, 3)$ and $f(1, 4)$.

8. $f(x, y) = x^2 + xy + 1$ **9.** $f(x, y) = \dfrac{x + 2y}{3x - y}$

10. $f(x, y) = xy \ln \dfrac{x}{y}$

In Exercises 11 and 12, find the first partial derivatives and evaluate each at the point $(10, -1)$.

11. $f(x, y) = 3x^2 + 9xy^2 - 2$ **12.** $f(x, y) = x\sqrt{x + y}$

In Exercises 13 and 14, find the critical points, relative extrema, and saddle points of the function.

13. $f(x, y) = 3x^2 + 4y^2 - 6x + 16y - 4$

14. $f(x, y) = 4xy - x^4 - y^4$

15. The production function for a company is given by

$$f(x, y) = 60x^{0.7}y^{0.3}$$

where x is the number of units of labor (at \$42 per unit) and y is the number of units of capital (at \$144 per unit). The total cost for labor and capital cannot exceed \$240,000. Use Lagrange multipliers to find the maximum production level for this manufacturer.

16. Find the least squares regression line for the points $(1, 2), (3, 3), (6, 4), (8, 6)$, and $(11, 7)$.

In Exercises 17 and 18, evaluate the double integral.

17. $\displaystyle\int_0^1 \int_x^1 (30x^2y - 1)\, dy\, dx$ **18.** $\displaystyle\int_0^{\sqrt{e-1}} \int_0^{2y} \dfrac{1}{y^2 + 1}\, dx\, dy$

19. Use a double integral to find the area of the region bounded by the graphs of $y = 3$ and $y = x^2 - 2x + 3$ (see figure).

20. Use a double integral to find the volume of the solid bounded by the graphs of $z = 8 - 2x, z = 0, y = 0, y = 3, x = 0$, and $x = 4$.

21. Find the average value of $f(x, y) = x^2 + y$ over the region defined by a rectangle with vertices $(0, 0), (1, 0), (1, 3)$, and $(0, 3)$.

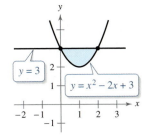

Figure for 19

Appendices

Appendices D, E, and F are located on the website that accompanies this text at *www.cengagebrain.com*.

A Precalculus Review

A.1 The Real Number Line and Order

- Represent, classify, and order real numbers.
- Use inequalities to represent sets of real numbers.
- Solve inequalities.
- Use inequalities to model and solve real-life problems.

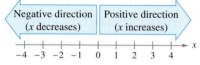

The Real Number Line

FIGURE A.1

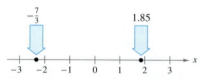

Every point on the real number line corresponds to one and only one real number.

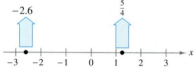

Every real number corresponds to one and only one point on the real number line.

FIGURE A.2

The Real Number Line

Real numbers can be represented with a coordinate system called the **real number line** (or x-axis), as shown in Figure A.1. The **positive direction** (to the right) is denoted by an arrowhead and indicates the direction of increasing values of x. The real number corresponding to a particular point on the real number line is called the **coordinate** of the point. As shown in Figure A.1, it is customary to label those points whose coordinates are integers.

The point on the real number line corresponding to zero is called the **origin.** Numbers to the right of the origin are **positive,** and numbers to the left of the origin are **negative.** The term **nonnegative** describes a number that is either positive or zero.

The importance of the real number line is that it provides you with a conceptually perfect picture of the real numbers. That is, each point on the real number line corresponds to one and only one real number, and each real number corresponds to one and only one point on the real number line. This type of relationship is called a **one-to-one correspondence** and is illustrated in Figure A.2.

Each of the four points in Figure A.2 corresponds to a real number that can be expressed as the ratio of two integers.

$$-2.6 = -\frac{13}{5} \qquad \frac{5}{4} \qquad -\frac{7}{3} \qquad 1.85 = \frac{37}{20}$$

Such numbers are called **rational.** Rational numbers have either terminating or infinitely repeating decimal representations.

Terminating Decimals	*Infinitely Repeating Decimals*
$\dfrac{2}{5} = 0.4$	$\dfrac{1}{3} = 0.333\ldots = 0.\overline{3}*$
$\dfrac{7}{8} = 0.875$	$\dfrac{12}{7} = 1.714285714285\ldots = 1.\overline{714285}$

Real numbers that are not rational are called **irrational,** and they cannot be represented as the ratio of two integers (or as terminating or infinitely repeating decimals). So, a decimal approximation is used to represent an irrational number. Some irrational numbers occur so frequently in applications that mathematicians have invented special symbols to represent them. For example, the symbols $\sqrt{2}$, π, and e represent irrational numbers whose decimal approximations are as shown. (See Figure A.3.)

$$\sqrt{2} \approx 1.4142135623$$
$$\pi \approx 3.1415926535$$
$$e \approx 2.7182818284$$

FIGURE A.3

---

*The bar indicates which digit or digits repeat infinitely.

Order and Intervals on the Real Number Line

One important property of the real numbers is that they are **ordered**: 0 is less than 1, -3 is less than -2.5, π is less than $\frac{22}{7}$, and so on. You can visualize this property on the real number line by observing that a is less than b if and only if a lies to the left of b on the real number line. Symbolically, "a is less than b" is denoted by the inequality $a < b$. For example, the inequality

$$\tfrac{3}{4} < 1$$

follows from the fact that $\frac{3}{4}$ lies to the left of 1 on the real number line, as shown in Figure A.4.

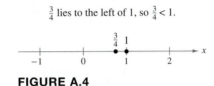

$\frac{3}{4}$ lies to the left of 1, so $\frac{3}{4} < 1$.

FIGURE A.4

When three real numbers a, x, and b are ordered such that $a < x$ and $x < b$, we say that x is **between** a and b and write

$a < x < b.$ *x is between a and b.*

The set of *all* real numbers between a and b is called the **open interval** between a and b and is denoted by (a, b). An interval of the form (a, b) does not contain the "endpoints" a and b. Intervals that include their endpoints are called **closed** and are denoted by $[a, b]$. Intervals of the form $[a, b)$ and $(a, b]$ are neither open nor closed. Figure A.5 shows the nine types of intervals on the real number line.

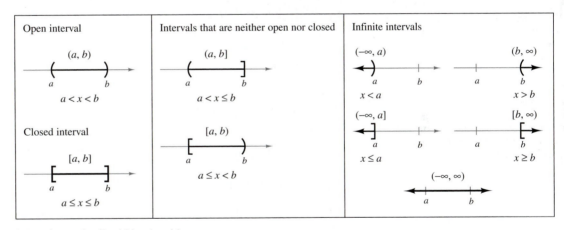

Intervals on the Real Number Line

FIGURE A.5

Note that a square bracket is used to denote "less than or equal to" ($\le$) or "greater than or equal to" ($\ge$). Furthermore, the symbols

∞ Positive infinity

and

$-\infty$ Negative infinity

denote **positive** and **negative infinity,** respectively. These symbols do not denote real numbers; they merely let you describe unbounded conditions more concisely. For instance, the interval $[b, \infty)$ is unbounded to the right because it includes *all* real numbers that are greater than or equal to b.

Solving Inequalities

In calculus, you are frequently required to "solve inequalities" involving variable expressions such as $3x - 4 < 5$. The number a is a **solution** of an inequality if the inequality is true when a is substituted for x. The set of all values of x that satisfy an inequality is called the **solution set** of the inequality. The following properties are useful for solving inequalities. (Similar properties are obtained when $<$ is replaced by $\leq$ and $>$ is replaced by $\geq$.)

STUDY TIP

Notice the differences between Properties 3 and 4. For example,

$$-3 < 4 \implies (-3)(2) < (4)(2)$$

and

$$-3 < 4$$
$$\implies (-3)(-2) > (4)(-2).$$

Properties of Inequalities

Let a, b, c, and d be real numbers.

1. Transitive property: $a < b$ and $b < c$ ➡ $a < c$

2. Adding inequalities: $a < b$ and $c < d$ ➡ $a + c < b + d$

3. Multiplying by a (positive) constant: $a < b$ ➡ $ac < bc, \quad c > 0$

4. Multiplying by a (negative) constant: $a < b$ ➡ $ac > bc, \quad c < 0$

5. Adding a constant: $a < b$ ➡ $a + c < b + c$

6. Subtracting a constant: $a < b$ ➡ $a - c < b - c$

Note that you *reverse the inequality* when you multiply by a negative number. For example, if $x < 3$, then $-4x > -12$. This principle also applies to division by a negative number. So, if $-2x > 4$, then $x < -2$.

Example 1   Solving an Inequality

Find the solution set of the inequality

$$3x - 4 < 5.$$

SOLUTION

$3x - 4 < 5$	Write original inequality.
$3x - 4 + 4 < 5 + 4$	Add 4 to each side.
$3x < 9$	Simplify.
$\dfrac{1}{3}(3x) < \dfrac{1}{3}(9)$	Multiply each side by $\frac{1}{3}$.
$x < 3$	Simplify.

So, the solution set is the interval $(-\infty, 3)$, as shown in Figure A.6. Once you have solved an inequality, it is a good idea to check some x-values in your solution set to see whether they satisfy the original inequality. You should also check some values outside your solution set to verify that they do *not* satisfy the inequality. For instance, Figure A.6 shows that when $x = 0$ or $x = 2$, the inequality is satisfied, but when $x = 4$, the inequality is not satisfied.

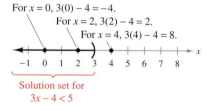

For $x = 0$, $3(0) - 4 = -4$.
For $x = 2$, $3(2) - 4 = 2$.
For $x = 4$, $3(4) - 4 = 8$.

Solution set for
$3x - 4 < 5$

FIGURE A.6

✓ Checkpoint 1

Find the solution set of the inequality

$$2x - 3 < 7.$$

In Example 1, all five inequalities listed as steps in the solution have the same solution set, and they are called **equivalent inequalities.**

The inequality in Example 1 involves a first-degree polynomial. To solve inequalities involving polynomials of higher degree, you can use the fact that a polynomial can change signs *only* at its real zeros (the real numbers that make the polynomial zero). Between two consecutive real zeros, a polynomial must be entirely positive or entirely negative. This means that when the real zeros of a polynomial are put in order, they divide the real number line into **test intervals** in which the polynomial has no sign changes. That is, if a polynomial has the factored form

$$(x - r_1)(x - r_2), \ldots, (x - r_n), \qquad r_1 < r_2 < r_3 < \cdots < r_{n-1} < r_n$$

then the test intervals are

$$(-\infty, r_1), \quad (r_1, r_2), \quad \ldots, \quad (r_{n-1}, r_n), \quad \text{and} \quad (r_n, \infty).$$

For example, the polynomial

$$x^2 - x - 6 = (x - 3)(x + 2)$$

can change signs only at $x = -2$ and $x = 3$. To determine the sign of the polynomial in the intervals $(-\infty, -2)$, $(-2, 3)$, and $(3, \infty)$, you need to test only *one value* in each interval.

Example 2   Solving a Polynomial Inequality

$$x^2 < x + 6 \qquad \text{Original inequality}$$
$$x^2 - x - 6 < 0 \qquad \text{Polynomial form}$$
$$(x - 3)(x + 2) < 0 \qquad \text{Factor.}$$

So, the polynomial $x^2 - x - 6$ has $x = -2$ and $x = 3$ as its zeros. You can solve the inequality by testing the sign of the polynomial in each of the intervals $(-\infty, -2)$, $(-2, 3)$, and $(3, \infty)$. In each interval, choose a representative x-value and evaluate the polynomial.

Interval	x-Value	Polynomial Value	Conclusion
$(-\infty, -2)$	$x = -3$	$(-3)^2 - (-3) - 6 = 6$	Positive
$(-2, 3)$	$x = 0$	$(0)^2 - (0) - 6 = -6$	Negative
$(3, \infty)$	$x = 4$	$(4)^2 - (4) - 6 = 6$	Positive

From this you can conclude that the inequality is satisfied for all x-values in $(-2, 3)$. This implies that the solution of the inequality $x^2 < x + 6$ is the interval $(-2, 3)$, as shown in Figure A.7. Note that the original inequality contains a "less than" symbol. This means that the solution set does not contain the endpoints of the test interval $(-2, 3)$.

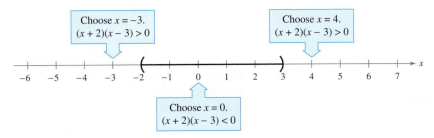

FIGURE A.7

✓ Checkpoint 2

Find the solution set of the inequality

$$x^2 > 3x + 10.$$

Application

Inequalities are frequently used to describe conditions that occur in business and science. For instance, the inequality

$$8.8 \le W \le 26.4$$

describes the typical weights W (in pounds) of adult rhesus monkeys. Example 3 shows how an inequality can be used to describe the production levels in a manufacturing plant.

Example 3   Production Levels

In addition to fixed overhead costs of $500 per day, the cost of producing x units of an item is $2.50 per unit. During the month of August, the total cost of production varied from a high of $1325 to a low of $1200 per day. Find the high and low *production levels* during the month.

SOLUTION   Because it costs $2.50 to produce one unit, it costs $2.5x$ to produce x units. Furthermore, because the fixed cost per day is $500, the total daily cost C (in dollars) of producing x units is

$$C = 2.5x + 500.$$

Now, because the cost ranged from $1200 to $1325, you can write the following.

$1200 \le$	$2.5x + 500$	≤ 1325	Write original inequality.
$1200 - 500 \le$	$2.5x + 500 - 500$	$\le 1325 - 500$	Subtract 500 from each part.
$700 \le$	$2.5x$	≤ 825	Simplify.
$\dfrac{700}{2.5} \le$	$\dfrac{2.5x}{2.5}$	$\le \dfrac{825}{2.5}$	Divide each part by 2.5.
$280 \le$	x	≤ 330	Simplify.

So, the daily production levels during the month of August varied from a low of 280 units to a high of 330 units, as shown in Figure A.8.

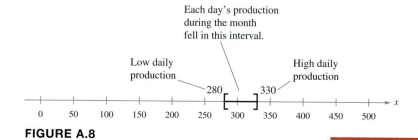

FIGURE A.8

✓ Checkpoint 3

Use the information in Example 3 to find the high and low production levels during the month of October, when the total cost of production varied from a high of $1500 to a low of $1000 per day.   ■

Exercises A.1

Classifying Real Numbers In Exercises 1–10, determine whether the real number is rational or irrational.

1. 0.25

2. -3678

3. $\dfrac{3\pi}{2}$

4. $3\sqrt{2} - 1$

5. $4.3\overline{451}$

6. $\dfrac{22}{7}$

7. $\sqrt[3]{64}$

8. $0.\overline{8177}$

9. $\sqrt[3]{60}$

10. $2e$

Checking Solutions In Exercises 11–14, determine whether each given value of x satisfies the inequality.

11. $5x - 12 > 0$

   (a) $x = 3$ (b) $x = -3$ (c) $x = \frac{5}{2}$

12. $x + 1 < \dfrac{x}{3}$

   (a) $x = 0$ (b) $x = 4$ (c) $x = -4$

13. $0 < \dfrac{x - 2}{4} < 2$

   (a) $x = 4$ (b) $x = 10$ (c) $x = 0$

14. $-1 < \dfrac{3 - x}{2} \le 1$

   (a) $x = 0$ (b) $x = 1$ (c) $x = 5$

Solving an Inequality In Exercises 15–28, solve the inequality. Then graph the solution set on the real number line. *See Examples 1 and 2.*

15. $x - 5 \ge 7$

16. $2x > 3$

17. $4x + 1 < 2x$

18. $2x + 7 < 3$

19. $4 - 2x < 3x - 1$

20. $x - 4 \le 2x + 1$

21. $-4 < 2x - 3 < 4$

22. $0 \le x + 3 < 5$

23. $\dfrac{3}{4} > x + 1 > \dfrac{1}{4}$

24. $-1 < -\dfrac{x}{3} < 1$

25. $\dfrac{x}{2} + \dfrac{x}{3} > 5$

26. $\dfrac{x}{2} - \dfrac{x}{3} > 5$

27. $2x^2 - x < 6$

28. $2x^2 + 1 < 9x - 3$

Writing Inequalities In Exercises 29–32, use inequality notation to describe the subset of real numbers.

29. **Earnings Per Share** A company expects its earnings per share E for the next quarter to be no less than $4.10 and no more than $4.25.

30. **Production** The estimated daily oil production p at a refinery is greater than 2 million barrels but less than 2.4 million barrels.

31. **Survey** According to a survey, the percent p of Americans who now conduct most of their banking transactions online is no more than 40%.

32. **Income** The net income I of a company is expected to be no less than $239 million.

33. **Physiology** The maximum heart rate of a person in normal health is related to the person's age by the equation

$$r = 220 - A$$

where r is the maximum heart rate (in beats per minute) and A is the person's age (in years). Some physiologists recommend that during physical activity, a sedentary person should strive to increase his or her heart rate to at least 60% of the maximum heart rate, and a highly fit person should strive to increase his or her heart rate to at most 90% of the maximum heart rate. Use inequality notation to express the range of the target heart rate for physical activity for a 20-year-old.

34. **Annual Operating Costs** A utility company has a fleet of vans. The annual operating cost C (in dollars) of each van is estimated to be

$$C = 0.35m + 2500$$

where m is the number of miles driven. What number of miles will yield an annual operating cost that is less than $13,000?

35. **Profit** The revenue for selling x units of a product is

$$R = 115.95x$$

and the cost of producing x units is

$$C = 95x + 750.$$

To obtain a profit, the revenue must be *greater than* the cost. For what values of x will this product return a profit?

36. **Sales** A doughnut shop sells a dozen doughnuts for $4.50. Beyond the fixed cost of $220 per day, it costs $2.75 for enough materials and labor to produce each dozen doughnuts. During the month of January, the daily profit varies between $60 and $270. Between what levels (in dozens) do the daily sales vary?

True or False? In Exercises 37 and 38, determine whether each statement is true or false, given $a < b$.

37. (a) $-2a < -2b$

   (b) $a + 2 < b + 2$

   (c) $6a < 6b$

   (d) $\dfrac{1}{a} < \dfrac{1}{b}$

38. (a) $a - 4 < b - 4$

   (b) $4 - a < 4 - b$

   (c) $-3b < -3a$

   (d) $\dfrac{a}{4} < \dfrac{b}{4}$

A.2 Absolute Value and Distance on the Real Number Line

- Find the absolute values of real numbers and understand the properties of absolute value.
- Find the distance between two numbers on the real number line.
- Define intervals on the real number line.
- Use intervals to model and solve real-life problems and find the midpoint of an interval.

Absolute Value of a Real Number

TECH TUTOR

Absolute value expressions can be evaluated on a graphing utility. When an expression such as $|3 - 8|$ is evaluated, parentheses should surround the expression, as in abs$(3 - 8)$.

Definition of Absolute Value

The **absolute value** of a real number a is

$$|a| = \begin{cases} a, & \text{if } a \geq 0 \\ -a, & \text{if } a < 0. \end{cases}$$

At first glance, it may appear from this definition that the absolute value of a real number can be negative, but this is not possible. For example, let $a = -3$. Then, because $-3 < 0$, you have

$$|a| = |-3| = -(-3) = 3.$$

The following properties are useful for working with absolute values.

Properties of Absolute Value

1. Multiplication: $|ab| = |a||b|$

2. Division: $\left|\dfrac{a}{b}\right| = \dfrac{|a|}{|b|}, \quad b \neq 0$

3. Power: $|a^n| = |a|^n$

4. Square root: $\sqrt{a^2} = |a|$

Be sure you understand the fourth property in this list. A common error in algebra is to imagine that by squaring a number and then taking the square root, you come back to the original number. But this is true only if the original number is nonnegative. For instance, if $a = 2$, then

$$\sqrt{2^2} = \sqrt{4} = 2$$

but if $a = -2$, then

$$\sqrt{(-2)^2} = \sqrt{4} = 2.$$

The reason for this is that (by definition) the square root symbol

$$\sqrt{}$$

denotes only the nonnegative root.

Distance on the Real Number Line

Consider two distinct points on the real number line, as shown in Figure A.9.

1. The **directed distance from a to b** is

$b - a$.

2. The **directed distance from b to a** is

$a - b$.

3. The **distance between a and b** is

$|a - b|$ or $|b - a|$.

In Figure A.9, note that because b is to the right of a, the directed distance from a to b (moving to the right) is positive. Moreover, because a is to the left of b, the directed distance from b to a (moving to the left) is negative. The distance *between* two points on the real number line can never be negative.

Directed distance
from a to b:

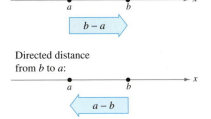

Directed distance
from b to a:

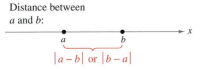

Distance between
a and b:

FIGURE A.9

Distance Between Two Points on the Real Number Line

The distance d between points x_1 and x_2 on the real number line is given by

$$d = |x_2 - x_1| = \sqrt{(x_2 - x_1)^2}.$$

Note that the order of subtraction with x_1 and x_2 does not matter because

$|x_2 - x_1| = |x_1 - x_2|$ and $(x_2 - x_1)^2 = (x_1 - x_2)^2$.

Example 1   **Finding Distance on the Real Number Line**

Determine the distance between -3 and 4 on the real number line. What is the directed distance from -3 to 4? What is the directed distance from 4 to -3?

SOLUTION   The distance between -3 and 4 is given by

$|-3 - 4| = |-7| = 7$ $|a - b|$

or

$|4 - (-3)| = |7| = 7$ $|b - a|$

as shown in Figure A.10.

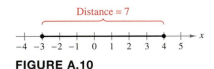

FIGURE A.10

The directed distance from -3 to 4 is

$4 - (-3) = 7.$ $b - a$

The directed distance from 4 to -3 is

$-3 - 4 = -7.$ $a - b$

 **Checkpoint 1**

Determine the distance between -2 and 6 on the real number line. What is the directed distance from -2 to 6? What is the directed distance from 6 to -2?

Intervals Defined by Absolute Value

Example 2   Defining an Interval on the Real Number Line

Find the interval on the real number line that contains all numbers that lie no more than two units from 3.

SOLUTION   Let x be any point in this interval. You need to find all x such that the distance between x and 3 is less than or equal to 2. This implies that

$$|x - 3| \le 2.$$

Requiring the absolute value of $x - 3$ to be less than or equal to 2 means that $x - 3$ must lie between -2 and 2. So, you can write

$$-2 \le x - 3 \le 2.$$

Solving this pair of inequalities, you have

$$-2 + 3 \le x - 3 + 3 \le 2 + 3$$
$$1 \le \quad x \quad \le 5. \qquad \text{Solution set}$$

So, the interval is $[1, 5]$, as shown in Figure A.11.

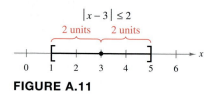

FIGURE A.11

✓ Checkpoint 2

Find the interval on the real number line that contains all numbers that lie no more than four units from 6.   ■

Two Basic Types of Inequalities Involving Absolute Value

Let a and d be real numbers, where $d > 0$.

$|x - a| \le d$ if and only if $a - d \le x \le a + d$.

$|x - a| \ge d$ if and only if $x \le a - d$ or $a + d \le x$.

Inequality	Interpretation	Graph
$|x - a| \le d$	All numbers x whose distance from a is less than or equal to d.	
$|x - a| \ge d$	All numbers x whose distance from a is greater than or equal to d.	

Be sure you see that inequalities of the form $|x - a| \ge d$ have solution sets consisting of two intervals. To describe the two intervals without using absolute values, you must use *two* separate inequalities, connected by an "or" to indicate union.

Application

 **Example 3**   **Quality Control**

A large manufacturer hired a quality control firm to determine the reliability of a product. Using statistical methods, the firm determined that the manufacturer could expect 0.35% ± 0.17% of the units to be defective. The manufacturer offers a money-back guarantee on this product. How much should be budgeted to cover the refunds on 100,000 units? (Assume that the retail price is $8.95.) Will the manufacturer have to establish a refund budget greater than $5000?

SOLUTION   Let r represent the percent of defective units (in decimal form). You know that r will differ from 0.0035 by at most 0.0017.

$$0.0035 - 0.0017 \le r \le 0.0035 + 0.0017$$
$$0.0018 \le r \le 0.0052 \qquad \text{Figure A.12(a)}$$

(a) Percent of defective units

Now, letting x be the number of defective units out of 100,000, it follows that $x = 100,000r$ and you have

$$0.0018(100{,}000) \le 100{,}000r \le 0.0052(100{,}000)$$
$$180 \le \quad x \quad \le 520. \qquad \text{Figure A.12(b)}$$

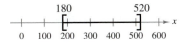

(b) Number of defective units

Finally, letting C be the cost of refunds, you have $C = 8.95x$. So, the total cost of refunds for 100,000 units should fall within the interval given by

$$180(8.95) \le 8.95x \le 520(8.95)$$
$$\$1611 \le \quad C \quad \le \$4654. \qquad \text{Figure A.12(c)}$$

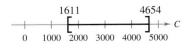

(c) Cost of refunds

FIGURE A.12

The manufacturer will *not* have to establish a refund budget greater than $5000.

✓ **Checkpoint 3**

Use the information in Example 3 to determine how much should be budgeted to cover refunds on 250,000 units.

In Example 3, the manufacturer should expect to spend between $1611 and $4654 for refunds. Of course, the safer budget figure for refunds would be the higher of these estimates. From a statistical point of view, however, the most representative estimate would be the average of these two extremes. Graphically, the average of two numbers is the **midpoint** of the interval with the two numbers as endpoints, as shown in Figure A.13.

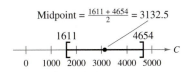

FIGURE A.13

Midpoint of an Interval

The **midpoint** of the interval with endpoints a and b is found by taking the average of the endpoints.

$$\text{Midpoint} = \frac{a + b}{2}$$

Exercises A.2

Finding Distance on the Real Number Line In Exercises 1–6, determine (a) the distance between a and b, (b) the directed distance from a to b, and (c) the directed distance from b to a. *See Example 1.*

1. $a = 126, b = 75$
2. $a = -126, b = -75$
3. $a = 9.34, b = -5.65$
4. $a = -2.05, b = 4.25$
5. $a = \frac{16}{5}, b = \frac{112}{75}$
6. $a = -\frac{18}{5}, b = \frac{61}{15}$

Describing Intervals Using Absolute Value In Exercises 7–18, use absolute values to describe the given interval (or pair of intervals) on the real number line.

7. $[-2, 2]$
8. $(-3, 3)$
9. $(-\infty, -2) \cup (2, \infty)$
10. $(-\infty, -3] \cup [3, \infty)$
11. $[2, 8]$
12. $(-7, -1)$
13. $(-\infty, 0) \cup (4, \infty)$
14. $(-\infty, 20) \cup (24, \infty)$
15. All numbers *less than* three units from 5
16. All numbers *more than* five units from 2
17. y is at most *two units* from a.
18. y is *less than* h units from c.

Solving an Inequality In Exercises 19–34, solve the inequality. Then graph the solution set on the real number line. *See Example 2.*

19. $|x| < 4$
20. $|2x| < 6$
21. $\left|\dfrac{x}{2}\right| > 3$
22. $|3x| > 12$
23. $|x - 5| < 2$
24. $|3x + 1| \geq 4$
25. $\left|\dfrac{x - 3}{2}\right| \geq 5$
26. $|2x + 1| < 5$
27. $|10 - x| > 4$
28. $|25 - x| \geq 20$
29. $|9 - 2x| < 1$
30. $\left|1 - \dfrac{2x}{3}\right| < 1$
31. $|x - a| \leq b, \ b > 0$
32. $|2x - a| \geq b, \ b > 0$
33. $\left|\dfrac{3x - a}{4}\right| < 2b, \ b > 0$
34. $\left|a - \dfrac{5x}{2}\right| > b, \ b > 0$

Finding a Midpoint In Exercises 35–40, find the midpoint of the given interval.

35. $[8, 24]$
36. $[7.3, 12.7]$
37. $[-6.85, 9.35]$
38. $[-4.6, -1.3]$
39. $\left[-\frac{1}{2}, \frac{3}{4}\right]$
40. $\left[\frac{5}{6}, \frac{5}{2}\right]$

41. **Stock Price** A stock market analyst predicts that over the next year, the price p of a stock will not change from its current price of \$33.15 by more than \$2. Use absolute values to write this prediction as an inequality.

42. **Production** The estimated daily production x at a refinery is given by

$$|x - 200{,}000| \leq 25{,}000$$

where x is measured in barrels of oil. Determine the high and low production levels.

43. **Manufacturing** The acceptable weights for a 20-ounce cereal box are given by

$$|x - 20| \leq 0.75$$

where x is measured in ounces. Determine the high and low weights for the cereal box.

44. **Weight** The American Kennel Club has developed guidelines for judging the features of various breeds of dogs. To not receive a penalty, the guidelines specify that the weights for male collies must satisfy the inequality

$$\left|\dfrac{w - 67.5}{7.5}\right| \leq 1$$

where w is the weight (in pounds). Determine the interval on the real number line in which these weights lie. *(Source: The American Kennel Club, Inc.)*

Budget Variance In Exercises 45–48, (a) use absolute value notation to represent the two intervals in which expenses must lie if they are to be within \$500 and within 5% of the specified budget amount and (b) using the more stringent constraint, determine whether the given expense is at variance with the budget restriction.

	Item	*Budget*	*Expense*
45.	Utilities	\$4750.00	\$5116.37
46.	Insurance	\$15,000.00	\$14,695.00
47.	Maintenance	\$20,000.00	\$22,718.35
48.	Taxes	\$7500.00	\$8691.00

49. **Quality Control** In determining the reliability of a product, a manufacturer determines that it should expect $0.05\% \pm 0.01\%$ of the units to be defective. The manufacturer offers a money-back guarantee on this product. How much should be budgeted to cover the refunds on 150,000 units? (Assume that the retail price is \$195.99.)

A.3 Exponents and Radicals

- Evaluate expressions involving exponents or radicals.
- Simplify expressions with exponents.
- Find the domains of algebraic expressions.

Expressions Involving Exponents or Radicals

Properties of Exponents

1. Whole-number exponents: $x^n = \underbrace{x \cdot x \cdot x \cdots x}_{n \text{ factors}}$

2. Zero exponent: $x^0 = 1, \quad x \neq 0$

3. Negative exponents: $x^{-n} = \dfrac{1}{x^n}, \quad x \neq 0$

4. Radicals (principal nth root): $\sqrt[n]{x} = a \implies x = a^n$

5. Rational exponents $(1/n)$: $x^{1/n} = \sqrt[n]{x}$

6. Rational exponents (m/n): $x^{m/n} = (x^{1/n})^m = \left(\sqrt[n]{x}\right)^m$

   $x^{m/n} = (x^m)^{1/n} = \sqrt[n]{x^m}$

7. Special convention (square root): $\sqrt[2]{x} = \sqrt{x}$

STUDY TIP

If n is even, then the principal nth root is positive. For example, $\sqrt{4} = +2$ and $\sqrt[4]{81} = +3$.

Example 1 Evaluating Expressions

	Expression	*x-Value*	*Substitution*
a.	$y = -2x^2$	$x = 4$	$y = -2(4^2) = -2(16) = -32$
b.	$y = 3x^{-3}$	$x = -1$	$y = 3(-1)^{-3} = \dfrac{3}{(-1)^3} = \dfrac{3}{-1} = -3$
c.	$y = (-x)^2$	$x = \dfrac{1}{2}$	$y = \left(-\dfrac{1}{2}\right)^2 = \dfrac{1}{4}$
d.	$y = \dfrac{2}{x^{-2}}$	$x = 3$	$y = \dfrac{2}{3^{-2}} = 2(3^2) = 18$

✓ **Checkpoint 1**

Evaluate $y = 4x^{-2}$ for $x = 3$.

Example 2 Evaluating Expressions

	Expression	*x-Value*	*Substitution*
a.	$y = 2x^{1/2}$	$x = 4$	$y = 2\sqrt{4} = 2(2) = 4$
b.	$y = \sqrt[3]{x^2}$	$x = 8$	$y = 8^{2/3} = (8^{1/3})^2 = 2^2 = 4$

✓ **Checkpoint 2**

Evaluate $y = 4x^{1/3}$ for $x = 8$.

Operations with Exponents

Operations with Exponents

1. Multiplying like bases: $\quad x^n x^m = x^{n+m}$ $\qquad$ Add exponents.

2. Dividing like bases: $\quad \dfrac{x^n}{x^m} = x^{n-m}$ $\qquad$ Subtract exponents.

3. Removing parentheses: $\quad (xy)^n = x^n y^n$

$$\left(\frac{x}{y}\right)^n = \frac{x^n}{y^n}$$

$$(x^n)^m = x^{nm}$$

4. Special conventions: $\quad -x^n = -(x^n), \quad -x^n \neq (-x)^n$

$$cx^n = c(x^n), \quad cx^n \neq (cx)^n$$

$$x^{n^m} = x^{(n^m)}, \quad x^{n^m} \neq (x^n)^m$$

Example 3   Simplifying Expressions with Exponents

Simplify each expression.

a. $2x^2(x^3)$ $\qquad$ **b.** $(3x)^2 \sqrt[3]{x}$ $\qquad$ **c.** $\dfrac{3x^2}{(x^{1/2})^3}$

d. $\dfrac{5x^4}{(x^2)^3}$ $\qquad$ **e.** $x^{-1}(2x^2)$ $\qquad$ **f.** $\dfrac{-\sqrt{x}}{5x^{-1}}$

SOLUTION

a. $2x^2(x^3) = 2x^{2+3} = 2x^5$ $\qquad\qquad$ $x^n x^m = x^{n+m}$

b. $(3x)^2 \sqrt[3]{x} = 9x^2 x^{1/3} = 9x^{2+(1/3)} = 9x^{7/3}$ $\qquad$ $x^n x^m = x^{n+m}$

c. $\dfrac{3x^2}{(x^{1/2})^3} = 3\left(\dfrac{x^2}{x^{3/2}}\right) = 3x^{2-(3/2)} = 3x^{1/2}$ $\qquad$ $(x^n)^m = x^{nm}, \dfrac{x^n}{x^m} = x^{n-m}$

d. $\dfrac{5x^4}{(x^2)^3} = \dfrac{5x^4}{x^6} = 5x^{4-6} = 5x^{-2} = \dfrac{5}{x^2}$ $\qquad$ $(x^n)^m = x^{nm}, \dfrac{x^n}{x^m} = x^{n-m}$

e. $x^{-1}(2x^2) = 2x^{-1}x^2 = 2x^{-1+2} = 2x$ $\qquad$ $x^n x^m = x^{n+m}$

f. $\dfrac{-\sqrt{x}}{5x^{-1}} = -\dfrac{1}{5}\left(\dfrac{x^{1/2}}{x^{-1}}\right) = -\dfrac{1}{5}x^{(1/2)+1} = -\dfrac{1}{5}x^{3/2}$ $\qquad$ $\dfrac{x^n}{x^m} = x^{n-m}$

✓ Checkpoint 3

Simplify each expression.

a. $3x^2(x^4)$

b. $(2x)^3 \sqrt{x}$

c. $\dfrac{4x^2}{(x^{1/3})^2}$

Note in Example 3 that one characteristic of simplified expressions is the absence of negative exponents. Another characteristic of simplified expressions is that sums and differences are written in *factored form*. To do this, you can use the **Distributive Property.**

$$abx^n + acx^{n+m} = ax^n(b + cx^m)$$

Study the next example carefully to be sure that you understand the concepts involved in the factoring process.

Example 4   Simplifying by Factoring

Simplify each expression by factoring.

a. $2x^2 - x^3$ **b.** $2x^3 + x^2$ **c.** $2x^{1/2} + 4x^{5/2}$ **d.** $2x^{-1/2} + 3x^{5/2}$

SOLUTION

a. $2x^2 - x^3 = x^2(2 - x)$

b. $2x^3 + x^2 = x^2(2x + 1)$

c. $2x^{1/2} + 4x^{5/2} = 2x^{1/2}(1 + 2x^2)$

d. $2x^{-1/2} + 3x^{5/2} = x^{-1/2}(2 + 3x^3) = \dfrac{2 + 3x^3}{\sqrt{x}}$

✓ Checkpoint 4

Simplify each expression by factoring.

a. $x^3 - 2x$

b. $2x^{1/2} + 8x^{3/2}$ ■

Many algebraic expressions obtained in calculus occur in unsimplified form. For instance, the two expressions shown in the following example are the result of an operation in calculus called *differentiation*. [The first is the derivative of $2(x + 1)^{3/2}(2x - 3)^{5/2}$, and the second is the derivative of $2(x + 1)^{1/2}(2x - 3)^{5/2}$.]

Example 5   Simplifying by Factoring

a. $3(x + 1)^{1/2}(2x - 3)^{5/2} + 10(x + 1)^{3/2}(2x - 3)^{3/2}$
$$= (x + 1)^{1/2}(2x - 3)^{3/2}[3(2x - 3) + 10(x + 1)]$$
$$= (x + 1)^{1/2}(2x - 3)^{3/2}(6x - 9 + 10x + 10)$$
$$= (x + 1)^{1/2}(2x - 3)^{3/2}(16x + 1)$$

b. $(x + 1)^{-1/2}(2x - 3)^{5/2} + 10(x + 1)^{1/2}(2x - 3)^{3/2}$
$$= (x + 1)^{-1/2}(2x - 3)^{3/2}[(2x - 3) + 10(x + 1)]$$
$$= (x + 1)^{-1/2}(2x - 3)^{3/2}(2x - 3 + 10x + 10)$$
$$= (x + 1)^{-1/2}(2x - 3)^{3/2}(12x + 7)$$
$$= \frac{(2x - 3)^{3/2}(12x + 7)}{(x + 1)^{1/2}}$$

✓ Checkpoint 5

Simplify the expression by factoring.

$$(x + 2)^{1/2}(3x - 1)^{3/2} + 4(x + 2)^{-1/2}(3x - 1)^{5/2}$$ ■

STUDY TIP

To check that a simplified expression is equivalent to the original expression, try substituting values for x into each expression.

Example 6 shows some additional types of expressions that can occur in calculus. [The expression in Example 6(d) is an antiderivative of $(x + 1)^{2/3}(2x + 3)$, and the expression in Example 6(e) is the derivative of $(x + 2)^3/(x - 1)^3$.]

Example 6   Factors Involving Quotients

Simplify each expression by factoring.

a. $\dfrac{3x^2 + x^4}{2x}$

b. $\dfrac{\sqrt{x} + x^{3/2}}{x}$

c. $(9x + 2)^{-1/3} + 18(9x + 2)$

d. $\dfrac{3}{5}(x + 1)^{5/3} + \dfrac{3}{4}(x + 1)^{8/3}$

e. $\dfrac{3(x + 2)^2(x - 1)^3 - 3(x + 2)^3(x - 1)^2}{[(x - 1)^3]^2}$

SOLUTION

a. $\dfrac{3x^2 + x^4}{2x} = \dfrac{x^2(3 + x^2)}{2x} = \dfrac{x^{2-1}(3 + x^2)}{2} = \dfrac{x(3 + x^2)}{2}$

b. $\dfrac{\sqrt{x} + x^{3/2}}{x} = \dfrac{x^{1/2}(1 + x)}{x} = \dfrac{1 + x}{x^{1-(1/2)}} = \dfrac{1 + x}{\sqrt{x}}$

c. $(9x + 2)^{-1/3} + 18(9x + 2) = (9x + 2)^{-1/3}[1 + 18(9x + 2)^{4/3}]$

$\qquad = \dfrac{1 + 18(9x + 2)^{4/3}}{\sqrt[3]{9x + 2}}$

d. $\dfrac{3}{5}(x + 1)^{5/3} + \dfrac{3}{4}(x + 1)^{8/3} = \dfrac{12}{20}(x + 1)^{5/3} + \dfrac{15}{20}(x + 1)^{8/3}$

$\qquad = \dfrac{3}{20}(x + 1)^{5/3}[4 + 5(x + 1)]$

$\qquad = \dfrac{3}{20}(x + 1)^{5/3}(4 + 5x + 5)$

$\qquad = \dfrac{3}{20}(x + 1)^{5/3}(5x + 9)$

e. $\dfrac{3(x + 2)^2(x - 1)^3 - 3(x + 2)^3(x - 1)^2}{[(x - 1)^3]^2}$

$\qquad = \dfrac{3(x + 2)^2(x - 1)^2[(x - 1) - (x + 2)]}{(x - 1)^6}$

$\qquad = \dfrac{3(x + 2)^2(x - 1 - x - 2)}{(x - 1)^{6-2}}$

$\qquad = \dfrac{-9(x + 2)^2}{(x - 1)^4}$

✓ Checkpoint 6

Simplify the expression by factoring.

$$\dfrac{5x^3 + x^6}{3x}$$

Domain of an Algebraic Expression

When working with algebraic expressions involving x, you face the potential difficulty of substituting a value of x for which the expression is not defined (does not produce a real number). For example, the expression $\sqrt{2x + 3}$ is *not defined* when $x = -2$ because

$$\sqrt{2(-2) + 3} = \sqrt{-1}$$

is not a real number.

The set of all values for which an expression is defined is called its **domain.** So, the domain of $\sqrt{2x + 3}$ is the set of all values of x such that $\sqrt{2x + 3}$ is a real number. In order for $\sqrt{2x + 3}$ to represent a real number, it is necessary that

$$2x + 3 \geq 0. \qquad \text{\color{red}Expression must be nonnegative.}$$

In other words, $\sqrt{2x + 3}$ is defined only for those values of x that lie in the interval $\left[-\frac{3}{2}, \infty\right)$, as shown in Figure A.14.

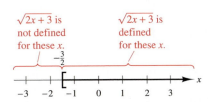

FIGURE A.14

Example 7 **Finding the Domain of an Expression**

Find the domain of each expression.

a. $\sqrt{3x - 2}$

b. $\dfrac{1}{\sqrt{3x - 2}}$

c. $\sqrt[3]{9x + 1}$

SOLUTION

a. The domain of $\sqrt{3x - 2}$ consists of all x such that

$$3x - 2 \geq 0 \qquad \text{\color{red}Expression must be nonnegative.}$$

which implies that $x \geq \frac{2}{3}$. So, the domain is $\left[\frac{2}{3}, \infty\right)$.

b. The domain of $1/\sqrt{3x - 2}$ is the same as the domain of $\sqrt{3x - 2}$, except that $1/\sqrt{3x - 2}$ is not defined when $3x - 2 = 0$. Because this occurs when $x = \frac{2}{3}$, the domain is $\left(\frac{2}{3}, \infty\right)$.

c. Because $\sqrt[3]{9x + 1}$ is defined for all real numbers, its domain is $(-\infty, \infty)$.

✓ **Checkpoint 7**

Find the domain of each expression.

a. $\sqrt{x - 2}$

b. $\dfrac{1}{\sqrt{x - 2}}$

c. $\sqrt[3]{x - 2}$

Exercises A.3

Evaluating Expressions   In Exercises 1–20, evaluate the expression for the given value of x. *See Examples 1 and 2.*

Expression	x-Value	Expression	x-Value
1. $-2x^3$	$x = 3$	**2.** $\dfrac{x^2}{3}$	$x = 6$
3. $4x^{-3}$	$x = 2$	**4.** $7x^{-2}$	$x = 5$
5. $\dfrac{1 + x^{-1}}{x^{-1}}$	$x = 3$	**6.** $x - 4x^{-2}$	$x = 3$
7. $3x^2 - 4x^3$	$x = -2$	**8.** $5(-x)^3$	$x = 3$
9. $6x^0 - (6x)^0$	$x = 10$	**10.** $\dfrac{1}{(-x)^{-3}}$	$x = 4$
11. $\sqrt[3]{x^2}$	$x = 27$	**12.** $\sqrt{x^3}$	$x = \frac{1}{9}$
13. $x^{-1/2}$	$x = 4$	**14.** $x^{-3/4}$	$x = 16$
15. $x^{-2/5}$	$x = -32$	**16.** $(x^{2/3})^3$	$x = 10$
17. $500x^{60}$	$x = 1.01$	**18.** $\dfrac{10{,}000}{x^{120}}$	$x = 1.1$
19. $\sqrt[3]{x}$	$x = -54$	**20.** $\sqrt[6]{x}$	$x = 325$

Simplifying Expressions with Exponents   In Exercises 21–30, simplify the expression. *See Example 3.*

21. $6y^{-2}(2y^4)^{-3}$ **22.** $z^{-3}(3z^4)$

23. $10(x^2)^2$ **24.** $(4x^3)^2$

25. $\dfrac{7x^2}{x^{-3}}$ **26.** $\dfrac{x^{-3}}{\sqrt{x}}$

27. $\dfrac{10(x + y)^3}{4(x + y)^{-2}}$ **28.** $\left(\dfrac{12s^2}{9s}\right)^3$

29. $\dfrac{3x\sqrt{x}}{x^{1/2}}$ **30.** $\left(\sqrt[3]{x^2}\right)^3$

Simplifying Radicals   In Exercises 31–36, simplify by removing all possible factors from the radical.

31. $\sqrt{8}$ **32.** $\sqrt[3]{\dfrac{16}{27}}$

33. $\sqrt[3]{54x^5}$ **34.** $\sqrt[4]{(3x^2y^3)^4}$

35. $\sqrt[3]{144x^9y^{-4}z^5}$ **36.** $\sqrt[4]{32xy^5z^{-8}}$

Simplifying by Factoring   In Exercises 37–44, simplify each expression by factoring. *See Examples 4, 5, and 6.*

37. $3x^3 - 12x$

38. $8x^4 - 6x^2$

39. $2x^{5/2} + x^{-1/2}$

40. $5x^{3/2} - x^{-3/2}$

41. $3x(x + 1)^{3/2} - 6(x + 1)^{1/2}$

42. $2x(x - 1)^{5/2} - 4(x - 1)^{3/2}$

43. $\dfrac{5x^6 + x^3}{3x^2}$

44. $\dfrac{(x + 1)(x - 1)^2 - (x - 1)^3}{(x + 1)^2}$

Finding the Domain of an Expression   In Exercises 45–52, find the domain of the expression. *See Example 7.*

45. $\sqrt{x - 4}$ **46.** $\sqrt{5 - 2x}$

47. $\sqrt{x^2 + 3}$ **48.** $\sqrt{4x^2 + 1}$

49. $\dfrac{1}{\sqrt[3]{x - 4}}$

50. $\dfrac{1}{\sqrt[3]{x + 4}}$

51. $\dfrac{\sqrt{x + 2}}{1 - x}$

52. $\dfrac{1}{\sqrt{2x + 3}} + \sqrt{6 - 4x}$

 Compound Interest   In Exercises 53–56, a certificate of deposit has a principal of P dollars and an annual percentage rate of r (expressed as a decimal) compounded n times per year. The balance A in the account is given by

$$A = P\left(1 + \frac{r}{n}\right)^N$$

where N is the number of compoundings. Use a graphing utility to find the balance in the account.

53. $P = \$10{,}000$, $r = 6.5\%$, $n = 12$, $N = 120$

54. $P = \$7000$, $r = 5\%$, $n = 365$, $N = 1000$

55. $P = \$5000$, $r = 5.5\%$, $n = 4$, $N = 60$

56. $P = \$8000$, $r = 7\%$, $n = 6$, $N = 90$

57. Period of a Pendulum   The period of a pendulum is

$$T = 2\pi\sqrt{\frac{L}{32}}$$

where T is the period (in seconds) and L is the length (in feet) of the pendulum. Find the period of a pendulum whose length is 4 feet.

58. Annuity   After n annual payments of P dollars have been made into an annuity earning an annual percentage rate of r compounded annually, the balance A is given by

$$A = P(1 + r) + P(1 + r)^2 + \cdots + P(1 + r)^n.$$

Rewrite this formula by completing the following factorization.

$$A = P(1 + r)(\quad\quad)$$

A.4 Factoring Polynomials

■ Use special products and factorization techniques to factor polynomials.
■ Use synthetic division to factor polynomials of degree three or more.
■ Use the Rational Zero Theorem to find the real zeros of polynomials.

Factorization Techniques

The **Fundamental Theorem of Algebra** states that every nth-degree polynomial

$$a_n x^n + a_{n-1} x^{n-1} + \cdots + a_1 x + a_0, \quad a_n \neq 0$$

has precisely n **zeros.** (The zeros may be repeated or imaginary.) The zeros of a polynomial in x are the values of x that make the polynomial zero. The problem of finding the zeros of a polynomial is equivalent to the problem of factoring the polynomial into linear factors.

Special Products and Factorization Techniques

Quadratic Formula

$$ax^2 + bx + c = 0 \implies x = \frac{-b \pm \sqrt{b^2 - 4ac}}{2a}$$

Example

$$x^2 + 3x - 1 = 0 \implies x = \frac{-3 \pm \sqrt{13}}{2}$$

Special Products

$$x^2 - a^2 = (x - a)(x + a)$$

$$x^3 - a^3 = (x - a)(x^2 + ax + a^2)$$

$$x^3 + a^3 = (x + a)(x^2 - ax + a^2)$$

$$x^4 - a^4 = (x - a)(x + a)(x^2 + a^2)$$

Examples

$$x^2 - 9 = (x - 3)(x + 3)$$

$$x^3 - 8 = (x - 2)(x^2 + 2x + 4)$$

$$x^3 + 64 = (x + 4)(x^2 - 4x + 16)$$

$$x^4 - 16 = (x - 2)(x + 2)(x^2 + 4)$$

Binomial Theorem

$$(x + a)^2 = x^2 + 2ax + a^2$$

$$(x - a)^2 = x^2 - 2ax + a^2$$

$$(x + a)^3 = x^3 + 3ax^2 + 3a^2x + a^3$$

$$(x - a)^3 = x^3 - 3ax^2 + 3a^2x - a^3$$

$$(x + a)^4 = x^4 + 4ax^3 + 6a^2x^2 + 4a^3x + a^4$$

$$(x - a)^4 = x^4 - 4ax^3 + 6a^2x^2 - 4a^3x + a^4$$

Examples

$$(x + 3)^2 = x^2 + 6x + 9$$

$$(x^2 - 5)^2 = x^4 - 10x^2 + 25$$

$$(x + 2)^3 = x^3 + 6x^2 + 12x + 8$$

$$(x - 1)^3 = x^3 - 3x^2 + 3x - 1$$

$$(x + 2)^4 = x^4 + 8x^3 + 24x^2 + 32x + 16$$

$$(x - 4)^4 = x^4 - 16x^3 + 96x^2 - 256x + 256$$

$$(x + a)^n = x^n + nax^{n-1} + \frac{n(n-1)}{2!}a^2x^{n-2} + \frac{n(n-1)(n-2)}{3!}a^3x^{n-3} + \cdots + na^{n-1}x + a^{n*}$$

$$(x - a)^n = x^n - nax^{n-1} + \frac{n(n-1)}{2!}a^2x^{n-2} - \frac{n(n-1)(n-2)}{3!}a^3x^{n-3} + \cdots \pm na^{n-1}x \mp a^n$$

Factoring by Grouping

$$acx^3 + adx^2 + bcx + bd = ax^2(cx + d) + b(cx + d)$$

$$= (ax^2 + b)(cx + d)$$

Example

$$3x^3 - 2x^2 - 6x + 4 = x^2(3x - 2) - 2(3x - 2)$$

$$= (x^2 - 2)(3x - 2)$$

---

* The factorial symbol ! is defined as follows: $0! = 1$, $1! = 1$, $2! = 2 \cdot 1 = 2$, $3! = 3 \cdot 2 \cdot 1 = 6$, $4! = 4 \cdot 3 \cdot 2 \cdot 1 = 24$, and so on.

Example 1   Applying the Quadratic Formula

Use the Quadratic Formula to find all real zeros of each polynomial.

a. $4x^2 + 6x + 1$

b. $x^2 + 6x + 9$

c. $2x^2 - 6x + 5$

SOLUTION

a. Using $a = 4$, $b = 6$, and $c = 1$, you can write

$$x = \frac{-b \pm \sqrt{b^2 - 4ac}}{2a} = \frac{-6 \pm \sqrt{6^2 - 4(4)(1)}}{2(4)}$$

$$= \frac{-6 \pm \sqrt{36 - 16}}{8}$$

$$= \frac{-6 \pm \sqrt{20}}{8}$$

$$= \frac{-6 \pm 2\sqrt{5}}{8}$$

$$= \frac{2(-3 \pm \sqrt{5})}{2(4)}$$

$$= \frac{-3 \pm \sqrt{5}}{4}.$$

So, there are two real zeros:

$$x = \frac{-3 - \sqrt{5}}{4} \approx -1.309 \quad \text{and} \quad x = \frac{-3 + \sqrt{5}}{4} \approx -0.191.$$

b. In this case, $a = 1$, $b = 6$, and $c = 9$, and the Quadratic Formula yields

$$x = \frac{-b \pm \sqrt{b^2 - 4ac}}{2a} = \frac{-6 \pm \sqrt{36 - 36}}{2} = -\frac{6}{2} = -3.$$

So, there is one (repeated) real zero: $x = -3$.

c. For this quadratic equation, $a = 2$, $b = -6$, and $c = 5$. So,

$$x = \frac{-b \pm \sqrt{b^2 - 4ac}}{2a} = \frac{6 \pm \sqrt{36 - 40}}{4} = \frac{6 \pm \sqrt{-4}}{4}.$$

Because $\sqrt{-4}$ is imaginary, there are no real zeros.

✓ **Checkpoint 1**

Use the Quadratic Formula to find all real zeros of each polynomial.

a. $2x^2 + 4x + 1$

b. $x^2 - 8x + 16$

c. $2x^2 - x + 5$

The zeros in Example 1(a) are irrational, and the zeros in Example 1(c) are imaginary. In both of these cases the quadratic is said to be **irreducible** because it cannot be factored into linear factors with rational coefficients. The next example shows how to find the zeros associated with *reducible* quadratics. In this example, factoring is used to find the zeros of each quadratic. Try using the Quadratic Formula to obtain the same zeros.

Recall that the zeros of a polynomial in x are the values of x that make the polynomial zero. To find the zeros, factor the polynomial into linear factors and set each factor equal to zero. For instance, the zeros of $(x - 2)(x - 3)$ occur when $x - 2 = 0$ and $x - 3 = 0$.

Example 2   Finding Real Zeros by Factoring

Find all the real zeros of each quadratic polynomial.

a. $x^2 - 5x + 6$ **b.** $x^2 - 6x + 9$ **c.** $2x^2 + 5x - 3$

SOLUTION

a. Because

$$x^2 - 5x + 6 = (x - 2)(x - 3)$$

the zeros are $x = 2$ and $x = 3$.

b. Because

$$x^2 - 6x + 9 = (x - 3)^2$$

the only zero is $x = 3$.

c. Because

$$2x^2 + 5x - 3 = (2x - 1)(x + 3)$$

the zeros are $x = \frac{1}{2}$ and $x = -3$.

✓ Checkpoint 2

Find all the real zeros of each quadratic polynomial.

a. $x^2 - 2x - 15$ **b.** $x^2 + 2x + 1$ **c.** $2x^2 - 7x + 6$ ■

Example 3   Finding the Domain of a Radical Expression

Find the domain of $\sqrt{x^2 - 3x + 2}$.

SOLUTION   Because

$$x^2 - 3x + 2 = (x - 1)(x - 2)$$

you know that the zeros of the quadratic are $x = 1$ and $x = 2$. So, you need to test the sign of the quadratic in the three intervals $(-\infty, 1)$, $(1, 2)$, and $(2, \infty)$, as shown in Figure A.15. After testing each of these intervals, you can see that the quadratic is negative in the center interval and positive in the outer two intervals. Moreover, because the quadratic is zero when $x = 1$ and $x = 2$, you can conclude that the domain of $\sqrt{x^2 - 3x + 2}$ is

$$(-\infty, 1] \cup [2, \infty). \qquad \text{Domain}$$

Values of $\sqrt{x^2 - 3x + 2}$

x	$\sqrt{x^2 - 3x + 2}$
0	$\sqrt{2}$
1	0
1.5	Undefined
2	0
3	$\sqrt{2}$

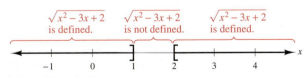

FIGURE A.15

✓ Checkpoint 3

Find the domain of

$$\sqrt{x^2 + x - 2}. \qquad ■$$

Factoring Polynomials of Degree Three or More

It can be difficult to find the zeros of polynomials of degree three or more. However, if one of the zeros of a polynomial is known, then you can use that zero to reduce the degree of the polynomial. For example, if you know that $x = 2$ is a zero of

$$x^3 - 4x^2 + 5x - 2$$

then you know that $(x - 2)$ is a factor, and you can use long division to factor the polynomial as shown.

$$x^3 - 4x^2 + 5x - 2 = (x - 2)(x^2 - 2x + 1)$$
$$= (x - 2)(x - 1)(x - 1)$$

As an alternative to long division, many people prefer to use **synthetic division** to reduce the degree of a polynomial.

Synthetic Division for a Cubic Polynomial

Given: $x = x_1$ is a zero of $ax^3 + bx^2 + cx + d$.

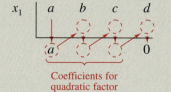

Coefficients for quadratic factor

Vertical pattern:
Add terms.

Diagonal pattern:
Multiply by x_1.

Performing synthetic division on the polynomial

$$x^3 - 4x^2 + 5x - 2$$

using the given zero, $x = 2$, produces the following.

$$
\begin{array}{r|rrrr}
2 & 1 & -4 & 5 & -2 \\
  &   & 2 & -4 & 2 \\
\hline
  & 1 & -2 & 1 & 0
\end{array}
$$

$(x - 2)(x^2 - 2x + 1) = x^3 - 4x^2 + 5x - 2$

When you use synthetic division, remember to take *all* coefficients into account—*even when some of them are zero.* For instance, when you know that $x = -2$ is a zero of $x^3 + 3x + 14$, you can apply synthetic division as shown.

$$
\begin{array}{r|rrrr}
-2 & 1 & 0 & 3 & 14 \\
   &   & -2 & 4 & -14 \\
\hline
   & 1 & -2 & 7 & 0
\end{array}
$$

$(x + 2)(x^2 - 2x + 7) = x^3 + 3x + 14$

STUDY TIP

The algorithm for synthetic division given above works *only* for divisors of the form $x - x_1$. Remember that $x + x_1 = x - (-x_1)$.

The Rational Zero Theorem

There is a systematic way to find the *rational* zeros of a polynomial. You can use the **Rational Zero Theorem** (also called the Rational Root Theorem).

Rational Zero Theorem

If a polynomial

$$a_n x^n + a_{n-1} x^{n-1} + \cdots + a_1 x + a_0$$

has integer coefficients, then every *rational* zero is of the form

$$x = \frac{p}{q}$$

where p is a factor of a_0, and q is a factor of a_n.

Example 4   Using the Rational Zero Theorem

Find all real zeros of the polynomial.

$$2x^3 + 3x^2 - 8x + 3$$

SOLUTION

$$(2)x^3 + 3x^2 - 8x + (3)$$

Factors of constant term: $\pm 1, \pm 3$

Factors of leading coefficient: $\pm 1, \pm 2$

The possible rational zeros are the factors of the constant term divided by the factors of the leading coefficient.

$$1, -1, 3, -3, \frac{1}{2}, -\frac{1}{2}, \frac{3}{2}, -\frac{3}{2}$$

By testing these possible zeros, you can see that $x = 1$ works.

$$2(1)^3 + 3(1)^2 - 8(1) + 3 = 2 + 3 - 8 + 3 = 0$$

Now, by synthetic division you have the following.

$$
\begin{array}{r|rrrr}
1 & 2 & 3 & -8 & 3 \\
  &   & 2 & 5 & -3 \\
\hline
  & 2 & 5 & -3 & 0 \\
\end{array}
$$

$$(x - 1)(2x^2 + 5x - 3) = 2x^3 + 3x^2 - 8x + 3$$

Finally, by factoring the quadratic

$$2x^2 + 5x - 3 = (2x - 1)(x + 3)$$

you have

$$2x^3 + 3x^2 - 8x + 3 = (x - 1)(2x - 1)(x + 3)$$

and you can conclude that the zeros are $x = 1$, $x = \frac{1}{2}$, and $x = -3$.

STUDY TIP

In Example 4, you can check that the zeros are correct by substituting into the original polynomial.

Check that $x = 1$ is a zero.
$$2(1)^3 + 3(1)^2 - 8(1) + 3$$
$$= 2 + 3 - 8 + 3$$
$$= 0$$

Check that $x = \frac{1}{2}$ is a zero.
$$2\left(\frac{1}{2}\right)^3 + 3\left(\frac{1}{2}\right)^2 - 8\left(\frac{1}{2}\right) + 3$$
$$= \frac{1}{4} + \frac{3}{4} - 4 + 3$$
$$= 0$$

Check that $x = -3$ is a zero.
$$2(-3)^3 + 3(-3)^2 - 8(-3) + 3$$
$$= -54 + 27 + 24 + 3$$
$$= 0$$

✓ Checkpoint 4

Find all real zeros of the polynomial.

$$2x^3 - 3x^2 - 3x + 2$$

Exercises A.4

Applying the Quadratic Formula   In Exercises 1–8, use the Quadratic Formula to find all real zeros of the second-degree polynomial. *See Example 1.*

1. $6x^2 - 7x + 1$
2. $8x^2 - 2x - 1$
3. $4x^2 - 12x + 9$
4. $9x^2 + 12x + 4$
5. $y^2 + 4y + 1$
6. $y^2 + 5y - 2$
7. $2x^2 + 3x - 4$
8. $3x^2 - 8x - 4$

Factoring Polynomials   In Exercises 9–18, write the second-degree polynomial as the product of two linear factors.

9. $x^2 - 4x + 4$
10. $x^2 + 10x + 25$
11. $4x^2 + 4x + 1$
12. $9x^2 - 12x + 4$
13. $3x^2 - 4x + 1$
14. $2x^2 - x - 1$
15. $3x^2 - 5x + 2$
16. $4x^2 + 19x + 12$
17. $x^2 - 4xy + 4y^2$
18. $x^2 - xy - 2y^2$

Factoring Polynomials   In Exercises 19–34, completely factor the polynomial.

19. $81 - y^4$
20. $x^4 - 16$
21. $x^3 - 8$
22. $y^3 - 64$
23. $y^3 + 64$
24. $z^3 + 125$
25. $x^3 - y^3$
26. $(x - a)^3 + b^3$
27. $x^3 - 4x^2 - x + 4$
28. $x^3 - x^2 - x + 1$
29. $2x^3 - 3x^2 + 4x - 6$
30. $x^3 - 5x^2 - 5x + 25$
31. $2x^3 - 4x^2 - x + 2$
32. $x^3 - 7x^2 - 4x + 28$
33. $x^4 - 15x^2 - 16$
34. $2x^4 - 49x^2 - 25$

Finding Real Zeros by Factoring   In Exercises 35–54, find all real zeros of the polynomial. *See Example 2.*

35. $x^2 - 5x$
36. $2x^2 - 3x$
37. $x^2 - 9$
38. $x^2 - 25$
39. $x^2 - 3$
40. $x^2 - 8$
41. $(x - 3)^2 - 9$
42. $(x + 1)^2 - 36$
43. $x^2 + x - 2$
44. $x^2 + 5x + 6$
45. $x^2 - 5x - 6$
46. $x^2 + x - 20$
47. $3x^2 + 5x + 2$
48. $2x^2 - x - 1$
49. $x^3 + 64$
50. $x^3 - 216$
51. $x^4 - 16$
52. $x^4 - 625$
53. $x^3 - x^2 - 4x + 4$
54. $2x^3 + x^2 + 6x + 3$

Finding the Domain of a Radical Expression   In Exercises 55–60, find the domain of the expression. *See Example 3.*

55. $\sqrt{x^2 - 4}$
56. $\sqrt{4 - x^2}$
57. $\sqrt{x^2 - 7x + 12}$
58. $\sqrt{x^2 - 8x + 15}$
59. $\sqrt{5x^2 + 6x + 1}$
60. $\sqrt{3x^2 - 10x + 3}$

Using Synthetic Division   In Exercises 61–64, use synthetic division to complete the indicated factorization.

61. $x^3 - 3x^2 - 6x - 2 = (x + 1)(\quad)$
62. $x^3 - 2x^2 - x + 2 = (x - 2)(\quad)$
63. $2x^3 - x^2 - 2x + 1 = (x + 1)(\quad)$
64. $x^4 - 16x^3 + 96x^2 - 256x + 256 = (x - 4)(\quad)$

Using the Rational Zero Theorem   In Exercises 65–74, use the Rational Zero Theorem to find all real zeros of the polynomial. *See Example 4.*

65. $x^3 - x^2 - 10x - 8$
66. $x^3 - 7x - 6$
67. $x^3 - 6x^2 + 11x - 6$
68. $x^3 + 2x^2 - 5x - 6$
69. $6x^3 - 11x^2 - 19x - 6$
70. $18x^3 - 9x^2 - 8x + 4$
71. $x^3 - 3x^2 - 3x - 4$
72. $2x^3 - x^2 - 13x - 6$
73. $4x^3 + 11x^2 + 5x - 2$
74. $3x^3 + 4x^2 - 13x + 6$

75. **Production Level**   The minimum average cost of producing x units of a product occurs when the production level is set at the (positive) solution of

$$0.0003x^2 - 1200 = 0.$$

How many solutions does this equation have? Find and interpret the solution(s) in the context of the problem. What production level will minimize the average cost?

76. **Profit**   The profit P (in dollars) from sales is given by

$$P = -200x^2 + 2000x - 3800$$

where x is the number of units sold per day (in hundreds). Determine the interval for x such that the profit will be greater than \$1000.

A.5 Fractions and Rationalization

- Simplify rational expressions.
- Add and subtract rational expressions.
- Simplify rational expressions involving radicals.
- Rationalize numerators and denominators of rational expressions.

Simplifying Rational Expressions

In this section, you will review operations involving fractional expressions such as

$$\frac{2}{x}, \quad \frac{x^2 + 2x - 4}{x + 6}, \quad \text{and} \quad \frac{1}{\sqrt{x^2 + 1}}.$$

The first two expressions have polynomials as both numerator and denominator and are called **rational expressions.** A rational expression is **proper** when the degree of the numerator is less than the degree of the denominator. For example,

$$\frac{x}{x^2 + 1}$$

is proper. If the degree of the numerator is greater than or equal to the degree of the denominator, then the rational expression is **improper.** For example,

$$\frac{x^2}{x^2 + 1} \quad \text{and} \quad \frac{x^3 + 2x + 1}{x + 1}$$

are both improper.

A fraction is in simplest form when its numerator and denominator have no factors in common aside from ±1. To write a fraction in simplest form, divide out common factors.

$$\frac{a \cdot c}{b \cdot c} = \frac{a}{b}, \quad c \neq 0$$

The key to success in simplifying rational expressions lies in your ability to factor polynomials. When simplifying rational expressions, be sure to factor each polynomial completely before concluding that the numerator and denominator have no common factors.

Example 1 Simplifying a Rational Expression

Write $\dfrac{12 + x - x^2}{2x^2 - 9x + 4}$ in simplest form.

SOLUTION

$$\frac{12 + x - x^2}{2x^2 - 9x + 4} = \frac{(4 - x)(3 + x)}{(2x - 1)(x - 4)} \qquad \text{Factor completely.}$$

$$= \frac{-(x - 4)(3 + x)}{(2x - 1)(x - 4)} \qquad (4 - x) = -(x - 4)$$

$$= -\frac{3 + x}{2x - 1}, \quad x \neq 4 \qquad \text{Divide out common factors.}$$

STUDY TIP

To simplify a rational expression, it may be necessary to change the sign of a factor by factoring out (-1), as shown in Example 1.

✓ Checkpoint 1

Write $\dfrac{x^2 + 8x - 20}{x^2 + 11x + 10}$ in simplest form.

Operations with Fractions

Operations with Fractions

1. Add fractions (find a common denominator):

$$\frac{a}{b} + \frac{c}{d} = \frac{a}{b}\left(\frac{d}{d}\right) + \frac{c}{d}\left(\frac{b}{b}\right) = \frac{ad}{bd} + \frac{bc}{bd} = \frac{ad + bc}{bd}, \quad b \neq 0, d \neq 0$$

2. Subtract fractions (find a common denominator):

$$\frac{a}{b} - \frac{c}{d} = \frac{a}{b}\left(\frac{d}{d}\right) - \frac{c}{d}\left(\frac{b}{b}\right) = \frac{ad}{bd} - \frac{bc}{bd} = \frac{ad - bc}{bd}, \quad b \neq 0, d \neq 0$$

3. Multiply fractions:

$$\left(\frac{a}{b}\right)\left(\frac{c}{d}\right) = \frac{ac}{bd}, \quad b \neq 0, d \neq 0$$

4. Divide fractions (invert and multiply):

$$\frac{a/b}{c/d} = \left(\frac{a}{b}\right)\left(\frac{d}{c}\right) = \frac{ad}{bc}, \quad b \neq 0, c \neq 0, d \neq 0$$

$$\frac{a/b}{c} = \frac{a/b}{c/1} = \left(\frac{a}{b}\right)\left(\frac{1}{c}\right) = \frac{a}{bc}, \quad b \neq 0, c \neq 0$$

5. Divide out common factors:

$$\frac{\acute{a}b}{\acute{a}c} = \frac{b}{c}, \quad a \neq 0, c \neq 0$$

$$\frac{ab + ac}{ad} = \frac{\acute{a}(b + c)}{\acute{a}d} = \frac{b + c}{d}, \quad a \neq 0, d \neq 0$$

Example 2   **Adding and Subtracting Rational Expressions**

Perform each indicated operation and simplify.

a. $x + \dfrac{1}{x}$ **b.** $\dfrac{1}{x + 1} - \dfrac{2}{2x - 1}$

SOLUTION

a. $x + \dfrac{1}{x} = \dfrac{x^2}{x} + \dfrac{1}{x}$ Write with common denominator.

$\qquad = \dfrac{x^2 + 1}{x}$ Add fractions.

b. $\dfrac{1}{x + 1} - \dfrac{2}{2x - 1} = \dfrac{(2x - 1)}{(x + 1)(2x - 1)} - \dfrac{2(x + 1)}{(x + 1)(2x - 1)}$

$\qquad = \dfrac{2x - 1 - 2x - 2}{2x^2 + x - 1}$

$\qquad = \dfrac{-3}{2x^2 + x - 1}$

✓ **Checkpoint 2**

Perform each indicated operation and simplify.

a. $x + \dfrac{2}{x}$ **b.** $\dfrac{2}{x + 1} - \dfrac{1}{2x + 1}$

In adding (or subtracting) fractions whose denominators have no common factors, it is convenient to use the following pattern.

$$\frac{a}{b} + \frac{c}{d} = \frac{a + c}{b d} = \frac{ad + bc}{bd}$$

For instance, in Example 2(b), you could have used this pattern as shown.

$$\frac{1}{x + 1} - \frac{2}{2x - 1} = \frac{(2x - 1) - 2(x + 1)}{(x + 1)(2x - 1)}$$

$$= \frac{2x - 1 - 2x - 2}{(x + 1)(2x - 1)}$$

$$= \frac{-3}{2x^2 + x - 1}$$

In Example 2, the denominators of the rational expressions have no common factors. When the denominators do have common factors, it is best to find the least common denominator before adding or subtracting. For instance, when adding

$$\frac{1}{x} \quad \text{and} \quad \frac{2}{x^2}$$

you can recognize that the least common denominator is x^2 and write

$$\frac{1}{x} + \frac{2}{x^2} = \frac{x}{x^2} + \frac{2}{x^2}$$ Write with common denominator.

$$= \frac{x + 2}{x^2}.$$ Add fractions.

This is further demonstrated in Example 3.

Example 3   Adding Rational Expressions

Add the rational expressions.

$$\frac{x}{x^2 - 1} + \frac{3}{x + 1}$$

SOLUTION   Because $x^2 - 1 = (x + 1)(x - 1)$, the least common denominator is $x^2 - 1$.

$$\frac{x}{x^2 - 1} + \frac{3}{x + 1} = \frac{x}{(x - 1)(x + 1)} + \frac{3}{x + 1}$$ Factor.

$$= \frac{x}{(x - 1)(x + 1)} + \frac{3(x - 1)}{(x - 1)(x + 1)}$$ Write with common denominator.

$$= \frac{x + 3(x - 1)}{(x - 1)(x + 1)}$$ Add fractions.

$$= \frac{x + 3x - 3}{(x - 1)(x + 1)}$$ Multiply.

$$= \frac{4x - 3}{x^2 - 1}$$ Simplify.

✓ **Checkpoint 3**

Add the rational expressions.

$$\frac{x}{x^2 - 4} + \frac{2}{x - 2}$$

■

Example 4　Subtracting Rational Expressions

Subtract the rational expressions.

$$\frac{1}{2(x^2 + 2x)} - \frac{1}{4x}$$

SOLUTION　In this case, the least common denominator is $4x(x + 2)$.

$$\frac{1}{2(x^2 + 2x)} - \frac{1}{4x} = \frac{1}{2x(x + 2)} - \frac{1}{2(2x)} \qquad \text{Factor.}$$

$$= \frac{2}{2(2x)(x + 2)} - \frac{x + 2}{2(2x)(x + 2)} \qquad \text{Write with common denominator.}$$

$$= \frac{2 - (x + 2)}{4x(x + 2)} \qquad \text{Subtract fractions.}$$

$$= \frac{2 - x - 2}{4x(x + 2)} \qquad \text{Remove parentheses.}$$

$$= \frac{-x}{4x(x + 2)} \qquad \text{Divide out common factor.}$$

$$= \frac{-1}{4(x + 2)}, \quad x \neq 0 \qquad \text{Simplify.}$$

✓ **Checkpoint 4**

Subtract the rational expressions.

$$\frac{1}{3(x^2 + 2x)} - \frac{1}{3x}$$

Example 5　Combining Three Rational Expressions

Perform the operations and simplify.

$$\frac{3}{x - 1} - \frac{2}{x} + \frac{x + 3}{x^2 - 1}$$

SOLUTION　Using the factored denominators $(x - 1)$, x, and $(x + 1)(x - 1)$, you can see that the least common denominator is $x(x + 1)(x - 1)$.

$$\frac{3}{x - 1} - \frac{2}{x} + \frac{x + 3}{x^2 - 1} = \frac{3(x)(x + 1)}{x(x + 1)(x - 1)} - \frac{2(x + 1)(x - 1)}{x(x + 1)(x - 1)} + \frac{(x + 3)(x)}{x(x + 1)(x - 1)}$$

$$= \frac{3(x)(x + 1) - 2(x + 1)(x - 1) + (x + 3)(x)}{x(x + 1)(x - 1)}$$

$$= \frac{3x^2 + 3x - 2x^2 + 2 + x^2 + 3x}{x(x + 1)(x - 1)}$$

$$= \frac{2x^2 + 6x + 2}{x(x + 1)(x - 1)}$$

$$= \frac{2(x^2 + 3x + 1)}{x(x + 1)(x - 1)}$$

✓ **Checkpoint 5**

Perform the operations and simplify.

$$\frac{4}{x} - \frac{2}{x^2} + \frac{4}{x + 3}$$

Expressions Involving Radicals

In calculus, the operation of differentiation tends to produce "messy" expressions when applied to fractional expressions. This is especially true when the fractional expressions involve radicals. When differentiation is used, it is important to be able to simplify these expressions in order to obtain more manageable forms. The expressions in Example 6 are the results of differentiation. In each case, note how much *simpler* the simplified form is than the original form.

Example 6 **Simplifying an Expression with Radicals**

Simplify each expression.

a. $\dfrac{\sqrt{x+1} - \dfrac{x}{2\sqrt{x+1}}}{x+1}$

b. $\left(\dfrac{1}{x+\sqrt{x^2+1}}\right)\left(1 + \dfrac{2x}{2\sqrt{x^2+1}}\right)$

SOLUTION

a. $\dfrac{\sqrt{x+1} - \dfrac{x}{2\sqrt{x+1}}}{x+1} = \dfrac{\dfrac{2(x+1)}{2\sqrt{x+1}} - \dfrac{x}{2\sqrt{x+1}}}{x+1}$ Write with common denominator.

$= \dfrac{\dfrac{2x+2-x}{2\sqrt{x+1}}}{\dfrac{x+1}{1}}$ Subtract fractions.

$= \dfrac{x+2}{2\sqrt{x+1}}\left(\dfrac{1}{x+1}\right)$ To divide, invert and multiply.

$= \dfrac{x+2}{2(x+1)^{3/2}}$ Multiply.

b. $\left(\dfrac{1}{x+\sqrt{x^2+1}}\right)\left(1 + \dfrac{2x}{2\sqrt{x^2+1}}\right) = \left(\dfrac{1}{x+\sqrt{x^2+1}}\right)\left(1 + \dfrac{x}{\sqrt{x^2+1}}\right)$

$= \left(\dfrac{1}{x+\sqrt{x^2+1}}\right)\left(\dfrac{\sqrt{x^2+1}}{\sqrt{x^2+1}} + \dfrac{x}{\sqrt{x^2+1}}\right)$

$= \left(\dfrac{1}{x+\sqrt{x^2+1}}\right)\left(\dfrac{x+\sqrt{x^2+1}}{\sqrt{x^2+1}}\right)$

$= \dfrac{1}{\sqrt{x^2+1}}$

✓ **Checkpoint 6**

Simplify each expression.

a. $\dfrac{\sqrt{x+2} - \dfrac{x}{4\sqrt{x+2}}}{x+2}$

b. $\left(\dfrac{1}{x+\sqrt{x^2+4}}\right)\left(1 + \dfrac{x}{\sqrt{x^2+4}}\right)$

Rationalization Techniques

In working with quotients involving radicals, it is often convenient to move the radical expression from the denominator to the numerator, or vice versa. For example, you can move $\sqrt{2}$ from the denominator to the numerator in the following quotient by multiplying by $\sqrt{2}/\sqrt{2}$.

Radical in Denominator	*Rationalize*	*Radical in Numerator*
$\dfrac{1}{\sqrt{2}}$	$\dfrac{1}{\sqrt{2}}\left(\dfrac{\sqrt{2}}{\sqrt{2}}\right)$	$\dfrac{\sqrt{2}}{2}$

This process is called **rationalizing the denominator.** A similar process is used to **rationalize the numerator.**

STUDY TIP

The success of the second and third rationalizing techniques stems from the following.

$$\left(\sqrt{a} - \sqrt{b}\right)\left(\sqrt{a} + \sqrt{b}\right) = a - b$$

Rationalizing Techniques

1. When the denominator is $\sqrt{a}$, multiply by $\dfrac{\sqrt{a}}{\sqrt{a}}$.

2. When the denominator is $\sqrt{a} - \sqrt{b}$, multiply by $\dfrac{\sqrt{a} + \sqrt{b}}{\sqrt{a} + \sqrt{b}}$.

3. When the denominator is $\sqrt{a} + \sqrt{b}$, multiply by $\dfrac{\sqrt{a} - \sqrt{b}}{\sqrt{a} - \sqrt{b}}$.

The same guidelines apply to rationalizing numerators.

Example 7   Rationalizing Denominators and Numerators

Rationalize the denominator or numerator.

a. $\dfrac{3}{\sqrt{12}}$ b. $\dfrac{\sqrt{x+1}}{2}$ c. $\dfrac{1}{\sqrt{5} + \sqrt{2}}$ d. $\dfrac{1}{\sqrt{x} - \sqrt{x+1}}$

SOLUTION

a. $\dfrac{3}{\sqrt{12}} = \dfrac{3}{2\sqrt{3}} = \dfrac{3}{2\sqrt{3}}\left(\dfrac{\sqrt{3}}{\sqrt{3}}\right) = \dfrac{3\sqrt{3}}{2(3)} = \dfrac{\sqrt{3}}{2}$

b. $\dfrac{\sqrt{x+1}}{2} = \dfrac{\sqrt{x+1}}{2}\left(\dfrac{\sqrt{x+1}}{\sqrt{x+1}}\right) = \dfrac{x+1}{2\sqrt{x+1}}$

c. $\dfrac{1}{\sqrt{5} + \sqrt{2}} = \dfrac{1}{\sqrt{5} + \sqrt{2}}\left(\dfrac{\sqrt{5} - \sqrt{2}}{\sqrt{5} - \sqrt{2}}\right) = \dfrac{\sqrt{5} - \sqrt{2}}{5 - 2} = \dfrac{\sqrt{5} - \sqrt{2}}{3}$

d. $\dfrac{1}{\sqrt{x} - \sqrt{x+1}} = \dfrac{1}{\sqrt{x} - \sqrt{x+1}}\left(\dfrac{\sqrt{x} + \sqrt{x+1}}{\sqrt{x} + \sqrt{x+1}}\right)$

$$= \dfrac{\sqrt{x} + \sqrt{x+1}}{x - (x+1)}$$

$$= -\sqrt{x} - \sqrt{x+1}$$

✓ Checkpoint 7

Rationalize the denominator or numerator.

a. $\dfrac{5}{\sqrt{8}}$ b. $\dfrac{\sqrt{x+2}}{4}$ c. $\dfrac{1}{\sqrt{6} - \sqrt{3}}$ d. $\dfrac{1}{\sqrt{x} + \sqrt{x+2}}$ ■

Exercises A.5

Simplifying a Rational Expression In Exercises 1–4, write the rational expression in simplest form. *See Example 1.*

1. $\dfrac{x^2 - 7x + 12}{x^2 + 3x - 18}$

2. $\dfrac{x^2 - 5x - 6}{x^2 + 11x + 10}$

3. $\dfrac{x^2 + 3x - 10}{2x^2 - x - 6}$

4. $\dfrac{3x^2 + 13x + 12}{x^2 - 4x - 21}$

Adding and Subtracting Rational Expressions In Exercises 5–16, perform the indicated operations and simplify. *See Examples 2, 3, 4, and 5.*

5. $\dfrac{x}{x - 2} + \dfrac{3}{x - 2}$

6. $\dfrac{5x + 10}{2x - 1} - \dfrac{2x + 10}{2x - 1}$

7. $x - \dfrac{3}{x}$

8. $3x + \dfrac{2}{x^2}$

9. $\dfrac{2}{x - 3} + \dfrac{5x}{3x + 4}$

10. $\dfrac{3}{3x - 1} - \dfrac{1}{x + 2}$

11. $\dfrac{2}{x^2 - 4} - \dfrac{1}{x - 2}$

12. $\dfrac{5}{x^2 - 9} + \dfrac{x}{x + 3}$

13. $\dfrac{x}{x^2 + x - 2} - \dfrac{1}{x + 2}$

14. $\dfrac{2}{x + 1} + \dfrac{3x - 2}{x^2 - 2x - 3}$

15. $\dfrac{2}{x^2 + 1} - \dfrac{1}{x} + \dfrac{1}{x^3 + x}$

16. $\dfrac{3}{x + 2} + \dfrac{3}{x - 2} + \dfrac{1}{x^2 - 4}$

Simplifying an Expression with Radicals In Exercises 17–28, simplify the expression. *See Example 6.*

17. $\dfrac{-x}{(x + 1)^{3/2}} + \dfrac{2}{(x + 1)^{1/2}}$

18. $2\sqrt{x}(x - 2) + \dfrac{(x - 2)^2}{2\sqrt{x}}$

19. $\dfrac{2 - t}{2\sqrt{1 + t}} - \sqrt{1 + t}$

20. $-\dfrac{\sqrt{x^2 + 1}}{x^2} + \dfrac{1}{\sqrt{x^2 + 1}}$

21. $\left(2x\sqrt{x^2 + 1} - \dfrac{x^3}{\sqrt{x^2 + 1}}\right) \div (x^2 + 1)$

22. $\left(\sqrt{x^3 + 1} - \dfrac{3x^3}{2\sqrt{x^3 + 1}}\right) \div (x^3 + 1)$

23. $\dfrac{(x^2 + 2)^{1/2} - x^2(x^2 + 2)^{-1/2}}{x^2}$

24. $\dfrac{x(x + 1)^{-1/2} - (x + 1)^{1/2}}{x^2}$

25. $\dfrac{\dfrac{\sqrt{x + 1}}{\sqrt{x}} - \dfrac{\sqrt{x}}{\sqrt{x + 1}}}{2(x + 1)}$

26. $\dfrac{\dfrac{2x^2}{3(x^2 - 1)^{2/3}} - (x^2 - 1)^{1/3}}{x^2}$

27. $\dfrac{-x^2}{(2x + 3)^{3/2}} + \dfrac{2x}{(2x + 3)^{1/2}}$

28. $\dfrac{-x}{2(3 + x^2)^{3/2}} + \dfrac{3}{(3 + x^2)^{1/2}}$

Rationalizing Denominators and Numerators In Exercises 29–42, rationalize the denominator or numerator and simplify. *See Example 7.*

29. $\dfrac{2}{\sqrt{10}}$

30. $\dfrac{3}{\sqrt{21}}$

31. $\dfrac{4x}{\sqrt{x - 1}}$

32. $\dfrac{5y}{\sqrt{y + 7}}$

33. $\dfrac{49(x - 3)}{\sqrt{x^2 - 9}}$

34. $\dfrac{10(x + 2)}{\sqrt{x^2 - x - 6}}$

35. $\dfrac{5}{\sqrt{14} - 2}$

36. $\dfrac{13}{6 + \sqrt{10}}$

37. $\dfrac{1}{\sqrt{6} + \sqrt{5}}$

38. $\dfrac{x}{\sqrt{2} + \sqrt{3}}$

39. $\dfrac{2}{\sqrt{x} + \sqrt{x - 2}}$

40. $\dfrac{10}{\sqrt{x} + \sqrt{x + 5}}$

41. $\dfrac{\sqrt{x + 2} - \sqrt{2}}{x}$

42. $\dfrac{\sqrt{x + 1} - 1}{x}$

43. **Installment Loan** The monthly payment M (in dollars) for an installment loan is given by the formula

$$M = P\left[\dfrac{r/12}{1 - \left(\dfrac{1}{(r/12) + 1}\right)^N}\right]$$

where P is the amount of the loan (in dollars), r is the annual percentage rate (in decimal form), and N is the number of monthly payments. Enter the formula into a graphing utility, and use it to find the monthly payment for a loan of \$10,000 at an annual percentage rate of 7.5% ($r = 0.075$) for 5 years ($N = 60$ monthly payments).

44. **Inventory** A retailer has determined that the cost C (in dollars) of ordering and storing x units of a product is

$$C = 6x + \dfrac{900{,}000}{x}.$$

(a) Write the expression for cost as a single fraction.

(b) Which order size should the retailer place: 240 units, 387 units, or 480 units? Explain your reasoning.

B Alternative Introduction to the Fundamental Theorem of Calculus

In this appendix, a summation process is used to provide an alternative development of the definite integral. It is intended that this supplement follow Section 5.3 in the text. If used, this appendix should replace the material preceding Example 2 in Section 5.4. Example 1 below shows how the area of a region in the plane can be approximated by the use of rectangles.

Example 1 Using Rectangles to Approximate the Area of a Region

Use the four rectangles shown in Figure B.1 to approximate the area of the region lying between the graph of

$$f(x) = \frac{x^2}{2}$$

and the x-axis, between $x = 0$ and $x = 4$.

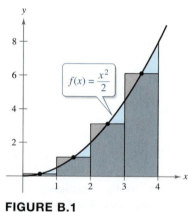

$$f(x) = \frac{x^2}{2}$$

FIGURE B.1

SOLUTION You can find the heights of the rectangles by evaluating the function f at each of the midpoints of the subintervals

$$[0, 1], \quad [1, 2], \quad [2, 3], \quad [3, 4].$$

Because the width of each rectangle is 1, the sum of the areas of the four rectangles is

$$S = \overbrace{(1)}^{\text{width}} \overbrace{f\left(\tfrac{1}{2}\right)}^{\text{height}} + \overbrace{(1)}^{\text{width}} \overbrace{f\left(\tfrac{3}{2}\right)}^{\text{height}} + \overbrace{(1)}^{\text{width}} \overbrace{f\left(\tfrac{5}{2}\right)}^{\text{height}} + \overbrace{(1)}^{\text{width}} \overbrace{f\left(\tfrac{7}{2}\right)}^{\text{height}}$$

$$= \frac{1}{8} + \frac{9}{8} + \frac{25}{8} + \frac{49}{8}$$

$$= \frac{84}{8}$$

$$= 10.5.$$

So, you can approximate the area of the region to be 10.5 square units.

> **STUDY TIP**
>
> The approximation technique used in Example 1 is called the *Midpoint Rule*. The Midpoint Rule is discussed further in Section 5.6.

The procedure shown in Example 1 can be generalized. Let f be a continuous function defined on the closed interval $[a, b]$. To begin, partition the interval into n subintervals, each of width

$$\Delta x = \frac{b - a}{n}$$

as shown.

$$a = x_0 < x_1 < x_2 < \cdots < x_{n-1} < x_n = b$$

In each subinterval $[x_{i-1}, x_i]$, choose an arbitrary point c_i and form the sum

$$S = f(c_1)\,\Delta x + f(c_2)\,\Delta x + \cdots + f(c_{n-1})\,\Delta x + f(c_n)\,\Delta x.$$

This type of summation is called a **Riemann sum** and is often written using summation notation, as shown below.

$$S = \sum_{i=1}^{n} f(c_i)\,\Delta x, \quad x_{i-1} \le c_i \le x_i$$

For the Riemann sum in Example 1, the interval is $[a, b] = [0, 4]$, the number of subintervals is $n = 4$, the width of each subinterval is $\Delta x = 1$, and the point c_i in each subinterval is its midpoint. So, you can write the approximation in Example 1 as

$$S = \sum_{i=1}^{n} f(c_i)\,\Delta x$$

$$= \sum_{i=1}^{4} f(c_i)(1)$$

$$= \frac{1}{8} + \frac{9}{8} + \frac{25}{8} + \frac{49}{8}$$

$$= \frac{84}{8}.$$

Example 2   Using a Riemann Sum to Approximate Area

Use a Riemann sum to approximate the area of the region bounded by the graph of

$$f(x) = -x^2 + 2x$$

and the x-axis, for $0 \le x \le 2$. In the Riemann sum, let $n = 6$ and choose c_i to be the left endpoint of each subinterval.

SOLUTION   Subdivide the interval $[0, 2]$ into six subintervals, each of width

$$\Delta x = \frac{2 - 0}{6}$$

$$= \frac{1}{3}$$

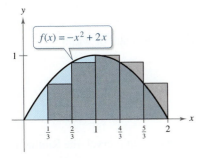

$f(x) = -x^2 + 2x$

FIGURE B.2

as shown in Figure B.2. Because c_i is the left endpoint of each subinterval, the Riemann sum is given by

$$S = \sum_{i=1}^{n} f(c_i)\,\Delta x$$

$$= \left[f(0) + f\left(\frac{1}{3}\right) + f\left(\frac{2}{3}\right) + f(1) + f\left(\frac{4}{3}\right) + f\left(\frac{5}{3}\right) \right]\left(\frac{1}{3}\right)$$

$$= \left[0 + \frac{5}{9} + \frac{8}{9} + 1 + \frac{8}{9} + \frac{5}{9} \right]\left(\frac{1}{3}\right)$$

$$= \frac{35}{27} \text{ square units}.$$

Example 2 illustrates an important point. If a function f is continuous and nonnegative over the interval $[a, b]$, then the Riemann sum

$$S = \sum_{i=1}^{n} f(c_i)\,\Delta x$$

can be used to approximate the area of the region bounded by the graph of f and the x-axis, between $x = a$ and $x = b$. Moreover, for a given interval, as the number of subintervals increases, the approximation to the actual area will improve. This is illustrated in the next two examples by using Riemann sums to approximate the area of a triangle.

Example 3   Approximating the Area of a Triangle

Use a Riemann sum to approximate the area of the triangular region bounded by the graph of

$$f(x) = 2x$$

and the x-axis, $0 \le x \le 3$. Use a partition of six subintervals and choose c_i to be the left endpoint of each subinterval.

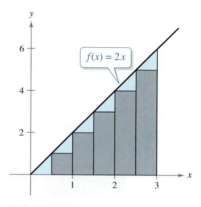

FIGURE B.3

SOLUTION   Subdivide the interval $[0, 3]$ into six subintervals, each of width

$$\Delta x = \frac{3 - 0}{6}$$

$$= \frac{1}{2}$$

as shown in Figure B.3. Because c_i is the left endpoint of each subinterval, the Riemann sum is given by

$$S = \sum_{i=1}^{n} f(c_i)\,\Delta x$$

$$= \left[f(0) + f\left(\frac{1}{2}\right) + f(1) + f\left(\frac{3}{2}\right) + f(2) + f\left(\frac{5}{2}\right) \right]\left(\frac{1}{2}\right)$$

$$= [0 + 1 + 2 + 3 + 4 + 5]\left(\frac{1}{2}\right)$$

$$= \frac{15}{2} \text{ square units.}$$

The approximations in Examples 2 and 3 are called **left Riemann sums,** because c_i was chosen to be the left endpoint of each subinterval. Using the right endpoints in Example 3, the **right Riemann sum** is $\frac{21}{2}$. Note that the exact area of the triangular region in Example 3 is

$$\text{Area} = \frac{1}{2}(\text{base})(\text{height})$$

$$= \frac{1}{2}(3)(6)$$

$$= 9 \text{ square units.}$$

So, the left Riemann sum gives an approximation that is less than the actual area, and the right Riemann sum gives an approximation that is greater than the actual area.

In Example 4, you will see that the approximation improves as the number of subintervals increases.

TECH TUTOR

Most graphing utilities are able to sum the first n terms of a sequence. Try using a graphing utility to verify the right Riemann sum in Example 3.

Example 4   Increasing the Number of Subintervals

Let $f(x) = 2x$, $0 \le x \le 3$. Use a graphing utility to determine the left and right Riemann sums for $n = 10$, $n = 100$, and $n = 1000$ subintervals.

SOLUTION   A graphing utility program for this problem is shown in Figure B.4. [Note that the function $f(x) = 2x$ is entered as Y1.]

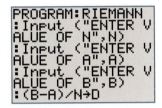
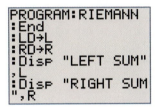

FIGURE B.4

Running this program for $n = 10$, $n = 100$, and $n = 1000$ gives the results shown in the table.

n	Left Reimann sum	Right Reimann sum
10	8.100	9.900
100	8.910	9.090
1000	8.991	9.009

From the results of Example 4, it appears that the Riemann sums are approaching the limit 9 as n approaches infinity. It is this observation that motivates the definition of a **definite integral.** In this definition, consider the partition of $[a, b]$ into n subintervals of equal width $\Delta x = (b - a)/n$, as shown.

$$a = x_0 < x_1 < x_2 < \cdots < x_{n-1} < x_n = b$$

Moreover, consider c_i to be an arbitrary point in the ith subinterval $[x_{i-1}, x_i]$. To say that the number of subintervals n approaches infinity is equivalent to saying that the width, Δx, of the subintervals approaches zero.

Definition of Definite Integral

If f is a continuous function defined on the closed interval $[a, b]$, then the **definite integral of f on $[a, b]$** is

$$\int_a^b f(x)\, dx = \lim_{\Delta x \to 0} \sum_{i=1}^n f(c_i)\, \Delta x$$

$$= \lim_{n \to \infty} \sum_{i=1}^n f(c_i)\, \Delta x.$$

If f is continuous and nonnegative on the interval $[a, b]$, then the definite integral of f on $[a, b]$ gives the area of the region bounded by the graph of f, the x-axis, and the vertical lines $x = a$ and $x = b$.

Evaluation of a definite integral by its limit definition can be difficult. However, there are times when a definite integral can be solved by recognizing that it represents the area of a common type of geometric figure.

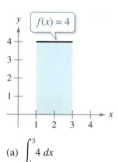

(a) $\displaystyle\int_1^3 4\, dx$

Rectangle

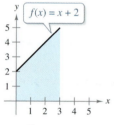

(b) $\displaystyle\int_0^3 (x + 2)\, dx$

Trapezoid

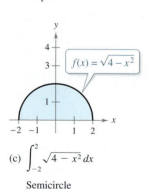

(c) $\displaystyle\int_{-2}^2 \sqrt{4 - x^2}\, dx$

Semicircle

FIGURE B.5

Example 5   **The Areas of Common Geometric Figures**

Sketch the region corresponding to each of the definite integrals. Then evaluate each definite integral using a geometric formula.

a. $\displaystyle\int_1^3 4\, dx$

b. $\displaystyle\int_0^3 (x + 2)\, dx$

c. $\displaystyle\int_{-2}^2 \sqrt{4 - x^2}\, dx$

SOLUTION   A sketch of each region is shown in Figure B.5.

a. The region associated with this definite integral is a rectangle of height 4 and width 2. Moreover, because the function $f(x) = 4$ is continuous and nonnegative on the interval $[1, 3]$, you can conclude that the area of the rectangle is given by the definite integral. So, the value of the definite integral is

$$\int_1^3 4\, dx = 4(2) = 8 \text{ square units.}$$

b. The region associated with this definite integral is a trapezoid with an altitude of 3 and parallel bases of lengths 2 and 5. The formula for the area of a trapezoid is $\frac{1}{2}h(b_1 + b_2)$, and so you have

$$\int_0^3 (x + 2)\, dx = \frac{1}{2}(3)(2 + 5)$$

$$= \frac{21}{2} \text{ square units.}$$

c. The region associated with this definite integral is a semicircle of radius 2. The formula for the area of a semicircle is $\frac{1}{2}\pi r^2$, and so you have

$$\int_{-2}^2 \sqrt{4 - x^2}\, dx = \frac{1}{2}\pi(2^2)$$

$$= 2\pi \text{ square units.}$$

For some simple functions, it is possible to evaluate definite integrals by the Riemann sum definition. In the next example, you will use the fact that the sum of the first n integers is given by the formula

$$1 + 2 + \cdots + n = \sum_{i=1}^{n} i = \frac{n(n + 1)}{2}$$ See Exercise 29.

to compute the area of the triangular region in Examples 3 and 4.

Example 6   Evaluating a Definite Integral by Its Definition

Evaluate $\displaystyle\int_{0}^{3} 2x \, dx$.

SOLUTION   Let

$$\Delta x = \frac{b - a}{n} = \frac{3}{n}$$

and choose c_i to be the right endpoint of each subinterval,

$$c_i = \frac{3i}{n}.$$

Then you have

$$\int_{0}^{3} 2x \, dx = \lim_{\Delta x \to 0} \sum_{i=1}^{n} f(c_i)\Delta x$$

$$= \lim_{n \to \infty} \sum_{i=1}^{n} 2\left(i\frac{3}{n}\right)\left(\frac{3}{n}\right)$$

$$= \lim_{n \to \infty} \frac{18}{n^2} \sum_{i=1}^{n} i$$

$$= \lim_{n \to \infty} \left(\frac{18}{n^2}\right)\left(\frac{n(n + 1)}{2}\right)$$

$$= \lim_{n \to \infty} \left(9 + \frac{9}{n}\right).$$

This limit can be evaluated in the same way that you calculated horizontal asymptotes in Section 3.6. In particular, as n approaches infinity, you see that $9/n$ approaches 0, and the limit above is 9. So, you can conclude that

$$\int_{0}^{3} 2x \, dx = 9.$$

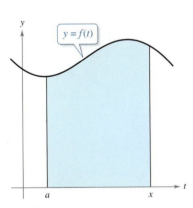

FIGURE B.6

From Example 6, you can see that it can be difficult to evaluate the definite integral of even a simple function by using Riemann sums. A computer can help in calculating these sums for large values of n, but this procedure would give only an approximation of the definite integral. Fortunately, the **Fundamental Theorem of Calculus** provides a technique for evaluating definite integrals using antiderivatives, and for this reason it is often thought to be the most important theorem in calculus. In the remainder of this appendix, you will see how derivatives and integrals are related via the Fundamental Theorem of Calculus.

To simplify the discussion, assume that f is a continuous nonnegative function defined on the interval $[a, b]$. Let $A(x)$ be the area of the region under the graph of f from a to x, as indicated in Figure B.6. The area under the shaded region in Figure B.7 is

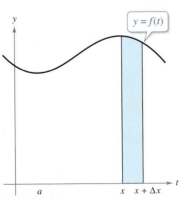

FIGURE B.7

$$A(x + \Delta x) - A(x).$$

If Δx is small, then this area is approximated by the area of the rectangle of height $f(x)$ and width Δx. So, you have

$$A(x + \Delta x) - A(x) \approx f(x)\, \Delta x.$$

Dividing by Δx produces

$$f(x) \approx \frac{A(x + \Delta x) - A(x)}{\Delta x}.$$

By taking the limit as Δx approaches 0, you can see that

$$f(x) = \lim_{\Delta x \to 0} \frac{A(x + \Delta x) - A(x)}{\Delta x}$$
$$= A'(x)$$

and you can establish the fact that the area function $A(x)$ is an antiderivative of f. Although it was assumed that f is continuous and nonnegative, this development is valid when the function f is simply continuous on the closed interval $[a, b]$. This result is used in the proof of the Fundamental Theorem of Calculus.

Fundamental Theorem of Calculus

If f is a continuous function on the closed interval $[a, b]$, then

$$\int_a^b f(x)\,dx = F(b) - F(a)$$

where F is any function such that $F'(x) = f(x)$.

PROOF   From the discussion above, you know that

$$\int_a^x f(x)\,dx = A(x)$$

and in particular,

$$A(a) = \int_a^a f(x)\,dx = 0$$

and

$$A(b) = \int_a^b f(x)\,dx.$$

If F is *any* antiderivative of f, then you know that F differs from A by a constant. That is,

$$A(x) = F(x) + C.$$

So,

$$\int_a^b f(x)\,dx = A(b) - A(a)$$
$$= [F(b) + C] - [F(a) + C]$$
$$= F(b) + C - F(a) - C$$
$$= F(b) - F(a).$$

You are now ready to continue Section 5.4, on page 341, just after the statement of the Fundamental Theorem of Calculus.

Exercises B

Using Rectangles to Approximate the Area of a Region   In Exercises 1 and 2, use the rectangles to approximate the area of the region. *See Example 1.*

1. $y = x + 1$

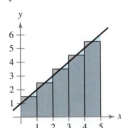

2. $y = 4 - x^2$

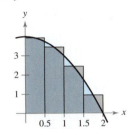

Using a Riemann Sum to Approximate Area   In Exercises 3–8, use the left Riemann sum and the right Riemann sum to approximate the area of the region using the indicated number of subintervals. *See Examples 2 and 3.*

3. $y = \sqrt{x}$

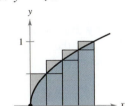

4. $y = \sqrt{x} + 1$

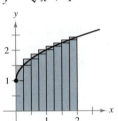

5. $y = \dfrac{1}{x}$

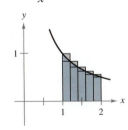

6. $y = \dfrac{1}{x - 2}$

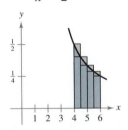

7. $y = \sqrt{1 - x^2}$

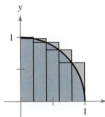

8. $y = \sqrt{x + 1}$

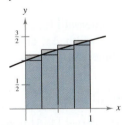

9. Comparing Riemann Sums   Consider a triangle of area 2 bounded by the graphs of $y = x$, $y = 0$, and $x = 2$.

(a) Sketch the graph of the region.

(b) Divide the interval $[0, 2]$ into n equal subintervals and show that the endpoints of the subintervals are

$$0 < 1\left(\frac{2}{n}\right) < \cdots < (n - 1)\left(\frac{2}{n}\right) < n\left(\frac{2}{n}\right).$$

(c) Show that the left Riemann sum is

$$S_L = \sum_{i=1}^{n} \left[(i - 1)\left(\frac{2}{n}\right)\right]\left(\frac{2}{n}\right).$$

(d) Show that the right Riemann sum is

$$S_R = \sum_{i=1}^{n} \left[i\left(\frac{2}{n}\right)\right]\left(\frac{2}{n}\right).$$

(e) Complete the table below.

n	5	10	50	100
Left sum, S_L				
Right sum, S_R				

(f) Show that $\lim\limits_{n \to \infty} S_L = \lim\limits_{n \to \infty} S_R = 2$.

10. Comparing Riemann Sums   Consider a trapezoid of area 4 bounded by the graphs of $y = x$, $y = 0$, $x = 1$, and $x = 3$.

(a) Sketch the graph of the region.

(b) Divide the interval $[1, 3]$ into n equal subintervals and show that the endpoints of the subintervals are

$$1 < 1 + 1\left(\frac{2}{n}\right) < \cdots < 1 + (n - 1)\left(\frac{2}{n}\right) < 1 + n\left(\frac{2}{n}\right).$$

(c) Show that the left Riemann sum is

$$S_L = \sum_{i=1}^{n} \left[1 + (i - 1)\left(\frac{2}{n}\right)\right]\left(\frac{2}{n}\right).$$

(d) Show that the right Riemann sum is

$$S_R = \sum_{i=1}^{n} \left[1 + i\left(\frac{2}{n}\right)\right]\left(\frac{2}{n}\right).$$

(e) Complete the table below.

n	5	10	50	100
Left sum, S_L				
Right sum, S_R				

(f) Show that $\lim\limits_{n \to \infty} S_L = \lim\limits_{n \to \infty} S_R = 4$.

Writing a Definite Integral   In Exercises 11–18, set up a definite integral that yields the area of the region. (Do not evaluate the integral.)

11. $f(x) = 3$

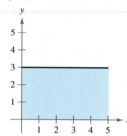

12. $f(x) = 4 - 2x$

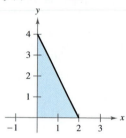

13. $f(x) = 4 - |x|$

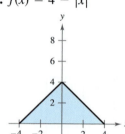

14. $f(x) = x^2$

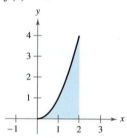

15. $f(x) = 4 - x^2$

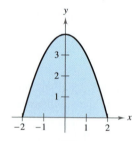

16. $f(x) = \dfrac{1}{x^2 + 1}$

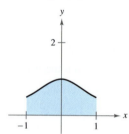

17. $f(x) = \sqrt{x + 1}$

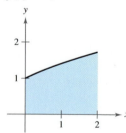

18. $f(x) = (x^2 + 1)^2$

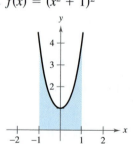

Finding Areas of Common Geometric Figures   In Exercises 19–28, sketch the region whose area is given by the definite integral. Then use a geometric formula to evaluate the integral ($a > 0, r > 0$). *See Example 5.*

19. $\displaystyle\int_0^3 4\,dx$

20. $\displaystyle\int_{-a}^a 4\,dx$

21. $\displaystyle\int_0^4 x\,dx$

22. $\displaystyle\int_0^4 \frac{x}{2}\,dx$

23. $\displaystyle\int_0^2 (2x + 5)\,dx$

24. $\displaystyle\int_0^5 (5 - x)\,dx$

25. $\displaystyle\int_{-1}^1 \left(1 - |x|\right)\,dx$

26. $\displaystyle\int_{-a}^a \left(a - |x|\right)\,dx$

27. $\displaystyle\int_{-3}^3 \sqrt{9 - x^2}\,dx$

28. $\displaystyle\int_{-r}^r \sqrt{r^2 - x^2}\,dx$

29. Proving a Sum Show that $\displaystyle\sum_{i=1}^n i = \frac{n(n + 1)}{2}$.

(*Hint:* Add the two sums below.)

$$S = 1 + 2 + 3 + \cdots + (n - 2) + (n - 1) + n$$

$$S = n + (n - 1) + (n - 2) + \cdots + 3 + 2 + 1$$

30. Evaluating a Definite Integral by Its Definition Use the Riemann sum definition and the result of Exercise 29 to evaluate the definite integrals.

(a) $\displaystyle\int_1^2 x\,dx$ (b) $\displaystyle\int_0^4 3x\,dx$

Comparing a Sum with an Integral   In Exercises 31 and 32, use the figure to fill in the blank with the symbol $<$, $>$, or $=$.

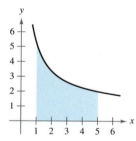

31. The interval $[1, 5]$ is partitioned into n subintervals of equal width Δx, and x_i is the left endpoint of the ith subinterval.

$$\sum_{i=1}^n f(x_i)\,\Delta x \quad \rule{1cm}{0.4pt} \quad \int_1^5 f(x)\,dx$$

32. The interval $[1, 5]$ is partitioned into n subintervals of equal width Δx, and x_i is the right endpoint of the ith subinterval.

$$\sum_{i=1}^n f(x_i)\,\Delta x \quad \rule{1cm}{0.4pt} \quad \int_1^5 f(x)\,dx$$

C Formulas

C.1 Differentiation and Integration Formulas

■ Use differentiation and integration tables to supplement differentiation and integration techniques.

Differentiation Formulas

1. $\dfrac{d}{dx}[cu] = cu'$

2. $\dfrac{d}{dx}[u \pm v] = u' \pm v'$

3. $\dfrac{d}{dx}[uv] = uv' + vu'$

4. $\dfrac{d}{dx}\left[\dfrac{u}{v}\right] = \dfrac{vu' - uv'}{v^2}$

5. $\dfrac{d}{dx}[c] = 0$

6. $\dfrac{d}{dx}[u^n] = nu^{n-1}u'$

7. $\dfrac{d}{dx}[x] = 1$

8. $\dfrac{d}{dx}[\ln u] = \dfrac{u'}{u}$

9. $\dfrac{d}{dx}[e^u] = e^u u'$

10. $\dfrac{d}{dx}[\sin u] = (\cos u)u'$

11. $\dfrac{d}{dx}[\cos u] = -(\sin u)u'$

12. $\dfrac{d}{dx}[\tan u] = (\sec^2 u)u'$

13. $\dfrac{d}{dx}[\cot u] = -(\csc^2 u)u'$

14. $\dfrac{d}{dx}[\sec u] = (\sec u \tan u)u'$

15. $\dfrac{d}{dx}[\csc u] = -(\csc u \cot u)u'$

Integration Formulas

Forms Involving u^n

1. $\displaystyle\int u^n \, du = \dfrac{u^{n+1}}{n+1} + C, \quad n \neq -1$

2. $\displaystyle\int \dfrac{1}{u} \, du = \ln|u| + C$

Forms Involving $a + bu$

3. $\displaystyle\int \dfrac{u}{a+bu} \, du = \dfrac{1}{b^2}(bu - a\ln|a+bu|) + C$

4. $\displaystyle\int \dfrac{u}{(a+bu)^2} \, du = \dfrac{1}{b^2}\left(\dfrac{a}{a+bu} + \ln|a+bu|\right) + C$

5. $\displaystyle\int \dfrac{u}{(a+bu)^n} \, du = \dfrac{1}{b^2}\left[\dfrac{-1}{(n-2)(a+bu)^{n-2}} + \dfrac{a}{(n-1)(a+bu)^{n-1}}\right] + C, \quad n \neq 1, 2$

6. $\displaystyle\int \dfrac{u^2}{a+bu} \, du = \dfrac{1}{b^3}\left[-\dfrac{bu}{2}(2a - bu) + a^2\ln|a+bu|\right] + C$

7. $\displaystyle\int \dfrac{u^2}{(a+bu)^2} \, du = \dfrac{1}{b^3}\left(bu - \dfrac{a^2}{a+bu} - 2a\ln|a+bu|\right) + C$

8. $\displaystyle\int \dfrac{u^2}{(a+bu)^3} \, du = \dfrac{1}{b^3}\left[\dfrac{2a}{a+bu} - \dfrac{a^2}{2(a+bu)^2} + \ln|a+bu|\right] + C$

9. $\displaystyle\int \dfrac{u^2}{(a+bu)^n} \, du = \dfrac{1}{b^3}\left[\dfrac{-1}{(n-3)(a+bu)^{n-3}} + \dfrac{2a}{(n-2)(a+bu)^{n-2}} - \dfrac{a^2}{(n-1)(a+bu)^{n-1}}\right] + C, \quad n \neq 1, 2, 3$

10. $\displaystyle\int \dfrac{1}{u(a+bu)} \, du = \dfrac{1}{a}\ln\left|\dfrac{u}{a+bu}\right| + C$

Integration Formulas   (continued)

11. $\displaystyle\int \frac{1}{u(a + bu)^2}\, du = \frac{1}{a}\left(\frac{1}{a + bu} + \frac{1}{a}\ln\left|\frac{u}{a + bu}\right|\right) + C$

12. $\displaystyle\int \frac{1}{u^2(a + bu)}\, du = -\frac{1}{a}\left(\frac{1}{u} + \frac{b}{a}\ln\left|\frac{u}{a + bu}\right|\right) + C$

13. $\displaystyle\int \frac{1}{u^2(a + bu)^2}\, du = -\frac{1}{a^2}\left[\frac{a + 2bu}{u(a + bu)} + \frac{2b}{a}\ln\left|\frac{u}{a + bu}\right|\right] + C$

Forms Involving $\sqrt{a + bu}$

14. $\displaystyle\int u^n \sqrt{a + bu}\, du = \frac{2}{b(2n + 3)}\left[u^n(a + bu)^{3/2} - na\int u^{n-1}\sqrt{a + bu}\, du\right]$

15. $\displaystyle\int \frac{1}{u\sqrt{a + bu}}\, du = \frac{1}{\sqrt{a}}\ln\left|\frac{\sqrt{a + bu} - \sqrt{a}}{\sqrt{a + bu} + \sqrt{a}}\right| + C, \quad a > 0$

16. $\displaystyle\int \frac{1}{u^n\sqrt{a + bu}}\, du = \frac{-1}{a(n - 1)}\left[\frac{\sqrt{a + bu}}{u^{n-1}} + \frac{(2n - 3)b}{2}\int \frac{1}{u^{n-1}\sqrt{a + bu}}\, du\right], \quad n \neq 1$

17. $\displaystyle\int \frac{\sqrt{a + bu}}{u}\, du = 2\sqrt{a + bu} + a\int \frac{1}{u\sqrt{a + bu}}\, du$

18. $\displaystyle\int \frac{\sqrt{a + bu}}{u^n}\, du = \frac{-1}{a(n - 1)}\left[\frac{(a + bu)^{3/2}}{u^{n-1}} + \frac{(2n - 5)b}{2}\int \frac{\sqrt{a + bu}}{u^{n-1}}\, du\right], \quad n \neq 1$

19. $\displaystyle\int \frac{u}{\sqrt{a + bu}}\, du = -\frac{2(2a - bu)}{3b^2}\sqrt{a + bu} + C$

20. $\displaystyle\int \frac{u^n}{\sqrt{a + bu}}\, du = \frac{2}{(2n + 1)b}\left(u^n\sqrt{a + bu} - na\int \frac{u^{n-1}}{\sqrt{a + bu}}\, du\right)$

Forms Involving $u^2 - a^2$, $a > 0$

21. $\displaystyle\int \frac{1}{u^2 - a^2}\, du = -\int \frac{1}{a^2 - u^2}\, du = \frac{1}{2a}\ln\left|\frac{u - a}{u + a}\right| + C$

22. $\displaystyle\int \frac{1}{(u^2 - a^2)^n}\, du = \frac{-1}{2a^2(n - 1)}\left[\frac{u}{(u^2 - a^2)^{n-1}} + (2n - 3)\int \frac{1}{(u^2 - a^2)^{n-1}}\, du\right], \quad n \neq 1$

Forms Involving $\sqrt{u^2 \pm a^2}$, $a > 0$

23. $\displaystyle\int \sqrt{u^2 \pm a^2}\, du = \frac{1}{2}\left(u\sqrt{u^2 \pm a^2} \pm a^2 \ln\left|u + \sqrt{u^2 \pm a^2}\right|\right) + C$

24. $\displaystyle\int u^2\sqrt{u^2 \pm a^2}\, du = \frac{1}{8}\left[u(2u^2 \pm a^2)\sqrt{u^2 \pm a^2} - a^4 \ln\left|u + \sqrt{u^2 \pm a^2}\right|\right] + C$

25. $\displaystyle\int \frac{\sqrt{u^2 + a^2}}{u}\, du = \sqrt{u^2 + a^2} - a\ln\left|\frac{a + \sqrt{u^2 + a^2}}{u}\right| + C$

26. $\displaystyle\int \frac{\sqrt{u^2 \pm a^2}}{u^2}\, du = \frac{-\sqrt{u^2 \pm a^2}}{u} + \ln\left|u + \sqrt{u^2 \pm a^2}\right| + C$

27. $\displaystyle\int \frac{1}{\sqrt{u^2 \pm a^2}}\, du = \ln\left|u + \sqrt{u^2 \pm a^2}\right| + C$

28. $\displaystyle\int \frac{1}{u\sqrt{u^2 + a^2}}\, du = \frac{-1}{a}\ln\left|\frac{a + \sqrt{u^2 + a^2}}{u}\right| + C$

29. $\displaystyle \int \frac{u^2}{\sqrt{u^2 \pm a^2}}\, du = \frac{1}{2}\left(u\sqrt{u^2 \pm a^2} \mp a^2 \ln\left|u + \sqrt{u^2 \pm a^2}\right|\right) + C$

30. $\displaystyle \int \frac{1}{u^2\sqrt{u^2 \pm a^2}}\, du = \mp \frac{\sqrt{u^2 \pm a^2}}{a^2 u} + C$

31. $\displaystyle \int \frac{1}{(u^2 \pm a^2)^{3/2}}\, du = \frac{\pm u}{a^2\sqrt{u^2 \pm a^2}} + C$

Forms Involving $\sqrt{a^2 - u^2}$, $a > 0$

32. $\displaystyle \int \frac{\sqrt{a^2 - u^2}}{u}\, du = \sqrt{a^2 - u^2} - a \ln\left|\frac{a + \sqrt{a^2 - u^2}}{u}\right| + C$

33. $\displaystyle \int \frac{1}{u\sqrt{a^2 - u^2}}\, du = \frac{-1}{a} \ln\left|\frac{a + \sqrt{a^2 - u^2}}{u}\right| + C$

34. $\displaystyle \int \frac{1}{u^2\sqrt{a^2 - u^2}}\, du = \frac{-\sqrt{a^2 - u^2}}{a^2 u} + C$

35. $\displaystyle \int \frac{1}{(a^2 - u^2)^{3/2}}\, du = \frac{u}{a^2\sqrt{a^2 - u^2}} + C$

Forms Involving e^u

36. $\displaystyle \int e^u\, du = e^u + C$

37. $\displaystyle \int u e^u\, du = (u - 1)e^u + C$

38. $\displaystyle \int u^n e^u\, du = u^n e^u - n \int u^{n-1} e^u\, du$

39. $\displaystyle \int \frac{1}{1 + e^u}\, du = u - \ln(1 + e^u) + C$

40. $\displaystyle \int \frac{1}{1 + e^{nu}}\, du = u - \frac{1}{n}\ln(1 + e^{nu}) + C$

Forms Involving ln u

41. $\displaystyle \int \ln u\, du = u(-1 + \ln u) + C$

42. $\displaystyle \int u \ln u\, du = \frac{u^2}{4}(-1 + 2 \ln u) + C$

43. $\displaystyle \int u^n \ln u\, du = \frac{u^{n+1}}{(n+1)^2}\left[-1 + (n+1)\ln u\right] + C, \quad n \ne -1$

44. $\displaystyle \int (\ln u)^2\, du = u\left[2 - 2 \ln u + (\ln u)^2\right] + C$

45. $\displaystyle \int (\ln u)^n\, du = u(\ln u)^n - n \int (\ln u)^{n-1}\, du$

Forms Involving sin u or cos u

46. $\displaystyle \int \sin u\, du = -\cos u + C$

47. $\displaystyle \int \cos u\, du = \sin u + C$

48. $\displaystyle \int \sin^2 u\, du = \frac{1}{2}(u - \sin u \cos u) + C$

49. $\displaystyle \int \cos^2 u\, du = \frac{1}{2}(u + \sin u \cos u) + C$

50. $\displaystyle \int \sin^n u\, du = -\frac{\sin^{n-1} u \cos u}{n} + \frac{n-1}{n} \int \sin^{n-2} u\, du$

51. $\displaystyle \int \cos^n u\, du = \frac{\cos^{n-1} u \sin u}{n} + \frac{n-1}{n} \int \cos^{n-2} u\, du$

52. $\displaystyle \int u \sin u\, du = \sin u - u \cos u + C$

53. $\displaystyle \int u \cos u\, du = \cos u + u \sin u + C$

54. $\displaystyle \int u^n \sin u\, du = -u^n \cos u + n \int u^{n-1} \cos u\, du$

Integration Formulas   (continued)

55. $\displaystyle\int u^n \cos u \, du = u^n \sin u - n\int u^{n-1} \sin u \, du$

56. $\displaystyle\int \frac{1}{1 \pm \sin u} \, du = \tan u \mp \sec u + C$

57. $\displaystyle\int \frac{1}{1 \pm \cos u} \, du = -\cot u \pm \csc u + C$

58. $\displaystyle\int \frac{1}{\sin u \cos u} \, du = \ln|\tan u| + C$

Forms Involving tan *u*, cot *u*, sec *u*, or csc *u*

59. $\displaystyle\int \tan u \, du = -\ln|\cos u| + C$

60. $\displaystyle\int \cot u \, du = \ln|\sin u| + C$

61. $\displaystyle\int \sec u \, du = \ln|\sec u + \tan u| + C$

62. $\displaystyle\int \csc u \, du = \ln|\csc u - \cot u| + C$

63. $\displaystyle\int \tan^2 u \, du = -u + \tan u + C$

64. $\displaystyle\int \cot^2 u \, du = -u - \cot u + C$

65. $\displaystyle\int \sec^2 u \, du = \tan u + C$

66. $\displaystyle\int \csc^2 u \, du = -\cot u + C$

67. $\displaystyle\int \tan^n u \, du = \frac{\tan^{n-1} u}{n-1} - \int \tan^{n-2} u \, du, \quad n \neq 1$

68. $\displaystyle\int \cot^n u \, du = -\frac{\cot^{n-1} u}{n-1} - \int \cot^{n-2} u \, du, \quad n \neq 1$

69. $\displaystyle\int \sec^n u \, du = \frac{\sec^{n-2} u \tan u}{n-1} + \frac{n-2}{n-1}\int \sec^{n-2} u \, du, \quad n \neq 1$

70. $\displaystyle\int \csc^n u \, du = -\frac{\csc^{n-2} u \cot u}{n-1} + \frac{n-2}{n-1}\int \csc^{n-2} u \, du, \quad n \neq 1$

71. $\displaystyle\int \frac{1}{1 \pm \tan u} \, du = \frac{1}{2}(u \pm \ln|\cos u \pm \sin u|) + C$

72. $\displaystyle\int \frac{1}{1 \pm \cot u} \, du = \frac{1}{2}(u \mp \ln|\sin u \pm \cos u|) + C$

73. $\displaystyle\int \frac{1}{1 \pm \sec u} \, du = u + \cot u \mp \csc u + C$

74. $\displaystyle\int \frac{1}{1 \pm \csc u} \, du = u - \tan u \pm \sec u + C$

C.2 Formulas from Business and Finance

■ Summary of business and finance formulas

Formulas from Business

Basic Terms

x = number of units produced (or sold)

p = price per unit

R = total revenue from selling x units

C = total cost of producing x units

$\overline{C}$ = average cost per unit

P = total profit from selling x units

Basic Equations

$$R = xp \qquad \overline{C} = \frac{C}{x} \qquad P = R - C$$

Typical Graphs of Supply and Demand Curves

Supply curves increase as price increases and demand curves decrease as price increases. The equilibrium point occurs when the supply and demand curves intersect.

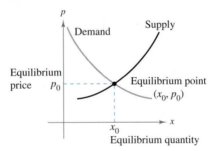

Demand Function: $p = f(x)$ = price required to sell x units

$$\eta = \frac{p/x}{dp/dx} = \text{price elasticity of demand}$$

(When $|\eta| < 1$, the demand is inelastic. When $|\eta| > 1$, the demand is elastic.)

Typical Graphs of Revenue, Cost, and Profit Functions

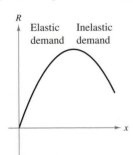

Revenue Function

The low prices required to sell more units eventually result in a decreasing revenue.

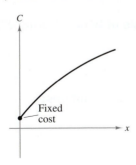

Cost Function

The total cost to produce x units includes the fixed cost.

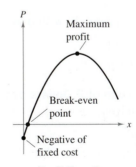

Profit Function

The break-even point occurs when $R = C$.

Formulas from Business   (continued)

Marginals

$\dfrac{dR}{dx}$ = marginal revenue ≈ the *extra* revenue from selling one additional unit

$\dfrac{dC}{dx}$ = marginal cost ≈ the *extra* cost of producing one additional unit

$\dfrac{dP}{dx}$ = marginal profit ≈ the *extra* profit from selling one additional unit

Marginal revenue

1 unit

Extra revenue for one unit

Revenue Function

Formulas from Finance

Basic Terms

P = amount of deposit r = interest rate

n = number of times interest is compounded per year

t = number of years A = balance after t years

Compound Interest Formulas

1. Balance when interest is compounded n times per year: $A = P\left(1 + \dfrac{r}{n}\right)^{nt}$

2. Balance when interest is compounded continuously: $A = Pe^{rt}$

Effective Rate of Interest

$$r_{eff} = \left(1 + \frac{r}{n}\right)^n - 1$$

Present Value of a Future Investment

$$P = \frac{A}{\left(1 + \dfrac{r}{n}\right)^{nt}}$$

Balance of an Increasing Annuity After *n* Deposits of *P* per Year for *t* Years

$$A = P\left[\left(1 + \frac{r}{n}\right)^{nt} - 1\right]\left(1 + \frac{n}{r}\right)$$

Initial Deposit for a Decreasing Annuity with *n* Withdrawals of *W* per Year for *t* Years

$$P = W\left(\frac{n}{r}\right)\left\{1 - \left[\frac{1}{1 + (r/n)}\right]^{nt}\right\}$$

Monthly Installment *M* for a Loan of *P* Dollars over *t* Years at *r*% Interest

$$M = P\left\{\frac{r/12}{1 - \left[\dfrac{1}{1 + (r/12)}\right]^{12t}}\right\}$$

Amount of an Annuity

$$e^{rT}\int_0^T c(t)e^{-rt}\,dt$$

$c(t)$ is the continuous income function in dollars per year and T is the term of the annuity in years.

Answers to Selected Exercises

Chapter 1

Section 1.1 *(page 8)*

Section 1.2 *(page 17)*

Section 1.3 *(page 27)*

Quiz Yourself *(page 28)*

1. (a) 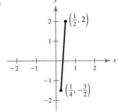 (b) $d = 3\sqrt{5}$
(c) Midpoint: $(0, -0.5)$

2. (a) 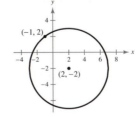 (b) $d = \sqrt{12.3125}$
(c) Midpoint: $\left(\frac{3}{8}, \frac{1}{4}\right)$

3. (a) 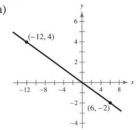 (b) $d = 6\sqrt{10}$
(c) Midpoint: $(-3, 1)$

4. $d_1 = \sqrt{5}$
$d_2 = \sqrt{45}$
$d_3 = \sqrt{50}$
$d_1^2 + d_2^2 = d_3^2$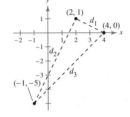

5. 9681.5 thousand

6. 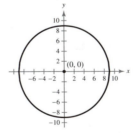 **7.**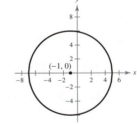

8.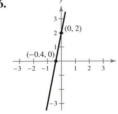

9. $x^2 + y^2 = 81$ **10.** $(x + 1)^2 + y^2 = 36$

11. $(x - 2)^2 + (y + 2)^2 = 25$ **12.** 4735 units

13. $y = \frac{1}{2}x - 3$

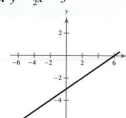

14. $y = 2x - 1$

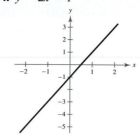

15. $y = -\frac{1}{3}x + 7$

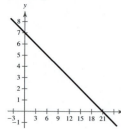

16. $y = -1.2x + 0.2$

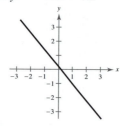

17. $x = -2$

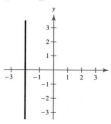

18. $y = 2$

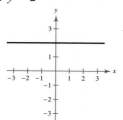

19. (a) $y = -0.25x - 4.25$ (b) $y = 4x - 17$

20. 2015: \$2,270,000; 2018: \$2,622,500

21. $C = 0.55x + 175$

22. (a) $S = 1600t + 20,200$ (b) \$44,200

Section 1.4   *(page 38)*

Skills Warm Up   *(page 38)*

1. 20   **2.** 10   **3.** $x^2 + x - 6$

4. $x^3 + 9x^2 + 26x + 30$ **5.** $\dfrac{1}{x}$ **6.** $\dfrac{2x - 1}{x}$

7. $y = -2x + 17$ **8.** $y = \frac{6}{5}x^2 + \frac{1}{5}$

9. $y = 3 \pm \sqrt{5 + (x + 1)^2}$ **10.** $y = \pm\sqrt{4x^2 + 2}$

11. $y = 2x + \dfrac{1}{2}$ **12.** $y = \dfrac{x^3}{2} + \dfrac{1}{2}$

Section 1.5   *(page 48)*

Skills Warm Up   *(page 48)*

1. $\frac{1}{3}x^2 + \frac{1}{6}x$ **2.** $x^2(x + 9)$ **3.** $x + 4$ **4.** $x + 6$

5. (a) 7   (b) $c^2 - 3c + 3$
    (c) $x^2 + 2xh + h^2 - 3x - 3h + 3$

6. (a) -4 (b) 10   (c) $3t^2 + 4$ **7.** h **8.** 4

9. Domain: $(-\infty, 0) \cup (0, \infty)$
    Range: $(-\infty, 0) \cup (0, \infty)$

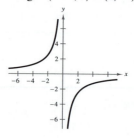

10. Domain: $[-5, 5]$
     Range: $[0, 5]$

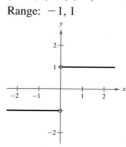

11. Domain: $(-\infty, \infty)$
     Range: $[0, \infty)$

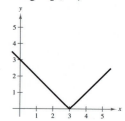

12. Domain:
     $(-\infty, 0) \cup (0, \infty)$
     Range: $-1, 1$

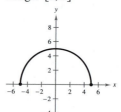

13. y is not a function of x.

14. y is a function of x.

Section 1.6   *(page 56)*

Skills Warm Up   *(page 56)*

1. $\dfrac{x + 4}{x - 8}$ **2.** $\dfrac{x + 1}{x - 3}$ **3.** $\dfrac{x + 2}{2(x - 3)}$ **4.** $\dfrac{x - 4}{x - 2}$

5. $x = 0, -7$ **6.** $x = -5, 1$ **7.** $x = -\frac{2}{3}, -2$

8. $x = 0, 3, -8$ **9.** 13   **10.** -1

Review Exercises for Chapter 1   *(page 61)*

1.

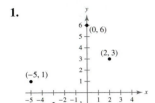

3. $\sqrt{29}$ **5.** $3\sqrt{2}$ **7.** $\dfrac{\sqrt{17}}{2}$ **9.** $(7, 4)$

11. $(-8, 6)$ **13.** $\left(\dfrac{5}{2}, \dfrac{2}{5}\right)$

15. $P = R - C$; The difference in the heights of the bars that represent revenue and cost is equal to the height of the bar that represents profit.

17. $(-2, 7), (-1, 8), (1, 5)$

19.

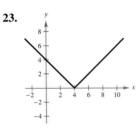

21.

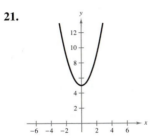

23.

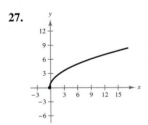

25.

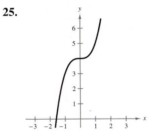

27.

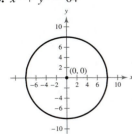

29. $\left(-\dfrac{3}{4}, 0\right), (0, -3)$ **31.** $(-4, 0), (2, 0), (0, -8)$

33. $x^2 + y^2 = 64$ **35.** $x^2 + y^2 = 9$

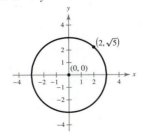

37. $(-2, 9)$ **39.** $(-1, -1), (0, 0), (1, 1)$

41. (a) $C = 10x + 200$
    $R = 14x$
    (b) 50 shirts

43. $p = \$46.40$
    $x = 5000$ units

45. Slope: -1
    y-intercept: $(0, 12)$

47. Slope: -3
    y-intercept: $(0, -2)$

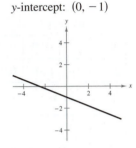

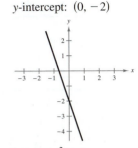

49. Slope: 0 (horizontal line)
    y-intercept: $\left(0, -\dfrac{5}{3}\right)$

51. Slope: $-\dfrac{2}{5}$
    y-intercept: $(0, -1)$

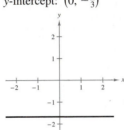

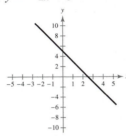

53. $\dfrac{6}{7}$ **55.** $\dfrac{20}{21}$

57. $y = -2x + 5$ **59.** $y = -4$

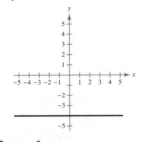

61. $y = 2x - 9$ **63.** $x = 5$

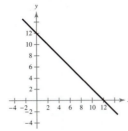

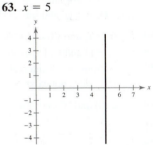

65. (a) $7x - 8y + 69 = 0$ (b) $2x + y = 0$
    (c) $2x + y = 0$ (d) $2x + 3y - 12 = 0$

67. (a) $x = -10p + 1070$ (b) 725 units   (c) 650 units

69. y is a function of x.   **71.** y is not a function of x.

73. Domain: $(-\infty, \infty)$ **75.** Domain: $[-1, \infty)$
    Range: $(-\infty, \infty)$ Range: $[0, \infty)$

77. Domain: $(-\infty, -4) \cup (-4, 3) \cup (3, \infty)$
    Range: $(-\infty, 0) \cup \left(0, \dfrac{1}{7}\right) \cup \left(\dfrac{1}{7}, \infty\right)$

79. (a) 7   (b) -11 (c) $3x + 7$

81. (a) $x^2 + 2x$ (b) $x^2 - 2x + 2$ (c) $2x^3 - x^2 + 2x - 1$
    (d) $\dfrac{1 + x^2}{2x - 1}$ (e) $4x^2 - 4x + 2$ (f) $2x^2 + 1$

CHAPTER 1

83.

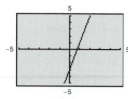

$f(x)$ is one-to-one.
$f^{-1}(x) = \frac{1}{4}(x + 3)$

85.

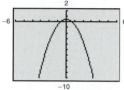

$f(x)$ does not have an inverse function.

87.

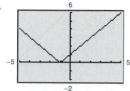

$f(x)$ does not have an inverse function.

89.

x	0.9	0.99	0.999	1	1.001	1.01	1.1
$f(x)$	0.6	0.96	0.996	?	1.004	1.04	1.4

$\lim\limits_{x\to1} (4x - 3) = 1$

91.

x	-0.1	-0.01	-0.001	0
$f(x)$	35.71	355.26	3550.71	?

x	0.001	0.01	0.1
$f(x)$	-3550.31	-354.85	-35.30

$\lim\limits_{x\to0} \dfrac{\sqrt{x + 6} - 6}{x}$ does not exist.

93. 8   **95.** 7   **97.** $-\frac{2}{5}$ **99.** Limit does not exist.
101. $-\frac{1}{4}$ **103.** $-\infty$ **105.** Limit does not exist.
107. 5   **109.** $3x^2 - 1$
111. $(-\infty, \infty)$; For any c on the real number line, $F(c)$ is defined, $\lim\limits_{x\to c} f(x)$ exists, and $\lim\limits_{x\to c} f(x) = f(c)$.
113. $(-\infty, -4)$ and $(-4, \infty)$; $f(-4)$ is undefined.
115. $(-\infty, -1)$ and $(-1, \infty)$; $f(-1)$ is undefined.
117. Continuous on all intervals $(c, c + 1)$, where c is an integer; $\lim\limits_{x\to c} f(c)$ does not exist.
119. $(-\infty, 0)$ and $(0, \infty)$; $\lim\limits_{x\to0} f(x)$ does not exist.
121. $a = 2$
123. (a)

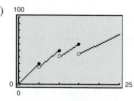

Explanations will vary. The function is defined for all values of x greater than zero. The function is discontinuous when $x = 5$, $x = 10$, and $x = 15$.
(b) $49.90

125. (a) $C(t) = \begin{cases} 1 + 0.1[\![t]\!], & t > 0, \ t \text{ not an integer} \\ 1 + 0.1[\![t - 1]\!], & t > 0, \ t \text{ an integer} \end{cases}$

(b)

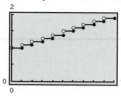

C is not continuous at $t = 1, 2, 3, \ldots$.

Test Yourself *(page 65)*

1. (a) $d = 5\sqrt{2}$ (b) Midpoint: $(-1.5, 1.5)$
(c) $m = -1$ (d) $y = -x$
(e)

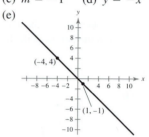

2. (a) $d = 2.5$ (b) Midpoint: $(1.25, 2)$
(c) $m = 0$ (d) $y = 2$
(e)

3. (a) $d = 2\sqrt{10}$ (b) Midpoint: $(-1, 2)$
(c) $m = \frac{1}{3}$ (d) $y = \frac{1}{3}x + \frac{7}{3}$
(e)

4. $(5.5, 53.45)$
5. $m = \frac{1}{5}$; $(0, -2)$

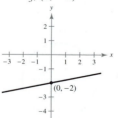

6. m is undefined; no y-intercept

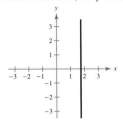

7. $m = -2.5$; $(0, 6.25)$

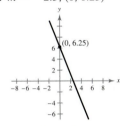

8. $y = -\frac{1}{4}x - \frac{29}{4}$ **9.** $y = \frac{5}{2}x - 4$

10. (a)

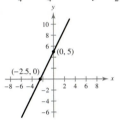

(b) Domain: $(-\infty, \infty)$
   Range: $(-\infty, \infty)$
(c) $f(-3) = -1; f(-2) = 1; f(3) = 11$
(d) The function is one-to-one.

11. (a)

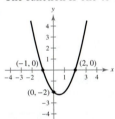

(b) Domain: $(-\infty, \infty)$
   Range: $(-2.25, \infty)$
(c) $f(-3) = 10; f(-2) = 4; f(3) = 4$
(d) The function is not one-to-one.

12. (a)

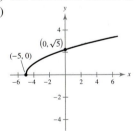

(b) Domain: $[-5, \infty)$
   Range: $[0, \infty)$
(c) $f(-3) = \sqrt{2}; f(-2) = \sqrt{3}; f(3) = 2\sqrt{2}$
(d) The function is one-to-one.

13. $f^{-1}(x) = \frac{1}{4}x - \frac{3}{2}$ **14.** $f^{-1}(x) = -\frac{1}{3}x^3 + \frac{8}{3}$
15. -1 **16.** Limit does not exist.   **17.** 2   **18.** $\frac{1}{6}$
19. $(-\infty, 4)$ and $(4, \infty)$; Explanations will vary. There is a discontinuity at $x = 4$, because $f(4)$ is not defined.
20. $(-\infty, 5]$; Explanations will vary.
21. $(-\infty, \infty)$; Explanations will vary.
22. (a) The model fits the data well. Explanations will vary.
   (b) 128,087.392 thousand (128,087,392)

Chapter 2

Section 2.1   *(page 76)*

Skills Warm Up   *(page 76)*

1. $x = 2$ **2.** $y = 2$ **3.** $y = -x + 2$

4. $y = 3x - 4$ **5.** $2x$ **6.** $3x^2$ **7.** $\dfrac{1}{x^2}$

8. $2x$ **9.** $(-\infty, \infty)$ **10.** $(-\infty, 1) \cup (1, \infty)$
11. $(-\infty, \infty)$ **12.** $(-\infty, 0) \cup (0, \infty)$

Section 2.2   *(page 86)*

Skills Warm Up   *(page 86)*

1. (a) 8   (b) 16   (c) $\frac{1}{2}$
2. (a) $\frac{1}{36}$ (b) $\frac{1}{32}$ (c) $\frac{1}{64}$

3. $4x(3x^2 + 1)$ **4.** $\frac{3}{2}x^{1/2}(x^{3/2} - 1)$ **5.** $\dfrac{1}{4x^{3/4}}$

6. $x^2 - \dfrac{1}{x^{1/2}} + \dfrac{1}{3x^{2/3}}$ **7.** $0, -\dfrac{2}{3}$
8. $0, \pm 1$ **9.** $-10, 2$ **10.** $-2, 12$

Section 2.3   *(page 97)*

Skills Warm Up   *(page 97)*

1. 3   **2.** -7 **3.** -3 **4.** 2.4
5. $y' = 8x - 2$ **6.** $y' = -9t^2 + 4t$
7. $s' = -32t + 24$ **8.** $y' = -32x + 54$
9. $A' = -\frac{3}{5}r^2 + \frac{3}{5}r + \frac{1}{2}$ **10.** $y' = 2x^2 - 4x + 7$
11. $y' = 12 - \dfrac{x}{2500}$ **12.** $y' = 74 - \dfrac{3x^2}{10,000}$

CHAPTER 2

Section 2.4   *(page 105)*

Skills Warm Up   *(page 105)*

1. $2(3x^2 + 7x + 1)$ **2.** $4x^2(6 - 5x^2)$

3. $8x^2(x^2 + 2)^3 + (x^2 + 4)$

4. $(2x)(2x + 1)[2x + (2x + 1)^3]$

5. $\dfrac{23}{(2x + 7)^2}$ **6.** $-\dfrac{x^2 + 8x + 4}{(x^2 - 4)^2}$

7. $-\dfrac{2(x^2 + x - 1)}{(x^2 + 1)^2}$ **8.** $\dfrac{4(3x^4 - x^3 + 1)}{(1 - x^4)^2}$

9. $\dfrac{4x^3 - 3x^2 + 3}{x^2}$ **10.** $\dfrac{x^2 - 2x + 4}{(x - 1)^2}$

11. 11 **12.** 0 **13.** $-\dfrac{1}{4}$ **14.** $\dfrac{17}{4}$

Quiz Yourself   *(page 106)*

1. $f(x) = 5x + 3$

$f(x + \Delta x) = 5x + 5\Delta x + 3$

$f(x + \Delta x) - f(x) = 5\Delta x$

$\dfrac{f(x + \Delta x) - f(x)}{\Delta x} = 5$

$\displaystyle\lim_{\Delta x \to 0} \dfrac{f(x + \Delta x) - f(x)}{\Delta x} = 5$

$f'(x) = 5$

$f'(-2) = 5$

2. $f(x) = \sqrt{x + 3}$

$f(x + \Delta x) = \sqrt{x + \Delta x + 3}$

$f(x + \Delta x) - f(x) = \sqrt{x + \Delta x + 3} - \sqrt{x + 3}$

$\dfrac{f(x + \Delta x) - f(x)}{\Delta x} = \dfrac{1}{\sqrt{x + \Delta x + 3} + \sqrt{x + 3}}$

$\displaystyle\lim_{\Delta x \to 0} \dfrac{f(x + \Delta x) - f(x)}{\Delta x} = \dfrac{1}{2\sqrt{x + 3}}$

$f'(x) = \dfrac{1}{2\sqrt{x + 3}}$

$f'(1) = \dfrac{1}{4}$

3. $f(x) = x^2 - 2x$

$f(x + \Delta x) = x^2 + 2x\Delta x + (\Delta x)^2 - 2x - 2\Delta x$

$f(x + \Delta x) - f(x) = 2x\Delta x + (\Delta x)^2 - 2\Delta x$

$\dfrac{f(x + \Delta x) - f(x)}{\Delta x} = 2x + \Delta x - 2$

$\displaystyle\lim_{\Delta x \to 0} \dfrac{f(x + \Delta x) - f(x)}{\Delta x} = 2x - 2$

$f'(x) = 2x - 2$

$f'(3) = 4$

4. $f'(x) = 0$ **5.** $f'(x) = 19$ **6.** $f'(x) = -6x$

7. $f'(x) = \dfrac{3}{x^{3/4}}$ **8.** $f'(x) = -\dfrac{8}{x^3}$ **9.** $f'(x) = \dfrac{1}{\sqrt{x}}$

10. $f'(x) = -\dfrac{5}{(3x + 2)^2}$ **11.** $f'(x) = -6x^2 + 8x - 2$

12. $f'(x) = 15x^2 + 26x + 14$ **13.** $f'(x) = \dfrac{-4(x^2 - 3)}{(x^2 + 3)^2}$

14.

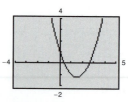

Average rate: 0

Instantaneous rates: $f'(0) = -3, f'(3) = 3$

15.

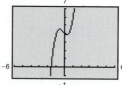

Average rate: 1

Instantaneous rates: $f'(-1) = 3, f'(1) = 7$

16.

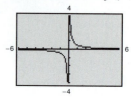

Average rate: $-\dfrac{1}{20}$

Instantaneous rates: $f'(2) = -\dfrac{1}{8}, f'(5) = -\dfrac{1}{50}$

17.

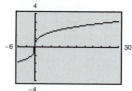

Average rate: $\dfrac{1}{19}$

Instantaneous rates: $f'(8) = \dfrac{1}{12}, f'(27) = \dfrac{1}{27}$

18. (a) $\$11.61$ (b) $\$11.63$

(c) The results are approximately equal.

19. $y = -4x - 6$ **20.** $y = 10x - 8$

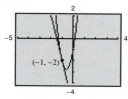   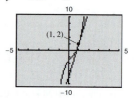

21. (a) $\dfrac{dS}{dt} = -0.40668t^2 + 3.7364t - 4.351$

(b) 2004: $\$4.08772/\text{yr}$

2007: $\$1.87648/\text{yr}$

2008: $-\$0.48732/\text{yr}$

Section 2.5   *(page 114)*

Skills Warm Up   *(page 114)*

1. $(1 - 5x)^{2/5}$ 2. $(2x - 1)^{3/4}$
3. $(4x^2 + 1)^{-1/2}$ 4. $(x - 6)^{-1/3}$
5. $x^{1/2}(1 - 2x)^{-1/3}$ 6. $(2x)^{-1}(3 - 7x)^{3/2}$
7. $(x - 2)(3x^2 + 5)$ 8. $(x - 1)(5\sqrt{x} - 1)$
9. $(x^2 + 1)^2(4 - x - x^3)$
10. $(3 - x^2)(x - 1)(x^2 + x + 1)$

Section 2.6   *(page 120)*

Skills Warm Up   *(page 120)*

1. $t = 0, \frac{3}{2}$ 2. $t = -2, 7$ 3. $t = -2, 10$

4. $t = \dfrac{9 \pm 3\sqrt{10{,}249}}{32}$ 5. $\dfrac{dy}{dx} = 6x^2 + 14x$

6. $\dfrac{dy}{dx} = 8x^3 + 18x^2 - 10x - 15$

7. $\dfrac{dy}{dx} = \dfrac{2x(x + 7)}{(2x + 7)^2}$ 8. $\dfrac{dy}{dx} = -\dfrac{6x^2 + 10x + 15}{(2x^2 - 5)^2}$

9. Domain: $(-\infty, \infty)$ 10. Domain: $[7, \infty)$

   Range: $[-4, \infty)$ Range: $[0, \infty)$

Section 2.7   *(page 126)*

Skills Warm Up   *(page 126)*

1. $y = x^2 - 2x$ 2. $y = \dfrac{x - 3}{4}$

3. $y = 1, x \neq -6$ 4. $y = -4, x \neq \pm\sqrt{3}$
5. $y = \pm\sqrt{5 - x^2}$ 6. $y = \pm\sqrt{6 - x^2}$ 7. $\frac{8}{3}$
8. $-\frac{1}{2}$ 9. $\frac{5}{7}$

Section 2.8   *(page 132)*

Skills Warm Up   *(page 132)*

1. $A = \pi r^2$ 2. $V = \frac{4}{3}\pi r^3$ 3. $S = 6s^2$
4. $V = s^3$ 5. $V = \frac{1}{3}\pi r^2 h$ 6. $A = \frac{1}{2}bh$

7. $-\dfrac{x}{y}$ 8. $\dfrac{2x - 3y}{3x}$ 9. $-\dfrac{2x + y}{x + 2}$

10. $-\dfrac{y^2 - y + 1}{2xy - 2y - x}$

Review Exercises for Chapter 2   *(page 137)*

1. -2 3. 0

5. Answers will vary. Sample answer:

   $t = 4$: slope $\approx$ \$290 million/yr; Sales were increasing by about \$290 million/yr in 2004.

   $t = 7$: slope $\approx$ \$320 million/yr; Sales were increasing by about \$320 million/yr in 2007.

7. Answers will vary. Sample answer:

   $t = 1$: slope $\approx$ 65 thousand visitors/month; The number of visitors to the national park is increasing at about 65,000 visitors per month in January.

   $t = 8$: slope $\approx$ 0 visitors/month; The number of visitors to the national park is neither increasing nor decreasing in August.

   $t = 12$: slope ≈ -1000 thousand visitors/month; The number of visitors to the national park is decreasing at about 1,000,000 visitors per month in December.

9. -3 11. -2 13. $\frac{1}{4}$ 15. -1 17. 9

19. $-x + 2$ 21. $\dfrac{1}{2\sqrt{x - 5}}$ 23. $-\dfrac{5}{x^2}$

25. All values except $x = 1$; The function is not defined at $x = 1$.
27. All values except $x = 0$; A function is not differentiable at a discontinuity.

29. 0 31. $3x^2$ 33. $8x$ 35. $\dfrac{8x^3}{5}$ 37. $8x^3 + 6x$

39. $2x + 6$ 41. -0.125 43. 5
45. (a) $y = 5x - 7$

   (b) and (c)

47. (a) $y = x - 1$

   (b) and (c)

49. (a) 2004: $m \approx 290$
       2007: $m \approx 320$

   (b) Results should be similar.

   (c) The slope shows the rate at which sales were increasing or decreasing in a particular year.

51.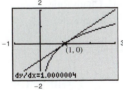

   Average rate of change: 4
   Instantaneous rate of change at $t = -3$: 4
   Instantaneous rate of change at $t = 1$: 4

53.

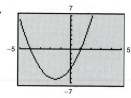

Average rate of change: 4
Instantaneous rate of change at $x = 0$: 3
Instantaneous rate of change at $x = 1$: 5

55. (a) -24 ft/sec   (b) $t = 1$: -8 ft/sec
$t = 3$: -72 ft/sec
(c) 5.14 sec
(d) -140.5 ft/sec

57. $\dfrac{dC}{dx} = 320$ **59.** $\dfrac{dC}{dx} = \dfrac{1.275}{\sqrt{x}}$

61. $\dfrac{dR}{dx} = -1.2x + 150$ **63.** $\dfrac{dR}{dx} = -12x^2 + 4x + 100$

65. $\dfrac{dP}{dx} = -0.0006x^2 + 12x - 1$

67. (a) \$9.95   (b) \$10
(c) Parts (a) and (b) differ by only \$0.05.

In Exercises 69–89, the differentiation rule(s) used may vary.
A sample answer is provided.

69. $15x^2(1 - x^2)$; Power Rule
71. $16x^3 - 33x^2 + 12x$; Product Rule
73. $\dfrac{3}{(x + 3)^2}$; Quotient Rule
75. $\dfrac{2(3 + 5x - 3x^2)}{(x^2 + 1)^2}$; Quotient Rule
77. $30x(5x^2 + 2)^2$; Chain Rule
79. $-\dfrac{1}{(x + 1)^{3/2}}$; Quotient Rule   **81.** $\dfrac{2x^2 + 1}{\sqrt{x^2 + 1}}$; Product Rule
83. $80x^4 - 24x^2 + 1$; Product Rule
85. $18x^5(x + 1)(2x + 3)^2$; Chain Rule
87. $x(x - 1)^4(7x - 2)$; Product Rule
89. $\dfrac{3(9t + 5)}{2\sqrt{3t + 1}(1 - 3t)^3}$; Quotient Rule
91. (a) $t = 1$: -6.63 $t = 3$: -6.5
$t = 5$: -4.33 $t = 10$: -1.36
(b)

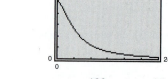

The rate of decrease is
approaching zero.

93. 6   **95.** $-\dfrac{120}{x^6}$ **97.** $\dfrac{35x^{3/2}}{2}$ **99.** $\dfrac{2}{x^{2/3}}$

101. (a) $s(t) = -16t^2 + 5t + 30$ (b) About 1.534 sec
(c) About -44.09 ft/sec   (d) -32 ft/sec²

103. $-\dfrac{2x + 3y}{3(x + y^2)}$ **105.** $\dfrac{2x - 8}{2y - 9}$ **107.** $y = \dfrac{1}{3}x + \dfrac{1}{3}$

109. $y = \dfrac{4}{3}x + \dfrac{2}{3}$

111. (a) 12π square inches per minute
(b) 40π square inches per minute

113. $\dfrac{1}{64}$ ft/min

Test Yourself   *(page 141)*

1. $f(x) = x^2 + 1$
$f(x + \Delta x) = x^2 + 2x\Delta x + \Delta x^2 + 1$
$f(x + \Delta x) - f(x) = 2x\Delta x + \Delta x^2$
$\dfrac{f(x + \Delta x) - f(x)}{\Delta x} = 2x + \Delta x$
$\displaystyle\lim_{\Delta x \to 0} \dfrac{f(x + \Delta x) - f(x)}{\Delta x} = 2x$
$f'(x) = 2x$
$f'(2) = 4$

2. $f(x) = \sqrt{x} - 2$
$f(x + \Delta x) = \sqrt{x + \Delta x} - 2$
$f(x + \Delta x) - f(x) = \sqrt{x + \Delta x} - \sqrt{x}$
$\dfrac{f(x + \Delta x) - f(x)}{\Delta x} = \dfrac{1}{\sqrt{x + \Delta x} + \sqrt{x}}$
$\displaystyle\lim_{\Delta x \to 0} \dfrac{f(x + \Delta x) - f(x)}{\Delta x} = \dfrac{1}{2\sqrt{x}}$
$f'(x) = \dfrac{1}{2\sqrt{x}}$
$f'(4) = \dfrac{1}{4}$

3. $f'(t) = 3t^2 + 2$ **4.** $f'(x) = 8x - 8$ **5.** $f'(x) = \dfrac{3\sqrt{x}}{2}$

6. $f'(x) = 3x^2 + 10x + 6$ **7.** $f'(x) = \dfrac{9}{x^4}$

8. $f'(x) = \dfrac{5 + x}{2\sqrt{x}} + \sqrt{x}$ **9.** $f'(x) = 36x^3 + 48x$

10. $f'(x) = -\dfrac{1}{\sqrt{1 - 2x}}$

11. $f'(x) = \dfrac{(10x + 1)(5x - 1)^2}{x^2} = 250x - 75 + \dfrac{1}{x^2}$

12. $y = 2x - 2$

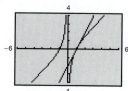

13. (a) \$18.69 billion/yr
(b) 2005: \$10.50 billion/yr
2008: \$14.95 billion/yr
(c) The annual sales of CVS Caremark from 2005 to 2008
increased on average by about \$18.69 billion/yr, and the
instantaneous rates of change for 2005 and 2008 are
\$10.50 billion/yr and \$14.95 billion/yr, respectively.

14. (a) $P = -0.016x^2 + 1460x - 715,000$ (b) \$1437.60

15. 0   **16.** $-\dfrac{3}{8(3 - x)^{5/2}}$ **17.** $-\dfrac{96}{(2x - 1)^4}$

18. $s(t) = -16t^2 + 30t + 75$ $s(2) = 71$ ft
$v(t) = -32t + 30$ $v(2) = -34$ ft/sec
$a(t) = -32$ $a(2) = -32$ ft/sec²

19. $\dfrac{dy}{dx} = -\dfrac{1 + y}{x}$ **20.** $\dfrac{dy}{dx} = -\dfrac{1}{y - 1}$ **21.** $\dfrac{dy}{dx} = \dfrac{x}{2y}$

22. (a) 3.75π cm³/min   (b) 15π cm³/min

Chapter 3

Section 3.1   *(page 151)*

Skills Warm Up   *(page 151)*

1. $x = 0, x = 8$ **2.** $x = 0, x = 24$ **3.** $x = \pm 5$
4. $x = 0$ **5.** $(-\infty, 3) \cup (3, \infty)$ **6.** $(-\infty, 1)$
7. $(-\infty, -2) \cup (-2, 5) \cup (5, \infty)$ **8.** $\left(-\sqrt{3}, \sqrt{3}\right)$
9. $x = -2$: -6 **10.** $x = -2$: 60
 $\quad\;\; x = 0$: 2 $\qquad\;\; x = 0$: -4
 $\quad\;\; x = 2$: -6 $\qquad\;\; x = 2$: 60
11. $x = -2$: $-\frac{1}{3}$ **12.** $x = -2$: $\frac{1}{18}$
 $\quad\;\;\; x = 0$: 1 $\qquad\;\; x = 0$: $-\frac{1}{8}$
 $\quad\;\;\; x = 2$: 5 $\qquad\;\; x = 2$: $-\frac{3}{2}$

Section 3.2   *(page 159)*

Skills Warm Up   *(page 159)*

1. $0, \pm\frac{1}{2}$ **2.** $-2, 5$ **3.** 1 **4.** $0, 125$
5. $-4 \pm \sqrt{17}$ **6.** $1 \pm \sqrt{5}$
7. Negative   **8.** Positive   **9.** Positive
10. Negative   **11.** Increasing   **12.** Decreasing

Section 3.3   *(page 167)*

Skills Warm Up   *(page 167)*

1. $f''(x) = 48x^2 - 54x$ **2.** $g''(s) = 12s^2 - 18s + 2$
3. $g''(x) = 56x^6 + 120x^4 + 72x^2 + 8$
4. $f''(x) = \dfrac{4}{9(x-3)^{2/3}}$ **5.** $h''(x) = \dfrac{190}{(5x-1)^3}$
6. $f''(x) = -\dfrac{42}{(3x+2)^3}$ **7.** $x = \pm\dfrac{\sqrt{3}}{3}$
8. $x = 0, 3$ **9.** $t = \pm 4$ **10.** $x = 0, \pm 5$

Section 3.4   *(page 173)*

Skills Warm Up   *(page 173)*

1. $x + \frac{1}{2}y = 12$ **2.** $2xy = 24$ **3.** $xy = 24$
4. $\sqrt{(x_2 - x_1)^2 + (y_2 - y_1)^2} = 10$
5. $x = -3$ **6.** $x = -\frac{2}{3}, 1$ **7.** $x = \pm 5$
8. $x = 4$ **9.** $x = \pm 1$ **10.** $x = \pm 3$

Quiz Yourself   *(page 174)*

1. Critical number: $x = 3$
 Increasing on $(3, \infty)$
 Decreasing on $(-\infty, 3)$

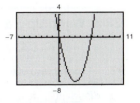

2. Critical numbers:
 $x = -4, x = 0$
 Increasing on
 $(-\infty, -4)$ and $(0, \infty)$
 Decreasing on $(-4, 0)$

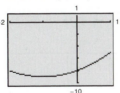

3. Critical numbers: $x = 5$,
 $x = -5$
 Increasing on $(-5, 5)$
 Decreasing on $(-\infty, -5)$
 and $(5, \infty)$

4. Relative minimum: $(0, -5)$
 Relative maximum: $(-2, -1)$
5. Relative minima: $(2, -13), (-2, -13)$
 Relative maximum: $(0, 3)$
6. Relative minimum: $(0, 0)$
7. Minimum: $(-1, -9)$ **8.** Minimum: $(3, -54)$
 Maximum: $(1, -5)$ $\quad$ Maximum: $(-3, 54)$

9. Minimum: $(0, 0)$
 Maximum: $(1, 0.5)$

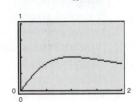

10. Point of inflection: $(2, -2)$
 Concave downward on $(-\infty, 2)$
 Concave upward on $(2, \infty)$
11. Points of inflection: $(-2, -80)$ and $(2, -80)$
 Concave downward on $(-2, 2)$
 Concave upward on $(-\infty, -2)$ and $(2, \infty)$
12. Relative minimum: $(1, 9)$
 Relative maximum: $(-2, 36)$
13. Relative minimum: $(3, 12)$
 Relative maximum: $(-3, -12)$
14. $\$120{,}000\ (x = 120)$ **15.** 50 ft by 100 ft
16. (a) Increasing from 2000 to early 2008
 $\qquad$ Decreasing from early 2008 to 2009
 $\quad$ (b) Early 2008; 2000

CHAPTER 3

Section 3.5 *(page 182)*

Skills Warm Up *(page 182)*

1. 1   **2.** $\frac{6}{5}$ **3.** 2   **4.** $\frac{1}{2}$

5. $\frac{dC}{dx} = 1.2 + 0.006x$ **6.** $\frac{dP}{dx} = 0.02x + 11$

7. $\frac{dP}{dx} = -1.4x + 7$ **8.** $\frac{dC}{dx} = 4.2 + 0.003x^2$

9. $\frac{dR}{dx} = 14 - \frac{x}{1000}$ **10.** $\frac{dR}{dx} = 3.4 - \frac{x}{750}$

Section 3.6 *(page 191)*

Skills Warm Up *(page 191)*

1. 3   **2.** 1   **3.** -11 **4.** 4   **5.** $-\frac{1}{4}$

6. -2 **7.** 0   **8.** 1

9. $\overline{C} = \frac{150}{x} + 3$ **10.** $\overline{C} = \frac{1900}{x} + 1.7 + 0.002x$

$\frac{dC}{dx} = 3$ $\frac{dC}{dx} = 1.7 + 0.004x$

11. $\overline{C} = 0.005x + 0.5 + \frac{1375}{x}$ **12.** $\overline{C} = \frac{760}{x} + 0.05$

$\frac{dC}{dx} = 0.01x + 0.5$ $\frac{dC}{dx} = 0.05$

Section 3.7 *(page 199)*

Skills Warm Up *(page 199)*

1. Vertical asymptote: $x = 0$
   Horizontal asymptote: $y = 0$
2. Vertical asymptote: $x = 2$
   Horizontal asymptote: $y = 0$
3. Vertical asymptote: $x = -3$
   Horizontal asymptote: $y = 40$
4. Vertical asymptotes: $x = 1, x = 3$
   Horizontal asymptote: $y = 1$
5. Decreasing on $(-\infty, -2)$
   Increasing on $(-2, \infty)$
6. Increasing on $(-\infty, -4)$
   Decreasing on $(-4, \infty)$
7. Increasing on $(-\infty, -1)$ and $(1, \infty)$
   Decreasing on $(-1, 1)$
8. Decreasing on $(-\infty, 0)$ and $(\sqrt[3]{2}, \infty)$
   Increasing on $(0, \sqrt[3]{2})$
9. Increasing on $(-\infty, 1)$ and $(1, \infty)$
10. Decreasing on $(-\infty, -3)$ and $(\frac{1}{3}, \infty)$
   Increasing on $(-3, \frac{1}{3})$

Section 3.8 *(page 205)*

Skills Warm Up *(page 205)*

1. $\frac{dC}{dx} = 0.18x$ **2.** $\frac{dC}{dx} = 0.15$

3. $\frac{dR}{dx} = 1.25 + 0.03\sqrt{x}$ **4.** $\frac{dR}{dx} = 15.5 - 3.1x$

5. $\frac{dP}{dx} = -\frac{0.01}{\sqrt[3]{x^2}} + 1.4$ **6.** $\frac{dP}{dx} = -0.04x + 25$

7. $\frac{dA}{dx} = \frac{\sqrt{3}}{2}x$ **8.** $\frac{dA}{dx} = 12x$ **9.** $\frac{dC}{dr} = 2\pi$

10. $\frac{dP}{dw} = 4$ **11.** $\frac{dS}{dr} = 8\pi r$ **12.** $\frac{dP}{dx} = 2 + \sqrt{2}$

13. $A = \pi r^2$ **14.** $A = x^2$

15. $V = x^3$ **16.** $V = \frac{4}{3}\pi r^3$

Review Exercises for Chapter 3 *(page 210)*

1. $x = 1$ **3.** $x = -3, x = 3$ **5.** $x = 1, x = \frac{7}{3}$

7. Critical number: $x = -\frac{1}{2}$
   Increasing on $\left(-\frac{1}{2}, \infty\right)$
   Decreasing on $\left(-\infty, -\frac{1}{2}\right)$
9. Critical numbers: $x = 0, x = 4$
   Increasing on $(0, 4)$
   Decreasing on $(-\infty, 0)$ and $(4, \infty)$
11. Critical number: $x = 1$
   Increasing on $(1, \infty)$
   Decreasing on $(-\infty, 1)$
13. The only critical number is $t \approx -10.85$. Any $t > -10.85$ produces a positive dR/dt, so the sales were increasing from 2004 to 2009.
15. Relative maximum: $(0, -2)$
   Relative minimum: $(1, -4)$
17. Relative minimum: $(8, -52)$
19. Relative maxima: $(-1, 1), (1, 1)$
   Relative minimum: $(0, 0)$
21. Relative maximum: $(0, 6)$
23. Relative maximum: $(0, 0)$
   Relative minimum: $(4, 8)$
25. Maximum: $(0, 6)$
   Minimum: $\left(-\frac{5}{2}, -\frac{1}{4}\right)$
27. Maxima: $(-2, 17), (4, 17)$
   Minima: $(-4, -15), (2, -15)$
29. Maximum: $(1, 1)$
   Minimum: $(9, -3)$
31. Maximum: $(1, 1)$
   Minimum: $(-1, -1)$
33. $r \approx 1.58$ in.

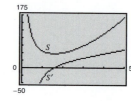

35. Concave upward on $(2, \infty)$
Concave downward on $(-\infty, 2)$

37. Concave upward on $\left(-\dfrac{2\sqrt{3}}{3}, \dfrac{2\sqrt{3}}{3}\right)$

Concave downward on $\left(-\infty, -\dfrac{2\sqrt{3}}{3}\right)$ and $\left(\dfrac{2\sqrt{3}}{3}, \infty\right)$

39. $(0, 0), (4, -128)$

41. $(0, 0), (1.0652, 4.5244), (2.5348, 3.5246)$

43. No relative extrema

45. Relative maximum: $\left(-\sqrt{3}, 6\sqrt{3}\right)$
Relative minimum: $\left(\sqrt{3}, -6\sqrt{3}\right)$

47. Relative maxima: $\left(-\dfrac{\sqrt{2}}{2}, \dfrac{1}{2}\right), \left(\dfrac{\sqrt{2}}{2}, \dfrac{1}{2}\right)$
Relative minimum: $(0, 0)$

49. $\left(50, 166\frac{2}{3}\right)$ **51.** $l = w = 15$ m **53.** 144 in.^{3}

55. $x = 900$ **57.** $x = 150$ **59.** (a) $24 (b) $8

61. (a) For $0 < x < 750$, $|\eta| > 1$ and the demand is elastic.
For $750 < x < 1500$, $|\eta| < 1$ and the demand is inelastic.
For $x = 750$, the demand has unit elasticity.
(b) From 0 to 750 units, revenue is increasing.
From 750 to 1500 units, revenue does not increase.

63. $x = -7, x = 0$ **65.** $x = -\frac{1}{2}$ **67.** $-\infty$

69. ∞ **71.** $y = \frac{2}{3}$ **73.** $y = 0$

75. (a) $\overline{C} = 0.75 + \dfrac{4000}{x}$

(b) $\overline{C}(100) = 40.75$
$\overline{C}(1000) = 4.75$

(c) The limit is 0.75. As more and more units are produced, the average cost per unit will approach $0.75.

77. (a) 20%: $62.5 million
50%: $250 million
90%: $2250 million
(b) The limit is ∞, meaning that as the percent gets very close to 100, the cost grows without bound.

79.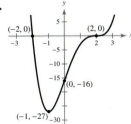

81.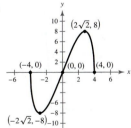

83.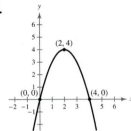

85.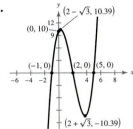

87.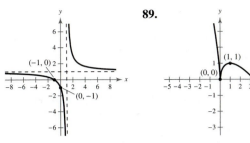

89.

91. (a)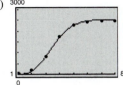

The model fits the data well.
(b) ≈ 2434 bacteria
(c) Answers will vary.

93. $dy = 0.08, \Delta y = 0.0802$ **95.** $dy = -2.1, \Delta y = -2.191$

97. $800 **99.** $15.25 **101.** $\approx$$4.52

103. $dy = 1.5x^2 \, dx$ **105.** $dy = 18x(3x^2 - 2)^2 \, dx$

107. $dy = -\dfrac{7}{(x + 5)^2} \, dx$

109. (a) $164
(b) $163.2, a difference of $0.80.

111. $B = 0.1\sqrt{5w}$
$\dfrac{dB}{dw} = \dfrac{0.05\sqrt{5}}{\sqrt{w}}$

$\Delta B \approx dB = \dfrac{0.05\sqrt{5}}{\sqrt{w}} \, dw$

$= \dfrac{0.05\sqrt{5}}{\sqrt{90}}(5)$

≈ 0.059 m^2

Test Yourself (page 214)

1. Critical number: $x = 0$
Increasing on $(0, \infty)$
Decreasing on $(-\infty, 0)$

2. Critical numbers: $x = -2, x = 2$
Increasing on $(-\infty, -2)$ and $(2, \infty)$
Decreasing on $(-2, 2)$

3. Critical number: $x = 5$
Increasing on $(5, \infty)$
Decreasing on $(-\infty, 5)$

4. Relative minimum: $(3, -14)$
Relative maximum: $(-3, 22)$

5. Relative minima: $(-1, -7)$ and $(1, -7)$
Relative maximum: $(0, -5)$

6. Relative maximum: $(0, 2.5)$

7. Minimum: $(-3, -1)$ **8.** Minimum: $(0, 0)$
Maximum: $(0, 8)$ Maximum: $(2.25, 9)$

9. Minimum: $\left(2\sqrt{3}, 2\sqrt{3}\right)$
Maximum: $(1, 6.5)$

10. Concave upward: $(2, \infty)$
Concave downward: $(-\infty, 2)$

11. Concave upward: $\left(-\infty, -\frac{2\sqrt{2}}{3}\right)$ and $\left(\frac{2\sqrt{2}}{3}, \infty\right)$

Concave downward: $\left(-\frac{2\sqrt{2}}{3}, \frac{2\sqrt{2}}{3}\right)$

12. No point of inflection

The graph is concave upward on its entire domain.

13. Concave upward: $(-\infty, -3)$ and $(3, \infty)$

Concave downward: $(-3, 3)$

Points of inflection: $(-3, -175)$ and $(3, -175)$

14. Relative minimum: $(6, -166)$

Relative maximum: $(-2, 90)$

15. Relative minimum: $(3, -97.2)$

Relative maximum: $(-3, 97.2)$

16. Vertical asymptote: $x = 5$

Horizontal asymptote: $y = 3$

17. Horizontal asymptote: $y = 2$

18. Vertical asymptote: $x = 1$

19. 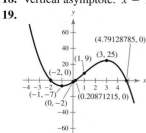   **20.**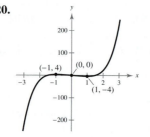

21. 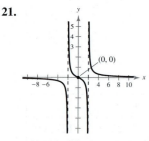   **22.** $dy = 10x\, dx$

23. $dy = \dfrac{-4}{(x+3)^2}\, dx$ **24.** $dy = 3(x+4)^2\, dx$

25. For $0 \le x < 350$, $|\eta| > 1$ and the demand is elastic.

For $350 < x \le 700$, $|\eta| < 1$ and the demand is inelastic.

For $x = 350$, the demand has unit elasticity.

Chapter 4

Section 4.1   *(page 220)*

> ### Skills Warm Up   *(page 220)*
>
> **1.** Horizontal shift to the left two units
> **2.** Reflection about the x-axis
> **3.** Vertical shift down one unit
> **4.** Reflection about the y-axis
> **5.** Horizontal shift to the right one unit
> **6.** Vertical shift up two units
> **7.** 125   **8.** 22.63   **9.** 9   **10.** $\frac{1}{125}$
> **11.** $\frac{1}{2}$ **12.** $\frac{25}{64}$ **13.** 5   **14.** $\frac{4}{3}$ **15.** $-9, 1$
> **16.** $2 \pm 2\sqrt{2}$ **17.** $1, -5$ **18.** $\frac{1}{2}, 1$

Section 4.2   *(page 227)*

> ### Skills Warm Up   *(page 227)*
>
> **1.** Continuous on $(-\infty, \infty)$
> **2.** Discontinuous at $x = \pm 2$
> **3.** Discontinuous at $x = \pm\sqrt{3}$
> **4.** Removable discontinuity at $x = 4$
> **5.** $y = 0$ **6.** $y = 0$ **7.** $y = 4$ **8.** $y = \frac{1}{2}$
> **9.** $y = \frac{3}{2}$ **10.** $y = 6$ **11.** $y = 0$ **12.** $y = 0$

Section 4.3   *(page 234)*

> ### Skills Warm Up   *(page 234)*
>
> **1.** $\frac{1}{2}e^x(2x^2 - 1)$ **2.** $\dfrac{e^x(x+1)}{x}$ **3.** $e^x(x - e^x)$
>
> **4.** $e^{-x}(e^{2x} - x)$ **5.** $-\dfrac{6}{7x^3}$ **6.** $6x - \dfrac{1}{6}$
>
> **7.** $6(2x^2 - x + 6)$ **8.** $\dfrac{t+2}{2t^{3/2}}$
>
> **9.** Relative maximum: $\left(-\dfrac{4\sqrt{3}}{3}, \dfrac{16\sqrt{3}}{9}\right)$
>
> Relative minimum: $\left(\dfrac{4\sqrt{3}}{3}, -\dfrac{16\sqrt{3}}{9}\right)$
>
> **10.** Relative maximum: $(0, 5)$
> Relative minima: $(-1, 4), (1, 4)$

Quiz Yourself   *(page 235)*

1. 1024   **2.** 216   **3.** 27   **4.** $\sqrt{15}$
5. e^7 **6.** $e^{11/3}$ **7.** e^6 **8.** e^3
9. 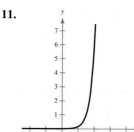   **10.**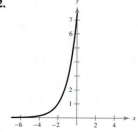

11. 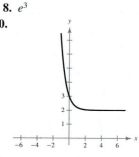   **12.**

13.
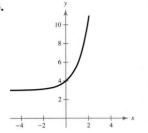

14.
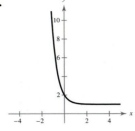

15. 22.69 grams   **16.** $31.06

17. (a) $3571.02   (b) $3572.83   (c) $3573.74

18. $10,379.21   **19.** $5e^{5x}$ **20.** e^{x-4} **21.** $5e^{x+2}$

22. $e^x(2 - x)$ **23.** $y = -2x + 1$

24.
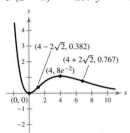

Relative maximum: $(4, 8e^{-2})$

Relative minimum: $(0, 0)$

Points of inflection:
$(4 - 2\sqrt{2}, 0.382),$
$(4 + 2\sqrt{2}, 0.767)$

Horizontal asymptote to the right: $y = 0$

Section 4.4   *(page 242)*

Skills Warm Up   *(page 242)*

1. $f^{-1}(x) = \frac{1}{5}x$ **2.** $f^{-1}(x) = x + 6$

3. $f^{-1}(x) = \dfrac{x - 2}{3}$ **4.** $f^{-1}(x) = \frac{4}{3}(x + 9)$

5. $x > -4$ **6.** Any real number x

7. $x < -1$ or $x > 1$ **8.** $x > 5$

9. $3462.03   **10.** $3374.65

Section 4.5   *(page 249)*

Skills Warm Up   *(page 249)*

1. $2\ln(x + 1)$ **2.** $\ln x + \ln(x + 1)$

3. $\ln x - \ln(x + 1)$ **4.** $3[\ln x - \ln(x - 3)]$

5. $\ln 4 + \ln x + \ln(x - 7) - 2\ln x$

6. $3\ln x + \ln(x + 1)$ **7.** $-\dfrac{y}{x + 2y}$

8. $\dfrac{3 - 2xy + y^2}{x(x - 2y)}$ **9.** $-12x + 2$ **10.** $-\dfrac{6}{x^4}$

Section 4.6   *(page 255)*

Skills Warm Up   *(page 255)*

1. $-\dfrac{1}{4}\ln 2$ **2.** $\dfrac{1}{5}\ln\dfrac{10}{3}$ **3.** $-\dfrac{\ln(25/16)}{0.01}$

4. $-\dfrac{\ln(11/16)}{0.02}$ **5.** $7.36e^{0.23t}$ **6.** $1.296e^{0.072t}$

7. $-33.6e^{-1.4t}$ **8.** $-0.025e^{-0.001t}$ **9.** 4

10. 12 **11.** $2x + 1$ **12.** $x^2 + 1$

Review Exercises for Chapter 4   *(page 260)*

1. (a) 16,384   (b) 117,649   (c) 0.0625   (d) 81

3.
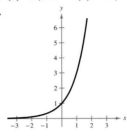

5.
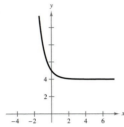

7.
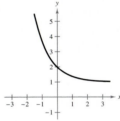

9. (a) 5894.39 (thousand)   (b) 6203.76 (thousand)

11. (a) $69,295.66   (b) $233,081.88

13. (a) e^{10} (b) $\dfrac{1}{e^2}$ (c) $e^{11/2}$ (d) $\dfrac{1}{e^8}$

15.

17.

19.

n	1	2	4	12
A	$1216.65	$1218.99	$1220.19	$1221.00

n	365	Continuous compounding
A	$1221.39	$1221.40

21.

n	1	2	4	12
A	4231.80	4244.33	4250.73	4255.03

n	365	Continuous compounding
A	4257.13	4257.20

23. b   **25.** (a) 6%   (b) 6.09%   (c) 6.14%   (d) 6.17%

27. $10,338.10

29. (a) $8276.81   (b) $7697.12   (c) $7500

31. (a) The model fits the data very well.

(b) $y = 116.85x + 111.1$

The linear model fits the data moderately well.
The exponential model is a better fit.

(c) Exponential: $4357.50 (million)
Linear: $1863.85 (million)

CHAPTER 4

33. (a) $P \approx 1049$ fish

(b)   13 months

(c) Yes, P approaches 10,000 fish as t approaches ∞.

35. $8xe^{x^2}$ **37.** $\dfrac{1 - 2x}{e^{2x}}$ **39.** $-\dfrac{10e^{2x}}{(1 + e^{2x})^2}$

41. $y = 3 - x$ **43.** $y = \dfrac{x}{e}$

45.

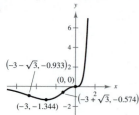

Relative minimum: $(-3, -1.344)$

Inflection points: $(0, 0)$, $(-3 + \sqrt{3}, -0.574)$,

and $(-3 - \sqrt{3}, -0.933)$

Horizontal asymptote: $y = 0$

47.

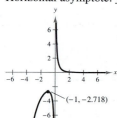

Relative maximum: $(-1, -2.718)$

Horizontal asymptote: $y = 0$

Vertical asymptote: $x = 0$

49. $e^{2.4849} \approx 12$ **51.** $\ln 4.4816 \approx 1.5$

53.

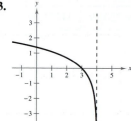

55.

57. $\ln x + \frac{1}{2}\ln(x - 1)$ **59.** $2 \ln x - 3 \ln(x + 1)$

61. $3[\ln(1 - x) - \ln 3 - \ln x]$

63. $\ln(2x^2 - x - 15)$ **65.** $4 \ln\left(\dfrac{x^5 - x^2}{x - 5}\right)$

67. 3   **69.** $e^3 \approx 20.09$ **71.** 1

73. $\dfrac{3 + \sqrt{13}}{2} \approx 3.3028$ **75.** $-\dfrac{\ln(0.25)}{1.386} \approx 1.0002$

77. $\frac{1}{2}(\ln 6 + 1) \approx 1.3959$ **79.** $\dfrac{\ln 1.1}{\ln 1.21} = 0.5$

81. (a) ≈ 28.07 years   (b) ≈ 27.75 years

(c) ≈ 27.73 years   (d) ≈ 27.73 years

83. (a) 75   (b) 65.34   (c) ≈ 11 months

85. $\dfrac{2}{x}$ **87.** $\dfrac{1}{x} + \dfrac{1}{x - 1} - \dfrac{1}{x - 2} = \dfrac{x^2 - 4x + 2}{x(x - 2)(x - 1)}$

89. 2   **91.** $\dfrac{1 - 3\ln x}{x^4}$ **93.** $\dfrac{4x}{3(x^2 - 2)}$

95. $\dfrac{2}{x} + \dfrac{1}{2(x + 1)}$ **97.** $\dfrac{1}{1 + e^x}$ **99.** 2   **101.** 0

103. 1.594   **105.** 1.500   **107.** $(2 \ln 5)5^{2x+1}$

109. $\dfrac{2}{(2x - 1)\ln 3}$ **111.** $\dfrac{-1}{\ln 10} \cdot \dfrac{1}{x} = -\dfrac{1}{x \ln 10}$

113.

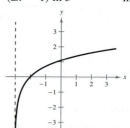

115.

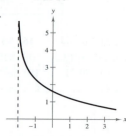

No relative extrema   No relative extrema

No points of inflection   No points of inflection

117. 2005: 256.4 (million)

2008: 160.25 (million)

119. $y = 3e^{-0.27465t}$ **121.** 5.19 g; 0.10 g

123. 20.18 g; 17.88 g   **125.** 2.47 g; 1.85 g

127. 8.66 yr, $1335.32, $4433.43

129. 2%, 34.66 yr, $24,730.82

131. (a) $D = 500e^{-0.38376t}$

(b) 107.72 milligrams per milliliter

Test Yourself *(page 264)*

1. 1   **2.** $\frac{1}{256}$ **3.** $e^{9/2}$ **4.** e^{12}

5.

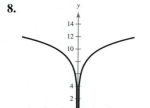

6.

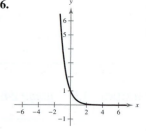

7.

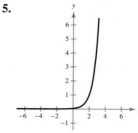

8.

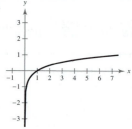

9.

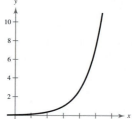

10.

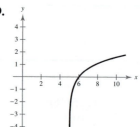

11. $\ln 3 - \ln 2$ **12.** $\frac{1}{2}\ln(x + y)$ **13.** $\ln(x + 1) - \ln y$

14. $\ln[y(x + 1)]$ **15.** $\ln\dfrac{x^3}{(x - 1)^2}$ **16.** $\ln\dfrac{xy^4}{\sqrt{(z + 4)}}$

17. $x \approx 3.197$ **18.** $x \approx 1.750$ **19.** $x \approx 58.371$

20. (a) 17.67 yr   (b) 17.36 yr   (c) 17.33 yr   (d) 17.33 yr

21. $-3e^{-3x}$ **22.** $7e^{x+2} + 2$

23. $\dfrac{2x}{3 + x^2}$ **24.** $\dfrac{2}{x(x + 2)}$

25. (a) \$2241.54 million   (b) \$138.30 million/yr

26. 59.4%   **27.** 39.61 yr

Chapter 5

Section 5.1 *(page 273)*

> **Skills Warm Up** *(page 273)*
>
> **1.** $x^{-1/2}$ **2.** $(2x)^{4/3}$ **3.** $5^{1/2}x^{3/2} + x^{5/2}$
> **4.** $x^{-1/2} + x^{-2/3}$ **5.** $(x + 1)^{5/2}$ **6.** $x^{1/6}$
> **7.** -12 **8.** -10 **9.** 14 **10.** 14

Section 5.2 *(page 281)*

> **Skills Warm Up** *(page 281)*
>
> **1.** $\frac{1}{2}x^4 + x + C$ **2.** $\frac{3}{2}x^2 + \frac{2}{3}x^{3/2} - 4x + C$
> **3.** $-\dfrac{1}{x} + C$ **4.** $-\dfrac{1}{6t^2} + C$ **5.** $\frac{4}{7}t^{7/2} + \frac{2}{5}t^{5/2} + C$
> **6.** $\frac{4}{5}x^{5/2} - \frac{2}{3}x^{3/2} + C$ **7.** $\dfrac{5x^3 - 4}{2x} + C$
> **8.** $\dfrac{-6x^2 + 5}{3x^3} + C$ **9.** $\frac{2}{5}\sqrt{x}(8x^2 + 15) + C$

Section 5.3 *(page 287)*

> **Skills Warm Up** *(page 287)*
>
> **1.** $x + 2 - \dfrac{2}{x + 2}$ **2.** $x - 2 + \dfrac{1}{x - 4}$
> **3.** $x + 8 + \dfrac{2x - 4}{x^2 - 4x}$ **4.** $x^2 - x - 4 + \dfrac{20x + 22}{x^2 + 5}$
> **5.** $\frac{1}{4}x^4 - \dfrac{1}{x} + C$ **6.** $\frac{1}{2}x^2 + 2x + C$
> **7.** $\frac{1}{2}x^2 - \dfrac{4}{x} + C$ **8.** $-\dfrac{1}{x} - \dfrac{3}{2x^2} + C$

Quiz Yourself *(page 288)*

1. $3x + C$ **2.** $5x^2 + C$ **3.** $-\dfrac{1}{4x^4} + C$

4. $\dfrac{x^3}{3} - x^2 + 15x + C$ **5.** $\dfrac{(6x + 1)^4}{4} + C$

6. $\dfrac{1}{50}(5x^2 - 2)^5 + C$ **7.** $\dfrac{(x^2 - 5x)^2}{2} + C$

8. $-\dfrac{1}{2(x^3 + 3)^2} + C$ **9.** $\frac{2}{15}(5x + 2)^{3/2} + C$

10. $f(x) = 8x^2 + 1$ **11.** $f(x) = 3x^3 + 4x - 2$

12. (a) $C = -0.03x^2 + 16x + 9.03$ (b) \$9.03   (c) \$509.03

13. $f(x) = \frac{2}{3}x^3 + x + 1$

14. (a) 1000 bolts   (b) About 8612 bolts

15. $e^{5x+4} + C$ **16.** $e^{x^3} + C$

17. $\frac{1}{2}e^{(x^2 - 6x)} + C$ **18.** $\ln|2x - 1| + C$

19. $-\frac{1}{8}\ln|3 - 8x| + C$ **20.** $\frac{1}{6}\ln(3x^2 + 4) + C$

21. (a) $S(t) = 13.16t^2 + 848.99\ln(t) + 2504.44$
    (b) \$5112.11 million

Section 5.4 *(page 298)*

> **Skills Warm Up** *(page 298)*
>
> **1.** $\frac{3}{2}x^2 + 7x + C$ **2.** $\frac{2}{5}x^{5/2} + \frac{4}{3}x^{3/2} + C$
> **3.** $\dfrac{1}{5}\ln|x| + C$ **4.** $-\dfrac{1}{6e^{6x}} + C$
> **5.** $C = 0.008x^{5/2} + 29,500x + C$
> **6.** $R = x^2 + 9000x + C$
> **7.** $P = 25,000x - 0.005x^2 + C$
> **8.** $C = 0.01x^3 + 4600x + C$

Section 5.5 *(page 305)*

> **Skills Warm Up** *(page 305)*
>
> **1.** $-x^2 + 3x + 2$ **2.** $-2x^2 + 4x + 4$
> **3.** $-x^3 + 2x^2 + 4x - 5$ **4.** $x^3 - 6x - 1$
> **5.** $(0, 4), (4, 4)$ **6.** $(1, -3), (2, -12)$
> **7.** $(-3, 9), (2, 4)$ **8.** $(-2, -4), (0, 0), (2, 4)$

Section 5.6 *(page 310)*

> **Skills Warm Up** *(page 310)*
>
> **1.** $\frac{1}{6}$ **2.** $\frac{3}{20}$ **3.** $\frac{7}{40}$ **4.** $\frac{13}{12}$ **5.** $\frac{61}{30}$ **6.** $\frac{53}{18}$
> **7.** $\frac{2}{3}$ **8.** $\frac{4}{7}$ **9.** 0 **10.** 5

Review Exercises for Chapter 5 *(page 315)*

1. $16x + C$ **3.** $\frac{3}{10}x^2 + C$ **5.** $x^3 + C$

7. $\frac{2}{3}x^3 + \frac{5}{2}x^2 + C$ **9.** $x^{2/3} + C$ **11.** $\frac{3}{7}x^{7/3} + \frac{3}{2}x^2 + C$

13. $\frac{4}{9}x^{9/2} - 2\sqrt{x} + C$ **15.** $6x^2 - 3$ **17.** $x^3 - 4x^2 + 15$

19. $s(t) = -16t^2 + 80t$
    5 seconds

21. $\frac{1}{4}(x + 4)^4 + C$ **23.** $\frac{1}{5}(5x + 1)^5 + C$

25. $x + 5x^2 + \frac{25}{3}x^3 + C$ or $\frac{1}{15}(1 + 5x)^3 + C_1$

27. $\dfrac{(3x^3 + 1)^3}{27} + C$ **29.** $\dfrac{-1}{12(2x^3 - 5)^2} + C$

31. $\frac{2}{5}\sqrt{5x - 1} + C$

33. (a) 30.54 board-feet   (b) 125.2 board-feet

35. $e^{4x} + C$ **37.** $-\frac{1}{5}e^{-5x} + C$

39. $\dfrac{7e^{3x^2}}{6} + C$ **41.** $\ln|x - 6| + C$

43. $\frac{2}{3}\ln|6x - 1| + C$ **45.** $-\frac{1}{3}\ln|1 - x^3| + C$

47.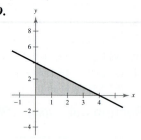

Area = 6

49.

Area = 8

51. (a) 13   (b) 7   (c) 11   (d) 50

53. $\frac{32}{3}$ **55.** 2 ln 2   **57.** $4e^{1/2} - 4$ **59.** 16

61. 0   **63.** 4   **65.** 2   **67.** 5 ln 3 ≈ 5.49

69. 1.899   **71.** Average value = 3; $x = 1$

73. Average value = $\frac{2}{3}(1 - e^3) \approx -12.724$;
$x = \ln\left[-\frac{1}{3}(1 - e^3)\right] \approx 1.850$

75. Average value = $\frac{2}{5}$; $x = \frac{25}{4}$ **77.** 0   **79.** 115.2

81. $17,492.94   **83.** Increases by $700.25   **85.** $520.54

87.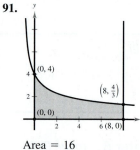

Area = $\frac{4}{9}$

89.

Area = $\frac{64}{3}$

91.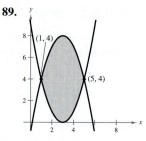

Area = 16

93.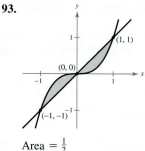

Area = $\frac{1}{2}$

95. Consumer surplus = 1417.5
Producer surplus = 202.5

97. Consumer surplus = 1250
Producer surplus = 1250

99. R_2; $84.5 million

101. $300 million

103. Approximation: 1.5
Actual area: 1.5

105. 2.625   **107.** 1.070

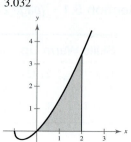

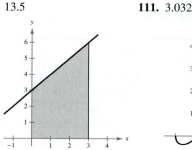

109. 13.5   **111.** 3.032

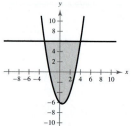

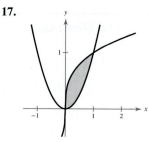

113. 9840 ft²

Test Yourself   *(page 319)*

1. $3x^3 - 2x^2 + 13x + C$ **2.** $\dfrac{(x + 1)^3}{3} + C$

3. $\dfrac{2(x^4 - 7)^{3/2}}{3} + C$ **4.** $\dfrac{10x^{3/2}}{3} - 12x^{1/2} + C$

5. $5e^{3x} + C$ **6.** $\frac{3}{4}\ln|4x - 1| + C$

7. $f(x) = 3x^2 - 5x - 2$ **8.** $f(x) = e^x + x$

9. 8   **10.** 18   **11.** $\frac{2}{3}$

12. $2\sqrt{5} - 2\sqrt{2} \approx 1.644$ **13.** $\frac{1}{4}(e^{12} - 1) \approx 40,688.4$

14. $\ln 6 \approx 1.792$

15. (a) $S(t) = 2240e^{0.1013t} + 4.3906$, $0 \le t \le 9$
(b) $3661.68 million

16.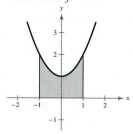

Area = $\frac{343}{6} \approx 57.167$

17.

Area = $\frac{5}{12}$

18. Consumer surplus = 20 million
Producer surplus = 8 million

19. Midpoint Rule: $\frac{63}{64} \approx 0.9844$
Exact area: 1

20. Midpoint Rule: $\frac{21}{8} = 2.625$
Exact area: $\frac{8}{3} = 2.\overline{6}$

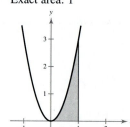

Chapter 6

Section 6.1   *(page 329)*

Skills Warm Up   *(page 329)*

1. $\dfrac{1}{x+1}$ 2. $\dfrac{2x}{x^2-1}$ 3. $3x^2e^{x^3}$

4. $-2xe^{-x^2}$ 5. $e^x(x^2+2x)$ 6. $e^{-2x}(1-2x)$

7. $\dfrac{64}{3}$ 8. $\dfrac{4}{3}$ 9. 36   10. 8

Section 6.2   *(page 335)*

Skills Warm Up   *(page 335)*

1. $x^2+8x+16$ 2. x^2-2x+1

3. $x^2+x+\dfrac{1}{4}$ 4. $x^2-\dfrac{2}{3}x+\dfrac{1}{9}$

5. $2e^x(x-1)+C$ 6. $x^3\ln x-\dfrac{x^3}{3}+C$

Quiz Yourself   *(page 336)*

1. $\dfrac{1}{5}xe^{5x}-\dfrac{1}{25}e^{5x}+C$ 2. $3x\ln x-3x+C$

3. $\dfrac{1}{2}x^2\ln x+x\ln x-\dfrac{1}{4}x^2-x+C$

4. $\dfrac{2}{3}x(x+3)^{3/2}-\dfrac{4}{15}(x+3)^{5/2}+C$

5. $\dfrac{x^2}{4}\ln x-\dfrac{x^2}{8}+C$ 6. $-\dfrac{1}{2}e^{-2x}\left(x^2+x+\dfrac{1}{2}\right)+C$

7. (a) 282,016 units   (b) 56,403 units

8. (a) \$784,000   (b) \$673,108.31

9. $\dfrac{1}{4}(2x-\ln|1+2x|)+C$ 10. $10\ln\left|\dfrac{x}{0.1+0.2x}\right|+C$

11. $\ln\left|x+\sqrt{x^2-16}\right|-\dfrac{\sqrt{x^2-16}}{x}+C$

12. $\dfrac{1}{2}\ln\left|\dfrac{\sqrt{4+9x}-2}{\sqrt{4+9x}+2}\right|+C$

13. $\dfrac{1}{4}[4x^2-\ln(1+e^{4x^2})]+C$ 14. $x^2e^{x^2+1}+C$

15. (a) \$84,281,126.52   (b) \$257,392,429.72

16. $\dfrac{8}{e}-4\approx-1.0570$ 17. $10\ln 2-\dfrac{15}{4}\approx 3.1815$

18. $\dfrac{64}{3}(\sqrt{2}-1)\approx 8.8366$ 19. $e-2\approx 0.7183$

20. $\dfrac{\sqrt{5}}{18}\approx 0.1242$ 21. $\dfrac{1}{4}\left(\ln\dfrac{17}{19}-\ln\dfrac{7}{9}\right)\approx 0.0350$

Section 6.3   *(page 343)*

Skills Warm Up   *(page 343)*

1. $\dfrac{2}{x^3}$ 2. $-\dfrac{96}{(2x+1)^4}$ 3. $-\dfrac{12}{x^4}$ 4. $6x-4$

5. $16e^{2x}$ 6. $e^{x^2}(4x^2+2)$ 7. $(3, 18)$

8. $(1, 8)$ 9. $n<-5\sqrt{10},\ n>5\sqrt{10}$

10. $n<-5, n>5$

Section 6.4   *(page 351)*

Skills Warm Up   *(page 351)*

1. 9   2. 3   3. $-\dfrac{1}{8}$ 4. Limit does not exist.

5. Limit does not exist.   6. -4

7. (a) $\dfrac{32}{3}b^3-16b^2+8b-\dfrac{4}{3}$ (b) $-\dfrac{4}{3}$

8. (a) $\dfrac{b^2-b-11}{(b-2)^2(b-5)}$ (b) $\dfrac{11}{20}$

9. (a) $\ln\left(\dfrac{5-3b^2}{b+1}\right)$ (b) $\ln 5\approx 1.609$

10. (a) $e^{-3b^2}(e^{6b^2}+1)$ (b) 2

Review Exercises for Chapter 6   *(page 356)*

1. $2\sqrt{x}\ln x-4\sqrt{x}+C$ 3. xe^x+C

5. $\dfrac{2}{15}(x-5)^{3/2}(3x+10)+C$ 7. $x^2e^{2x}-xe^{2x}+\dfrac{1}{2}e^{2x}+C$

9. $3e^2-\dfrac{3(e^2-1)}{2}\approx 12.584$ 11. $16-20e^{-1/4}\approx 0.4240$

13. \$90,634.62   15. \$865,958.50

17. (a) \$1,200,000   (b) \$1,052,649.52

19. $\dfrac{1}{54}(9x^2-12x+8\ln|3x+2|)+C$

21. $\dfrac{x}{2}\sqrt{x^2-16}-8\ln\left(\sqrt{x^2-16}+x\right)+C$

23. $\dfrac{1}{9}\left(\dfrac{2}{2+3x}+\ln|2+3x|\right)+C$

25. $\sqrt{x^2+25}-5\ln\left|\dfrac{5+\sqrt{x^2+25}}{x}\right|+C$

27. $\dfrac{1}{4}\ln\left|\dfrac{x-2}{x+2}\right|+C$ 29. $\dfrac{2}{3}(x-2)\sqrt{1+x}+C$

31. $2\sqrt{1+x}+\ln\left|\dfrac{\sqrt{1+x}-1}{\sqrt{1+x}+1}\right|+C$

33. (a) 0.675   (b) 0.290

35. Exact: $\dfrac{2}{3}\approx 0.6667$ 37. Exact: $\dfrac{3}{8}=0.375$
    Trapezoidal: 0.7050   Trapezoidal: 0.3786
    Simpson's: 0.6715   Simpson's: 0.3751

39. Exact: $2-2e^{-2}\approx 1.7293$ 41. (a) 0.741   (b) 0.737
    Trapezoidal: 1.7652
    Simpson's: 1.7299

43. (a) 0.305   (b) 0.289   45. (a) 2.961   (b) 2.936

47. (a) $|E|\le\dfrac{e^4}{6}\approx 9.0997$ (b) $|E|\le\dfrac{e^4}{90}\approx 0.6066$

49. (a) $n=214$ (b) $n=2$

51. Converges, $-\dfrac{1}{4}$ 53. Diverges   55. Diverges

57. $A\approx 4$ 59. $A=1$ 61. \$266,666.67   63. No

Test Yourself   *(page 358)*

1. $xe^{x+1}-e^{x+1}+C$ 2. $3x^3\ln x-x^3+C$

3. $-3x^2e^{-x/3}-18xe^{-x/3}-54e^{-x/3}+C$

4. (a) $\approx\$6494.47$ million   (b) $\approx\$811.81$ million

5. $\dfrac{1}{4}\left(\dfrac{7}{7+2x}+\ln|7+2x|\right)+C$ 6. $x^3-\ln(1+e^{x^3})+C$

7. $-\dfrac{2}{75}(2-5x^2)\sqrt{1+5x^2}+C$ 8. $-1+\dfrac{3}{2}\ln 3\approx 0.6479$

9. $8\dfrac{2}{3}$ 10. $4\ln\left[3(\sqrt{17}-4)\right]+\sqrt{17}-5\approx -4.8613$

11. Exact: 18.0
   Trapezoidal: 18.28
12. Simpson's Rule: 41.3606; Exact: 41.1711
13. Converges; $\frac{1}{3}$ **14.** Converges; 12   **15.** Diverges
16. (a) \$498.75   (b) Plan B, because \$149 < \$498.75.

Chapter 7

Section 7.1   *(page 365)*

Skills Warm Up   *(page 365)*

1. $2\sqrt{5}$ **2.** 5   **3.** 8   **4.** 8   **5.** (4, 7)
6. (1, 0)   **7.** (0, 3)   **8.** (−1, 1)
9. $(x - 2)^2 + (y - 3)^2 = 4$
10. $(x - 1)^2 + (y - 4)^2 = 25$

Section 7.2   *(page 373)*

Skills Warm Up   *(page 373)*

1. (4, 0), (0, 3)   **2.** $\left(-\frac{4}{3}, 0\right)$, (0, −8)
3. (1, 0), (0, −2)   **4.** (−5, 0), (0, −5)
5. $x^2 + y^2 + z^2 = \frac{1}{4}$ **6.** $x^2 + y^2 + z^2 = 4$

Section 7.3   *(page 379)*

Skills Warm Up   *(page 379)*

1. 11   **2.** −16   **3.** 7   **4.** 4   **5.** $(-\infty, \infty)$
6. $(-\infty, -3) \cup (-3, 0) \cup (0, \infty)$
7. $[5, \infty)$ **8.** $\left(-\infty, -\sqrt{5}\right] \cup \left[\sqrt{5}, \infty\right)$
9. 55.0104   **10.** 6.9165

Section 7.4   *(page 388)*

Skills Warm Up   *(page 388)*

1. $\dfrac{x}{\sqrt{x^2 + 3}}$ **2.** $-6x(3 - x^2)^2$ **3.** $e^{2t+1}(2t + 1)$
4. $\dfrac{e^{2x}(2 - 3e^{2x})}{\sqrt{1 - e^{2x}}}$ **5.** $-\dfrac{2}{3 - 2x}$ **6.** $\dfrac{3(t^2 - 2)}{2t(t^2 - 6)}$
7. $-\dfrac{10x}{(4x - 1)^3}$ **8.** $-\dfrac{(x + 2)^2(x^2 + 8x + 27)}{(x^2 - 9)^3}$
9. $f'(2) = 8$ **10.** $g'(2) = \frac{7}{2}$

Section 7.5   *(page 395)*

Skills Warm Up   *(page 395)*

1. (3, 2)   **2.** (11, 6)   **3.** (1, 4)   **4.** (4, 4)
5. (5, 2)   **6.** (3, −2)   **7.** (0, 0), (−1, 0)
8. (−2, 0), (2, −2)

9. $\dfrac{\partial z}{\partial x} = 12x^2$ $\dfrac{\partial^2 z}{\partial y^2} = -6$

   $\dfrac{\partial z}{\partial y} = -6y$ $\dfrac{\partial^2 z}{\partial x \partial y} = 0$

   $\dfrac{\partial^2 z}{\partial x^2} = 24x$ $\dfrac{\partial^2 z}{\partial y \partial x} = 0$

10. $\dfrac{\partial z}{\partial x} = 10x^4$ $\dfrac{\partial^2 z}{\partial y^2} = -6y$

   $\dfrac{\partial z}{\partial y} = -3y^2$ $\dfrac{\partial^2 z}{\partial x \partial y} = 0$

   $\dfrac{\partial^2 z}{\partial x^2} = 40x^3$ $\dfrac{\partial^2 z}{\partial y \partial x} = 0$

11. $\dfrac{\partial z}{\partial x} = 4x^3 - \dfrac{\sqrt{xy}}{2x}$ $\dfrac{\partial^2 z}{\partial y^2} = \dfrac{\sqrt{xy}}{4y^2}$

   $\dfrac{\partial z}{\partial y} = -\dfrac{\sqrt{xy}}{2y} + 2$ $\dfrac{\partial^2 z}{\partial x \partial y} = -\dfrac{\sqrt{xy}}{4xy}$

   $\dfrac{\partial^2 z}{\partial x^2} = 12x^2 + \dfrac{\sqrt{xy}}{4x^2}$ $\dfrac{\partial^2 z}{\partial y \partial x} = -\dfrac{\sqrt{xy}}{4xy}$

12. $\dfrac{\partial z}{\partial x} = 4x - 3y$ $\dfrac{\partial^2 z}{\partial y^2} = 2$

   $\dfrac{\partial z}{\partial y} = 2y - 3x$ $\dfrac{\partial^2 z}{\partial x \partial y} = -3$

   $\dfrac{\partial^2 z}{\partial x^2} = 4$ $\dfrac{\partial^2 z}{\partial y \partial x} = -3$

13. $\dfrac{\partial z}{\partial x} = y^3 e^{xy^2}$ $\dfrac{\partial^2 z}{\partial y^2} = 4x^2 y^3 e^{xy^2} + 6xy e^{xy^2}$

   $\dfrac{\partial z}{\partial y} = 2xy^2 e^{xy^2} + e^{xy^2}$ $\dfrac{\partial^2 z}{\partial x \partial y} = 2xy^4 e^{xy^2} + 3y^2 e^{xy^2}$

   $\dfrac{\partial^2 z}{\partial x^2} = y^5 e^{xy^2}$ $\dfrac{\partial^2 z}{\partial y \partial x} = 2xy^4 e^{xy^2} + 3y^2 e^{xy^2}$

14. $\dfrac{\partial z}{\partial x} = e^{xy}(xy + 1)$ $\dfrac{\partial^2 z}{\partial y^2} = x^3 e^{xy}$

   $\dfrac{\partial z}{\partial y} = x^2 e^{xy}$ $\dfrac{\partial^2 z}{\partial x \partial y} = xe^{xy}(xy + 2)$

   $\dfrac{\partial^2 z}{\partial x^2} = ye^{xy}(xy + 2)$ $\dfrac{\partial^2 z}{\partial y \partial x} = xe^{xy}(xy + 2)$

Quiz Yourself   *(page 396)*

1. (a)

2. (a)

(b) 3   (c) $\left(0, \frac{5}{2}, 1\right)$ (b) $2\sqrt{35}$ (c) (2, 2, −1)

3. (a)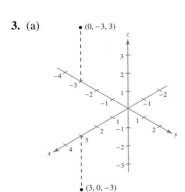

(b) $3\sqrt{6}$ (c) $\left(\frac{3}{2}, -\frac{3}{2}, 0\right)$

4. $(x - 2)^2 + (y + 1)^2 + (z - 3)^2 = 16$

5. $(x - 1)^2 + (y - 4)^2 + (z + 2)^2 = 11$

6. Center: $(4, 1, 3)$; radius: 7

7. 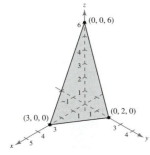   **8.**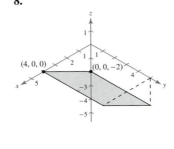

9. 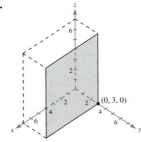   **10.** Ellipsoid

11. Hyperboloid of two sheets   **12.** Elliptic paraboloid

13. $f(1, 0) = 1$ **14.** $f(1, 0) = 2$

$f(4, -1) = -5$ $f(4, -1) = 3\sqrt{7}$

15. $f(1, 0) = 0$

$f(4, -1) = \ln 6 \approx 1.79$

16. (a) Between $30°$ and $50°$ (b) Between $40°$ and $80°$

(c) Between $70°$ and $90°$

17. $f_x = 2x - 3$; $f_x(-2, 3) = -7$

$f_y = 4y - 1$; $f_y(-2, 3) = 11$

18. $f_x = \dfrac{y(3 + y)}{(x + y)^2}$; $f_x(-2, 3) = 18$

$f_y = \dfrac{-2xy - y^2 - 3x}{(x + y)^2}$; $f_y(-2, 3) = 9$

19. $f_x = 3x^2 e^{2y}$; $f_x(-2, 3) = 12e^6 \approx 4841.15$

$f_y = 2x^3 e^{2y}$; $f_y(-2, 3) = -16e^6 \approx -6454.86$

20. $f_x = \dfrac{2}{2x + 7y}$; $f_x(-2, 3) = \dfrac{2}{17} \approx 0.118$

$f_y = \dfrac{7}{2x + 7y}$; $f_y(-2, 3) = \dfrac{7}{17} \approx 0.412$

21. Critical point: $(1, 0)$

Relative minimum:

$(1, 0, -3)$

22. Critical points: $(0, 0)$, $\left(\frac{4}{3}, \frac{4}{3}\right)$

Relative maximum:

$\left(\frac{4}{3}, \frac{4}{3}, \frac{59}{27}\right)$

Saddle point: $(0, 0, 1)$

23. $x = 80$, $y = 20$; $20,000$

Section 7.6   *(page 402)*

Skills Warm Up   *(page 402)*

1. $\left(\frac{7}{8}, \frac{1}{12}\right)$ **2.** $\left(-\frac{1}{24}, -\frac{7}{8}\right)$ **3.** $\left(\frac{55}{12}, -\frac{25}{12}\right)$

4. $\left(\frac{22}{23}, -\frac{3}{23}\right)$ **5.** $\left(\frac{5}{3}, \frac{1}{3}, 0\right)$ **6.** $\left(\frac{14}{19}, -\frac{10}{19}, -\frac{32}{57}\right)$

7. $f_x = 2xy + y^2$ **8.** $f_x = 50y^2(x + y)$

$f_y = x^2 + 2xy$ $f_y = 50y(x + y)(x + 2y)$

9. $f_x = 3x^2 - 4xy + yz$ **10.** $f_x = yz + z^2$

$f_y = -2x^2 + xz$ $f_y = xz + z^2$

$f_z = xy$ $f_z = xy + 2xz + 2yz$

Section 7.7   *(page 408)*

Skills Warm Up   *(page 408)*

1. 5.0225 **2.** 0.0189

3. $S_a = 2a - 4 - 4b$ **4.** $S_a = 8a - 6 - 2b$

$S_b = 12b - 8 - 4a$ $S_b = 18b - 4 - 2a$

5. 15 **6.** 42 **7.** $\frac{25}{12}$ **8.** 14 **9.** 31 **10.** 95

Section 7.8   *(page 415)*

Skills Warm Up   *(page 415)*

1. 1 **2.** 6 **3.** 42 **4.** $\frac{1}{2}$ **5.** $\frac{19}{4}$

6. $\frac{16}{3}$ **7.** $\frac{1}{7}$ **8.** 4 **9.** $\ln 5$ **10.** $\ln(e - 1)$

11. $\dfrac{e}{2}(e^4 - 1)$ **12.** $\dfrac{1}{2}\left(1 - \dfrac{1}{e^2}\right)$

13. 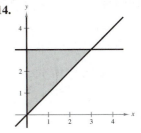   **14.**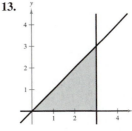

15. 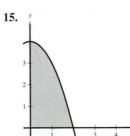   **16.**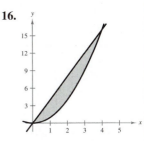

CHAPTER 7

Section 7.9   (page 422)

Skills Warm Up   (page 422)

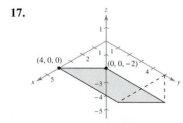

5. 1   **6.** 6   **7.** $\frac{1}{3}$ **8.** $\frac{40}{3}$ **9.** $\frac{28}{3}$ **10.** $\frac{7}{6}$

Review Exercises for Chapter 7   (page 427)

1.

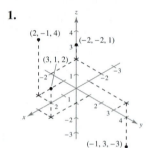

3. $\sqrt{65}$ **5.** $(-1, 4, 6)$ **7.** $x^2 + (y - 1)^2 + z^2 = 25$

9. $(x - 2)^2 + (y + 2)^2 + (z + 3)^2 = 9$

11. Center: $(4, -2, 3)$; radius: 7

13.   **15.**

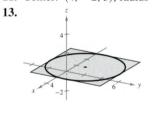

17.

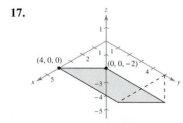

19. Sphere   **21.** Ellipsoid   **23.** Elliptic paraboloid

25. Top half of a circular cone

27. (a) 18   (b) 0   (c) -245 (d) -32

29. Domain: all points (x, y) inside or on the circle $x^2 + y^2 = 1$

Range: $[0, 1]$

31. Domain: all points (x, y)

Range: $(0, \infty)$

33. The level curves are lines of slope $-\frac{2}{5}$.

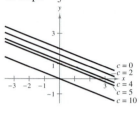

35. The level curves are hyperbolas.

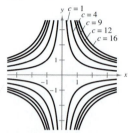

37. (a) No; the precipitation increments are 7.99 in., 9.99 in., 9.99 in., 9.99 in., and 19.99 in.

(b) Increase the number of level curves to correspond to smaller increments of precipitation.

39. (a) \$2.49

(b) Because $\frac{\partial z}{\partial y} = 0.060 > \frac{\partial z}{\partial x} = 0.046$, y has the greater influence.

41. $f_x = 2xy + 3y + 2$

$f_y = x^2 + 3x - 5$

43. $z_x = \dfrac{2x}{y^2}$

$z_y = \dfrac{-2x^2}{y^3}$

45. $f_x = \dfrac{5}{5x + 4y}$

$f_y = \dfrac{4}{5x + 4y}$

47. $f_x = ye^x + e^y$

$f_y = xe^y + e^x$

49. $w_x = yz^2$

$w_y = xz^2$

$w_z = 2xyz$

51. (a) -9 (b) -6 **53.** (a) -2 (b) -2

55. $f_{xx} = 6$

$f_{yy} = 12y$

$f_{xy} = f_{yx} = -1$

57. $f_{xx} = f_{yy} = f_{xy} = f_{yx} = \dfrac{-1}{4(1 + x + y)^{3/2}}$

59. $f_{xx} = 10yz^3$ $f_{yx} = 1 + 10xz^3$ $f_{zx} = 30xyz^2$

$f_{xy} = 1 + 10xz^3$ $f_{yy} = -18yz$ $f_{zy} = 15x^2z^2 - 9y^2$

$f_{xz} = 30xyz^2$ $f_{yz} = 15x^2z^2 - 9y^2$ $f_{zz} = 30x^2yz$

61. (a) $C_x(500, 250) = 99.50$

$C_y(500, 250) = 140$

(b) Downhill skis; this is determined by comparing the marginal costs for the two models of skis at the production level (500, 250).

63. Critical point: $(0, 0)$

Relative minimum: $(0, 0, 0)$

65. Critical point: $(-2, 3)$

Saddle point: $(-2, 3, 1)$

67. Critical points: $(0, 0)$, $\left(\frac{1}{6}, \frac{1}{12}\right)$

Relative minimum: $\left(\frac{1}{6}, \frac{1}{12}, -\frac{1}{432}\right)$

Saddle point: $(0, 0, 0)$

69. Critical points: $(1, 1), (-1, -1). (1, -1), (-1, 1)$

Relative minimum: $(1, 1, -2)$

Relative maximum: $(-1, -1, 6)$

Saddle points: $(1, -1, 2), (-1, 1, 2)$

71. $x_1 = 50, x_2 = 200$

73. At $(3, 6)$, the relative maximum is 36.

75. At $(2, 2)$, the relative minimum is 8.

77. At $\left(\frac{4}{3}, \frac{2}{3}, \frac{4}{3}\right)$, the relative maximum is $\frac{32}{27}$.

79. $x_1 = 378$ units; $x_2 = 623$ units

81. $y = \frac{60}{59}x - \frac{15}{59}$

83. (a) $y = -2.6x + 347$
 (b) 126 cameras
 (c) About $56.54

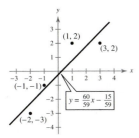

85. 1   **87.** $\frac{7}{4}$ **89.** $\frac{32}{3}$ **91.** $\frac{9}{2}$ **93.** 20   **95.** 8

97. $\frac{4096}{9}$ **99.** 3   **101.** $5700   **103.** About $155.69/\text{ft}^2$

12. $f_x = (x + y)^{1/2} + \dfrac{x}{2(x + y)^{1/2}}$; $f_x(10, -1) = \dfrac{14}{3}$

$f_y = \dfrac{x}{2(x + y)^{1/2}}$; $f_y(10, -1) = \dfrac{5}{3}$

13. Critical point: $(1, -2)$; Relative minimum: $(1, -2, -23)$

14. Critical points: $(0, 0), (1, 1), (-1, -1)$
 Saddle point: $(0, 0, 0)$
 Relative maxima: $(1, 1, 2), (-1, -1, 2)$

15. About 128,613 units   **16.** $y = 0.52x + 1.4$

17. $\frac{3}{2}$ **18.** 1   **19.** $\frac{4}{3}$ units^2 **20.** 48   **21.** $\frac{11}{6}$

Test Yourself   *(page 431)*

1. (a)

2. (a)

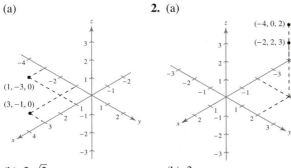

 (b) $2\sqrt{2}$ (b) 3

 (c) $(2, -2, 0)$ (c) $(-3, 1, 2.5)$

3. (a)

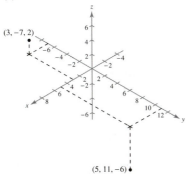

 (b) $14\sqrt{2}$

 (c) $(4, 2, -2)$

4. Center: $(10, -5, 5)$; radius: 5

5. Hyperboloid of one sheet

6. Elliptic cone   **7.** Hyperbolic paraboloid

8. $f(3, 3) = 19$ **9.** $f(3, 3) = \frac{3}{2}$
 $f(1, 4) = 6$ $f(1, 4) = -9$

10. $f(3, 3) = 0$
 $f(1, 4) = 4 \ln \frac{1}{4} \approx -5.5$

11. $f_x = 6x + 9y^2$; $f_x(10, -1) = 69$
 $f_y = 18xy$; $f_y(10, -1) = -180$

Answers to Checkpoints

Chapter 1

Checkpoints for Section 1.1

1 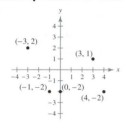　**2**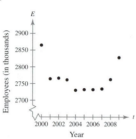

3 5　　**4** $d_1 = \sqrt{20}, d_2 = \sqrt{45}, d_3 = \sqrt{65}$　　**5** 25 yd
$$d_1^2 + d_2^2 = 20 + 45 = 65 = d_3^2$$
6 $(-2, 5)$　　**7** \$87.5 billion
8 $(-1, -4), (1, -2), (1, 2), (-1, 0)$

Checkpoints for Section 1.2

1 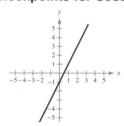　**2**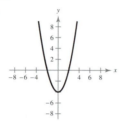

3 x-intercepts: $(3, 0), (-1, 0)$, y-intercept: $(0, -3)$
4 $(x + 2)^2 + (y - 1)^2 = 25$

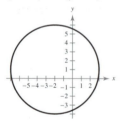

5 12,500 units　　**6** 4 million units at \$122/unit
7 The projection obtained from the model is \$10,814.3 million, which is less than the *Value Line* projection.

Checkpoints for Section 1.3

1 (a)　　　　(b)

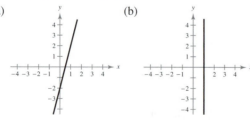

(c)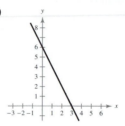

2 Yes, $\frac{27}{312} \approx 0.08654 > \frac{1}{12} = 0.08\overline{3}$.
3 The y-intercept $(0, 1500)$ tells you that the original value of the copier is \$1500. The slope of $m = -300$ tells you that the value decreases by \$300/yr.
4 (a) 2　(b) $-\frac{1}{2}$　(c) 0　　**5** $y = 2x + 4$
6 $y = 0.52t - 2.67$; \$2.01
7 (a) $y = \frac{1}{2}x$　(b) $y = -2x + 5$
8 $V = -1375t + 12,000$

Checkpoints for Section 1.4

1 (a) Yes, $y = x - 1$.　(b) No, $y = \pm\sqrt{4 - x^2}$.
　　(c) No, $y = \pm\sqrt{2 - x}$.　(d) Yes, $y = x^2$.
2 (a) Domain: $[-1, \infty)$; Range: $[0, \infty)$
　　(b) Domain: $(-\infty, \infty)$; Range: $[0, \infty)$
3 $f(0) = 1, f(1) = -3, f(4) = -3$
　　No, f is not one-to-one.
4 (a) $x^2 + 2x\,\Delta x + (\Delta x)^2 + 3$　(b) $2x + \Delta x, \Delta x \neq 0$
5 (a) $2x^2 + 5$　(b) $4x^2 + 4x + 3$
6 (a) $f^{-1}(x) = 5x$　(b) $f^{-1}(x) = x + 6$
7 $f^{-1}(x) = \sqrt{x - 2}$
8 　　　$f(x) = x^2 + 4$
　　　　　$y = x^2 + 4$
　　　　　$x = y^2 + 4$
　　　　$x - 4 = y^2$
　　$\pm\sqrt{x - 4} = y$

Checkpoints for Section 1.5

1 6　　**2** (a) 4　(b) Does not exist　(c) 4
3 (a) 5　(b) 6　(c) 25　(d) -2　　**4** 5　　**5** 12
6 7　　**7** $\frac{1}{4}$　　**8** (a) -1　(b) 1　　**9** 1
10 $\lim\limits_{x \to 1^-} f(x) = 18$ and $\lim\limits_{x \to 1^+} f(x) = 20$

　　$\lim\limits_{x \to 1^-} f(x) \neq \lim\limits_{x \to 1^+} f(x)$

11 Does not exist

Checkpoints for Section 1.6

1 (a) f is continuous on the entire real line.
　　(b) f is continuous on the entire real line.
　　(c) f is continuous on the entire real line.
2 (a) f is continuous on $(-\infty, 1)$ and $(1, \infty)$.
　　(b) f is continuous on $(-\infty, 2)$ and $(2, \infty)$.
　　(c) f is continuous on the entire real line.
3 f is continuous on $[2, \infty)$.　　**4** f is continuous on $[-1, 5]$.

5

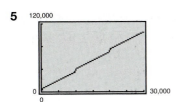

6 $A = 10,000(1 + 0.0075)^{[\![4t]\!]}$

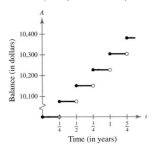

Chapter 2

Checkpoints for Section 2.1

1 3

2 For the months to the left of July on the graph, the tangent lines have positive slopes. For the months to the right of July, the tangent lines have negative slopes. The average daily temperature is increasing prior to July and decreasing after July.

3 4   **4** 2

5 $m = 8x$
     At $(0, 1)$, $m = 0$.
     At $(1, 5)$, $m = 8$.

6 $2x - 5$ **7** $-\dfrac{4}{t^2}$

Checkpoints for Section 2.2

1 (a) 0   (b) 0   (c) 0   (d) 0

2 (a) $4x^3$ (b) $-\dfrac{3}{x^4}$ (c) $2w$

3 $f'(x) = 3x^2$
     $m = f'(-1) = 3$;
     $m = f^{-1}(0) = 0$;
     $m = f^{-1}(1) = 3$

4 (a) $8x$ (b) $\dfrac{8}{\sqrt{x}}$ **5** (a) $\dfrac{1}{4}$ (b) $-\dfrac{2}{5}$

6 (a) $-\dfrac{9}{2x^3}$ (b) $-\dfrac{9}{8x^3}$ **7** (a) $\dfrac{\sqrt{5}}{2\sqrt{x}}$ (b) $\dfrac{1}{4x^{3/4}}$

8 (a) $4x + 5$ (b) $4x^3 - 2$ **9** -1

10 $y = -x + 2$ **11** $\$0.34/\text{yr}$

Checkpoints for Section 2.3

1 (a) $0.5\overline{6}$ mg/ml/min   (b) 0 mg/ml/min
     (c) -1.5 mg/ml/min

2 (a) -16 ft/sec   (b) -48 ft/sec   (c) -80 ft/sec

3 When $t = 1.75$, $h'(1.75) = -56$ ft/sec.
     When $t = 2$, $h'(2) = -64$ ft/sec.

4 (a) $\frac{3}{2}$ sec   (b) -32 ft/sec

5 When $x = 100$, $\dfrac{dP}{dx} = \$16/\text{unit}$.
     Actual gain = $\$16.06$

6 $p = -0.057x + 23.82$, $R = -0.057x^2 + 23.82x$

7 Revenue: $R = 2000x - 4x^2$
     Marginal revenue: $\dfrac{dR}{dx} = 2000 - 8x$; $\$0/\text{unit}$

8 $\dfrac{dP}{dx} = \$1.44/\text{unit}$
     Actual increase in profit $\approx \$1.44$

Checkpoints for Section 2.4

1 $-27x^2 + 12x + 24$ **2** $\dfrac{2x^2 - 1}{x^2}$

3 (a) $18x^2 + 30x$ (b) $12x + 15$ **4** $-\dfrac{22}{(5x - 2)^2}$

5 $y = \frac{8}{25}x - \frac{4}{5}$;

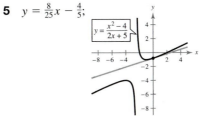

6 $\dfrac{-3x^2 + 4x + 8}{x^2(x + 4)^2}$ **7** (a) $\dfrac{2}{5}x + \dfrac{4}{5}$ (b) $3x^3$

8

t	0	1	2	3	4	5	6	7
$\dfrac{dP}{dt}$	0	-50	-16	-6	-2.77	-1.48	-0.88	-0.56

As t increases, the rate at which the blood pressure drops decreases.

Checkpoints for Section 2.5

1 (a) $u = g(x) = x + 1$, $y = f(u) = \dfrac{1}{\sqrt{u}}$
     (b) $u = g(x) = x^2 + 2x + 5$, $y = f(u) = u^3$

2 $6x^2(x^3 + 1)$ **3** $4(2x + 3)(x^2 + 3x)^3$ **4** $y = \frac{1}{3}x + \frac{8}{3}$

5 $-\dfrac{8}{(2x + 1)^2}$ **6** $\dfrac{x(3x^2 + 2)}{\sqrt{x^2 + 1}}$ **7** $-\dfrac{12(x + 1)}{(x - 5)^3}$

8 About $\$3.48/\text{yr}$

Checkpoints for Section 2.6

1 $f'(x) = 18x^2 - 4x$, $f''(x) = 36x - 4$, $f'''(x) = 36$, $f^{(4)}(x) = 0$

2 18   **3** $\dfrac{120}{x^6}$

4 Height = 144 ft
     Velocity = 0 ft/sec
     Acceleration = -32 ft/sec^2

5 -9.8 m/sec^2

6

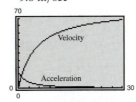

Acceleration approaches zero.

Checkpoints for Section 2.7

1 $-\dfrac{2}{x^3}$

2 (a) $12x^2$ (b) $6y\dfrac{dy}{dx}$ (c) $1 + 5\dfrac{dy}{dx}$ (d) $y^3 + 3xy^2\dfrac{dy}{dx}$

3 $\dfrac{dy}{dx} = -\dfrac{x-2}{y-1}$ **4** $\dfrac{3}{4}$ **5** $\dfrac{5}{9}$

6 $\dfrac{dx}{dp} = -\dfrac{2}{p^2(0.002x+1)}$

Checkpoints for Section 2.8

1 9   **2** $12\pi \approx 37.7\ \text{ft}^2/\text{sec}$

3 \$1500/day   **4** \$28,400/wk

Chapter 3

Checkpoints for Section 3.1

1 $f'(x) = 4x^3$

  $f'(x) < 0$ if $x < 0$; therefore, f is decreasing on $(-\infty, 0)$.

  $f'(x) > 0$ if $x > 0$; therefore, f is increasing on $(0, \infty)$.

2 $\dfrac{dF}{dt} = -1.5348t + 2.872 < 0$ when $3 \le t \le 8$, which implies

  that the consumption of fresh fruit was decreasing from 2003 through 2008.

3 $x = \frac{1}{2}$

4 Increasing on $(-\infty, -2)$ and $(2, \infty)$

  Decreasing on $(-2, 2)$

5 Increasing on $(0, \infty)$

  Decreasing on $(-\infty, 0)$

6 Increasing on $(-\infty, -1)$ and $(1, \infty)$

  Decreasing on $(-1, 0)$ and $(0, 1)$

7 Because $f'(x) = -3x^2 = 0$ when $x = 0$ and because f is decreasing on $(-\infty, 0) \cup (0, \infty)$, f is decreasing on $(-\infty, \infty)$.

8 $(0, 3000)$

Checkpoints for Section 3.2

1 Relative maximum at $(-1, 5)$

  Relative minimum at $(1, -3)$

2 Relative minimum at $(3, -27)$

3 Relative maximum at $(1, 1)$

  Relative minimum at $(0, 0)$

4 Absolute maximum at $(0, 10)$

  Absolute minimum at $(4, -6)$

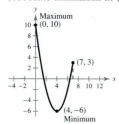

5

x (units)	24,000	24,200	24,300	24,400
P (profit)	\$24,760	\$24,766	\$24,767.50	\$24,768

x (units)	24,500	24,600	24,800
P (profit)	\$24,767.50	\$24,766	\$24,760

Checkpoints for Section 3.3

1 (a) $f'' = -4$; because $f''(x) < 0$ for all x, f is concave downward for all x.

  (b) $f''(x) = \dfrac{1}{2x^{3/2}}$; because $f''(x) > 0$ for all $x > 0$, f is concave upward for all $x > 0$.

2 Because $f''(x) > 0$ for $x < -\dfrac{2\sqrt{3}}{3}$ and $x > \dfrac{2\sqrt{3}}{3}$, f is concave upward on $\left(-\infty, -\dfrac{2\sqrt{3}}{3}\right)$ and $\left(\dfrac{2\sqrt{3}}{3}, \infty\right)$. Because $f''(x) < 0$ for $-\dfrac{2\sqrt{3}}{3} < x < \dfrac{2\sqrt{3}}{3}$, f is concave downward on $\left(-\dfrac{2\sqrt{3}}{3}, \dfrac{2\sqrt{3}}{3}\right)$.

3 f is concave upward on $(-\infty, 0)$.

  f is concave downward on $(0, \infty)$.

  Point of inflection: $(0, 0)$

4 f is concave upward on $(-\infty, 0)$ and $(1, \infty)$.

  f is concave downward on $(0, 1)$.

  Points of inflection: $(0, 1)$, $(1, 0)$

5 Relative minimum: $(3, -26)$

6 Point of diminishing returns: $x = \$150$ thousand

Checkpoints for Section 3.4

1

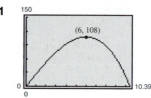

  Maximum volume $= 108$ in.^{3}

2 $\left(\sqrt{\tfrac{1}{2}}, \tfrac{7}{2}\right)$ and $\left(-\sqrt{\tfrac{1}{2}}, \tfrac{7}{2}\right)$ **3** 8 in. by 12 in.

Checkpoints for Section 3.5

1 125 units yield a maximum revenue of \$1,562,500.

2 400 units   **3** \$6.25/unit   **4** \$4.00

5 Demand is elastic when $0 < x < 144$.

  Demand is inelastic when $144 < x < 324$.

  Demand is of unit elasticity when $x = 144$.

Checkpoints for Section 3.6

1 (a) $\displaystyle\lim_{x \to 2^-} \dfrac{1}{x-2} = -\infty$ (b) $\displaystyle\lim_{x \to 2^+} \dfrac{1}{x-2} = \infty$

  (c) $\displaystyle\lim_{x \to -3^-} \dfrac{-1}{x+3} = -\infty$ (d) $\displaystyle\lim_{x \to -3^+} \dfrac{-1}{x+3} = \infty$

2 $x = 0, x = 4$ **3** $x = 3$

4 $\displaystyle\lim_{x \to 2^-} \dfrac{x^2-4x}{x-2} = \infty$; $\displaystyle\lim_{x \to 2^+} \dfrac{x^2-4x}{x-2} = -\infty$

5 2

6 (a) $y = 0$ (b) $y = \frac{1}{2}$ (c) No horizontal asymptote

7 $C = 0.75x + 25,000$

  $\overline{C} = 0.75 + \dfrac{25,000}{x}$

  $\displaystyle\lim_{x \to \infty} \overline{C} = \$0.75/\text{unit}$

8 No, the cost function is not defined at $p = 100$, which implies that it is not possible to remove 100% of the pollutants.

Checkpoints for Section 3.7

1

	$f(x)$	$f'(x)$	$f''(x)$	Shape of graph
x in $(-\infty, -1)$		$-$	$+$	Decreasing, concave upward
$x = -1$	-32	0	$+$	Relative minimum
x in $(-1, 1)$		$+$	$+$	Increasing, concave upward
$x = 1$	-16	$+$	0	Point of inflection
x in $(1, 3)$		$+$	$-$	Increasing, concave downward
$x = 3$	0	0	$-$	Relative maximum
x in $(3, \infty)$		$-$	$-$	Decreasing, concave downward

2

	$f(x)$	$f'(x)$	$f''(x)$	Shape of graph
x in $(-\infty, 0)$		$-$	$+$	Decreasing, concave upward
$x = 0$	5	0	0	Point of inflection
x in $(0, 2)$		$-$	$-$	Decreasing, concave downward
$x = 2$	-11	$-$	0	Point of inflection
x in $(2, 3)$		$-$	$+$	Decreasing, concave upward
$x = 3$	-22	0	$+$	Relative minimum
x in $(3, \infty)$		$+$	$+$	Increasing, concave upward

3

	$f(x)$	$f'(x)$	$f''(x)$	Shape of graph
x in $(-\infty, 0)$		$+$	$-$	Increasing, concave downward
$x = 0$	0	0	$-$	Relative maximum
x in $(0, 1)$		$-$	$-$	Decreasing, concave downward
$x = 1$	Undef.	Undef.	Undef.	Vertical asymptote
x in $(1, 2)$		$-$	$+$	Decreasing, concave upward
$x = 2$	4	0	$+$	Relative minimum
x in $(2, \infty)$		$+$	$+$	Increasing, concave upward

4

	$f(x)$	$f'(x)$	$f''(x)$	Shape of graph
x in $(-\infty, -1)$		$+$	$+$	Increasing, concave upward
$x = -1$	Undef.	Undef.	Undef.	Vertical asymptote
x in $(-1, 0)$		$+$	$-$	Increasing, concave downward
$x = 0$	-1	0	$-$	Relative maximum
x in $(0, 1)$		$-$	$-$	Decreasing, concave downward
$x = 1$	Undef.	Undef.	Undef.	Vertical asymptote
x in $(1, \infty)$		$-$	$+$	Decreasing, concave upward

5

	$f(x)$	$f'(x)$	$f''(x)$	Shape of graph
x in $(0, 1)$		$-$	$+$	Decreasing, concave upward
$x = 1$	-4	0	$+$	Relative minimum
x in $(1, \infty)$		$+$	$+$	Increasing, concave upward

Checkpoints for Section 3.8

1 $dy = 0.32$; $\Delta y = 0.32240801$ **2** $dR = \$22$; $\Delta R = \$21$

3 $dP = \$10.96$; $\Delta P = \$10.98$

4 (a) $dy = 12x^2\, dx$ (b) $dy = \frac{2}{3}\, dx$

(c) $dy = (6x - 2)\, dx$ (d) $dy = -\dfrac{2}{x^3}\, dx$

Chapter 4

Checkpoints for Section 4.1

1 (a) 243   (b) 3   (c) 64   (d) 8   (e) $\frac{1}{2}$ (f) $\sqrt{10}$

2 (a) 5.453×10^{-13} (b) 1.621×10^{-13}

(c) 2.629×10^{-14}

3 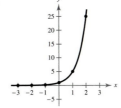   **4**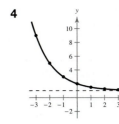

Checkpoints for Section 4.2

1

x	-2	-1	0	1	2
$g(x)$	$e^2 \approx 7.389$	$e \approx 2.718$	1	$\dfrac{1}{e} \approx 0.368$	$\dfrac{1}{e^2} \approx 0.135$

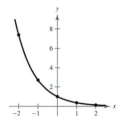

2 After 0 h, $y = 1.25$ g.
After 1 h, $y \approx 1.338$ g.
After 10 h, $y \approx 1.498$ g.

$$\lim_{t \to \infty} \frac{1.50}{1 + 0.2e^{-0.5t}} = 1.50 \text{ g}$$

3 (a) \$4870.38   (b) \$4902.71
(c) \$4918.66   (d) \$4919.21
All else being equal, the more often interest is compounded, the greater the balance.

4 (a) 7%   (b) 7.12%   (c) 7.19%   (d) 7.23%
5 \$16,712.90

Checkpoints for Section 4.3

1 At $(0, 2)$, the slope is 2. At $(1, 2e)$, the slope is $2e$.
2 (a) $3e^{3x}$

(b) $-\dfrac{6x^2}{e^{2x^3}}$

(c) $8xe^{x^2}$

(d) $-\dfrac{2}{e^{2x}}$

3 (a) 0
(b) $3e^{3x+1}$
(c) $xe^x(x + 2)$
(d) $\frac{1}{2}(e^x - e^{-x})$

(e) $\dfrac{e^x(x - 2)}{x^3}$

(f) $e^x(x^2 + 2x - 1)$

4
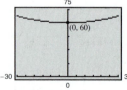

5 \$18.39/unit (80,000 units); \$1,471,517.77
6
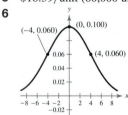

Points of inflection: $(-4, 0.060)$, $(4, 0.060)$

Checkpoints for Section 4.4

1

x	-1.5	-1	-0.5	0	0.5	1
$f(x)$	-0.693	0	0.405	0.693	0.916	1.099

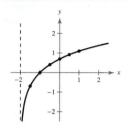

2 (a) 3   (b) $x + 1$
3 (a) $\ln 2 - \ln 5$ (b) $\frac{1}{3}\ln(x + 2)$
(c) $\ln x - \ln 5 - \ln y$ (d) $\ln x + 2\ln(x + 1)$

4 (a) $\ln x^4 y^3$ (b) $\ln \dfrac{x + 1}{(x + 3)^2}$ **5** (a) $\ln 6$ (b) $5\ln 5$

6 (a) e^4 (b) e^3 **7** 7.9 yr

Checkpoints for Section 4.5

1 $\dfrac{1}{x}$ **2** (a) $\dfrac{2x}{x^2 - 4}$ (b) $x(1 + 2\ln x)$ (c) $\dfrac{2\ln x - 1}{x^3}$

3 $\dfrac{1}{3(x + 1)}$ **4** $\dfrac{2}{x} + \dfrac{x}{x^2 + 1}$ **5** $y = 4x - 4$

6 Relative minimum: $(2, 2 - 2\ln 2) \approx (2, 0.6137)$

7 $\dfrac{dp}{dt} = -1.3\%/\text{mo}$

The average score would decrease at a greater rate than the model in Example 7.

8 (a) 4   (b) -2 (c) -5 (d) 3
9 (a) 2.322   (b) 2.631   (c) 3.161   (d) -0.5

10
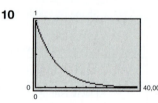
As time increases, the derivative approaches 0. The rate of change of the amount of carbon isotopes is proportional to the amount present.

Checkpoints for Section 4.6

1 About 2113.7 yr   **2** $y = 25e^{0.6931t}$
3 $r = \frac{1}{8}\ln 2 \approx 0.0866$ or 8.66%   **4** About 12.42 mo

Chapter 5

Checkpoints for Section 5.1

1 (a) $5x + C$ (b) $-r + C$ (c) $2t + C$ **2** $\frac{5}{2}x^2 + C$

3 (a) $-\dfrac{1}{x} + C$ (b) $\dfrac{3}{4}x^{4/3} + C$

4 (a) $\frac{1}{2}x^2 + 4x + C$ (b) $x^4 - \frac{5}{2}x^2 + 2x + C$
5 $\frac{2}{3}x^{3/2} + 4x^{1/2} + C$
6 General solution: $F(x) = 2x^2 + 2x + C$
Particular solution: $F(x) = 2x^2 + 2x + 4$
7 $s(t) = -16t^2 + 32t + 48$. The ball hits the ground 3 seconds after it is thrown, with a velocity of -64 feet per second.
8 $C = -0.01x^2 + 28x + 12.01$
$C(200) = \$5212.01$

Checkpoints for Section 5.2

1 (a) $\dfrac{(x^3 + 6x)^3}{3} + C$ (b) $\dfrac{2}{3}(x^2 - 2)^{3/2} + C$

2 $\dfrac{1}{36}(3x^4 + 1)^3 + C$ **3** $\dfrac{1}{9}(x^3 - 3x)^3 + C$

4 $2x^9 + \dfrac{12}{5}x^5 + 2x + C$ **5** $\dfrac{5}{3}(x^2 - 1)^{3/2} + C$

6 $-\dfrac{1}{3}(1 - 2x)^{3/2} + C$ **7** $\dfrac{1}{3}(x^2 + 4)^{3/2} + C$

8 About \$34,068

Checkpoints for Section 5.3

1 (a) $3e^x + C$ (b) $e^{5x} + C$ (c) $e^x - \dfrac{x^2}{2} + C$

2 $\dfrac{1}{2}e^{2x+3} + C$ **3** $2e^{x^2} + C$

4 (a) $2 \ln|x| + C$ (b) $\ln|x^3| + C$ (c) $\ln|2x + 1| + C$

5 $\dfrac{1}{4}\ln|4x + 1| + C$ **6** $\dfrac{3}{2}\ln(x^2 + 4) + C$

7 (a) $4x - 3 \ln|x| - \dfrac{2}{x} + C$ (b) $2\ln(1 + e^x) + C\,dx$

(c) $\dfrac{x^2}{2} + x + 3 \ln|x + 1| + C$

Checkpoints for Section 5.4

1 $\dfrac{1}{2}(3)(12) = 18$

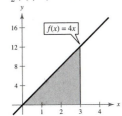

2 $\dfrac{22}{3}$ units^2 **3** 68

4 (a) $\dfrac{1}{4}(e^4 - 1) \approx 13.3995$ (b) $-\ln 5 + \ln 2 \approx -0.9163$

5 $\dfrac{13}{2}$ **6** (a) About \$14.18   (b) \$141.79

7 \$13.70   **8** (a) $\dfrac{2}{5}$ (b) 0   **9** About \$12,295.62

Checkpoints for Section 5.5

1 $\dfrac{8}{3}$ units^2

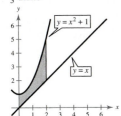

2 $\dfrac{32}{3}$ units^2 **3** $\dfrac{9}{2}$ units^2

4 $\dfrac{253}{12}$ units^2

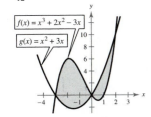

5 Consumer surplus: 40
Producer surplus: 20

6 The company can save \$47.52 million.

Checkpoints for Section 5.6

1 $\dfrac{37}{8}$ units^2 **2** 0.436 unit^2 **3** 5.642 units^2

4 About 1.463

Chapter 6

Checkpoints for Section 6.1

1 $\dfrac{1}{2}xe^{2x} - \dfrac{1}{4}e^{2x} + C$ **2** $\dfrac{x^2}{2}\ln x - \dfrac{1}{4}x^2 + C$

3 $x \ln 2x - x + C$ **4** $e^x(x^3 - 3x^2 + 6x - 6) + C$

5 $e - 2$ **6** \$538,145   **7** \$721,632.08

Checkpoints for Section 6.2

1 $\dfrac{2}{3}(x - 4)\sqrt{2 + x} + C$ (Formula 19)

2 $\sqrt{x^2 + 16} - 4 \ln\left|\dfrac{4 + \sqrt{x^2 + 16}}{x}\right| + C$ (Formula 25)

3 $\dfrac{1}{4}\ln\left|\dfrac{x - 2}{x + 2}\right| + C$ (Formula 21)

4 $\dfrac{1}{3}[1 - \ln(1 + e) + \ln 2] \approx 0.12663$ (Formula 39)

5 $x(\ln x)^2 + 2x - 2x \ln x + C$ (Formula 44)

6 About 18.2%

Checkpoints for Section 6.3

1 3.2608   **2** 3.1956   **3** 1.154

Checkpoints for Section 6.4

1 (a) Converges; $\dfrac{1}{2}$ (b) Diverges

2 1   **3** $\dfrac{1}{2}$ **4** 0.0013 or 0.13%

5 No, you do not have enough money to start the scholarship
fund because you need \$125,000.   (\$125,000 > \$120,000)

Chapter 7

Checkpoints for Section 7.1

1

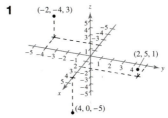

2 $2\sqrt{6}$ **3** $\left(-\dfrac{5}{2}, 2, -2\right)$

4 $(x - 4)^2 + (y - 3)^2 + (z - 2)^2 = 25$

5 $(x - 1)^2 + (y - 3)^2 + (z - 2)^2 = 38$

6 Center: $(-3, 4, -1)$; radius: 6

7 $(x + 1)^2 + (y - 2)^2 = 16$

Checkpoints for Section 7.2

1 x-intercept: $(4, 0, 0)$;
y-intercept: $(0, 2, 0)$;
z-intercept: $(0, 0, 8)$

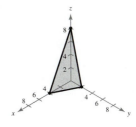

2 xy-trace: circle, $x^2 + y^2 = 1$
yz-trace: hyperbola, $y^2 - z^2 = 1$
xz-trace: hyperbola, $x^2 - z^2 = 1$
$z = 3$ trace: circle, $x^2 + y^2 = 10$
Hyperboloid of one sheet

3 (a) Elliptic paraboloid   (b) Elliptic cone

Checkpoints for Section 7.3

1 (a) 0   (b) $\frac{9}{4}$

2 Domain: $x^2 + y^2 \le 9$
Range: $0 \le z \le 3$

3 For each value of c, the equation $f(x, y) = c$ is a circle (or point) in the xy-plane.

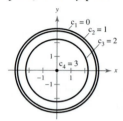

4 $f(1500, 1000) \approx 127{,}542$ units
$f(1000, 1500) \approx 117{,}608$ units
x, person-hours, has a greater effect on production.

5 (a) $M = \$421.60/\text{mo}$
(b) Total paid $= (30 \times 12) \times 421.60$
$= \$151{,}776$

Checkpoints for Section 7.4

1 $\dfrac{\partial z}{\partial x} = 4x - 8xy^3$

$\dfrac{\partial z}{\partial y} = -12x^2y^2 + 4y^3$

2 $f_x(x, y) = 2xy^3$; $f_x(1, 2) = 16$
$f_y(x, y) = 3x^2y^2$; $f_y(1, 2) = 12$

3 (a) $f_x(1, -1, 49) = 8$ (b) $f_y(1, -1, 49) = -18$

4 Substitute product relationship

5 $\dfrac{\partial w}{\partial x} = xy + 2xy \ln(xz)$

$\dfrac{\partial w}{\partial y} = x^2 \ln xz$

$\dfrac{\partial w}{\partial z} = \dfrac{x^2y}{z}$

6 $f_{xx} = 8y^2$
$f_{yy} = 8x^2 + 8$
$f_{xy} = 16xy$
$f_{yx} = 16xy$

7 $f_{xx} = 0$ $f_{xy} = e^y$ $f_{xz} = 2$
$f_{yx} = e^y$ $f_{yy} = xe^y + 2$ $f_{yz} = 0$
$f_{zx} = 2$ $f_{zy} = 0$ $f_{zz} = 0$

Checkpoints for Section 7.5

1 $f(-8, 2) = -64$: relative minimum
2 $f(0, 0) = 1$: relative maximum
3 $f(0, 0) = 0$: saddle point
4 $P(3.11, 3.81) = \$744.81$ maximum profit
5 $V\left(\frac{4}{3}, \frac{2}{3}, \frac{8}{3}\right) = \frac{64}{27}$ units^3

Checkpoints for Section 7.6

1 $V\left(\frac{4}{3}, \frac{2}{3}, \frac{8}{3}\right) = \frac{64}{27}$ units^3 **2** $f(187.5, 50) \approx 13{,}474$ units
3 About 26,740 units
4 $P(3.35, 4.26) = \$758.08$ maximum profit

Checkpoints for Section 7.7

1 For $f(x)$, $S = 10$. For $g(x)$, $S = 0.76$.
The quadratic model is a better fit.
2 $f(x) = \frac{6}{5}x + \frac{23}{10}$
3 $y = 16{,}194.4t + 132{,}405$
About 342,932,200 subscribers

Checkpoints for Section 7.8

1 (a) $\frac{1}{4}x^4 + 2x^3 - 2x - \frac{1}{4}$ (b) $\ln|y^2 + y| - \ln|2y|$

2 $\frac{25}{2}$ **3** $\displaystyle\int_2^4 \int_1^5 dx\, dy = 8$ **4** $\frac{4}{3}$

5 (a)

(b) $\displaystyle\int_0^4 \int_0^{x/2} dy\, dx$

(c) $\displaystyle\int_0^2 \int_{2y}^4 dx\, dy = 4 = \int_0^4 \int_0^{x/2} dy\, dx$

$R: 0 \le y \le 2$
$\quad 2y \le x \le 4$

6 $\displaystyle\int_{-1}^3 \int_{x^2}^{2x+3} dy\, dx = \dfrac{32}{3}$

Checkpoints for Section 7.9

1 $\frac{16}{3}$ **2** $e - 1$ **3** $\frac{176}{15}$
4 Integration by parts   **5** 3

Appendix A

Checkpoints for Section A.1

1 $x < 5$ or $(-\infty, 5)$
2 $x < -2$ or $x > 5$; $(-\infty, -2) \cup (5, \infty)$
3 $200 \le x \le 400$; so the daily production levels during the month varied between a low of 200 units and a high of 400 units.

Checkpoints for Section A.2

1 8; 8; -8 **2** $2 \le x \le 10$
3 $\$4027.50 \le C \le \$11{,}635$

Checkpoints for Section A.3

1 $\frac{4}{9}$ **2** 8   **3** (a) $3x^6$ (b) $8x^{7/2}$ (c) $4x^{4/3}$
4 (a) $x(x^2 - 2)$ (b) $2x^{1/2}(1 + 4x)$
5 $\dfrac{(3x - 1)^{3/2}(13x - 2)}{(x + 2)^{1/2}}$ **6** $\dfrac{x^2(5 + x^3)}{3}$
7 (a) $[2, \infty)$ (b) $(2, \infty)$ (c) $(-\infty, \infty)$

Checkpoints for Section A.4

1 (a) $\dfrac{-2 \pm \sqrt{2}}{2}$ (b) 4   (c) No real zeros
2 (a) $x = -3$ and $x = 5$ (b) $x = -1$
(c) $x = \frac{3}{2}$ and $x = 2$
3 $(-\infty, -2] \cup [1, \infty)$ **4** $-1, \frac{1}{2}, 2$

Checkpoints for Section A.5

1 $\dfrac{x-2}{x+1}$, $x \neq -10$ **2** (a) $\dfrac{x^2+2}{x}$ (b) $\dfrac{3x+1}{(x+1)(2x+1)}$

3 $\dfrac{3x+4}{(x+2)(x-2)}$ **4** $-\dfrac{x+1}{3x(x+2)}$

5 $\dfrac{2(4x^2+5x-3)}{x^2(x+3)}$ **6** (a) $\dfrac{3x+8}{4(x+2)^{3/2}}$ (b) $\dfrac{1}{\sqrt{x^2+4}}$

7 (a) $\dfrac{5\sqrt{2}}{4}$ (b) $\dfrac{x+2}{4\sqrt{x+2}}$ (c) $\dfrac{\sqrt{6}+\sqrt{3}}{3}$

     (d) $\dfrac{\sqrt{x+2}-\sqrt{x}}{2}$

Answers to Tech Tutors

Tech Tutor

Section 1.3   *(page 29)*

The lines appear perpendicular in the setting $-9 \le x \le 9$ and $-6 \le y \le 6$.

Section 1.6   *(page 61)*

Most calculators set in connected mode will join the two branches of the graph with a nearly vertical line near $x = 2$. This line is not part of the graph.

Section 4.5   *(page 287)*

Answers will vary.

Section 6.3   *(page 395)*

1.46265

Index

Basic Differentiation Rules

1. $\dfrac{d}{dx}[cu] = cu'$

2. $\dfrac{d}{dx}[u \pm v] = u' \pm v'$

3. $\dfrac{d}{dx}[uv] = uv' + vu'$

4. $\dfrac{d}{dx}\left[\dfrac{u}{v}\right] = \dfrac{vu' - uv'}{v^2}$

5. $\dfrac{d}{dx}[c] = 0$

6. $\dfrac{d}{dx}[u^n] = nu^{n-1}u'$

7. $\dfrac{d}{dx}[x] = 1$

8. $\dfrac{d}{dx}[\ln u] = \dfrac{u'}{u}$

9. $\dfrac{d}{dx}[e^u] = e^u u'$

10. $\dfrac{d}{dx}[\log_a u] = \dfrac{u'}{(\ln a)u}$

11. $\dfrac{d}{dx}[a^u] = (\ln a)a^u u'$

12. $\dfrac{d}{dx}[\sin u] = (\cos u)u'$

13. $\dfrac{d}{dx}[\cos u] = -(\sin u)u'$

14. $\dfrac{d}{dx}[\tan u] = (\sec^2 u)u'$

15. $\dfrac{d}{dx}[\cot u] = -(\csc^2 u)u'$

16. $\dfrac{d}{dx}[\sec u] = (\sec u \tan u)u'$

17. $\dfrac{d}{dx}[\csc u] = -(\csc u \cot u)u'$

Basic Integration Formulas

1. $\displaystyle\int kf(u)\,du = k\int f(u)\,du$

2. $\displaystyle\int [f(u) \pm g(u)]\,du = \int f(u)\,du \pm \int g(u)\,du$

3. $\displaystyle\int du = u + C$

4. $\displaystyle\int a^u\,du = \left(\dfrac{1}{\ln a}\right)a^u + C$

5. $\displaystyle\int e^u\,du = e^u + C$

6. $\displaystyle\int \ln u\,du = u(-1 + \ln u) + C$

7. $\displaystyle\int \sin u\,du = -\cos u + C$

8. $\displaystyle\int \cos u\,du = \sin u + C$

9. $\displaystyle\int \tan u\,du = -\ln|\cos u| + C$

10. $\displaystyle\int \cot u\,du = \ln|\sin u| + C$

11. $\displaystyle\int \sec u\,du = \ln|\sec u + \tan u| + C$

12. $\displaystyle\int \csc u\,du = -\ln|\csc u + \cot u| + C$

13. $\displaystyle\int \sec^2 u\,du = \tan u + C$

14. $\displaystyle\int \csc^2 u\,du = -\cot u + C$

Trigonometric Identities

Pythagorean Identities

$\sin^2\theta + \cos^2\theta = 1$

$\tan^2\theta + 1 = \sec^2\theta$

$\cot^2\theta + 1 = \csc^2\theta$

Sum or Difference of Two Angles

$\sin(\theta \pm \phi) = \sin\theta\cos\phi \pm \cos\theta\sin\phi$

$\cos(\theta \pm \phi) = \cos\theta\cos\phi \mp \sin\theta\sin\phi$

$\tan(\theta \pm \phi) = \dfrac{\tan\theta \pm \tan\phi}{1 \mp \tan\theta\tan\phi}$

Double Angle

$\sin 2\theta = 2\sin\theta\cos\theta$

$\cos 2\theta = 2\cos^2\theta - 1 = 1 - 2\sin^2\theta$

Reduction Formulas

$\sin(-\theta) = -\sin\theta$

$\cos(-\theta) = \cos\theta$

$\tan(-\theta) = -\tan\theta$

$\sin\theta = -\sin(\theta - \pi)$

$\cos\theta = -\cos(\theta - \pi)$

$\tan\theta = \tan(\theta - \pi)$

Half Angle

$\sin^2\theta = \tfrac{1}{2}(1 - \cos 2\theta)$

$\cos^2\theta = \tfrac{1}{2}(1 + \cos 2\theta)$

ALGEBRA

Quadratic Formula:

If $p(x) = ax^2 + bx + c$, $a \neq 0$ and $b^2 - 4ac \geq 0$, then the real zeros of p are $x = \left(-b \pm \sqrt{b^2 - 4ac}\right)/2a$.

Example

If $p(x) = x^2 + 3x - 1$, then $p(x) = 0$ if

$$x = \frac{-3 \pm \sqrt{13}}{2}.$$

Special Factors:

$x^2 - a^2 = (x - a)(x + a)$

$x^3 - a^3 = (x - a)(x^2 + ax + a^2)$

$x^3 + a^3 = (x + a)(x^2 - ax + a^2)$

$x^4 - a^4 = (x - a)(x + a)(x^2 + a^2)$

$x^4 + a^4 = \left(x^2 + \sqrt{2}ax + a^2\right)\left(x^2 - \sqrt{2}ax + a^2\right)$

$x^n - a^n = (x - a)(x^{n-1} + ax^{n-2} + \cdots + a^{n-1})$, for n odd

$x^n + a^n = (x + a)(x^{n-1} - ax^{n-2} + \cdots + a^{n-1})$, for n odd

$x^{2n} - a^{2n} = (x^n - a^n)(x^n + a^n)$

Examples

$x^2 - 9 = (x - 3)(x + 3)$

$x^3 - 8 = (x - 2)(x^2 + 2x + 4)$

$x^3 + 4 = \left(x + \sqrt[3]{4}\right)\left(x^2 - \sqrt[3]{4}x + \sqrt[3]{16}\right)$

$x^4 - 4 = \left(x - \sqrt{2}\right)\left(x + \sqrt{2}\right)(x^2 + 2)$

$x^4 + 4 = (x^2 + 2x + 2)(x^2 - 2x + 2)$

$x^5 - 1 = (x - 1)(x^4 + x^3 + x^2 + x + 1)$

$x^7 + 1 = (x + 1)(x^6 - x^5 + x^4 - x^3 + x^2 - x + 1)$

$x^6 - 1 = (x^3 - 1)(x^3 + 1)$

Exponents and Radicals:

$a^0 = 1$, $a \neq 0$

$a^{-x} = \dfrac{1}{a^x}$

$a^x a^y = a^{x+y}$

$\dfrac{a^x}{a^y} = a^{x-y}$

$(a^x)^y = a^{xy}$

$(ab)^x = a^x b^x$

$\left(\dfrac{a}{b}\right)^x = \dfrac{a^x}{b^x}$

$\sqrt{a} = a^{1/2}$

$\sqrt[n]{a} = a^{1/n}$

$\sqrt[n]{a^m} = a^{m/n} = \left(\sqrt[n]{a}\right)^m$

$\sqrt[n]{ab} = \sqrt[n]{a}\,\sqrt[n]{b}$

$\sqrt[n]{\left(\dfrac{a}{b}\right)} = \dfrac{\sqrt[n]{a}}{\sqrt[n]{b}}$

Algebraic Errors to Avoid:

$\dfrac{a}{x + b} \neq \dfrac{a}{x} + \dfrac{a}{b}$

(To see this error, let $a = b = x = 1$.)

$\sqrt{x^2 + a^2} \neq x + a$

(To see this error, let $x = 3$ and $a = 4$.)

$a - b(x - 1) \neq a - bx - b$

[Remember to distribute negative signs. The equation should be $a - b(x - 1) = a - bx + b$.]

$\dfrac{\left(\dfrac{x}{a}\right)}{b} \neq \dfrac{bx}{a}$

[To divide fractions, invert and multiply. The equation should be

$$\frac{\left(\dfrac{x}{a}\right)}{b} = \frac{\left(\dfrac{x}{a}\right)}{\left(\dfrac{b}{1}\right)} = \left(\frac{x}{a}\right)\left(\frac{1}{b}\right) = \frac{x}{ab}.]$$

$\sqrt{-x^2 + a^2} \neq -\sqrt{x^2 - a^2}$

(The negative sign cannot be factored out of the square root.)

$\dfrac{\not{a} + bx}{\not{a}} \neq 1 + bx$

(This is one of many examples of incorrect dividing out. The equation should be

$$\frac{a + bx}{a} = \frac{a}{a} + \frac{bx}{a} = 1 + \frac{bx}{a}.)$$

$\dfrac{1}{x^{1/2} - x^{1/3}} \neq x^{-1/2} - x^{-1/3}$

(This error is a more complex version of the first error.)

$(x^2)^3 \neq x^5$

[This equation should be $(x^2)^3 = x^2 x^2 x^2 = x^6$.]